T0230413

Lecture Notes in Artificial Intelligence 592

Subseries of Lecture Notes in Computer Science
Edited by J. Siekmann

Lecture Notes in Computer Science
Edited by G. Goos and J. Hartmanis

A. Voronkov (Ed.)

Logic Programming

First Russian Conference on Logic Programming
Irkutsk, Russia, September 14-18, 1990
Second Russian Conference on Logic Programming
St. Petersburg, Russia, September 11-16, 1991
Proceedings

Springer-Verlag
Berlin Heidelberg New York
London Paris Tokyo
Hong Kong Barcelona
Budapest

Series Editor

Jörg Siekmann
University of Saarland
German Research Center for Artificial Intelligence (DFKI)
Stuhlsatzenhausweg 3, W-6600 Saarbrücken 11, FRG

Volume Editor

Andrei Voronkov
European Computer-Industry Research Centre (ECRC)
Arabellastraße 17, W-8000 München 81, FRG
and
International Laboratory of Intelligent Systems (SINTEL)
Universitetski Prospect 4, 630090 Novosibirsk 90, Russia

CR Subject Classification (1991): F.4.1, I.2.3

ISBN 3-540-55460-2 Springer-Verlag Berlin Heidelberg New York
ISBN 0-387-55460-2 Springer-Verlag New York Berlin Heidelberg

Printed in Germany

Typesetting: Camera ready by author
Printing and binding: Druckhaus Beltz, Hemsbach/Bergstr.
45/3140-543210 - Printed on acid-free paper

Preface

The Russian Conferences on Logic Programming were organised with the aim of bringing together researchers from the Russian and the international logic programming communities. The first conference was planned to be held on the shore of Lake Baikal. However due to some local problems it was held September 14–18, 1990, in Irkutsk — a pleasant city in the Eastern part of Siberia. The number of participants was 71 from the Soviet Union and 11 from the other countries. The second conference was held September 11–16, 1991, on the board the ship "Michail Lomonosov", named after the founder of the Russian Academy of Sciences. The ship started from St.Petersburg and sailed along the River Neva, Lake Ladoga and Lake Onega. This time there were 125 participants from the former Soviet Union and 32 from other countries.

This volume contains the selected papers presented to these two Russian Conferences on Logic Programming.

The idea to organise the conference on the ship proved very successful. The next conference will be held on the same ship July 15–20, 1992 in the famous period of white nights. For several reasons it has been decided to change the name of the conference to LPAR — Logic Programming and Automated Reasoning.

I wish to thank all the people who did a lot to organise these conferences in the time when Russia was in total disorder. Special thanks are due to Victor Durasov and Nelya Dulatova, who made it possible to rearrange the first conference in one day. Further thanks are due to Tania Rybina, Yuri Shcheglyuk, Yulya Mantsivoda, Lena Deriglazova, Maxim Bushuev, Vladimir Bechbudov, Lena Shemyakina and Andrei Mantsivoda for the first conference. For the second conference special thanks are due to Eugene Dantsin, Robert Freidson, Per Bilse and Valeri Shatrov. Further thanks are due to Robert Kowalski, Cheryl Anderson, Herve Gallaire, Tania Rybina, Oleg Gussikhin, Nikolai Ilinski, Arkadi Tompakov, George Selvais, Yuri Shestov, Michael Simuni, Vladislav Valkovski, Oleg Alekseev and Edward Yanchevsky.

Munich, March 1992

Andrei Voronkov

Conference Sponsors

RCLP'90

International Laboratory of Intelligent Systems (SINTEL)

RCLP'91

Association for Logic Programming

SRI Opyt

Per Gregers Bilse

Prolog Development Center A/S

Applied Logic Systems Inc.

Logic Programming Associates Ltd.

SICS Sweden

St. Peterburg Institute of Electrical Engineering

Conference Organizers:

RCLP'90:

International Laboratory of Intelligent Systems (SINTEL)

Irkutsk State University

RCLP'91

Russian Association for Logic Programming

Association for Logic Programming

Eurobalt Inc.

International Laboratory of Intelligent Systems (SINTEL)

St. Peterburg Institute of Electrical Engineering

IRI Inc.

Contents

Real-time memory management for Prolog

Yves Bekkers, Lucien Ungaro

INRIA / IRISA
Av. du Général Leclerc
35042 Rennes-Cedex
France

Tel : (33) 99 84 71 00
E-mail : bekkers@irisa.fr, ungaro@irisa.fr

Abstract

This paper relates a long experiment on implementing real time garbage collectors for Prolog. First, the main peculiarities of Prolog memory management are briefly reviewed. The relation between non-determinism and garbage collection are explained. Early-reset and variable shunting are presented. Attributed variables, a new type of data for implementing Prolog extensions and realizing a value trail mechanism is introduced. Then, a realtime garbage collection algorithm, taking these aspects into account, is entirely presented. The synchronizing problems, including those linked to non-determinism, are discussed in details. Finally, two concrete implementations are described.

Key words : Prolog, garbage collector, realtime, implementation, abstract machine, early reset, variable shunting, attributed variable, virtual backtracking

1 Peculiarity of Prolog memory management

The efficiency of Prolog systems is due to the trailing mechanism which allows the representation of choice points without copying them. For implementing a complete garbage collection, it is necessary to find which objects belong to a choice point representation. Such a GC must interpret the trailing information [Bekkers84a], [Bekkers84b].

1.1 Early reset of variable and variable shunting

The main idea is to watch variables and interpret their correct binding through the different choice points. The following cases can be distinguished :

Keeping all information - If a variable is accessible in its bound and unbound state, the variable and its binding must be kept.

Early reset of variables - If a variable is only accessible in its unbound state then its binding is useless. In this case the variable should be unbound in order to loose access to his binding [Appleby88], [Schimpf90], [Bruynooghe84], [Pittomvils85], [Barklund87a].

Variable shunting - If a variable is only accessible in its bound state then only its binding is useful and the variable itself is useless. In this case one should replace any occurrence of the variable by its binding value, [Huitouze90].

In §4.2.4.1, a detailled implementation of these mechanisms is given.

1.2 Attributed variables

For implementing extensions to Prolog, we have designed a new type of variable called attributed variable [Huitouze88], [Huitouze90], [Brisset91]. It is like a variable with an extra term attached to it, its attribute. The main property of this object is that the attribute is only accessible when the variable is unbound. Attributed variables take all their flavor with variable shunting. The space occupied by useless attributes is automatically reclaimed by variable shunting.

Attributed variables are used in implementing PROLOGII `freeze` and `dif` primitives; it can also be used to implement other kind of constraints or *value trail* mechanisms such as described in [Carlsson87], [Turk86], [Barklund87b], [Toura88], [Neumerkel90].

1.3 The abstract machine MALI

MALI is an abstract machine which has been designed to encapsulate the memory management of Prolog systems [Bekkers86], [Bekkers88]. It offers a set of commands for creating, accessing, modifying Prolog objects such as constructed terms, logical variables, ... the backtrack stack itself is managed by MALI. In top of that, MALI offers an automatic memory management.

Many different implementations of MALI have been experimented, some with a serial GC, some with a realtime GC, some entirely in software (in C), some microprogrammed on a specialized hardware to be inserted into PCs. The same Prolog system, written in C, compatible with PrologII [Colmerauer82], uses any of these implementations.

2 The state of a Prolog system

As usual, the state a Prolog system is summarized into three informations, the current goal statement, the backtrack stack and the trail. With MALI, the dynamic space is organized as a single heap managed by the garbage collector. This must be opposed to the WAM architecture where the dynamic space is split into several spaces, stacks and heap, each subject to its own specific management.

2.1 A tagged pointer scheme

Our runtime system uses a tagged pointer scheme with an elementary type of value called a `WORD`. A `WORD` contains an information field and a tag field. The tag specifies the type of the represented value which may be atom, list, tuple, variable, trail, level, etc ....

```
typedef struct {
    TAG tag ;
    INFO info ;
} WORD ;
```

From the point of view of the Garbage collector there are two types of tags, those indicating a pointer, and the others indicating small atomic values. In the two implementations of MALI that we are presenting here, the size of a referenced object, including tuples, is given by tags (see description of tuples later).

2.2 Representing Prolog terms

Each Prolog term, such as atom, cons, tuple, variable, is designated by a `WORD`.

atom : the tag is `T_atom`, the information is a bit-coding of the value of the atomic constant.

cons : the tag is `T_cons`, the information is a pointer to a `CONS` structure.

```
typedef struct {
    WORD left ;
    WORD right ;
} CONS ;
```

T_cons

left

right

tuple : the tag is `T_tuple(i)`, the information is a pointer to an array of *i* words, one `WORD` per component.

variable : the tag is `T_var`, the information is a pointer to a `VAR` structure. The `binding` field holds the variable binding and a special tag `T_free` means that the variable is *unbound*. The `age` field is a reference to the choice point which was at the top of backtrack stack when the variable was created. The tag of an age is `T_age`.

The `age` field is used by the garbage collector for implementing the variable shunting mechanism.

```
typedef struct {
  WORD binding ;
  WORD age ;
} VAR ;
```

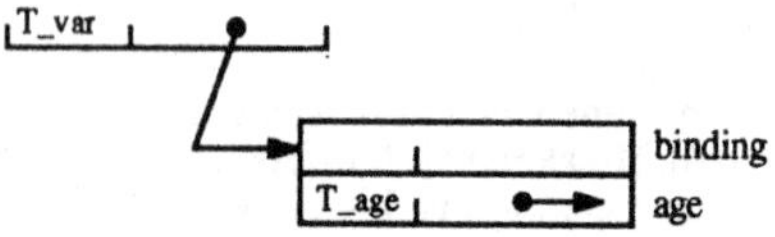

T_var | binding | T_age | age

attributed variable : the tag is `T_vara`, the information is a pointer to a VARA structure. This structure contains a supplementary field, the `attribute`, which is a term.

```
typedef struct {
    WORD binding ;
    WORD age ;
    WORD attribute ;
} VARA ;
```

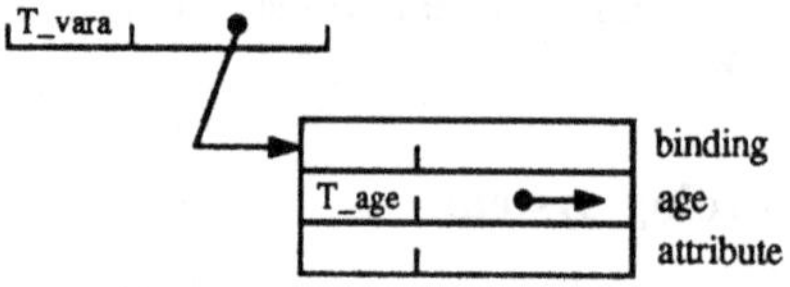

T_vara | binding | T_age | age | attribute

2.3 Representing the active goal statements

The active goal statements are those in the choice point stack plus the current one.

Current goal statement : it is a list of goals held in a register G. Each goal is a term as previously described.

Backtrack stack : it is represented as a linked list of choice points. Each choice point is a structure called a LEVEL containing four fields :

```
typedef struct {
    WORD goal ;      /* the goal statement (a list of goals) */
    WORD clause ;    /* the clause (pointer to program, not relevant to the GC) */
    WORD trail ;     /* the trail (a list of bindings) */
    WORD next ;      /* the link to the next choice point */
} LEVEL ;
```

The top of that stack, is held in a register S.

The trail : each choice point contains a list of references to bound variables. It records such bindings that have to be undone on backtracking to recover the next choice point. In MALI it is a linked list of trail elements :

```
typedef struct {
    WORD var ;
    WORD next ;
} TRAIL ;
```

The current list of bindings is held in a register T.

Notice that the first element of the current trail is an *empty* element, the grey element in figure 1. It can be seen as a trail element created in advance. This element makes algorithm for updating the trail, figure 8, more uniform by putting the content of register T into memory location. [Barklund87a] and [Schimpf90] have met the same problem, their solution was to create a dummy choice point.

Registers G, S and T contain WORDs which give access to the useful data, these are the roots for the marking algorithm.

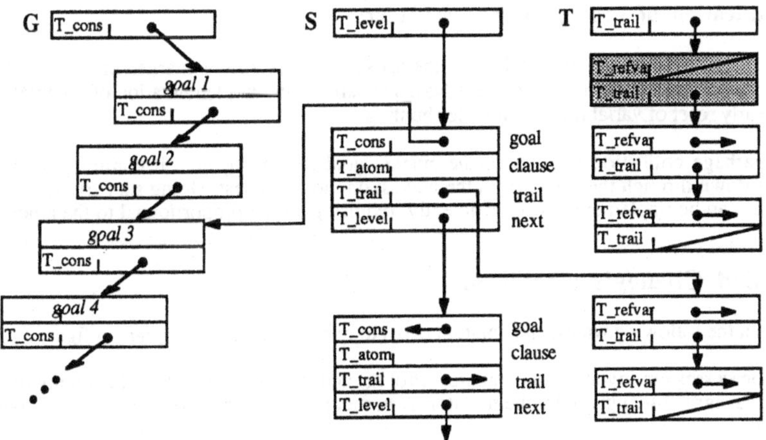

figure 1 : Goal statements, choice points and trail

3 Choosing the garbage collector

We took two constraints for the design of our garbage collector. First, it must work in real-time. Second, it must be well suited for Prolog, that is it must implement variable shunting and early reset of variables.

3.1 An algorithm for real-time garbage collectors

We have experimented two implementations of real time GC which are discussed later in §5. For both implementations, we have chosen Cheney's [Cheney70] copying algorithm as it has been recognized [Baker78], [Lieberman83] as being very well suited for real-time garbage collection.

Recall that Cheney's algorithm copies useful cells from a *fromspace* to a *tospace*, figure 2. It uses two indexes, a *visit* index *M* which points to the queue of copied objects not yet visited and a *copy* index *C* which points to the beginning of the free area where objects are copied.

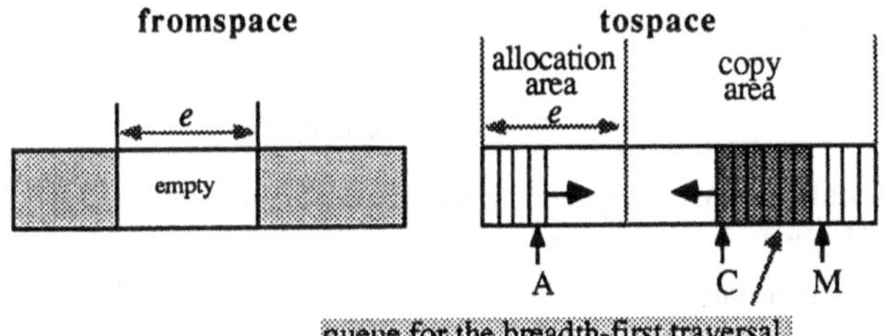

figure 2 : Visit and copy indexes for Cheney's algorithm

New objects are allocated in the allocation area in which index *A* gives the last allocated word. Allocation and copy areas grow in opposite directions. We have decided that C and M decrease towards lower addresses.

The size of the allocation area is bound by the size, *e*, of the previous empty area. Prolog system is suspended when reaching this boundary, until the current GC *batch* is finished. This is necessary to avoid deadlock on memory resource between Prolog system and its GC. At the end of a collecting batch, the allocation boundary is opened up to the C index until the next batch is started. A suspension of Prolog system corresponds to an abnormal functioning, in which the collector doesn't satisfy, in real time, Prolog system needs.

In the allocation area, objects are stacked as they are created. Hence, on backtracking this area may be instantly reclaimed. The instant reclaiming in the allocation area has proved to be sufficient for real-time GC, see discussion in §6.2.

3.2 Stratification of Prolog garbage collectors

Goal statements, the current one and those in the backtrack stack, can be marked independently. Doing so, the correct bindings is interpreted [Bekkers84a]. This allows the implementation of complete GCs which performs early reset of variables and variable shunting.

The garbage collection starts with the current goal statement, then it continues with active goal statements down through the stack, from the newest one to the oldest. Doing so, early reset and shunting of variables can be implemented very efficiently, requiring a time proportionnal to the amount of usefull objects.

4 Stratified Cheney's algorithm

Notation : in the following we use the notation (*F-L*) to refer line *L* of the program in figure *F*.

Two processes are involved, the mutator (Prolog) and the collector. At the beginning the collector is waiting for a signal (4-2) given by the mutator. A particular command, `Reduce`, figure 3, possibly starts the collector.

4.1 Starting the garbage collector

Between each resolution step, the state of Prolog is summarized within the three registers G, S and T ; Prolog executes the command `Reduce` to signal the collector that a batch can be started (3-7) with the three mentioned registers as roots (3-4). The command `Reduce` has no effect when the collector is working or when it is not enabled (3-2).

```
LEVEL * SM ; WORD * TM ;

1    Command Reduce() {
2        if (waiting(StartGC) && CollectorEnabled()){
3            "exchange from/to spaces" ;
4            G=NewVersion(G); S=NewVersion(S); T=Copy(T);
5            SM = S.info ;
6            TM = &(((TRAIL*) T.info)->next) ;
7            signal(StartGC)
8        }
9    }
```

figure 3 : The command Reduce

The collector is only *enabled* if a certain percentage of the memory is used, this is to avoid slowing down the mutator, see §6.1.

Memory overflow detection

One of the problems with realtime garbage collectors is to decide when memory overflow has occurred. If there is not enough memory the mutator must be suspended until the current collector batch is finished. Then if there is not yet enough memory, a new collector batch must be completed. If there is still not enough memory then the system is running out of memory.

It is impossible to start a collector batch at any time because roots of access are not always summarized in the intended registers. We have solved this synchronization problem with the `Reduce` command which defines the correct instants for starting the GC. Prolog is suspended inside this command if *there is not enough memory* to continue. With this solution, one has to decide of a maximum amount of memory which can be allocated between two executions of the command `Reduce`. Prolog is suspended if the free memory is smaller than this amount. For Prolog, this amount can be bounded by the size of the largest clause.

4.2 The garbage collector

4.2.1 The main loop of the collector

For each choice point, the collector copies its goal statement using Cheney's algorithm (4-4). Then, the trail is updated before going on with the next choice point. This is done by the procedure `UnTrailOneElt` (4-5).

For synchronizing reasons explained in §4.2.4, only one element of the trail is processed at a time ; the procedure is also in charge to prime the copy of the next level when the end of a segment trail is reached and to signal the end of the backtrack stack.

```
while (true) {
     wait(StartGC) ;
     while (true) {
          CopyLevel() ;
          if (UnTrailOneElt() != EndOfLevels) exit
     }
}
```

figure 4 : main loop of the collector

4.2.2 Copying a goal statement

Copying a level, figure 5, consists in a simple loop to update the references within the segment of memory situated between indexes C and M. The loop ends when C=M. Of course, this process might involve copying new information in the copy area, this is done by the procedure `NewVersion`, figure 6.

Notice that not all words are visited (5-4). The procedure `AllowedVisit`, not detailed here, successes only if M does not point to the attribute of a bound variable or to the goal statement of a `LEVEL`. According to the stratification principle :

- the attribute of a bound variable is copied while updating its trail element (9-9), see §4.2.4.2,
- the goal statement of a level is copied when the collector primes the copying of this level (8-6).

```
1    CopyLevel() {
2         while (M != C) {
3              M = M - 1 ;
4              if (AllowedVisit)
5                   exclusion (CopyObj) *M = NewVersion(*M)
6         }
7    }
```

figure 5 : copying an entire goal statement

4.2.3 Getting the new version of an object

The `NewVersion` procedure, figure 6, gives the reference to a new version of an object, either by copying the object (6-16), or by using its forward reference (6-12).

This procedure is also in charge of shunting variables (6-14) when they have been marked "shunted" (9-12).

```
1    WORD NewVersion(W)
2    WORD W ;
3    {
4         switch W.tag {
5         case T_atom, T_free,
6               T_trail, T_refvar, T_refvara :
7              return(W) ;
8         case T_cons, T_var, T_vara, T_level, T_age :
9              if (InToSpace(W.info)
10                  return(W)
11             elseif (Forwarded(W))
12                  return((W.info)->info)
13             elseif (IsShuntedVar(W))
14                  return(NewVersion(((VAR*) W.info)->binding))
15             else
16                  return(Copy(W))
17                  }
18        }
19   }
```

figure 6 : give the new version of an object

Critical section for copying objects : the mutator and the collector are both involved in copying objects. Hence, the presence of the critical section `CopyObj`, (5-5), (8-5) for the collector and (11-3), (12-15) for the mutator. The exclusion also includes the update of the `WORD` pointed to by index M.

```
WORD Copy(W)
WORD W ;
{    int Size = SizeOfObject(W.tag) ;
     if (W.info == nil) return(W) ;
     C = C-Size ;
     CopyBlock(Size, W.info, C) ;
     *(W.info).tag=T_forward; *(W.info).info=C ;
     W.info = C ;
     return(W)
}
```

figure 7 : Copy and forward an object

4.2.4 Updating the trail

Once a goal statement has been copied, the trail segment it holds is copied and updated. Early reset is applied and shunted variables are marked *shunted.* The procedure tackles only one element at a time, the iteration for a trail section is done by the loop of the collector (4-3).

```
1    EOL UnTrailOneElt()) {
2         exclusion (Trail) {
3              if (SM = nil) return EndOfLevels ;
4              if (TM->info = nil) {
5                   exclusion (CopyObj)
6                        SM->goal = NewVersion(SM->goal) ;
7                   SM = (SM->next).info ;
8                   TM = (SM->trail).info
9              }
10             else UpdateTrailElt() ;
11             return NotEndOfLevels
12        }
13   }
```

figure 8 : visiting a trail section

Critical section for manipulating trailing information : the trail may be pushed down by a `Backtrack` command, therefore, updating a trail element may not be done in parallel with such a command. Hence, the presence of the critical section `Trail`, (8-2) for the collector and (12-2) for the mutator.

The synchronizing problem happens when the mutator backtracks to a choice point which has not yet been visited by the collector (12-14). In that case the collector must stop its untrailing process, because trail elements are no more significant. When such a *deep backtrack* occurs (12-14), the backtrack stack seen by the collector is forced to shrink (TM and SM registers are updated (12-19)) and the `CopyLevel` process is primed with the roots of the recovered level (12-16). After the `Backtrack` command, the next iteration of the collector loop (4-3) will see a nonempty queue between C and M and will pursue the copying of this level even if it was currently tackling a trail section.

4.2.4.1 Implementation of early reset and variable shunting

When updating a trail element three cases must be considered :

(1) useful binding (the variable has already been copied) :

(1.1) useful free variable (the variable is older than the next choice point) :
the trail element is copied to be kept in the trail; for an attributed variable a new version of its attribute is obtained, figure 10;

(1.2) useless free variable (variable is younger than the next choice point) :
the trail element is discarded and the variable is marked *shunted*;

(2) useless binding (the variable has not been copied):
the trail element is discarded and the variable is reset *free.*

```
1    UpdateTrailElt()
2    {    WORD RefVar ; LEVEL * AgeOfVar ;
3         RefVar = (TM->info)->var ;
4         if (Forwarded(RefVar)) { /*(1)*/
5              RefVar.info = (RefVar.info)->info ;
6              AgeOfVar = (((VAR*) RefVar.info)->age).info
7              If (AgeOfVar != SM) {    /*(1.1)*/
8                   exclusion (Copy) TM* = Copy(TM*) ;
9                   UpdateAttribute(RefVar)
10             }
11             else {                   /*(1.2)*/
12                  SetShunted(RefVar);
13                  TM* = (TM->info)->next
14             }
15        }
16        else {                   /*(2)*/
17             ((VAR* RefVar.info)->binding).tag = free;
18             TM* = (TM->info)->next
19        }
20   }
```

figure 9 : visiting a trail element

Attributes of shunted variables (9-11) does not need to be updated, this is because such attributes will never be accessed.

4.2.4.2 Copying attributes with the correct binding environment

The attributes of bound variables are not updated in a standard way. This is to avoid copying an attribute with an incorrect binding environment. Attributes are copied while visiting the trail (9-9).

```
UpdateAttribute(RefVar)
WORD RefVar ;
{WORD *RefAttrib ;
     if (RefVar.tag == T_refvara)
     exclusion (CopyObj) {
          RefAttrib=&(((VARA*)RefVar.info)->attribute);
          *RefAttrib = NewVersion(*RefAttrib)
     }
}
```

figure 10 : updating the attribute of a bound VARA

4.3 Examples of commands

To illustrate the interactions between the collector and the mutator, some examples of commands are given, an access command (`Left`), the backtrack command (`Backtrack`) and a binding command (`BindVar`).

4.3.1 How the mutator is involved in copying objects

The command `Left` returns the left component of a binary "cons" term. In case this component is a reference to an object still in the fromspace, the command applies the `NewVersion` procedure. Therefore, Prolog system never gets any reference into the fromspace and the allocation area will never contain any reference to the fromspace. Of course, the NewVersion procedure may enforce the copy of an object.

```
Command WORD Left(Cons)
WORD Cons ;
{    exclusion (CopyObj)
          return(NewVersion((CONS*) Cons.info)->left)) ;
}
```

figure 11 : Accessing a cons, the command `Left`

4.3.2 Synchronizing problems while backtracking

The command `Backtrack` retrieves the top of the backtrack stack. Due to concurrent use of the trail, this command is executed under the `Trail` exclusion. Variables belonging to the current trail section are reset and registers G, S and T are updated. If the collector has not yet started the copy of the recovered choice point, (12-14) this is what we have called a *deep backtrack* condition, §4.2.4, the backtrack stack seen by the collector is forced to shrink (12-19).

```
1    Command Backtrack() {
2         exclusion (Trail) {
3              TRAIL * TT = (((TRAIL*) T.info)->next).info ;
4              while (TT != nil) {
5                   VAR * Refvar = (TT->var).info ;
6                   (RefVar->binding).tag = free ;
7                   TT := TT->next ;
8              } ;
9              LEVEL * ChoicePoint = S.info ;
10             G = ChoicePoint->goal ;
11             Clause = ChoicePoint->clause ;
12             S = ChoicePoint-> next ;
13             ((TRAIL*) T.info)->next = ChoicePoint->trail
14             if (ChoicePoint == SM) {
15                  exclusion (CopyObj) {
16                       G = NewVersion(G) ;
17                       S = NewVersion(S) ;
18                  }
19                  SM = S.info ;
20                  TM = &(((TRAIL*) T.info)->next)
21             }
22        }
23   }
```

figure 12 : Backtracking operation, the command `Backtrack`

4.3.3 An example of allocation : binding a variable

This command `BindVar` creates an empty trail element requiring two words in the allocation area. At the beginning, there is a check (13-4) to see if there is enough space in the allocation area. The test is meant to always succeed because the system is supposed to be suspended inside the command `Reduce` if space is insufficient (see §4.1).

```
1    Command BindVar(Var, Term)
2    WORD Var, Term ;
3    { WORD * NewT ;
4         CheckAllocation(2);
5         ((VAR *)Var.info)->binding = Term ;
6         ((T.info)->var).info = Var.info
7         (++A)->tag = T_refvar ; /* creating a new empty */
8         A->info = nil ;         /* trail element     */
9         NewT = A ;
10        (++A)->tag = T_trail ;
11        A->info = T.info ;
12        T.info = NewT.info ;
13   }
```

figure 13 : Binding a variable, the command `BindVar`

5 Implementations of Realtime garbage collectors

We have implemented two kinds of realtime garbage collectors, a pseudo-parallel and a parallel one.

5.1 pseudo-parallel garbage collectors

The pseudo-parallel GC emulates two processes on a single processor. The first process is the Prolog system and the second is MALI garbage collector. The code for both processes is written in C. A corroutining mechanism provides sequencing between the two processes. Garbage collection is done incrementally : the smallest grain of work is either an iteration of the `CopyLevel` loop (5-2) or an execution of procedure `UnTrailOneElt` (figure 8). These steps of garbage collection may be done each time a mutator command requires memory allocation. Therefore the garbage collecting time is spread along Prolog interpretation time, but the two processes never work really in parallel.

In this implementation, the `Copy` and `Trail` exclusions are naturally realized by the chosen step of garbage collection.

5.2 parallel garbage collectors

To implement our parallel GC, we have chosen a private memory architecture, figure 14 : one processor, MALI, supports the garbage collector together with the commands and has exclusive access to Prolog memory state. The other processor, the host, supports Prolog interpretation. The two processors communicate via a small shared memory. Processor MALI is a specially microprogrammed processor board to be installed in an IBM PC compatible computer.

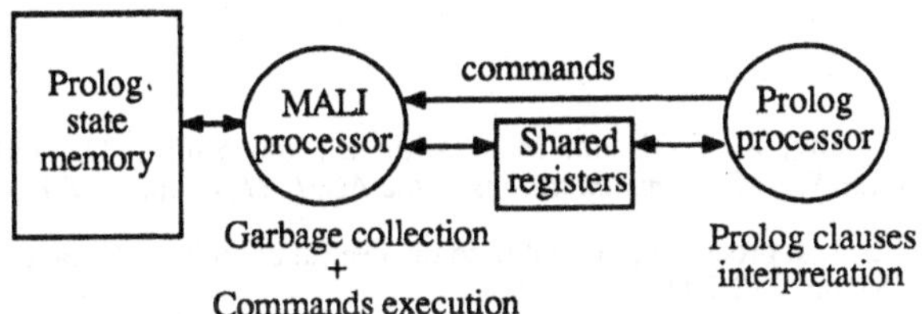

figure 14 : a MALI processor architecture

The GC process runs as a background task. Commands sent by the host are processed by the mutator as interruptions.

The `Copy` exclusion is realized in the GC code by disabling interruptions of the micromachine. Commands are delayed during the copy of objects, but this latency is reasonably short, see §6.1. The `Trail` exclusion is implemented with boolean variables plus a hardware mechanism to suspend and restart a command. The time spent by the collector to process a trail element is long enough to justify such a selective mechanism for implementing the `Trail` exclusion.

6 Discussion and results

The parallel garbage collector has been implemented on a PC AT with a 5MHz 80286 interpreting Prolog and a 6MHz microprogrammed processor built around an AMD 2916 supporting MALI. Compared with a software version of MALI implemented in C on the same PC having a serial garbage collector, this parallel version showed a speedup factor of 2. This result is essentially due to the fact that the commands, which roughtly represents 50% of the total execution time with the software version of MALI, are executed on the specialized hardware which is about 10 time faster for that job.

The pseuso-parallel collector has been written in C and is portable with Prolog on any computer. The relative speed of the collector with respect to the mutator is tuned by giving the number of collector steps executed for each allocated word. The best execution time which has been observed with this implementation is 30% longer than that of the serial collector implementation, and it was obtained for a speed ratio less than one step per allocated word. Generally, for programs requiring 70% or less of the memory, a real-time behavior is obtained for a collector speed between 1 and 2 steps per allocated word. As expected, speed degradation increases with collector speed, because useless collecting work is done.

6.1 Interferences between a parallel collector and the mutator

When the parallel collector is running the speed of the mutator is somewhat reduced by two agents.

- The latency of execution of the commands is increased because the collector must complete the atomic action it is currently doing. In our implementation, this latency has been measured to be about 10% of the command execution time.
- As we have said earlier, the mutator may be forced to copy objects. The time spent for these copies has been measured indirectly and represents between 10% to 15% of the command execution time.

Therefore, the command execution time is around 20% slower when the collector is running. In our system, due to the specialized implementation of commands, MALI commands represents only 10% to 20% of the total execution time of Prolog. So the overall degradation induced by the parallel collector is only 3% to 4% of the total system performances.

6.2 Instant reclaiming on backtracking and realtime garbage collector

In a conventional Prolog system, space is immediately reclaimed on backtracking. With our GC, this reclaiming is not entirely applicable because the order of objects in memory is modified by Cheney's copying technique. On backtracking, we can only reclaim memory in the allocation area, where objects are still located in creation order.

The expected normal functioning, also called stable functioning, of such a collector is when the collector is fast enough to avoid suspensions of the mutator for allocation reasons. In our system, the immediate reclaiming in the allocation area has proved to be amply sufficient to slow down apparent consumption of memory so that the garbage collector follows in real time the mutator needs.

7 References

Appleby88. Appleby, K., Carlsson, M., Haridi, S., and Sahlin, D. Garbage collection for Prolog Based on the WAM. *Communications of the ACM 31*, 6 (June 1988), pp. 719-741.

Baker78. Baker, H.G. List processing in real time on a serial computer. *Communications of the ACM 21*, 4 (April 1978), pp. 280-294.

Barklund87a. Barklund, J. A Garbage collection Algorithm for Tricia. Tech. Rept. 37B, UPMAIL, Uppsala University, Sweeden, December, 1987.

Barklund87b. Barklund, J. and Millroth, H. Hash Tables in Logic Programming. In *Proceedings of the International conference on Logic Programming, ICLP87*, Lassez, J.L., Melbourne, Austria, MIT Press, May 1987, pp. 411-427.

Bekkers84a. Bekkers, Y., Canet, B., Ridoux, O., and Ungaro, L. A short note on garbage collection in Prolog interpreters. *Logic Programming News letters* , 5 (1984).

Bekkers84b. Bekkers, Y., Canet, B., Ridoux, O., and Ungaro, L. A memory management machine for Prolog Interpreters. In *Proceedings of the second international Logic Programming conference*, Uppsala, Sweden, July 1984, pp. 343-353.

Bekkers86. Bekkers, Y., Canet, B., Ridoux, O., and Ungaro, L. A memory with a real-time garbage collector for implementing logic programming languages. In *Proceedings of the second International symposium on Logic programming*, IEEE, Salt-Lake City, Utah, 1986, pp. 258-265.

Bekkers88. Bekkers, Y., Canet, B., Ridoux, O., and Ungaro, L. MALI, A memory for implementing logic programming languages. In *Programming of future generation computers*, K. Fuchi, M.N., Tokyo, Japan, Elsevier Sciences Publishers BV North Holland, 1988, pp. 25-34.

Brisset91. Brisset, P. and Ridoux, O. Naive reverse can be linear. In *Proceedings of the international conference on Logic programming*, Paris, France, June 1991.

Bruynooghe84. Bruynooghe, M. Garbage collection in prolog interpreters. In *Implementations of Prolog*, J. Campbell, E.H., 1984, pp. 259-267.

Carlsson87. Carlsson, M. Freeze, Indexing and other Implementation Issues in the WAM. In *Proceedinings of the 4th International Conference on Logic Programming, ICLP87,* Lassez, J.L., Melbourne, Australia, MIT Press, May 1987, pp. 40-58.

Cheney70. Cheney, C.J. A nonrecursive list compacting Algorithm. *Communications of the ACM 13*, 11 (November 1970), pp. 677-678.

Colmerauer82. Colmerauer, A. PrologII : Manuel de référence et modèle théorique. Tech. Rept. GIA, Université d'Aix-Marseille II, 1982.

Huitouze88. le Huitouze, S. *Mise en œuvre de PrologII/MALI*, Ph.D. dissertation, Université de Rennes I, Décembre 1988.

Huitouze90. le Huitouze, S. A new data structure for implementing Extensions to Prolog. In *Proceeding PLILP'90, LNCS 456,* 1990, pp. 136-150.

Lieberman83. Lieberman, H. and Hewitt, C. A real-time garbage collector based on the lifetimes of Objects. *Communications of the ACM 26*, 6 (June 1983), pp. 419-429.

Neumerkel90. Neumerkel, U. Extensible Unification by Metastructures. In *Proceeding of META90,* Leuven, Belgium, May 1990, pp. 352-363.

Pittomvils85. Pittomvils, E., Bruynooghe, M., and Willems, Y.D. Towards a real time garbage collector for Prolog. In *Proceedings of the IEEE Symposium on Logic Programming,* Boston, MA, July 1985, pp. 185-198.

Schimpf90. Schimpf, J. Garbage collection for Prolog based on twin Cells. In *Proceedings of the 2nd NACLP workshop on Logic Programming architecture and implementations,* Austin, Texas, November 1990.

Toura88. Touraïvane, M. *La récupération de mémoire dans les machines non déterministes*, Ph.D. dissertation, Université Aix-Marseille II, Faculté des Sciences de Luminy, Marseille, France, November 1988.

Turk86. Turk, A.K. Compiler Optimisations for the WAM. In *Proceedinings of the 3rd International Conference on Logic Programming, ICLP86,* London, UK, July 1986, pp. 657-662.

A process semantics of logic programs

M. BELMESK
LIFIA-Institut IMAG
46, Félix Viallet 38031 Grenoble Cedex FRANCE

Abstract

The notions of compositionality and equivalence are fundamental questions in programming language semantics. We focus on these notions and study the semantics of logic programs in the setting of a graph model. We represent a logic program by a graph model, we derive some semantics related to the well-known classic semantics of logic programs (success set, computed answer substitution set and finite failure set). Furthermore we consider the set of partial computations, and prove that is compatible with the set of logic consequences of the program. A simulation equivalence with silent (or invisible) steps on these graphs is also considered, we show that the subsumption equivalence defined on logic programs is compatible with this τ-simulation equivalence. Finally, we prove that the τ-simulation equivalence is a congruence w.r.t two graph combining operators which are the counterpart of the union and hiding operators respectively defined on logic programs.

1 Introduction

In logic programs there are many methods of giving semantics to programs. One of the most attractive features of logic programming paradigm based on Horn clause logic is the equivalence of its operational, model-theoretic and fixpoint semantics. These last years, many extensions or variations of pure HCL have been proposed in the literature which extend the classical kernel of logic programming with new features like concurrency, synchronization and flow control.

The classic semantics of logic programming language is unsatisfactory for these notions, it does not allow to characterize different aspects of the program behavior from a single model. So, several models are necessary to capture distinct aspects of logic programs. Furthermore the classic semantics does not exhibit sufficient compositionality, and to deal with single program is not satisfactory, above all, when the issues of concurrency are of concern. Another fundamental question concerns the relation between programs. This

relation is formulated in terms of equivalence relations. Indeed, each selected semantics, depending on the chosen notion of observation, induces an equivalence relation.
In [1], an algebra of logic programs and its compositional semantics was proposed. Also, other authors [2,3,4] have investigated notions of compositionality and equivalences in logic programing style semantics.

The development of these notions of semantics models, compositionality and equivalence of logic programs is the subject of this paper. We develop a behavioral view of logic programs in the setting of a graph model. This allows us to link the field of logic programming and some techniques borrowed from the process description language and leads to a more informative semantics. Therefore, from the graph model many distinct semantics aspects can be selected depending on what is considered observable.

In this paper, from the graph model, we derive some semantics related to the well-known classic semantics of logic programs (success set, computed answer substitution set, finite failure set). Several equivalences can be deduced, in the logic programming style as in [5,6] and process semantic style. Furthermore, we consider the set of partial computations, and prove that is related to the set of logical consequences of a program. we consider also a simulation relation with silent (or invisible) steps, defined on graphs and show that the subsumption equivalence defined on logic programs is compatible with this τ-simulation. Two graph combining operators are considered, they are the counterpart of the union and hiding predicate operators defined on logic programs.We show that the τ-simulation equivalence is a congruence w.r.t these operators.

The rest of this paper is organized as follows. We present some standard concepts in section 2. Section 3 is devoted to a presentation of an algebra and some equivalences on logic programs. In section 4, we define an algebraic model for HCL programs based on transition system and derive three semantics. Sections 5 and 6 define a graph algebra, a τ-simulation equivalence and its relation with the subsumption equivalence and its congruence properties of the τ-simulation w.r.t the operators.

2 Basic notions

A *logic program* is a finite set of definite clauses. A *definite clause is* a clause of the form H :- $B_1,\dots,B_n$ ($n \geq 0$), where H and B_i are atoms. An *atom* is an expression $p(t_1,\dots,t_k)$ $(k \geq 0)$, where p is a predicate symbol and $t_1,\dots, t_k$ are terms. *Terms* are constructed as usual from individual variables, constants and function symbols. A *substitution* is a simultaneous assignment of terms to variables $(x_1/t_1,\dots, x_n/t_n)$ with distinct x_i and each x_i is different from its corresponding t_i. Subst stands for the set of substitutions and *id* for the empty substitution. A substitution θ is said to be *more general* than a substitution β, notation $\theta \leq \beta$, iff there exists a substitution γ such that $\beta = \theta\gamma$. The relation $\leq$ is a preorder and let $\approx$ be the associated equivalence relation (variance). If a substitution θ has the form $(x_1/y_1,...x_n/y_n)$ where all the y_i are new and distinct variables, it is called a renaming substitution. We denote by *var (E)* the set of variables occuring in E, where E is any syntactic object. If θ is a substitution and x a set of variables, the restriction of θ to the variables of x is denoted by $\theta_{/x}$, and the domain of θ is denoted by $dom(\theta)$. A *computed answer substitution* θ for the program P and the goal g is the substitution obtained by restricting the composition of $\theta_1\ \theta_2...\theta_n$ to the variables of g, where $\theta_1\ \theta_2...\theta_n$ is the sequence of *mgu's* used in an *SLD-refutation* of program P. Every instantiation of the answer substitution is a *correct answer substitution*. More definitions of logic programming language terms can be found in [10].

3 Algebra and equivalence of logic programs

Let V be a finite global vocabulary, in which all programs and goals are written, *PRED* be the set of predicate symbols of V and ε a particular predicate symbol. Atom stands for the set of atoms which can be constructed from the predicate and function symbols occuring in V. Π denotes the class of programs .

3.1 Algebra of logic programs

We introduce operators acting on logic programs, this turns Π into an algebra. The operators we describe constitute a subset of those defined in [11].

Definition 1 *(union)*
For any program P_1 and $P_2 \in \Pi$, $P_1 \cup P_2$ is the program obtained as the union of the clauses of P_1 and P_2 . ❑

Definition 2 *(retraction of predicate)*
For any $P_1 \in \Pi$ and $t \in PRED$, $P_1 \backslash t$ is the program obtained from P_1 by removing all the clauses having as a head, the predicate symbol t. ❑
This operation can be generalized to a set of predicates.

3.2 Equivalence of logic programs

In this section we investigate a notion of equivalence for logic programs is based on the syntax, well- known as the subsumption equivalence [12].

Definition 3 *(clause subsumption)*
Let c_1 and c_2 be the following definite clauses $G{:}\text{-}\ D_1,...,D_n$ and $H{:}\text{-}\ C_1,...,C_m$. c_1 is *subsumed* by c_2, notation $c_1 \ll c_2$, if there exists a substitution δ such that $G = H\delta$ and $\{C_1\delta,...,C_m\delta\} \subseteq \{D_1,...,D_n\}$. ❑

This definition can be generalized to programs as follows.

Definition 4 *(program' s subsumption)*
Let P_1 and P_2 be two logic programs. P_1 is *subsumed* by P_2, notation $P_1 \ll P_2$, if for every clause c_2 of P_2, there exists a clause c_1 of P_1 such that $c_1 \ll c_2$. ❑

Definition 5 *(subsumption equivalence)*
Let P_1 and P_2 be two logic programs. P_1 and P_2 are *subsumption equivalent*, notation $P_1 \approx_{sub} P_2$, iff $P_1 \ll P_2$ and $P_2 \ll P_1$. ❑

The equivalence $\approx_{sub}$ can be used as a semantic equivalence, provided that is a congruence with respect to operators previously defined.
Generaly, an equivalence relation $\approx$ is a congruence w.r.t a set Σ of operators, if for every $f \in \Sigma$: $p_i \approx q_i$ for $i = 1, ...,n;$ implies $f(p_1,..., p_n) \approx f(q_1,..., q_n)$

It is obvious that $\approx_{sub}$ is a congruence w.r.t union and retraction operators, that is :

Proposition 1
For any $P, P_1, P_2 \in \Pi$ and $t \in PRED$, $P_1 \approx_{sub} P_2$ implies
- $P_1 \cup P \approx_{sub} P_2 \cup P$ and
- $P_1 \backslash t \approx_{sub} P_2 \backslash t$ ❑

Several other equivalences can be used to characterize the relationship between logic programs, most of them are defined in [12] and [15]. Here, we only recall those related to the success and failure sets.
The success set of a program P, denoted by SSg is the set of all ground atomic formulas A such that, there exists a refutation of A in P. We denote by $SSng$ the set of atoms g which have a refutation with a computed answer substitution . Finally, a program can be characterized by the set of its finite failure, that is, those atoms having only failed

derivations, we denote this set by FF and by FF^k the set of ground atoms which are finitely failed by depth k [10]..

Definition 6

Let P be a logic program and A a ground atom. A is *finitely failed* in P, denoted $A \in FF$, if and only if, for some $k \geq 0$, A is failed by depth k, denoted $A \in FF^k$, defined as follows : $A \in FF^0$ iff, for each clause B :- $B_1, ..., B_n$ in P and each substitution θ, $A \neq B\theta$. $A \in FF^{k+1}$ iff, for every clause B;- $B_1,..., B_n$ in P, if $B\theta = A$ then there exists i such that $1 \leq i \leq k$ and $Bi_\theta \in FF^i$.

Definition 7

The *finite failure* set FF of a program P is defined by $FF = \bigcup_{k \geq 0} FF^k$. ☐

4 An Algebraic model for Horn clause logic program

To define the behaviors of a logic program, we employ a transition system which mimics the operational behavior of a logic program.We consider the set $Subst = \{\alpha, \beta, ..., \gamma\}$ of (visible) actions. $\tau \notin$ Subst denotes a silent or invisible action. A transition sytem is a directed edge-labeled graph with labels from $Act = \{Subst \times PRED\} \cup \{\tau \times PRED\}$.

We need a function *pred: Atom ----> PRED*, which takes an atom and gives the predicate symbol of this atom.

In the rest of this paper, we use transition system and graph to designate the same object.

4.1 Transition system

For any logic program P, we associate a transition system G *defined as follows :*

Definition 8

Given a program P, the associated transition system is $G = (Q, Act, \longrightarrow, Goal)$ where $Q = \{<S, \theta> / S \subseteq Atom \text{ or } S = nil \text{ and } \theta \in Subst\}$ is the set of configurations or agents.

$Act = \{Subst \times PRED\} \cup \{\tau \times PRED\}$. *and*

$\longrightarrow \subseteq Q \times (\{Subst \times PRED\} \cup \{\tau \times PRED\}) \times Q$ is defined as the smallest relation such that

- $< A_1,...,A_i,...,A_n ; \theta > \xrightarrow[pred(A_i)]{\tau} < (A_1,...,A_{i-1},B_1,...,B_m,A_{i+1},...,A_n)\sigma, \theta\sigma >$ for any renamed clause H:-$B_1,...,B_m$ of P and σ such that $A_i\sigma = H\sigma$
- $< \emptyset, \theta > \xrightarrow[\varepsilon]{\theta\alpha} < nil, \theta\alpha >$ for all $\alpha \in Subst$. Intuitively, the agent $< \emptyset, \theta >$ executes the action $\theta\alpha$ and then behaves like $< nil, \theta\alpha >$.

- $< nil, \theta >$ has no transition. This agent $< nil, \theta >$ is necessary because it distinguishes successful termination of the system from unsuccessful one.

Goal $\subseteq Q$. □

The behavior of a logic program P with a goal g is represented by the rooted graph $(G,< g,id >)$ where $< g,id >$ is its root. A *computation* of $(G, < g,id >)$ is a finite or infinite maximal sequence $S_0 \xrightarrow[x_1]{a_1} S_1 \xrightarrow[x_2]{a_2} S_2 \ldots$ where $S_0 = < g,id >$. The set of computations of $(G,< g,id >)$ is given by :

$$seq(G,< g,id >) = \{(S_i \xrightarrow[x_i]{a_i} S_{i+1})_{i<\alpha} \ / \ \alpha<\omega , S0 = < g,id > \ and\ Si = <gi,\theta>\}$$

Let $C_1 = S_0 \xrightarrow[x_1]{a_1} S_1 \xrightarrow[x_2]{a_2} S_2 \ldots S_i \xrightarrow[x_i]{a_i} S_{i+1}$ be a finite sequence that is $S_i +1 = <nil, \theta >$, we denote by $|C_1|$ the length of C_1.

Definition 9

The operational semantics of a graph G is defined by $[G]_{SEQ} = \bigcup_{k=1\ldots n} seq(G,< g_k, id >)$.□

Definition 10

Any prefix of a computation is called a *partial computation*, the set of partial computations is

$$[G]_{pc} = \{s = (< g,id > \Rightarrow < g',\theta >) \ / \ \exists s' \in [G]_{seq} , s \prec s'\}$$

where $s \prec s'$ means that s is a prefix of s'. □

4.2 Success and failure sets

Let g be a unit goal, from the operational semantics defined before, we can derive some semantics which recover the classic semantics proposed in the literature. Depending on the choosen observation (ground, non-ground atomic formulas...) we obtain the following sets :

$$[G]_g = \{g \ / < g,id > \overset{\theta}{\Rightarrow} <nil, \theta> \in [G]_{SEQ}, g \text{ is a ground atom }\}$$

$$[G]_{ng} = \{<g,\theta_{|var(g)} > / < g,id > \Rightarrow < \emptyset,\theta > \in [G]_{pc}, g \text{ is an atom }\}$$

$$[G]_{FF} = \{g \ / < g,id > \overset{\theta}{\Rightarrow} < nil, \theta > \notin [G]_{SEQ} \text{ for all } \theta \text{ and } g \text{ a ground atom}\}$$

The first set $[G]_g$ denotes the success set in the case of the standard semantics of logic programs. The notion of success set we consider with $[G]_{ng}$ is introduced in [15] and extends the standard one given in [10] by allowing non-ground atoms (modulo variance) in the Herbrand interpretations. This set capture the operational notion of computed answer substitution.The last set $[G]_{FF}$ denotes the set of finite failure, that is, the atoms that give rise to only failing computations.

Proposition 2

Let P be a logic programs and $G$the associated graph, then

1) $[G]_g = SSg$

2) $[G]_{ng} = SSng$

3) $[G]_{FF} = FF$

Proof : We give a sketch of the proof of (2) and (3).

Case (2):

For the $(\Rightarrow)$ direction, it suffices to show by induction on the length of the refutation, that for every refutation of a goal g with a computed answer substitution $\theta_{/var(g)}$ there exists a successful computation from the configuration $< g,id >$ to $< \emptyset,\theta >$.

The $(\Leftarrow)$ direction can be proved by showing that each successful computation in the graph corresponds to a refutation of the goal g. By construction, we know that each transition is equivalent to a single resolution step. A refutation starts in the configuration $< g,id >$ and reaches the configuration $< \emptyset,\theta >$ which corresponds to the empty goal and the computed answer substitution $\theta_{/var(g)}$.

Case (3):

Let g be a ground atom, we prove by induction on the length of the sequence ($C = < g,id > \Rightarrow < g', \theta >$ for all θ where $g \in [G]_{FF}$ and $g' \neq nil$) that if $|C| = n$ then $g \in FF^n$ and the converse.

We only prove the $(\Rightarrow)$ direction.

For the base case, if $|C| = 0$, then $C = < g, id >$ and there is no transition from the configuration $< g,id >$. By construction, we know that there is no substitution θ and no clause H:- $B_1,...,B_m$ $m \geq 0$ in P such that $g\theta = H\theta$, it is obvious that $g \in FF^0$.

Now assume the result holds for all $i \leq n$.

Let $C =< g,id > \Rightarrow < g', \theta>$ be a sequence of $[G]_{SEQ}$ such that $|C| = n$ with $g' \neq nil$.. C can be write as $<g,\theta> \xrightarrow[x]{\tau} <g1, \theta 1> \Rightarrow < g', \theta>$. By construction and induction hypothesis, there exists an atom $B \in g1$ such that $B \in FF^{n-1}$and by definition of finite failure then $g \in FF^n$. ❑

5 Graph algebra and equivalence

5.1 Graph algebra

We introduce two operators acting on graphs. The *asynchronous parallel composition* and the *hiding* operators which respectively are the counterpart of the union and retraction combining operators of logic programs, this turns **G** into an algebra.

Definition 11 *(asynchronous parallel composition)*

For any graph $G_1 = (Q, Act, \longrightarrow_1, Goal)$, $G_2 = (Q, Act, \longrightarrow_2, Goal)$ in **G**,

$G_1 \,||\, G_2$ is the graph $(Q, Act, \longrightarrow_1 \cup \longrightarrow_2, Goal)$.
$G_1 \,||\, G_2$ is G_1 "in parallel" with G_2 and a step of $G_1 \,||\, G_2$ is a step of G_1 or a step of G_2. ❑

Definition 12 *(hiding action)*
For any graph $G = (Q, Act, \longrightarrow, Goal)$ and $t \in PRED$, $G \setminus t$ is the graph

$$(Q, Act, \longrightarrow_{\setminus t}, Goal)$$

where $p \xrightarrow[x]{\tau} q$ is in $\longrightarrow_{\setminus t}$ iff $x \neq t$.
$G \setminus t$ is G where all *t-step* have been removed, so, some nodes may be unreachable. ❑

This operation can be generalized to a set of actions $A \subseteq Act$.

5.2 Graph Equivalences

Each method of giving a semantics to a program induces a, possibly different, equivalence relation on programs. Therefore, all the semantics defined before induce an equivalence relation which can be used to compare logic programs. Here, in one hand we relate the equivalence induced by the equality of the set of partial computations and the set of logical consequences of logic programs.Furthermore, by taking into account the interpretation of τ as indicative of silent move, we define a notion of τ- simulation and relates the induced equivalence with the subsumption equivalence defined in the section 2.

Definition 13 *(pc-equivalence)*
Two graphs G_1 and G_2 are *pc-equivalent*, notation $G_1 \equiv_{pc} G_2$, iff

$$[G_1]_{PC} = [G_2]_{PC}$$ ❑

Proposition 3
Let g be a unit goal, P a program and G the associated graph,
$< g,id > \Rightarrow < g',\theta > \in [G]_{PC}$ implies $g\theta :\!-\, g'$ is a logical consequence of P.

Proof:
Let us sketch the proof (by induction on the length k of the computation).
$(k = 1)$: in this case there is a transition $< g,id > \xrightarrow[x]{\tau} < g',\theta >$. g is a unit goal, this transition corresponds to a single resolution step. That is, there exists a clause $H\!:\!-B_1,...,B_n$ in the program P such that $g\theta = H\theta$, $g' = (B_1,...,B_n)\theta$ and $pred(g) = x$. The clause $g\theta\!:\!-\, g'$ is an instance of the clause $H\!:\!-B_1,...,B_n$. Since $P \models H\!:\!-\, B_1,...,B_n$, again by the standard logic we have $P \models (H\!:\!-\, B_1,...,B_n)\theta$ and $P \models g\theta :\!-\, g'$, then $g\theta :\!-\, g'$ is a logic consequence of P.

$(k \geq n)$: in this case, assume that for all partial computation $<g,id> \Rightarrow <g1,\theta'>$ of length $k \leq n$ we have $g\theta'$:- g_1 as logic consequence of P. We have to prove that:

$<g,id> \Rightarrow <g1,\theta'> \xrightarrow[x]{\tau} <g',\theta'\sigma>$ implies $g\theta'\sigma$:- g'. By inductive hypothesis we have $P \models g\theta'$:- g_1 and a gain $P \models (g\theta'$:- $g_1)\sigma$, but we have $P \models g_1\sigma$:- g' therefore $P \models g\theta'\sigma$:-g'. ❑

Corollary 1

Let P_1, P_2 two logic programs and G_1, G_2 the associated graphs, then

$$G_1 \equiv_{pc} G_2 \ \ implies \ \ P_1 \simeq_{log} P_2.$$ ❑

where $\simeq_{log}$ denotes the equivalence between programs having the same logical consequences. The proof is straighforward from the proposition 3.

The notion of simulation has been explored in number of places e.g [13,14], here we consider a relation of τ- simulation.

Definition 14 *(τ-simulation)*

Given two rooted graph $G_1 = (P, Act, \longrightarrow_1, p_0)$, $G_2 = (Q, Act, \longrightarrow_2, q_0)$ a relation $R \subseteq P \times Q$ is a τ-simulation (written R: $G_1 \sqsubseteq_\tau G_2$) iff

- $p_0 \ R \ q_0$ and
- whenever $p \ R \ q$ and $p \xrightarrow[x]{a} p'$, then
  - $a = \tau$ and $p' R \ q$ or
  - there exists a transition $q \xrightarrow[x]{a} q'$ with $p' \ R \ q'$

We write $G_1 \sqsubseteq_\tau G_2$ when $R : G_1 \sqsubseteq_\tau G_2$ for some R and say that G_2 is simulated by G_1. G_1 and G_2 are τ-similar, notation $G_1 \leftrightarrow_\tau G_2$, iff there exists two relations R_1, R_2 such that $R_1 : G_1 \sqsubseteq_\tau G_2$ *and* $R2$: $G_2 \sqsubseteq_\tau G_1$. It is well-known that $\sqsubseteq_\tau \subseteq P \times Q$ is a preorder.

❑

We use this definition for our graph model.

Definition 15

Let P_1 and P_2 be two logic programs and $G_1 = <Q, Act, \longrightarrow_1, Goal>$, $G_2 = <Q, Act, \longrightarrow_2, Goal>$ the graphs associated to P_1 and P_2 respectively. We have:

$G_1 \sqsubseteq_\tau G_2$ *iff* $\forall <s, id> \in Goal, \ (G_1, <s, id>) \sqsubseteq_\tau (G_2, <s, id>)$ ❑

Example: Let P_1 and P_2 be the following programs

P_1: p(x):- q(x) r(x)
q(0):-
r(0) :-

P_2: *p(x):- q(x)*
q(0):-
r(0): -

and $(G_1, < p(x),id >)$, $(G_2, < p(x),id >)$ the following graphs

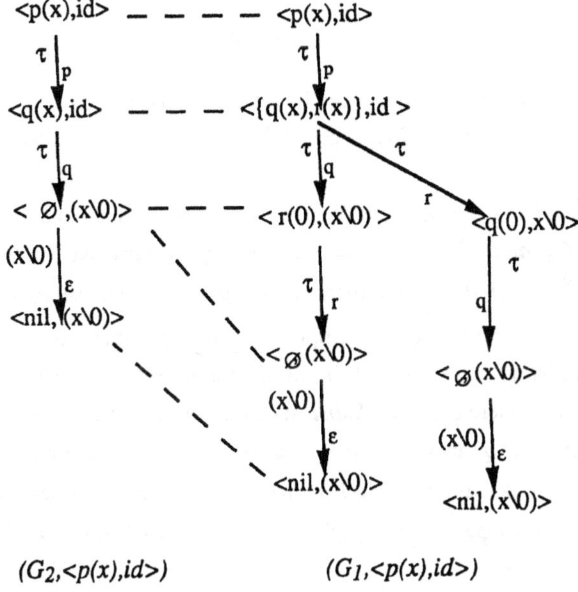

$(G_2,<p(x),id>)$ $(G_1,<p(x),id>)$

We have $(G_1,< p(x),id >) \sqsubseteq_\tau (G_2,< p(x),id >)$. The dotted lines represent the τ-simulation relation between the nodes of $(G_1,<p(x)>)$ and $(G_2,<p(x),id>)$. In order to have a clear graphical representation, we have only shown the relation with a single branch of $(G_1,<p(x),id>)$ and the node which can be reached from $< p(x),id >)$.
Before we investigate the link between this τ- simulation and the subsumption relation, we define a particular relation, we use in the proof of some results.

Definition 16

Let R be a binary relation $R \subseteq Q \times Q$ and X a set of variables.
$< p,\theta > R < q, \varphi >$ if $\exists \rho \in Subst$ such that :

1) $q\rho \subseteq p$
2) $(\varphi\,\rho)_{|X} \approx \theta_{|X}$ where $\approx$ is the variance relation.
3) $< p,\theta > R < nil,\theta >$ is always true for all p. □

Proposition 4

R is a pre-order

Proof : We have to check

1) R is reflexive: $< p,\theta > R < p,\theta >$ is trivial.

2) R is transitive: $< p,\theta > R <q, \varphi>$ and $< q, \varphi > R < r, \zeta >$ implies that there exists $\rho_1,\rho_2 \in Subst$ such that $q\rho_1 \subseteq p$, and $(\varphi\rho_1)_{|x} \approx \theta_{|x}$, furthermore $r\rho_2 \subseteq q$ *and* $(\zeta\rho_2)_{|x} \approx \varphi_{|x}$. Then it suffices to take $\rho_3 = \rho_2\rho_1$ to have $r\rho_3 \subseteq p$ and $(\zeta\rho_3)_{|x} \approx \theta_{|x}$, it follows that $< p,\theta > R < r, \zeta >$. □

Theorem 1

Let P_1 and P_2 be programs and G_1, G_2 be the associated graphs.
$P_1 \ll P_2$ implies $\forall < s, id > \in Goal, (G_1, < s, id>) \sqsubseteq_\tau (G_2, < s, id >)$

Proof :

Let $v = var(s)$, if $P_1 \ll P_2$, we have $\forall c_2 \in P_2, \exists c_1 \in P_1$ such that $r_1 \ll r_2$. Let c_1 and c_2 be $H{:}\text{-}\ C_1,...,C_n$ *and* $G{:}\text{-}\ D_1,...,D_m$ respectively. Assume we have $< p, \theta> R < q, \varphi >$ and $< p,\theta > \xrightarrow[x]{a} < p',\theta' >$, then there exists $p_1 \in p$, a clause $c_1 \in P_1$ and $\sigma \in Subst$ such that $H\sigma = p_1\sigma$. From the subsumption hypothesis, there exists $\psi \in Subst$ such that $H = G\psi$ and $\{D_1\psi,...,D_m\psi\} \subseteq \{C_1,...,C_n\}$ with $dom(\psi) \subseteq var(G)$ so we have $H\sigma = G\psi\sigma = p_1\sigma$ and also there exists $\rho \in Subst$ such that $q\rho \subseteq p$. There are two cases:

<u>Case1:</u> There is $q_1 \in q$ and $q_1\rho = p_1$, so we have $H\sigma = G\psi\sigma = p_1\sigma = q_1\rho\sigma$, since $(dom(\rho) \cap var(G) = \emptyset)$ and $(dom(\psi) \cap var(q_1) = \emptyset)$ this implies $G\rho\psi\sigma = q_1\rho\ \psi\sigma$, therefore there exists $<q,\varphi> \xrightarrow[x]{a} <q', \varphi'>$ where $q' = (q - \{q_1\} \cup \{D_1,...,D_m\})\rho\psi\sigma$ and $\varphi' = \varphi\rho\psi\sigma$.

Now, we show $<p', \theta'> R <q', \varphi'>$. In the transition $< p,\theta > \xrightarrow[x]{a} < p',\theta'>$, $p' = (p - \{p_1\} \cup \{C_1,...,C_n\})\sigma$ and $\theta' = \theta\sigma$. We have $q\rho \subseteq p$ and $(dom(\psi) \cap var(p) = \emptyset)$, so $q\rho\psi\sigma \subseteq p\ \sigma$ furtheremore $\{D_1\psi,...,D_m\psi\} \subseteq \{C_1,...,C_n\}$, if we take $\rho' = id$ then we have $q'\rho' \subseteq p'$. From $<p,\theta> R <q,\varphi>$, we have also $(\varphi\rho)_{|v} \approx \theta_{|v}$ and $dom(\psi) \cap v = \emptyset$, *so* $(\varphi\rho\psi\sigma\rho')_{|v} \approx \theta\sigma_{|v}$ *i-e* $(\varphi'\rho')_{|v} \approx \theta'_{|v}$.

<u>Case2</u>: There is no substitution ρ and atom $q_1 \in q$ such that $q_1\rho = p_1$, $a = \tau$ we show that $< p', \theta' > R < q, \varphi >$. We know that $< p, \theta > R < q, \varphi >$ so there exists ρ such that $q\rho \subseteq p$, $(\varphi\rho)_{|v} \approx \theta_{|v}$, and $(p\text{-}\{p_1\})\ \sigma \subseteq p'$, $q\rho\sigma \subseteq (p\text{-}\{p_1\})\sigma \subseteq p'$ so it suffices to take $\rho' = \rho\sigma$ to have $(\varphi\rho')_{|v} \approx \theta'_{|v}$. □

Proposition 5

Let P_1, P_2 be two logic programs and G_1, G_2 the associated graphs then $\forall < s, id > \in Goal, (G_1, < s, id >) \sqsubseteq_\tau (G_2, < s,id >)$ does not imply $P_1 \ll P_2$.

Proof: As proof we give a counter-example. Let P_1 be the program $p(x) :\text{-}\ q(f(x))$ and P_2 be the program $p(x) :\text{-}\ q(g(x))$, we have $(G_1,< p(x), id >) \sqsubseteq_\tau (G_2, < p(x),id >)$ however

$P_1 \ll P_2$ is not true. □

Corollary 2

$P_1 \approx_{sub} P_2$ implies $G_1 \leftrightarrow_\tau G_2$.

Proof:

Two applications of the theorem 1 gives the required result. ❑

6–Congruence properties

In this section we investigate the congruence properties of τ-simulation w.r.t operators.

6.1- τ- simulation equivalence and parallel composition

$\leftrightarrow_\tau$ is a congruence w.r.t asynchronous composition. For this purpose it suffices to show the following proposition:

Proposition 6

For any $G_1, G_2, G_3 \in G$, $G_1 \sqsubseteq_\tau G_2$ implies $G_1 || G_3 \sqsubseteq_\tau G_2 || G_3$

Proof:

We prove that $\forall <s, id> \in Goal, (G_2 || G_3), (G_1 || G_3, <s, id>) \sqsubseteq_\tau (G_2 || G_3, <s, id>)$. Let $v = var(s)$ and R be the τ-simulation between the nodes of G_1 and G_2. In order to establish $G_1 || G_3 \sqsubseteq_\tau G_2 || G_3$, we exhibit a τ-simulation R' between the nodes of $G_1 || G_3$ and $G_2 || G_3$. Let R' be the relation given by the definition 14.

Now assume that $p R' q$ and that $<p,\theta> \xrightarrow[x]{a} <p',\theta'>$ occurs in $G_1 || G_3$. From the construction of $G_1 || G_3$, the transition $<p,\theta> \xrightarrow[x]{a} <p',\theta'>$ is a transition of G_1 or G_3.

<u>Case 1</u>: $<p,\theta> \xrightarrow[x]{a} <p',\theta'>$ is a transition of G_1. This case is obvious since $G_1 \sqsubseteq_\tau G_2$.

<u>Case 2</u>: $<p,\theta> \xrightarrow[x]{a} <p',\theta'>$ is a transition of G_3. Assume $<p,\theta> R' <q,\varphi>$ and $<p,\theta> \xrightarrow[x]{a} <p',\theta'>$, in this case the transition is due to a step resolution where the atom p_1 is resolved with a clause $H\text{:-}C_1,\ldots,C_n$ whose head is unifiable with p_1, let σ be an unifier. Now $<p,\theta> R' <q,\varphi>$ implies there exists ρ such that $q\rho \subseteq p$ and $\varphi\rho_{/v} \approx \theta_{/v}$, if $p_1 \notin q\rho$ then G_2 have no transition corresponding to $<p,\theta> \xrightarrow[x]{a} <p',\theta'>$ $p' = (p-\{p1\} \cup \{(C_1,\ldots C_n)\})\sigma$ and $\theta' = \theta\sigma$ so $<p',\theta'> R' <q,\varphi>$ is true. Now if $p_1 \in q\rho$ we have a transition $<q,\varphi> \xrightarrow[x]{a} <q',\varphi'>$ of G_3 which is due to the step resolution with he clause $H\text{:-}C_1,\ldots,C_n$ with $q' = (q-\{p1\} \cup \{(C_1,\ldots C_n)\})\sigma$, $\varphi' = \varphi\sigma$ it suffices to take $\rho' = id$ to have $<p',\theta'> R' <q',\varphi'>$ then obviously R' is a τ-simulation. ❑

6.2 τ- simulation equivalence and hiding action

$\leftrightarrow_\tau$ is a congruence w.r.t action hiding. For this purpose it suffices to show the following proposition:

Proposition 7

For any $G_1 \in G$ and $t \in PRED$, $G_1 \sqsubseteq_\tau G_2$ implies $G_1\backslash t \sqsubseteq_\tau G_2 \backslash t$.

Proof: By using the same relation R' the proof is direct. ❑

Conclusion and future work

We have proposed a graph model to represent logic programs and to study some equivalence relations and congruence properties of some graph combining operators. We have shown some relationships between equivalences induced by semantics in logic programs style and equivalences on graph models. In a future work we believe that further analysis is required to define first a graph model which is less abstract and to extend the graph algebra by considering other operators (like synchronous parallel composition which will represent the intersection of logic programs) and to study other equivalences like the colored trace equivalence (which coïncide with the bisimulation equivalence [16]) since the graph we define is really a colored graph.

References

[1] A. Brogi, P. Mancarella, D. Pedreschi, F. Turini. Composition operators for logic theories. Computational logic. Symp.proceedings. Brussels, nov 1990.

[2] H. Gaifman, E. Shapiro. Fully abstract compositional semantics for logic programs. 6th annual ACM symp. of POPL. Jan. 1989.

[3] H. Gaifman, E. Shapiro. Proof theory and semantics of logic programs. 4th IEEE annual symp. on LICS. pp 50-62. 1989.

[4] M.J.Maher. Semantics of logic programs. Ph.D.dissertation, University of Melbourne, 1985.

[5] A. Corradini. An algebraic semantics for transition systems and logic programing. Ph.D Thesis, Dipermento di informatica, università di Pisa, Dec,1989.

[6] A. Corradini, U. Montanari. Towards a process semantics in logic programming style. LNCS 415

[7] R.Gerth, M.Codish, Y.Lichtenstein and E. Shapiro. Fully abstract denotational semantics for Flat concurrent Prolog. 3th IEEE annual symp. on LICS. pp 320- 335. 1988.

[8] J.Kok. Specialization in logic programming: from Horn Clause Logic to Prolog and Concurrent Prolog. LNCS 430.

[9] V.A. Saraswat. The concurrent logic programming language CP : Definition and operational semantics. 4th annual ACM symp. on POPL. pp 49- 62. 1987.

[10] J.W.Lloyd. Foundations of logic programming. Springer Verlag, New York. 1987.

[11] P.Mancarella, D. Pedreschi. An algebra of logic programs. In Proc. of 5th int. conf Symp. of logic Programming, 1988.

[12] M.J.Maher. Equivalences of logic programs. In J. Minker (ed.), Foundations of Deductive Databases and Logic programming, pp.627-658, Morgan Kaufmann Publishers, Los Altos. 1987.

[13] D.M.R.Park. Concurrency and automata on infinite sequences. In Proc. 5th GI conf LNCS 104.pp 167- 183.

[14] M.Hennessy , R. Milner. Algebraic laws for nondeterminism and concurrency. Journal of the ACM, 32(1). pp 137-161. 1985.

[15] M.Falaschi, G. Levi, M. Martelli, C. Palamidessi. Declarative Modeling of the Operational Behaviour of Logic Languages. In Proc. 5th Int. Conf. Symp. on Logic Programming, Seattle, MIT Press, pp. 993-1005, 1988.

[16] R.J. Van Glabbeeck . Comparative Concurrency Semantics and Refinement of Actions. Ph.D dissertation, 1990.

Logical Operational Semantics of Parlog
Part II: Or-Parallelism *

Egon Börger
Dip. di Informatica
C.so Italia 40
I-56100 PISA
boerger@dipisa.di.unipi.it

Elvinia Riccobene
Dip. di Matematica
V.le Andrea Doria 6
I-95125 CATANIA
riccobene@mathct.cineca.it

Abstract

This paper refines the definition of a complete, mathematical semantics for the parallel logic programming language PARLOG provided in [3] by giving an explicit *Evolving Algebras* formalization of the OR-Parallellism in Parlog. In particular we extend the algebras of [3] by new rules which describe the dynamics of the crucial *candidate clause* search in Parlog which was left abstract in [3].

1 Signature extension of Parlog Algebras

Due to space limitations we skip the Introduction and refer the reader for motivation and basic definition to [1], [2], [3] of the latter of which this paper is a direct sequel. The abstract function *candidate-clause(database,node,lit)*, which in [3] realizes the whole Parlog reduction process of a literal *lit* yielding as result the body of the selected clause and a unifying substitution of its head and *lit*, receives here an explicit descripition by *Evolving Algebras* rules. To this purpose we extend the previous Parlog Algebras, introducing new universes and functions and modifying the existing ones.

Due to space restrictions we can spell out here only a few new definitions and crucial rules, refering the reader for a full account to [4]. In the full paper [4] we prove the correctness of this *candidate-clause* specification w.r.t. its definition in [3] by showing that our extended Parlog algebras developed here are "conservative" over the Parlog algebras of [3].

When the Parlog computation system begins the reduction process of a given procedure call *lit*, it comes into the so-called **test-commit-output-spawn** phase. In the test phase

*The second author has been partially supported by "Progetto Finalizzato Sistemi Informatici e Calcolo Parallelo" of CNR, under Grant n.90.00671.69.

the system tries to find a clause which satisfies the **candidate clause** condition amoung those defining the procedure which comes in the form of

$$procdef(database, lit) = \underbrace{C_{1_1}.C_{1_2}.\cdots.C_{1_{m_1}}}_{S_1};\cdots\cdots;\underbrace{C_{n_1}.C_{n_2}.\cdots.C_{n_{m_n}}}_{S_n}$$

The *parclauses* S_i are searched through sequentially (*seq-search*). Each *parclause* is a sequence $C_1.C_2.\cdots.C_m$ of *guarded clauses* which are searched through in parallel (*or-search*). The *try-clause* computation, whether a clause is a *candidate clause* for *lit*, checks whether the clause is an unguarded clause for which **input matching** [1] of *lit* and *head* succeeds, or wheter it is a guarded clause for which: a) the *input-match* computation (unification of *lit* and *head*) succeeds with a substitution s; b) the *guard-eval* computation of the clause guard by the program succeeds with a substitution s'; c) the substitutions s and s' are consistent. (*seq-search, or-search, try-clause, input-match, guard-eval* are new tags.)

When a *candidate clause* is found, the calling literal *lit* commits to it (**commit** phase) - interrupting the search for (other) candidate clauses -; output unification is performed between the output mode argument terms of *lit* and those of the selected clause (**output** phase); the *lit* computation is reduced to the evaluation of the *candidate clause*'s body under the computed output substitution (**spawn** phase).

For an explicit description of this *candidate clause* search, we imagine each *or-par* node as root of a computation subtree performing the **test-commit-output** phase and of another (later created) subtree to perform the spawn phase.

During the *input-match* computation a node may come into a new mode *suspended*, namely when there is a unifier for the calling literal and the clause head which however tries to bind a variable occurring (in a term) in an *input* argument position of the call. During the spawn phase we need a function *sub* : **Node** $\rightarrow$ **Sub** which yields a substitution which is known at the given node, but which is not transparent to the main system. In the *or-par* node subtree description, the function *goal* will be used to pass from parent to child the calling literal *lit* which is responsable for the *candidate clause* search. The same function will be used to report from child to parent the *body* of the identified *candidate clause* for the calling literal.

Some other new mostly self explanatory functions will be introduced where needed in the rules.

2 The *or-par* node operation

The transition rules of Part I (see [3]) that perform the *or-par* node operation, are replaced by those that we are describing in this section.

When the *or-par* node has mode *starting*, it has just been created as child of a node

[1]**Input matching** (of a literal *lit* and a term *t*) is defined as unification of *lit* and *t* in which no variable is substituted which occurs in an *input argument* of *lit* (with respect to that program).

In contrast to input matching one speaks of **output unification** to refer to a unification of two terms which appear in an *output mode argument* (of a literal w.r.t. a given program). This façon de parler stresses that for the unification of terms occurring in *output mode arguments* there is no restriction on the direction of the bindings (from goal to clause head (*output*) or viceversa (*input*)).

labelled *and-seq* (see [3]): *lit* is the procedure call which has to be reduced now by starting the *candidate clause* search through its $procdef(database, lit)$, followed later by the computation of the selected body (if any). For this purpose the **or-par starting rule** passes the control to a newly created child labeled *seq-search* whose subtree manages the search of a *candidate clause* through the parclauses $S_1, S_2, \cdots, S_n$ of $procdef(database, lit)$ that are tried in sequence.

If the child *seq-search* is *reporting* with *failure*, no *candidate clause* has been found and by the **or-par failure rule** the *or-par* node becomes *reporting* with *failure*.

If the child is *reporting* with *success*, a *candidate clause* has been found and the system has commited to it (by effecting the corresponding output substitution). Therefore the body of this *candidate clause* is reported by the *goal* function of *seq-search* node.

The subsequent step, triggered by the following **or-par spawn rule**, is the computation of the selected clause body: a new *or-par* child node is created to perform the body computation. *Seq-search* node becomes *dormant*.

or – par spawn rule

If tag(p) = or-par & mode(p) = waiting & tag(child(p)) = seq-search
& mode(child(p)) = reporting & res(child(p)) = success
then
Create p-*subtree of child* temp
with tag $\tilde{\text{tag}}$(goal(child(p))), *mode* starting *and*
goal(temp):= goal(child(p)),
mode(child(p)):= dormant
end Create

If the body computation ends with *failure*, by the **or-par failure rule**, the *or-par* node becomes *reporting* with *failure*; if the body computation ends with *success*, by the **or-par success rule** the *or-par* node becomes *reporting* with *success*.

3 The *seq-search* node operation

A node labelled *seq-search* with goal *lit* manages the search of a *candidate clause* through the ";-"sequence of blocks of clauses that form the procedure defining *lit*.

When the node *seq-search* is *starting* and $procdef(database, lit) = nil$, by the **seq-search failure rule 1** *failure* is reported.

When the node *seq-search* is *starting* and $procdef(database, lit) \neq nil$, by the **seq-search starting rule** we create a child labeled *or-search* which becomes *starting*. By the function *parcl* the first *parclause* of those defining the procedure of the calling literal (which have to be tried in sequence searching a *candidate clause*) is passed to the child *or-search* together with goal *lit*, and a function *seqbloc* at the node *seq-search* is set to the remaining *parclause*.

If the search performed by the child labeled *or-search* fails, by the **seq-search continuation rule** a new child *or-search* is created with the same function for the next *parclause* in *seqbloc*; the previous node *or-search* becomes *dormant*.

If all *parclauses* have been tried and no *candidate clause* has been found, the *seq-search* node reports *failure* by the **seq-search failure rule 2.**

If the subcomputation of one *or-search* child node ends with *success* (namely a *candidate clause* has been found and - through committing to it - the *sub* substitution has been extended by the **output substitution**), by the **seq-search success rule** the node *seq-search* becomes *reporting* and reports to ist parent the body of the *candidate clause* by the *goal* function, thus preparing the start of the spawn phase.

The *seq-search* node rules are similar to those of our *and-seq* nodes (see [3]) and to the description of SEQ node for OCCAM [11] and are therefore skipped here.

4 The *or-search* node operation

A node *or-search* performs the parallel search of a *candidate clause* through a *parclause*, say $C_1.C_2.\cdots.C_m$.

When a node labelled *or-search* becomes *starting*, by the **or-search starting rule** as many *or-search*'s children labelled *try-clause* are created, as there are clauses in the given *parclause*. (The number of these clauses is computed by the function *length* applied to [.]-*decomp* evaluated on *parcl(p)*.)

Each child *try-clause* receives the calling literal as *goal* and one clause (renamed corresponding to the node) of the parent's *parclause*. All children get mode *ready*. When at least one (but maybe more than one) child *try-clause* ends its subcomputation with *success*, one computed *candidate body* together with its unifying substitution is selected by the waiting *or-search* parent all of which children's computations are aborted (by changing their mode into *dormant*); the *or-search* node becomes *reporting* thus realizing the commit phase of the current call. The current "transparent" substitution *sub* is extended by the, up to now "non-transparent", unifier's bindings for output variables reported by the *sub* function on the selected child node *try-clause*. We note that the following transition rule is performed if the guard condition *allowed*(p) = 1 is true. *Allowed* is an external function (see [9]) having value 1 on at most one node p at any moment. The role of this function is to avoid that the current value of *sub* could be simultaneously updated by more than one node p performing parallel processes. Using this function we simulate Parlog's approach to the commit/output access to shared information by parallel processes.

or – search success rule

```
Let  succ-resp(q)  (parent(q) = p & mode(q) = reporting & res(q) = success)
If tag(p) = or-search & mode(p) = waiting & for some q succ-resp(q)
& allowed(p) = 1
then
      Let choice ε q succ-resp(q),
            report from p-subtree with success
               saving goal from child choice
            sub:= join(sub, sub(choice))
```

where Hilbert's ε-operator is used as (external) choice function. This reflects the decision taken by the Parlog design to consider the *or-search* as non deterministic (implementation defined) language feature. (The *reporting* abbreviation is the obvious extension of the *reporting* abbreviation in [3].)

If all *try-clause* children end their subcomputation with *failure*, no *candidate clause* is in the initially given *parclause*. By the **or-search failure rule** the node *or-search* reports *failure*.

5 The *try-clause* node operation

The subcomputation of a node p labeled *try-clause* consists of verifying whether the associated clause *clause(p)* is a *candidate clause* for the calling literal *goal(p)* or not.

To be *candidate* the clause must satisfy the **input matching** and the **guard evaluation** conditions. These are tried in parallel, controlled by two newly created p-children labelled *input-match* resp. *guard-eval*. By the function *sub* we refer to ("non-transparent") substitutions attached to both nodes: in *sub* on node p labeled $input - match$ is coded the substitution satisfying the **input matching**, whereas the *sub* on $guard - eval$ node is the result of clause guard computation. In order that the clause is *candidate*, these last two substitutions have to be consistent. If they are and if the clause is selected by the **or-search success rule**, the union of these substitutions will extend the "transparent" substitution *sub*.

By the **try-clause starting rule** the child *input-match* receives from the parent - via *goal* function - the literal to be reduced, whereas the child *guard-eval* receives the guard of the clause. The function *sub* for the "non-transparent" bindings of the guard evaluation will be initialized to *empty* value for the child *guard-eval*; this is in accordance with the fact that the guard evaluation is done in parallel with the attempt to input matching and therefore without knowing about the previously computed variable bindings. It will also help to control the empty guard case.

If one of the two children fails its subcomputation, by the **try-clause failure rule 1** the *try-clause* node reports *failure* (because one of the two conditions, **input matching** or **guard evaluation**, has failed). There is another possibility for failure of a node labelled *try-clause*, namely if both children report *success* but their computed substitutions are inconsistent. In this case the **try-clause failure rule 2** checks the consistency of two substitutions using a function *cons* : $\mathbf{Sub} \times \mathbf{Sub} \rightarrow \{0, 1\}$ which says whether given substitutions are consistent or not. If these two substitutions computed by the children of a *try-clause* node are consistent, by the **try-clause success rule** the node*try-clause* becomes *reporting* with *success* and updates the value of the function *sub* by the union of the two children's substitutions (this is the **output substitution** relative to this clause) and, via the *goal* function, the body of the *candidate clause* by which the calling literal will be replaced (if this clause will be selected in the *or-search success rule*).

6 The *input-matching* node operation

A node labelled *input-match* is a leaf node of the Parlog computation "tree".

When the control arrives to an *input-match* node, we try to compute **input matching** between the head of the given clause (*clhead(clause(parent(p)))*) and the calling literal under the current substitution (*subres*(*goal*(*p*), *sub*)).

If there is no unification at all between *head* and*lit*, then by the **input-match failure rule** the node *input-match* becomes *reporting* with *failure*.

If there is some unifier *unif*, then we must control that it satisfies the *input matching* condition, i.e. that the set *domain*(*unif*) of variables bound by *unif* has empty intersection with the set *in_var*(*lit*, *db*) of variables occurring in input positions of *lit* w.r.t. the underlying program *db*. If the intersection is empty, by the **input-match success rule** the node *input-match* becomes *reporting* with *success* and updates the *empty* value of *sub* function by the unifying substitution.

If the intersection is not empty, then by the following rule the node *input-match* changes its mode into *suspended* and the *input* variables of *lit* that the unifier tries to bind are put into a set *susp_var(p,db)* of suspended variables:

input – match suspension rule

```
Let lit subres(goal(p),sub)
    head clhead(clause(parent(p)))
If tag(p) = input-matching & mode(p) = starting
& unify(lit,head) ≠ nil
& domain(unify(lit,head)) ∩ in_var(lit,db) ≠ ∅
then
      mode(p):= suspended,
      susp_var(p,db):= in_var(lit,db) ∩ domain(unify(lit,head)).
```

If and when one variable of *susp_var(p,db)* is bound by one of its producer processes - which means that it enters into the "transparent" substitution **sub** -, the node *input-match* starts to work again (its mode changes from *suspended* to *starting*) in its attempt to find an **input matching**.

input – match re – starting rule

```
If tag(p) = input-match & mode(p) = suspended
& susp_var(p,db) ∩ domain(sub) ≠ ∅
then
      mode(p):= starting,
```

7 The *guard-eval* node operation

In Parlog, guards of clauses are required to be "safe", i.e. a guard evaluation never attempts to bind a variable that occurs in an input mode argument position of the calling literal. This *guard safety* property has to be assumed by the programmer or may be considered at compile time. In our description we assume that guards are *safe*.

If the current clause has no guard, by the **guard-eval empty rule** the node becomes *reporting* with *success*; the *empty* substitution attached to the node will have no contribution to the clause output substitution (which in this case will be defined by the unifying substitution between clause head and calling literal).

If the clause guard is not empty, the *guard-eval* node has to start a (sub)computation which performes the **guard evaluation**. It consists of computing a substitution satisfying the guard of the given clause. During the guard evaluation some variable bindings will be computed. These bindings are not immediately "transparent" to the system: they become so only if and when the relative clause has been chosen as *candidate*. The guard variable bindings are part of the **output substitution** by which the "transparent" substitution *sub* will be extended if the clause will be selected by the **or-search success rule**. Thus we must consider the substitution attached to the *guard-eval* node (initialized to *empty* by the **try-clause starting rule**) as "transparent" for its entire subcomputation, but not for the whole system. We refer to it during the subcomputation by *sub*(*curr_geval*(*p*)) where *curr_geval* points to the nearest *guard-eval*'s subtree root (in ascending order). (Note that guard evaluation may be nested).

The *guard-eval* node works in the same way as the Parlog "tree" *root* having the renamed clause guard as its query. Therefore its subcomputation is performed by the rules seen so far using *sub*(*curr_geval*(*p*)) instead of *sub*. To express this difference in the use of our rules, we adapt from [5] a function $what\_to\_do : \mathbf{Node} \rightarrow \{run, guard\_test\}$ where we indicate by *run* that we are in run phase (i.e. using the "transparent" substitution *sub*) and by *guard_test* that we are in guard evaluation phase (i.e. using the "non-transparent" substitution *sub*(*curr_geval*(*p*)) instead of *sub*). *what_to_do* has to be initialized to the value *run* on the *root* of the Parlog "tree" (see [3]). The operation of a node labeled *guard-eval* is therefore as follows.

By the **guard-eval starting rule** the node labeled *guard-eval* (in *starting* mode) creates a child in *starting* mode to which the new query is passed by the function *goal*.

The label of that child is computed by the function $\tilde{tag}$ on *goal(p)* and *what_to_do* is updated to *guard_test*.

The *guard-eval* node receives the control again when its subcomputation ends. If the latter ends with *success*, then by the **guard-eval success rule** the node changes its mode into *reporting* and keeps stored - for use of its parent (*try-clause*) - the value of the computed substitution (satisfying the **guard evaluation**) coded in *sub(p)*. Its child becomes *dormant*.

If the child's computation ends with *failure* - there is no extension of the current substitution which satisfies the **guard evaluation** -, by the **guard-eval failure rule** the node *guard-eval* becomes *reporting* with *failure*.

To perform the guard evaluation the whole set of previous transition rules (included those presented in Part I (see [3])) are used. To distinguish if a rule is used in *run* or *guard_test* phase, we need to refine some rules. We add all to *starting* rules, *or-par spawn rule* and *seq-search continuation rule* to pass the value of *what_to_do* from parent to child, and in all rules refering to *sub* we add the condition "what_to_do(p):= run" among the guards. We also duplicate these rules replacing "run" by "guard_test" and "sub" by "sub(curr_geval(p))".

In the **input-match re-starting rule** for "guard_test" we have to assume that

$$\mathsf{sub}(\mathsf{curr\_geval}(p)) := \mathsf{join}(\mathsf{sub}(\mathsf{curr\_geval}(p)), \mathsf{sub}(\mathsf{sibling}(\mathsf{curr\_geval}(p))))$$

to assure that a possibly successful input matching unifier may help to unlock variables suspended during the guard evaluation of the sibling node. Suppose that a guard of the clause C has the form $p(x_i),q(x_i,y_i)$ and x_i is an input variable for both predicates p(x) and q(x,y) (but not input variable for the calling literal). When the guard evaluation starts, it could suspend and the only way to remove this state will be to consider the possible bound of x_i computed by the unifier between the head of the clause C and the calling literal. For this purpose it will be necessary to consider the join of *sub* at the *curr_geval(p)* and of the computed unifier (which is coded in *sub* of the *input − match* node associated to the *guard − eval* node, i.e. coded in *sub(sibling(curr_geval(p)))*.

References

[1] E.Börger, 1990 *A Logic Operational Semantics of full Prolog. Part I. Selection Core and Control*, CSL'89 3rd Workshop on Computer Science Logic, Springer LNCS 440, pp. 36-64.

[2] E.Börger, 1990 *A Logic Operational Semantics of full Prolog. Part II. Built-in Predicates for Database Manipulations*, MFCS'90 Mathematical Foundation of Computer Science (Ed. B.Rovan), Springer LNCS 452, pp. 1-14.

[3] E.Börger & E.Riccobene, 1991 *Logical Operational Semantics of Parlog. Part I: And-Parallelism*, PDK'91 International Workshop on Processing Declarative Knowledge, Symbolic Computation, Springer Lecture Notes in Computer Science (to appear).

[4] E.Börger & E.Riccobene, 1991 *A formal specification of PARLOG*, Dagstuhl-Seminar *Semantics of Programming Languages and Model Theory* (Eds. M.Droste, Y.Gurevich) (in preparation).

[5] E.Börger & D.Rosenzweig, 1991, *WAM Algebras - A Mathematical Study of Implementation. Part II*, Technical Report, CSE-TR-88-91, pp. 21-?, Dept. of EECS, University of Michigan, Ann Arbor (see these Proceedings).

[6] T.Conlon, *Programming in Parlog*, Addison Wesley 1989.

[7] T.Conlon & S.Gregory, *Hands on MacPARLOG 2.0 A User's Guide*, PLP Ltd 1990.

[8] S.Gregory, *Parallel Logic Programming in PARLOG*, Addison Wesley 1989.

[9] Y.Gurevich, 1991, *Dynamic Algebras. A Tutorial Introduction*, EATCS Bulletin 43, February 1991.

[10] Y.Gurevich, 1988, *Logic and Challenge of Computer Science*, Trends in Theoretical Computer Science (Ed. E.Börger), Computer Science Press, pp. 1-57

[11] Y.Gurevich & L.S.Moss, 1990, *Algebraic Operational Semantics and Occam*, CSL'89 3rd Workshop on Computer Science Logic (Eds. E.Börger, H.Kleine Büning, M.M.Richter), Springer LNCS 440, pp. 176-192.

WAM Algebras—A Mathematical Study of Implementation Part 2

Egon Börger
Dip. di Informatica
Universita di Pisa
Cso Italia 40
I-56100 PISA
boerger@dipisa.di.unipi.it

Dean Rosenzweig
FSB
University of Zagreb
Salajeva 5
YU-41000 Zagreb
dean.rosenzweig@uni-zg.ac.mail.yu

November 21, 1991

Abstract

In direct sequel to [Boerger,Rosenzweig 91a], term representing algebras are introduced, providing an evolving algebra account of WAM representation of terms and their unification. Prolog algebras of the preceeding paper are adapted to term representation, yielding, upon optimization, a mathematical reconstruction of the full WAM, and a proof of its correctness wrt to abstract Prolog algebras, given that the compiler satisfies a set of (explicitly stated) sufficient conditions.

Introduction

This paper is a direct sequel to [Boerger,Rosenzweig 91a] (referred to as Part I below), and both the introductory remarks and the reference list should be viewed as appended to those of Part I.

Term representing algebras, providing an evolving algebra account of WAM representation of terms and their (run time and compile time) unification are introduced in Section 1.

Prolog algebras of Part I are adapted to term representation in Section 2, with a discussion of *environment trimming* and the *last call optimization*.

Warren's *classification of variables* is derived in Section 3 by further optimization. *Switching instructions* are introduced, allowing a concrete notion of clause indexing. The resulting evolving algebras deserve to be called *WAM algebras.*

The methodology of correctness preserving stepwise refinement of evolving algebras allows to view the whole construction as providing a *mathematical correctness proof of the full WAM wrt abstract Prolog algebras* Part I starts with (cf. Main Theorem in 3.3). We did not have to prescribe a specific compiler—the cumulative assumptions we introduce along the way provide a set of (provably) sufficient conditions for the compiler to be correct in the above sense. It may need stressing that we provide a proof in the usual sense of (informal) mathematics, and not a derivation in any formal deductive system. Our attitude here is preformal rather then antiformal—useful deductive systems for evolving algebras have yet to be developed.

With the analysis completed, an introductory remark from Part I may be reemphasized:

> One is almost tempted to speak of a rational reconstruction of the WAM. It seems to us that the mathematical beauty of the WAM gets revealed in the process.

We join [Ait-Kaci 90] in saying, from a different context, that

> ...this understanding can make one appreciate all the more David H.D.Warren's feat, as he conceived this architecture in his mind as one whole.

The paper is a somewhat revised version of the technical report [Boerger,Rosenzweig 91b]. To conserve some space, we skip the treatment of any built-in predicates and control constructs, but see eg. [Boerger,Rosenzweig 91b] for a discussion of database operations and their implementation in a compatible framework.

In the transition rules we shall sometimes use an update of form

seq $i = t_1, \ldots, t_2$
  updates(i)
endseq

which is executed by evaluating t_1, t_2 to numbers, say n_1, n_2, and triggering sequential execution of $updates(n_1), \ldots, updates(n_2)$.

1 Term representing algebras

WAM representation of terms will be modelled by the notion of a term representing algebra, i.e. a pointer algebra

$$(DATAAREA, PO + MEMORY; top, bottom; +, -; val),$$

where *PO* (for *Prolog Objects*) is a universe supplied with functions

$$\begin{array}{rcl} type & : & PO \rightarrow \{Ref, Struct, Funct\} \\ ref & : & PO \rightarrow DATAAREA + ATOM \times ARITY \end{array}$$

where $ARITY = \{0, \ldots, maxarity\}$ and *MEMORY* is a universe containing *DATAAREA*, to be elaborated and used below. Its role in the codomain of val is to enable storage of pure pointers in *DATAAREA*. A careful reader who knows the WAM will notice that types of constants and lists are missing from the codomain of *type*. Indeed, we are skipping special representation of constants, lists (and anonymous variables) in order to conserve some space—hoping that the reader will be able to fill that in as a straightforward excercise, looking at [Warren 83, Ait-Kaci 90].

We shall abbreviate $type(val(l)), ref(val(l))$ as $type(l), ref(l)$ respectively, for $l \in DATAAREA$. Assignment of values, (very much) distinct from assignment of pointers, will be abbreviated as

$$l_1 \leftarrow l_2 \equiv val(l_1) := val(l_2).$$

Further abbreviations that will be used often are

$$\begin{array}{rcl} l \leftarrow \langle T, R \rangle & \equiv & type(l) := T \\ & & ref(l) := R \\ unbound(l) & \equiv & (type(l) = Ref \mathbin{\&} ref(l) = l) \\ mk\_ref(l) & \equiv & \langle Ref, l \rangle \\ mk\_unbound(l) & \equiv & l \leftarrow mk\_ref(l). \end{array}$$

We shall freely use the (partial) arithmetics and ordering induced on *DATAAREA* by the successor function + ; they are partial since we do not assume of *DATAAREA* to be Archimedean—cf. discussion in Section 2. A term representing algebra comes supplied with (partial) functions

$$\begin{array}{rcl} deref & : & DATAAREA \rightarrow DATAAREA \\ term & : & DATAAREA \rightarrow TERM \end{array}$$

such that $deref(l)$ follows the reference chain from l, while $term(l)$ reconstructs the term represented at 'location' l. More precisely, we assume

$$deref(l) = \begin{cases} deref(ref(l)) & \text{if } type(l) = Ref \ \& \ NOT(unbound(l)) \\ l & \text{otherwise} \end{cases}$$

$$term(l) = \begin{cases} mk\_var(l) & \text{if } unbound(l) \\ term(deref(l)) & \text{if } type(l) = Ref \ \& \ NOT(unbound(l)) \\ f(a_1,\ldots,a_n) & \text{if } type(l) = Struct \end{cases}$$

where, in the last case, $ref(ref(l)) = \langle f, n\rangle$, and $term(ref(l) + i) = a_i$, $1 \leq i \leq n$. Of mk_var we assume to associate a unique Prolog variable to an arbitrary location in $DATAAREA$.

Example 1. The term $s(Y_2, t(Y_1))$ would for instance be represented by a location l such that

$$\begin{array}{lll} l = \langle\, Struct, l_1 \,\rangle & l_1 = \langle\, Funct, \langle s, 2\rangle \,\rangle & l_1{+} = \langle\, Ref, l_1{+} \,\rangle \\ l_1{+}{+} = \langle\, Struct, l_2 \,\rangle & l_2 = \langle\, Funct, \langle t, 1\rangle \,\rangle & l_2{+} = \langle\, Ref, l_2{+} \,\rangle \end{array}$$

The reader may draw a picture and verify $term(l)$.

1.1 The heap, the push-down list and unification

Prolog structures will be represented on the $HEAP$, a subalgebra of $DATAAREA$

$$(HEAP; h, boh; +, -; val)$$

to be used as a stack, with $str \in HEAP$ the subterm (or structure) pointer to be used for navigating through substructures. The active part of the $HEAP$ will be abbreviated as $heap \equiv \{l \in HEAP \mid boh \leq l < h\}$ (its finiteness follows from boh being the initial value of h), and locations $l \in heap$ such that $type(l) = Ref$ will be referred to as *heap variables*.

Goal arguments will be represented in another subdomain $AREGS$ of $DATAAREA$, disjoint from the $HEAP$, on which we shall never use $+, -$; we assume just a function $x : \mathcal{N} \rightarrow AREGS$, $x_i \equiv x(i)$.

For unification we shall use a dedicated stack, the *pushdown list*, represented by a pointer algebra

$$(PDL, DATAAREA; pdl, nil; +, -, ; ref')$$

which may but need not be seen as embedded in $DATAAREA$, with $left \equiv ref'(pdl)$, $right \equiv ref'(pdl-)$.

In the unification algorithm below we shall use an abstract update $bind(l_1, l_2)$, of which we assume

WAM Assumption 1. For any $l, l_1, l_2 \in DATAAREA$ such that

- $unbound(l_1)$
- the variable $mk\_var(l_1)$ does not occur in $term(l_2)$,

if $term, term'$ are the values of $term(l)$ before and after the execution of $bind(l_1, l_2)$ respectively, and s is the substitution associating $term(l_2)$ to $mk\_var(l_1)$, then $term' = subres(term, s)$.

The mathematical (nondeterministic) unification algorithm [Lloyd 84] should fail on attempt to bind a variable to a term in which it occurs—unification is then said to be STO (subject to *occur-check*). Had we chosen to model such *unification with occur-check*, we would simply require of $bind(l_1, l_2)$ to trigger *backtracking* whenever the second condition of WAM Assumption 1 is violated, i.e. when $mk\_var(l_1)$ occurs in $term(l_2)$.

We choose however to allow for the usual practice of Prolog implementations, which, for pragmatic reasons, generally skip the occur check, and to adhere to the draft standard proposal [North,Scowen 91], which refrains from specifying the behaviour of systems upon STO unification by considering it as 'implementation dependent'. Hence we make no assumption whatsoever on STO binding. This decision is necessarily reflected in Unification Lemma and Pure Prolog Theorem below, which must refrain from stating anything for this case.

Unification will be triggered by setting a special 0–ary function *what_to_do* to *Unify*, given that the terms to be unified have already been pushed to *PDL*. We then have $what\_to\_do \in \{Run, Unify\}$, and the following Unification Rule:

```
if OK & what_to_do = Unify
thenif pdl = nil
then what_to_do := Run
elsif unbound(deref(left))
then bind(deref(left), deref(right))
     pdl := pdl − −
elsif unbound(deref(right))
then bind(deref(right), deref(left))
     pdl := pdl − −
elsif same_funcs(deref(left), deref(right))
then seq
        pdl := pdl − −
        seq i = arity(ref(deref(left))), ..., 1
          ref'(pdl+) := ref(deref(right)) + i
          ref'(pdl + +) := ref(deref(left)) + i
          pdl := pdl + +
        endseq
     endseq
else backtrack
     what_to_do := Run
```

where $same\_funcs(l_1, l_2)$ stands for

$$funct(ref(l_1)) = funct(ref(l_2)) \ \& \ arity(ref(l_1)) = arity(ref(l_2)).$$

Unification will be invoked exclusively by executing *unify* update, where

```
unify(l1, l2)  ≡  ref'(nil+) := l2
                  ref'(nil + +) := l1
                  pdl := nil + +
                  what_to_do := Unify
```

The above has the flavour of assembly code for calling a (recursive) subroutine, pushing the arguments to a stack, with setting *what_to_do* to *Unify* playing the role of 'jump to subroutine', and resetting it to *Run* that of 'return from subroutine'. Of other rules in the sequel, 'running the main program', we shall tacitly assume to contain the guard OK & $what\_to\_do = Run$.

Then we have

Unification Lemma. If $term(l_1), term(l_2) \in TERM$, then the effect of executing $unify(l_1, l_2)$, for any $l \in DATAAREA$ such that $term(l) \in TERM$, is as follows:

- if s is the most general unifier of $term(l_1), term(l_2)$, the new value of $term(l)$, when $what\_to\_do$ is set to Run again, will be the result of applying s to its old value;
- if the abstract unification algorithm fails without invoking the occur-check, $backtrack$ update will be executed.

Proof by induction over cumulative size of all left terms, i.e. over the value of $|term(ref'(nil{+}{+}))| + \ldots + |term(ref'(pdl - -))| + |term(ref'(pdl))|$.

Had we assumed of $bind$ to involve the occur-check, we could drop the non-STO limitations from our lemma.

If the reader prefers a different unification algorithm, and hence finds our description to be overspecific, he is welcome to consider the Unification Lemma as an additional WAM Assumption, assuming further that the $unify$ update binds variables only by invoking the abstract $bind$ update— all results (and proofs) below will remain valid.

1.2 Putting

Here we develop (almost) WAM code for constructing body goals. We shall use

$$(CODEAREA, p, bottom; +, -; code),$$

the pointer algebra from Part I, with $MEMORY \supseteq DATAAREA + CODEAREA$. The universe $INSTR$ of instructions is assumed to contain all instructions of form

$$\begin{array}{lll} put\_value(y_n, x_j), & put\_structure(f, a, x_i), & unify\_value(x_n), \\ unify\_value(y_n), & get\_value(y_n, x_j), & get\_structure(f, a, x_j), \\ unify\_variable(x_n), & & \end{array}$$

with $n, j, i \in \mathcal{N}$, $f \in ATOM$, $a \in ARITY$, $y_n \in DATAAREA$, and will be tacitly extended with any further instructions occurring below.

To define the sequence of putting instructions corresponding to a body goal, we rely on the notion of *term normal form* of first order logic, which comes in two variants, corresponding to analysis and synthesis of terms. We assume that all variables occurring in the input term are of form Y_n, and a sufficient supply of X variables, different from all Y's.

$$\begin{aligned} nf(X_i = Y_n) &= [X_i = Y_n] \\ nf(Y_i = Y_n) &= [\,] \\ nf_s(X_i = f(s_1, \ldots, s_m)) &= flatten([\, nf_s(Z_1 = s_1), \ldots, nf_s(Z_m = s_m), \\ &\qquad X_i = f(Z_1, \ldots, Z_m)]) \\ nf_a(X_i = f(s_1, \ldots, s_m)) &= flatten([\, X_i = f(Z_1, \ldots, Z_m), \\ &\qquad nf_a(Z_1 = s_1), \ldots, nf_a(Z_m = s_m)]), \end{aligned}$$

where Z_i is a fresh X variable if s_i is not a variable, s_i otherwise.

Example 2. For instance, we might have

$$\begin{aligned} nf_s(X_2 = s(Y_2, t(Y_1)) &= flatten([[\,], nf_s(X_3 = t(Y_1)), X_2 = s(Y_2, X_3)]) \\ &= [X_3 = t(Y_1), X_2 = s(Y_2, X_3)] \\ nf_a(X_2 = s(Y_2, t(Y_1)) &= flatten([X_2 = s(Y_2, X_3), [\,], nf_a(X_3 = t(Y_1))] \\ &= [X_2 = s(Y_2, X_3), X_3 = t(Y_1)] \end{aligned}$$

The *put_instr* of a normalized equation will be according to the following table, where j stands for an arbitrary 'top level' index (corresponding to input $X_i = t$ for term normalization), and k for a 'non top level' index (corresponding to an auxiliary variable introduced by normalization itself):

$$
\begin{array}{rcl}
X_j = Y_n & \rightarrow & [\,put\_value(y_n, x_j)\,] \\
X_k = Y_n & \rightarrow & [\,unify\_value(y_n)\,] \\
X_i = f(Z_1, \ldots, Z_a) & \rightarrow & [\,put\_structure(f, a, x_i), \\
 & & \quad unify\_value(z_1), \ldots, unify\_value(z_a)\,]
\end{array}
$$

with $y_i \in DATAAREA$, $x_i \in AREGS$, and z_i is lowercase version of Z_i.

The function *put_code* is defined by flattening the result of mapping *put_instr* along $nf_s(X_i = t)$. It is auxiliary to *put_seq*, specifying how a body goal is to be compiled:

$$put\_seq(g(s_1, \ldots, s_m) = flatten(\,[\,put\_code(X_1 = s_1), \ldots, put\_code(X_m = s_m)\,]\,)$$

with 'top level' $j = 1, \ldots, m$. The reader knowledgeable about the WAM will notice that this is not quite the WAM code, but wait till section 3.

Example 3. A body goal of form $p(Y_1, s(Y_2, t(Y_1)))$ would for instance generate the following code:

$$
\begin{array}{l}
put\_seq(p(Y_1, s(Y_2, t(Y_1)))) = [\,put\_value(y_1, x_1), put\_structure(t, 1, x_3), unify\_value(y_1), \\
\qquad\qquad\qquad\qquad\qquad put\_structure(s, 2, x_2), unify\_value(y_2), unify\_value(x_3)\,]
\end{array}
$$

Putting code will be 'executed' by the following rules

if $code(p) = put\_value(l, x_j)$
then $x_j \leftarrow l$
succeed

if $code(p) = unify\_value(l)$
& $mode = Write$
then $h \leftarrow l$
$h := h+$
succeed

if $code(p) = put\_structure(f, a, x_i)$
then $h \leftarrow \langle\, Struct, h+ \rangle$
$h+ \leftarrow \langle\, Funct, \langle f, a \rangle \rangle$
$x_i \leftarrow \langle\, Struct, h+ \rangle$
$h := h++$
$mode := Write$
succeed

Then we have

Putting lemma. If all variables occurring in Prolog literal $g(t_1, \ldots, t_m)$ are among $\{Y_1, \ldots, Y_l\}$, and if $y_n \in DATAAREA$ with $term(y_n) \in TERM$, s the substitution associating every Y_n with $term(y_n)$, $n = 1, \ldots, l$, and X_i fresh pairwise distinct variables, $1 = 1, \ldots, m$,

then the effect of executing (setting p to) $load(put\_seq(g(t_1, \ldots, t_m))$ is that $term(x_i)$ gets value of $subres(t_i, s)$, $i = 1, \ldots, m$.

Proof by induction over cumulative size of terms. Note that substructure representing X_k variables, generated by nf_s, always get 'instantiated', i.e. $term(x_k)$ gets a value in $TERM$, by *put_structure*, before being 'used' by *unify_value*. Uninstantiated variables inside structures get represented on the heap. In case of unbound Y_n, associating it to $mk\_var(y_n)$ amounts to renaming.

The reader might verify the effect of executing the code of Example 3.

1.3 Getting

The compilation of clause head, *compile time unification*, will again be specified by an abstract function *get_seq*, associating a sequence of instructions to a Prolog literal, relying on term normal forms. The *get_instr* of a normalized equation is defined by a table similar to that for *put_instr*, under the same conventions.

$$\begin{array}{rcl} X_j = Y_n & \rightarrow & [\,get\_value(y_n, x_j)\,] \\ X_k = Y_n & \rightarrow & [\,unify\_value(y_n)\,] \\ X_i = f(Z_1, \ldots, Z_a) & \rightarrow & [\,get\_structure(f, a, x_i), \\ & & \quad unify(z_1), \ldots, unify(z_a)\,] \end{array}$$

where $unify(y_n) = unify\_value(y_n)$, $unify(x_k) = unify\_variable(x_k)$.

The function *get_code* is defined by flattening the result of mapping *get_instr* along $nf_a(X_i = t)$. It is auxiliary to *get_seq*, specifying how a clause head is to be compiled:

$$get\_seq(g(s_1, \ldots, s_m) = flatten([\,get\_code(X_1 = s_1), \ldots, get\_code(X_m = s_m)\,])$$

with 'top level' $j = 1, \ldots, m$.

Example 4. A clause head of form $p(Y_1, s(Y_2, t(Y_1)))$ might generate the following code.

$$\begin{array}{rl} get\_seq(p(Y_1, s(Y_2, t(Y_1)))) = [& get\_value(y_1, x_1), get\_structure(s, 2, x_2), unify\_value(y_2) \\ & unify\_variable(x_3), get\_structure(t, 1, x_3), unify\_value(y_1)\,] \end{array}$$

The code sequence will be executed by the following rules.

if $code(p) = get\_value(l, x_j)$
then $unify(l, x_j)$
   succeed

if $code(p) = unify\_value(l)$
  & $mode = Read$
then $unify(l, str)$
   $str := str+$
   succeed

if $code(p) = unify\_variable(l)$
  & $mode = Read$
then $l \leftarrow str$
   $str := str+$
   succeed

if $code(p) = get\_structure(f, a, x_i)$
thenif $type(deref(x_i)) = Ref$
then $deref(x_i) \leftarrow \langle\, Struct, h+ \,\rangle$
   $trail(deref(x_i))$
   $h \leftarrow \langle\, Struct, h+ \,\rangle$
   $h+ \leftarrow \langle\, Funct, \langle f, a, \rangle\,\rangle$
   $h := h++$
   $mode := Write$
   succeed
elsif $type(deref(x_i)) = Struct$
   & $right\_func(f, a, deref(x_i))$
then $str := ref(deref(x_i))+$
   $mode := Read$
   succeed
else *backtrack*

where $right\_func(f, a, x_i)$ stands for

$$funct(ref(deref(x_i))) = f \;\;\&\;\; arity(ref(deref(x_i))) = a.$$

Then we have

Getting Lemma. If all variables occurring in a Prolog literal $g(t_1, \ldots, t_m)$ are among $\{Y_1, \ldots, Y_l\}$, and if further $y_n \in DATAAREA$ with $unbound(y_n)$, $n = 1, \ldots, l$, X_i a fresh variable with $x_i \in DATAAREA$ and $term(x_i) \in TERM$, $1 \leq i \leq m$, and

a) s is the unifying substitution of t_i and $term(x_i)$, $1 \leq i \leq m$, or

b) such unification fails without invoking the occur-check,

then the effect of executing (setting program pointer p to) $load(get\_seq(g(t_1, \ldots, t_m))$ for any $l \in DATAAREA$ with $term(l) = t \in TERM$ (before execution), is respectively

a) $term(l)$ gets the value of $subres(t, s)$ (up to renaming of Y's), or

b) backtracking.

Proof by induction over cumulative size of terms, relying on Unification Lemma. Note that substructure descriptors X_k, generated by nf_a, always get 'instantiated', i.e. $term(x_k)$ gets a value in $TERM$, by *unify_variable*, before being 'used' by *get_structure*.

The reader might verify the effect of executing the code of Example 4.

Since the heap will, in the sequel, turn out to be the most persistent of all data areas (subdomains of $DATAAREA$), we shall wish to uphold the following constraint:

Heap Variables Constraint. No heap variable points outside the heap, i.e. for any $l \in heap$, if $type(l) = Ref$, then $ref(l) \in heap$.

If we assume that the abstract *bind* (and hence *unify*) update does not violate the constraint, a simple examination of our rules suffices to see that the only instruction which might violate it is $unify\_value(y_n)$ in *Write* mode, with y_n a reference pointing outside the heap. If we call any such y_n (i.e. occurrence of Y_n) *local*, we can enforce the constraint by using a special instruction *unify_local_value*, triggering a run-time check for locality, with

```
if code(p) = unify_local_value(l)
   & mode = Write
then succeed
     if local(deref(l))
     then mk_heap_var(deref(l))
          trail(deref(l))
     else h ← deref(l)
          h := h+
```

where

$$
\begin{aligned}
local(l) &\equiv unbound(l) \ \&\ NOT(l \in heap) \\
mk\_heap\_var(l) &\equiv mk\_unbound(h) \\
& \quad\ l \leftarrow mk\_ref(h) \\
& \quad\ h := h+
\end{aligned}
$$

with a rule for *Read* mode of the same form as that for *unify_value*. The update $trail(deref(l))$ is of no import for Putting and Getting lemmas (and the Heap Variables lemma below), and thus can be regarded as a noop here; it will become significant in the next section.

The effect of *unify_local_value*, in terms of terms and substitutions constructed, is obviously the same as that of *unify_value* (up to substitution ordering), hence its usage in *put_code* function preserves the Putting and Getting lemmas. However, it preserves the Heap Variables Constraint as well, hence, given

Working Assumption. The *bind* update preserves the Heap Variables Constraint.

we have

Heap Variables Lemma. If *put_code* and *get_code* functions generate *unify_local_value* instead of *unify_value* for all occurrences of local variables, then the execution of *put_seq*,*get_seq* preserves the Heap Variables Constraint.

Proof is straightforward, up to the problem of recognizing occurrences of potentially local variables in Prolog terms. It will have to suffice to say here that this will include all occurrences of variables (within structures in body goals) for which the compiler cannot determine to be previously (in the given clause) allocated on the heap.

Example 5. In the clause $a(X) :- b(f(X))$ the second occurrence of X is local.

The Heap Variables Constraint will then be preserved under

Compiler Assumption 1. The assumptions of Heap Variables Lemma are satisfied.

The Working Assumption is temporary, since it will become a theorem when we discuss binding more carefully, in 2.1.3. below.

2 Term representing Prolog algebras

Time has come to connect the Prolog algebras of Part I to term representing algebras. The *STACK* will have to be adapted, with the most notable change being the representation of substitutions, shared between environments and the *TRAIL*—a new stack used to record binding history.

2.1 The stack and the trail—compiling pure Prolog

The *STACK* of Part I will now take the form of a subalgebra of *DATAAREA*

$$(STACK; tos(b,e), bos; +, -; val) \quad b, e, ct \in STACK,$$

disjoint from *HEAP* and *AREGS*, with the old decorating functions to be defined using $val, +, -$. The currently active finite part of the *STACK* will be abbreviated as $stack \equiv \{l \in STACK \mid bos \leq l \leq tos(b,e)\}$ (finiteness follows from bos being the initial value of $tos(b,e)$).

2.1.1 Environments

The most notable difference from the model of Part I consists in the environments holding the variables $y_1, \ldots, y_n$, where n is recorded in the code for the last *call* executed, and accessible via $cp-$ (cf. Pure Prolog Theorem below). The information stored in the environment ('decorating functions') will be accessible by the function *val* applied to fixed offsets from e. We namely define functions ce, cp, y_i, tos

$$\begin{aligned}
ce(l) &\equiv l+1 \\
cp(l) &\equiv l+2 \\
y_i &\equiv e+2+i \quad 1 \leq i \leq stack_offset(cp) \\
y_i(l) &\equiv l+2+i \quad 1 \leq i \leq stack_offset(val(cp(l))) \\
stack_offset(l) &\equiv n \quad \text{when } code(l-) = call(f,a,n) \\
tos(b,e) &\equiv \begin{cases} e+2+stack_offset(cp) & \text{if } b \leq e \\ b & \text{otherwise} \end{cases}
\end{aligned}$$

providing, so to speak, definitions for the abstract decorating functions of Part I—what used to be say $cp'(e)$ is now reconstructible as $val(cp(e))$. Current environment may then be depicted as

...	ce	cp	y_1	...	y_n

with e pointing below $ce \equiv ce(e)$.

Any $y_i(l)$ for l in the ce-chain will be called a ***stack variable*** in the sequel. Environments will be created and discarded by the ***environment-handling rules***, i.e. by executing homonymous instructions.

```
if code(p) = allocate(n)
then e := tos(b, e)
     val(ce(tos(b, e))) := e
     val(cp(tos(b, e))) := cp
     seq i = 1,...,n
       mk_unbound(y_i(tos(b, e)))
     endseq
     succeed
```

```
if code(p) = deallocate
then e := val(ce(e))
     cp := val(cp(e))
     succeed
```

Note that executing *allocate*, temporarily, includes initializing all new variables with *unbound*. This is a simplifying hypothesis to be dropped later.

2.1.2 Choicepoints

The decorating functions of a choicepoint will likewise be reconstructible by applying val to fixed offsets from b, with the addition of the current goal being represented by its arguments (the contents of argument registers), since its functor is accessible via $val(cp(b))-$, pictorially:

...	x_n	...	x_1	e	cp	b	p	tr	h

with b pointing at the top. To be completely formal, we have

$$\begin{array}{lll} h(l) \equiv l & tr(l) \equiv l - 1 & p(l) \equiv l - 2 \\ b(l) \equiv l - 3 & cp(l) \equiv l - 4 & e(l) \equiv l - 5 \\ x_i(l) \equiv l - 5 - i & hb \equiv val(h(b)) & \end{array}$$

Other layouts are of course possible, this is the original form of [Warren 83].

Rewriting the choicepoint handling rules of Part I in this framework (*try_me_else* | *try_me*, *retry_me_else* | *retry_me*, *trust_me* | *trust*) should by now be a straightforward excercise for the reader, noting that the instructions may now obtain an additional argument—the arity n.

2.1.3 Binding, trailing and unbinding

Here we take a closer look at how the variables are bound and how these bindings are recorded, i.e. at the *bind* update. For the case of variable-to-variable binding, we need explicit assumptions about 'memory layout', in order to avoid the problem of 'dangling pointers', as the stack and the heap are going to grow and shrink at in general different rates.

We shall extend the ordering $<$, defined in terms of $+$ on the *heap* and the *stack* (and in terms of x on *AREGS*), assuming

WAM Assumption 2. *HEAP* $<$ *STACK* $<$ *AREGS*.

This assumption forces us to view *DATAAREA* as an ordered non-Archimedean structure (remember that *heap* and *stack* are finite at all times). A more realistic Archimedean *DATAAREA* would force us to consider resource limitations and garbage collection at this level, complicating every correctness statement with a qualification 'given sufficient resources'. We prefer to separate our concerns, and (possibly) treat resource limitations separately, as a refinement of the present framework.

Given such an ordering, we further assume

WAM Assumption 3. If $unbound(l_1)$ & $unbound(l_2)$, then the effect of $bind(l_1, l_2)$ is to bind the higher location to the lower one.

Note that WAM Assumptions 2,3 imply our Working Assumption from section 1.3. Further, we assume of *bind* to record variable bindings, to be undone in case of backtracking, on the *TRAIL*, i.e. on a further pointer algebra

$$(TRAIL, DATAAREA; tr, botr; +, -; ref'')$$

to be used as a stack. More precisely, we assume

WAM Assumption 4. Whenever *bind* binds a location l, it executes the update

$$\begin{aligned} trail(l) \quad \equiv \quad & ref''(tr) := l \\ & tr := tr+ \end{aligned}$$

The *TRAIL* essentially records the substitutions, stored in Part I explicitly in choicepoints. Since, whenever there is an alternative state of computation, $p(b)$ will point to an instruction of *retry, trust* family, the *backtrack* update must, except for passing control to $p(b)$, reconstruct the old substitution by undoing all bindings trailed after the current choicepoint was pushed, i.e.

$$\begin{aligned} backtrack \quad \equiv \quad & p := val(p(b)) \\ & \text{seq } l = tr-, \ldots, tr(b) \\ & \quad mk\_unbound(ref''(l)) \\ & \text{endseq} \end{aligned}$$

Since, by executing the instruction pointed at by $p(b)$, *stack* will shrink to b and *heap* to hb, the variables residing above these points will become irrelevant, and recording them on the *TRAIL* turns out to be spurious. An implementation could thus optimize by executing $trail(l)$ only under the condition $l \in heap$ & $l < hb$ or $l \in stack$ & $l < b$. This is however an optimization step without any semantical import, which we may disregard (simplifying thus somewhat the proof of Pure Prolog Theorem below).

WAM Assumptions 3,4 would become theorems had we *defined* the *bind* update as

$$\begin{aligned} & \text{if } NOT(unbound(l_2)) \text{ or } l_1 > l_2 \\ & \text{then } l_1 \leftarrow l_2 \\ & \qquad trail(l_1) \\ & \text{elsif } l_2 > l_1 \\ & \text{then } l_2 \leftarrow l_1 \\ & \qquad trail(l_2) \end{aligned}$$

Since we do not wish to exclude other, say occur-checking, implementations, we keep the assumptions. They ensure the

Stack Variables Property. Every stack variable l points either to the heap or to a lower location of the stack, i.e $ref(l) \in heap$ or $ref(l) \in stack$ & $ref(l) < l$.

Proof by inspection of rules—they all preserve the property, given WAM Assumptions 2,3. In case of $get\_value(l, x_j)$ note that x_j should have been put there by an instruction executed before the current environment was allocated.

Although special instructions for constants could violate the letter of the property, i.e. l could get a literal constant [Warren 83, Ait-Kaci 90], this would not affect the uses we have for it.

2.1.4 Pure Prolog Theorem

At this point we may collect our assumption about the compiler in

Compiler Assumption 2. Assume functions *compile, load, unload* like in Part I, but now

$$\begin{aligned} compile(H :- G_1, \ldots, G_n) \quad = \quad & flatten([\, allocate(r), get\_seq(H), \\ & call\_seq(G_1), \ldots call\_seq(G_n), \\ & deallocate, proceed\,]) \end{aligned}$$

where $call\_seq(g(s_1, \ldots, s_k)) \equiv flatten([\,put\_seq(g(s_1, \ldots, s_k)), call(g, k, r)\,])$, with $\{Y_1, \ldots, Y_r\}$ being all variables occurring in the clause.

Function *procdef* will now be typed by $ATOM \times ARITY \rightarrow CODEAREA$, suppressing its database argument. Minding that the *call* instructions now have three arguments as above, and *execute* only the first two of them, it is a straightforward excercise to rewrite the rules query success, *proceed*, *call* | *execute* of Part I in this framework.

At this point we can describe a mapping $\mathcal{F}$, mapping states of our current algebra to states of Part I Prolog algebra, and (sequences of) our current rules to those of Part I Prolog algebra. In view however of our limitation to non-STO unification only (cf. discussion in 1.1.), we have, for the sake of proof, to *constrain* the Part I Prolog algebra by considering its abstract *unify* function as undefined in case unification is STO. This constraint is not meant as a modification of our model—it just expresses an attitude: we do not care what happens in that case, and we do not claim anything about it.

Setting up a correspondence with Prolog algebras of Part I, we shall presently ignore cutpoints *ct, cct, ct'* (since they are irrelevant for pure Prolog) and variable indexes *vi, cvi, vi'* (since their role of unique renaming is obviously taken over by offsets on the stack and the heap; if the reader cares about full detail, he can easily reintroduce them in our framework and extend the rules to handle the indexes like in Part I - they would never be used, except for reconstruction of a full Part I state).

Marking all names of universes and functions of Part I with an index 1, assume a simple minded $compile_1$,

$$compile_1(H :- G_1, \ldots, G_n) \quad = \quad [allocate, unify(H), call(G_1), \ldots, call(G_n), deallocate, proceed]$$

This defines the obvious (partial) function

$$codepointer : CODEAREA \rightarrow CODEAREA_1$$

which may be left undefined inside loaded *get_sequences* and *call_sequences*.

The rest of the correspondence will then be given by partial functions

$$\begin{aligned} subst &: TRAIL \rightarrow SUBST_1 \\ goal &: STACK \rightarrow TERM \\ choicepoint &: STACK \rightarrow STATE_1 \\ env &: STACK \rightarrow ENV_1 \end{aligned}$$

and an auxiliary function

$$term : DATAAREA \times TRAIL \rightarrow TERM_1$$

which can be defined inductively as follows, taking an object of $STATE$ ot ENV to be given by its g, s, p, cp, e, b or cp', e values, respectively.

$term(l, l_t)$ yields the value $term(l)$ would take after having unwound the trail to l_t

$subst(l_t)$ associates $mk\_var(ref''(l))$ to $term(ref''(l), l_t)$, $botr \leq l < tr$

$$\begin{aligned} goal(lb) &= g(term(x_1(lb), val(tr(lb))), \ldots, term(x_m(lb), val(tr(lb)))), \\ &\quad \text{where } code(val(cp(lb))-) = call(g, m, r); \\ choicepoint(lb) &= \langle\, goal(lb), \\ &\quad subst(val(tr(lb))), \\ &\quad codepointer(val(p(lb))), \\ &\quad codepointer(val(cp(lb))), \\ &\quad env(val(e(lb))), \\ &\quad choicepoint(val(b(lb)))\,\rangle \\ env(le) &= \langle\, codepointer(val(cp(le))), env(val(ce(le)))\,\rangle \end{aligned}$$

The 0–ary functions of Part I Prolog algebra are defined likewise:

$$\begin{aligned} g_1 &= g(term(x_1), \ldots, term(x_m)) \text{ where } code(cp-) = call(g, m, r) \\ s_1 &= subst(tr) \\ (c)p_1 &= codepointer((c)p) \\ e_1 &= env(e) \\ b_1 &= choicepoint(b) \end{aligned}$$

$\mathcal{F}$ preserves initial and stopping states (with *stop* value) and the substitution computed at the given state. With by now rather obvious mapping of rulesequences, straightforward (though somewhat tedious) verification of cases yields

Pure Prolog Theorem. For any significant term-representing Prolog algebra A and sequence of rule invocations R (which does not entail STO unification) the following diagram commutes

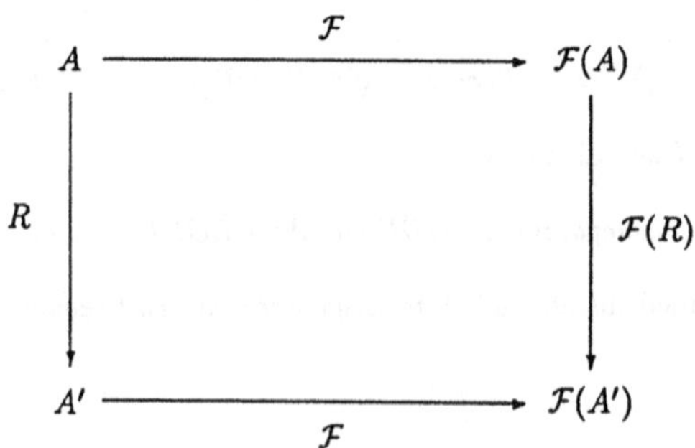

$\mathcal{F}$ is thus a functor (on categories we shall not describe here). Since it preserves initial and stopping states, we have

Corollary. An initial term-representing Prolog algebra A_{init} eventually reaches a stopping A_{stop} (without STO binding) iff $\mathcal{F}(A_{init})$ reaches $\mathcal{F}(A_{stop})$ (without STO unification).

2.2 Environment trimming and the last call optimization

Wishing to discard (i.e. recycle) stack space as soon as possible, we want any stack variable to get superfluous as soon as the *put_seq* for the goal containing its last occurrence in the clause is executed. A compiler could then be clever with variable numbering and trim the environment on the fly by generating a decreasing sequence of environment sizes (in third arguments of *call* instructions, cf. [Ait-Kaci 90]. According to [Warren 83], '... variables are arranged in their environment in such a way that they can be discarded as soon as they are no longer *needed*.' (our italics).

Let us then say, for now, that a variable, occurring before or in a body goal G_i of a clause $H :- G_1, \ldots, G_n$, $1 \leq i < n$, is *needed* at G_i if it occurs after G_i as well, i.e. in some G_l, $l > i$.

(This definition is going to change in Section 3, as we optimize the design extending the instruction set. The form of other definitions below, relying on that of being needed, will however remain intact.)

Environment trimming will consist in $call\_seq(g(t_1, \ldots, t_m))$ ending in $call(g, m, r)$, where r need not be the environment size any more—it will suffice that all variables needed at or after corresponding body goal occurrence $g(t_1, \ldots, t_m)$ are among $\{Y1, \ldots, Yr\}$, which is our **Compiler Assumption 3.**

Correctness of environment trimming (i.e. preservation of Putting Lemma across indexing instructions, *call*, *allocate*) would be ensured by the following property.

Argument Registers Property. If Y_k is no more needed in a body goal occurrence G, then after executing $put\_seq(G)$, for any x_i affected by that *put_seq*, the computation of $term(x_i)$ does not meet y_k.

In view of the Heap Variables Lemma and Stack Variables Property, a simple examination of the rules shows that the property could only be violated by $put\_value(y_n, x_j)$, with Y_n no more needed, and $ref(y_n) > e$. Corresponding occurences of Y_n in the clause are termed *unsafe* [Warren 83, Ait-Kaci 90]. The first of the two conditions amounts to Y_n occurring in an argument position of a body goal in which it is no more needed.

The second condition would be excluded had Y_n occurred in the clause head (y_n would then be set by the *get_seq* to point to a previously existing object, represented below e) or in a structure (y_n would be set to point to the *heap*, by a *unify* instruction). Thus the following definition.

An occurence of a variable Y_n in an argument position of a body goal, in which it is no more needed, is *unsafe* if Y_n does not previously occur in the clause head or in a structure.

Example 6. The second occurence of Y in the clause $has\_a(X) :- generate(X, Y), test(Y).$ is unsafe.

Since such an occurrence may nevertheless point below e, by having been previously bound, a run-time check would be appropriate, replacing $deref(y_n)$ by a new heap variable, but only if strictly necessary. The *compile* function should thus be modified to emit a new instruction,

$put\_unsafe\_value(y_n, x_j)$, including such a check, instead of $put\_value(y_n, x_j)$, for each unsafe occurrence of Y_n. The new instruction is 'executed' by the rule

$$\begin{array}{l} \textbf{if } code(p) = put\_unsafe\_value(y_n, x_j) \\ \textbf{then } succeed \\ \quad \textbf{if } deref(y_n) > e \\ \quad \textbf{then } mk\_heap\_var(deref(y_n)) \\ \quad\quad trail(deref(y_n)) \\ \quad\quad x_j \leftarrow mk\_ref(h)) \\ \quad \textbf{else } x_j \leftarrow deref(y_n) \end{array}$$

We can thus correctly assume environment trimming, reestablishing the Argument Registers Property, given

Compiler Assumption 2'. The *put_seq* function generates *put_unsafe_value* instead of *put_value* for any unsafe variable occurrence.

Since no variable is needed at the last goal, however, the environment will by then be trimmed down to continuation pointer, like in Part I, and the argument for the correctness of the ***last call optimization*** will be the same as there—namely, executing $call(g, a, 0)$, *deallocate*, *proceed* will, in view of the Argument Registers Property, have the same effect as *deallocate*, $execute(g, a)$. We may thus correctly assume LCO as well, modifying further the compile function so as given by

Compiler Assumption 4. If *compile* function performs the last call optimization, it is done in a way equivalent to replacing the *call_seq* for the last body goal by its *execute_seq*, where

$$execute\_seq(g(t_1, \ldots, t_m)) = flatten([put\_seq(g(t_1, \ldots, t_m), deallocate, execute(g, m)])$$

The discussion of this section is summed up in

Trimming Lemma. Environment trimming and LCO preserve the Pure Prolog Theorem.

3 WAM Algebras

In case of a fact or of a chain-rule, i.e. a void or a singleton body, no variable will ever be needed. However, we cannot (yet) say that *allocate* and $get\_seq(H)$ commute, or that *deallocate* is inverse to *allocate*, like in Part I, obviating the need for an environment. Before that we have to introduce special instructions, enabling us to handle the nowhere needed, ***temporary***, variables, without accessing an environment at all.

Temporary variables and related optimizations will be introduced in 3.1, 3.2. By then, the only feature of the classical WAM of [Warren 83], still lacking in our Prolog algebras (apart from handling of constants, lists and anonymous variables, which the reader can easily fill in for himself), will be the ***switching*** instructions, allowing the compiler to realize a concrete notion of clause indexing (hitherto understood as being abstractly encoded by the *procdef* function, cf. discussion in Part I). Switching instructions and related compiler assumptions are introduced in 3.3., making our algebras finally worthy of the above title.

3.1 Temporary variables

A variable occurring in a clause is ***temporary*** if it is not needed in any body goal. A variable which is not temporary is ***permanent***.

This may seem to be a weird way of saying that a temporary variable 'does not occur in more than one goal in the body, counting the head of the clause as part of the first goal' [Warren 83]. In Example 6, for instance, X is temporary while Y is permanent. Note however that our definition is parameterized by the notion of being ***needed***, which we are going to optimize in 3.2, making the above equivalent to Warren's full definition. [Ait-Kaci 90] has also attempted to derive Warren's original definition by optimizing the more naive notion above, but without properly adapting the notions of environment trimming and unsafe variables, to be later patched up by a notion of 'delayed trimming'. We have the three definitions, of ***temporary variables, environment trimming*** and ***unsafe variables***, coupled by the notion of being *needed* (and controlled by Argument Registers Property), so that optimization preserves correctness in a quasi automatic way.

The very notion of a temporary variable arises by optimization, intended to allow us not to *allocate* it in the environment. The Argument Registers Property tells us that permanent variables serve as communication channels between goals in the same body, leaving to temporary variables the role of mere descriptors of goal structure, akin to X_k's generated by term normal forms. The compiler will thus represent them likewise with *AREGS*. If we, for temporary variables, replace Y_n, y_n everywhere by fresh X_i, x_i, all the variable handling instructions (*get_value*, *put_value*, *unify_value*, *unify_local_value*) will work fine for them, given the Argument Registers Property, if the x_i's are properly initialized to objects on the *heap* or on the *stack*.

We shall then initialize each temporary variable on the fly, at its first occurrence, taking care that it gets the same value a permanent variable (initialized previously to *unbound* by *allocate*) would get in its place. We thus modify the *compile* function so as to emit *get_variable*(x_i, x_j), *unify_variable*(x_i), *put_variable*(x_i, x_j) for the first occurrence of a temporary variable X_i in a head argument, structure argument, body goal argument, respectively, which is our **Compiler Assumption 5.** Since temporary variables need (and should) not be trailed, new instructions will be 'executed' by the rules

if *code*$(p) =$ *get_variable*(l, x_j)
then $l \leftarrow x_j$
succeed

if *code*$(p) =$ *unify_variable*(l)
& *mode* $=$ *Write*
then *mk_heap_var*(l)
succeed

if *code*$(p) =$ *put_variable*(x_i, x_j)
then *mk_heap_var*(x_i)
$x_j \leftarrow$ *mk_ref*(h)
succeed

Note that our old rule for *unify_variable* in *Read* mode works fine for fresh x_i as well. By inspection of rules we get

Initialization Lemma. Instructions *get_variable*(l, x_j), *unify_variable*(l), *put_variable*(l, x_j) are equivalent (up to trailing) to, given $l > e$, initializing l to *unbound* and executing *get_value*(l, x_j), *unify_local_value*(l), *put_unsafe_value*(l, x_j) respectively.

It is usual to note that *get_variable*(x_i, x_i) and *put_value*(x_i, x_i) have the effect of *succeed*, permitting the compiler to be clever about numbering of temporary variables, confusing them with proper argument registers in order to minimize data movement ('peep-hole optimization', [Ait-Kaci 90]).

In case of a fact or a chain rule there will be no permanent variables, and *allocate* would just store *cp* on the stack. Now we can say that *allocate* would commute with *get_seq* and that *deallocate* would be its inverse, permitting us not to *allocate* at all, satisfying

Compiler Assumption 6. A fact or a chain-rule may be compiled to, respectively,

$$flatten(\,[get\_seq(Fact), proceed]\,)$$
$$flatten(\,[get\_seq(Head), pure\_execute\_sequence(Goal)]\,),$$

where *pure_execute_sequence* is like *execute_sequence*, but without *deallocate*.

3.2 Trading heap for stack or what is needed

By our definitions a temporary variable, first occurring as an argument of a body goal, would always occupy a fresh heap cell. Since the stack, in view of LCO and environment trimming (and determinacy detection), may retract much more often than the heap, trading in a persistent heap location for a volatile stack location may be considered as optimization.

A one goal variable first occurring as argument of a body goal may thus be better classified as permanent after all (excluding of course variables occurring in the last goal, if we are to preserve LCO). Since our treatment of permanent variables would however make it unsafe, nothing would be gained (and a stack location would be wasted) unless we reconsider that treatment.

A simple solution is not to make it unsafe—i.e. to let it live (be needed) a little longer by protecting it from being trimmed. A modified definition of being needed could adjust everything—variable classification, notion of unsafe and environment trimming.

A variable occurring before or in a body goal G_i of a clause $H :- G_1, \ldots, G_n$, $1 \leq i < n$, is *needed* at G_i if it occurs after G_i as well, i.e. in some G_l, $l > i$, or if its first occurence in the clause is an argument position of G_i.

With this definition of being needed, a one-goal variable which first occurs in an argument position of a body goal (not the last one), will be needed there, hence will not be unsafe, will not be trimmed till after the goal is called, and will be permanent. The Argument Registers Property is preserved by definition, a heap cell is traded in for a stack cell, and the definition of a temporary variable becomes equivalent to the classical one:

> A temporary variable is a variable that has its first occurrence in the head or in a structure or in the last goal, and that does not occur in more than one goal in the body, where the head of the clause is counted as part of the first goal. [Warren 83]

Example 7.[Ait-Kaci 90] In the clause $a :- b(X, X), c.$ the variable X is by the latest definition permanent.

One final step of optimization will bring our model to complete compliance with WAM compilation of clauses: it doesn't make sense (any more) to initialize a permanent variable to unbound at *allocate*, just in order to bind it to something else at its first occurrence. Permanent variables can be initialized on the fly, like temporary ones.

The *allocate* rule will thus lose its initialization update. The cases of a permanent variable occurring first in the head or in a structure are handled correctly (we have verified that already) by the existing *get_variable* and *unify_variable* rules. An attentive reader will have noticed that they bind a permanent variable without trailing it, but this has no semantical consequences (cf. discussion in 2.1.3). The only consequence would be to our proof of Pure Prolog Theorem, where the definition of *subst* should now take into account all permanent variables, whether trailed or not.

A permanent variable first occurring in an argument position of a body goal by (the latest) definition cannot be unsafe there, so we still need an instruction equivalent to $put\_value(y_n, x_j)$ with $unbound(y_n)$. It is usually called $put\_variable(y_n, x_j)$, enabling us to formulate our **Compiler Assumption 7** by extending Compiler Assumption 5 to permanent variables.

$$\begin{array}{l} \textbf{if } code(p) = put\_variable(y_n, x_j) \\ \textbf{then } mk\_unbound(y_n) \\ \quad x_j \leftarrow mk\_ref(y_n) \\ \quad succeed \end{array}$$

The above discussion may then be summed up as

Classification Lemma. Warren's classification of variables and on_the_fly initialization preserve the Pure Prolog Theorem.

For the record, we might list the final code for the clauses of examples 5,6,7.

$a(X) :- b(f(X))$.	$has\_a(X) :-$ $generate(X,Y), test(Y)$.	$a :- b(X,X), c$.
$get\_variable(x_2, x_1)$	$allocate(1)$	$allocate(1)$
$put\_structure(f, 1, x_1)$	$put\_variable(y_1, x_2)$	$put\_variable(y_1, x_1)$
$unify\_local\_value(x_2)$	$call(generate, 2, 1)$	$put\_value(y_1, x_2)$
$execute(b, 1)$	$put\_unsafe\_value(y_1, x_1)$	$call(b, 2, 1)$
	$deallocate$	$deallocate$
	$execute(test, 1)$	$execute(c, 0)$

3.3 Switching

Hitherto we have assumed of *procdef* to return a pointer to a *try_me_else, retry_me_else, trust_me* or a *try, retry, trust* chain (or a single clause) containing (at least) all candidate clauses for the given goal, cf. discussion in Part I. [Warren 83] prescribes a specific way of arranging dedicated instructions, *switch_on_term, switch_on_constant, switch_on_structure* in order to achieve a reasonable selection. In view of different methods of clause indexing, present in the literature and in implementations, we shall just describe the building blocks, the ***switching instructions***, assuming of the compiler to arrange them in such a way as to select at least all candidate clauses. We allow switching on any goal argument. Since we are skipping special representation of lists and constants, we leave it to the reader to extend *switch_on_term* with *Ll* and *Lc* arguments, to add *switch_on_constant* [Warren 83, Ait-Kaci 90] and to refine the ***chain*** function below.

The universe of instructions is then extended by

$$\{\, switch\_on\_term(i, Lv, Ls), switch\_on\_structure(i, N, T) \mid i, N \in \mathcal{N},\ Lv, Ls, T \in CODEAREA \,\}$$

where the new instructions are 'executed' by rules

$$\begin{array}{l}
\textbf{if } code(p) = switch_on_term(i, Lv, Ls) \\
\textbf{thenif } type(deref(x_i)) = Ref \\
\textbf{then } p := Lv \\
\textbf{elsif } type(deref(x_i)) = Struct \\
\textbf{then } p := Ls
\end{array}$$

$$\begin{array}{l}
\textbf{if } code(p) = switch_on_structure(i, N, T) \\
\textbf{then } p := hash(T, N, funct(ref(deref(x_i))), arity(ref(deref(x_i))))
\end{array}$$

with hash table access modelled by the function

$$hash : CODEAREA \times \mathcal{N} \times ATOM \times ARITY \rightarrow CODEAREA$$

to be thought of as returning, for $hash(T, N, f, a)$, the codepointer associated to f, a in the hash table of size N located at T.

Our assumptions about switching instructions generated by the compiler may be expressed by using an auxiliary function $chain : CODEAREA \rightarrow CODEAREA^*$ such that

$$\begin{array}{lll}
chain(Ptr) & = & \textbf{if } code(Ptr) = switch_on_term(i, Lv, Ls) \\
& & \textbf{then } (\ \textbf{if } type(deref(x_i)) = Ref \\
& & \qquad \textbf{then } chain(Lv) \\
& & \qquad \textbf{else } chain(Ls)\ ) \\
& & \textbf{elsif } code(Ptr) = switch_on_structure(i, N, T) \\
& & \textbf{then } chain(hash(T, N, funct(ref(deref(x_i))), arity(ref(deref(x_i))))) \\
& & \textbf{elsif } (\ code(Ptr) = try_me_else(N, n) \\
& & \qquad \textbf{or } code(Ptr) = retry_me_else(N, n)\ ) \\
& & \textbf{then } [\, Ptr \mid flatten([\, chain(Ptr+), chain(N)\,])\,] \\
& & \textbf{elsif } code(Ptr) \in \{\, trust(C), Proceed, Execute(g, m)\,\} \\
& & \textbf{then } [\, Ptr\,] \\
& & \textbf{else } [\, Ptr \mid chain(Ptr+)\,]
\end{array}$$

The precise assumption is then

Compiler Assumption 8. Mapping *code* along $chain(procdef(f, a))$ yields, with the same values of *AREGS*, the same sequence of instructions as hitherto assumed of $unload(procdef(f, a))$.

Given Compiler Assumption 8, inspection of rules (compared to definition of *chain*) suffices to establish

Switching Lemma. Switching preserves the Pure Prolog Theorem.

Example 8. Given the Prolog program

$$\begin{array}{l}
p(f_1(X)) :- q_1(X). \\
p(f_2(X)) :- q_2(X). \\
p(Other) :- default(Other).
\end{array}$$

and the compilation of clauses of form

$$\begin{array}{lll}
C_i+ : & get_structure(f_i, 1, x_1) & \\
& unify_variable(x_2) & \\
& put_value(x_2, x_1) & \\
& execute(q_i, 1) & i = 1, 2 \\
& & \\
C_3+ : & execute(default, 1) &
\end{array}$$

our assumptions (Compiler Assumption 8, and assumptions on *procdef* from Part I) would allow at least the following two layouts of switching instructions.

	$try\_me\_else(C_3, 1)$		$switch\_on\_term(1, C_1, L)$
	$switch\_on\_term(1, C_1, L)$	$L:$	$switch\_on\_structure(1, 2, T)$
$L:$	$switch\_on\_structure(1, 2, T)$		
		$L_1:$	$try(C_1+, 1)$
$C_1:$	$try\_me\_else(C_2, 1)$		$trust(C_3+, 1)$
	...		
		$L_2:$	$try(C_2+, 1)$
$C_2:$	$trust\_me(1)$		$trust(C_3+, 1)$
	...		
		$C_1:$	$try\_me\_else(C_2, 1)$
$C_3:$	$trust\_me(1)$		...
	...		
		$C_2:$	$retry\_me\_else(C_3, 1)$
			...
		$C_3:$	$trust\_me(1)$
			...
	with		with
	$hash(T, 2, f_i, 1) = C_i+$		$hash(T, 2, f_i, 1) = L_i$

Putting the lemmata together, we obtain, given WAM Assumptions 1,2,3,4 and Compiler Assumptions 1,2,2',3,4,5,6,7,8, the final result.

Main Theorem. The WAM is correct and complete wrt to Prolog (as defined by Prolog algebras of Part I, constrained to non-STO unification) in the sense of Pure Prolog Theorem.

References

[Ait-Kaci 90] Aït-Kaci,H. *The WAM: A (Real) Tutorial.* PRL Research Report 5, Digital Equipment Corporation, Paris Research Laboratory 1990.

[Boerger,Rosenzweig 91a] Börger, E. & Rosenzweig D. *From Prolog Algebras Towards WAM— A Mathematical Study of Implementation.* In: CSL '90, 4th Workshop on Computer Science Logic (Eds. E.Börger, H.Kleine Büning,M.M.Richter,W.Schönfeld), Springer LNCS 533, 1991., pp. 31-66

[Boerger,Rosenzweig 91b] Börger, E. & Rosenzweig D. *An Analysis of Prolog Database Views and Their Uniform Implementation* CSE-TR-88-91, University of Michigan, Ann Arbor, Michigan, 1991.

[Boerger,Rosenzweig 91c] Börger, E. & Rosenzweig D. *WAM Algebras—A Mathematical Study of Implementation, Part 2.* CSE-TR-89-91, University of Michigan, Ann Arbor, Michigan, 1991.

[Lloyd 84] Lloyd, J.W. *Foundations of Logic Programming,* Springer, Berlin-Heidelberg-New York-Tokyo 1984.

[North,Scowen 91] North, N.D. & Scowen, R.S. ISO/IEC JTCI SC22 WG 17 N.72, 1991.

[Warren 83] Warren, D.H.D. *An Abstract Prolog Instruction Set,* Technical Note 309, Artifficial Intelligence Center, SRI International 1983.

ABDUCTIVE SYSTEMS FOR NON-MONOTONIC REASONING

A.G. Bondarenko
Program Systems Institute
Pereslavle-Zalessky 152140 USSR

1. Introduction

This paper presents a new universal framework for formalizing Non-Monotonic reasoning. The notion of Abductive System is a generalization of the notion of Abductive Framework, proposed in [2]. In comparison with existing universal approaches to Non-Monotonic reasoning (such as the *Logic of Grounded Knowledge* [3], or *Non-monotonic Rule Systems* [4]), the formalism, presented here has some advantages:

- it combines preferential semantics with fixed-point approach;
- it is simple;
- it is general enough.

In this paper it is shown, that Default Logic [6], Autoepistemic Logic [5], the Logic of Grounded Knowledge [3], Non-Monotonic Modal Logic [7], and Stable Model Semantics for Logic Programming [1] can be expressed in terms of Abductive Systems.

The paper also presents two new Non-Monotonic logics:

- the Logic of Knowledge and Beliefs (KBL);
- Bimodal Non-Monotonic Logic (BNL).

KBL is a formalization of reasoning of a person about his own knowledge and beliefs. BNL is a modification of the Logic of Grounded Knowledge [3], and has some advantages in comparison with it.

2. Abductive systems

Definition 1: An *abductive system* is a guadruple $\langle L,\vdash,D,C\rangle$ where L is a formal language with negation "$\neg$";
$\vdash \subseteq 2^L \times L$ is a provability relation, defined by a deductive system;
$D \subseteq L$ is an abductive set;
$C \subseteq 2^L$ is a set of allowed extensions.

Let $T \subseteq L$; $\varphi,\varphi_1...\varphi_n \in L$. Introduce abbreviations:
1) $T \vdash \bot$ iff there is $\varphi \in L$, such that $T \vdash \varphi$ and $T \vdash \neg\varphi$.

In this case T is called *inconsistent*.

2) $T \nvdash \bot$, iff it isn't true that $T \vdash \bot$.

In this case T is called *consistent*.

3) $T, \varphi_1 ... \varphi_n \vdash \psi$ iff $T \cup \{\varphi_1 ... \varphi_n\} \vdash \psi$

4) $Th(T) = \{ \varphi \mid T \vdash \varphi \}$.

Fix an abductive system $\Sigma = \langle L, \vdash, D, C \rangle$.

Definition 2: Let $E, T \subseteq L$. Then E is an *extension* of T in Σ, iff

1) $E = Th (T \cup \{\varphi \mid E, \varphi \nvdash \bot, \varphi \in D\})$; and

2) $E \in C$.

A set of formulas E is called a *maximal set for* T in Σ, iff E is a maximal consistent set of the form $Th(T \cup Q)$, where $Q \subseteq D$.

Theorem 1. Let $T, E \subseteq L$, $T \nvdash \bot$. Then E is an extension of T in Σ, iff

1) E is a *maximal set for* T in Σ; and

2) $E \in C$.

The proof of this theorem is obvious.

Introduce abbreviations:

1) $T \vdash^1 \varphi$, iff either there is an extension of T, including φ, or T has has no extensions.

2) $T \vdash^2 \varphi$, iff φ is a member of all extensions of T.

Definition 3: A quadruple $\langle M, \models, \ll, Q \rangle$ is a *semantical structure for* a formal language L iff M is a nonempty set; $\models \subseteq M \times L$; $\ll \subseteq M \times M$ is a reflexive and transitive relation on M; and $Q \subseteq 2^M$.

Let $T \subseteq L$; $\varphi \in L$; m, m_1, $m_2 \in M$; $H \subseteq M$. Introduce abbreviations:

1) $m \not\models \varphi$ iff it isn't true that $m \models \varphi$;

2) $m \models T$ iff for any $\psi \in T$, $m \models \psi$ holds;

3) $T \models \varphi$ iff for any $m \in M$: $m \models T$ imply $m \models \varphi$;

4) $D(m) = \{\varphi \mid m \models \varphi, \varphi \in D\}$;

5) $H \models T$ iff for any $m \in H$, $m \models T$ holds;

6) $Mod(T) = \{m \mid m \models T \}$.

7) $m_1 \backsim m_2$ iff $m_1 \ll m_2$ and $m_2 \ll m_1$;

8) $m_1 < m_2$ iff $m_1 \ll m_2$ and not $m_1 \backsim m_2$;

9) $e(m,T) = \{m' \mid m' \backsim m, m' \models T \}$.

Let $S = \langle M, \models, \ll, Q \rangle$ be a semantical structure for L.

Elements of M are called *interpretations*.

A formula φ is *true* in an interpretation m, iff $m \models \varphi$. An interpretation m is a *model* of T, iff $m \models T$.

Definition 4: Let $T \subseteq L$, $m \subseteq M$. An interpretation m is called a *maximal model* of T in S, iff

1) $m \models T$;

2) there is no other $m' \in M$, such that $m' \models T$ and $m < m'$.

An interpretation m is called a *preferred model* of T in S, iff

1) m is a *maximal model* of T in S; and

2) $e(m,T) \in Q$.

If m is a preferred model of T, then e(m,T) is called a *preferred class of models* of T.

Definition 5: Let $\Sigma = \langle L, \vdash, D, C \rangle$ be an abductive system. A quadruple $S = \langle M, \models, <<, Q \rangle$ is a *formal semantics* for Σ, iff

1) S a semantical structure for L;

2) for any $m \in M$, $\varphi \in L$: $m \models \neg\varphi$ implies $m \not\models \varphi$;

3) for any $T \subseteq L$, $\varphi \in L$: $T \vdash \varphi$ iff $T \models \varphi$;

4) for any $m_1, m_2 \in M$: $m_1 << m_2$ iff $D(m_1) \subseteq D(m_2)$;

5) $Q = \{H \mid$ there is $E \in C$, such that $H = Mod(E)\}$.

Fix an abductive system $\Sigma = \langle L, \vdash, D, C \rangle$, and a formal semantics $S = \langle M, \models, <<, Q \rangle$ for it.

It is not hard to see, that a set of formulas T is consistent, iff it has a model, and if $C = 2^L$, then any consistent set of formulas T has a preferred model.

Theorem 2. Let $E, T \subseteq L$, $T \not\vdash \bot$. Then:

(a). if E is an extension of T then any model of E is a preferred model of T;

(b). if m is a preferred model of T then there is E' - an extension of T, such that $m \models E'$;

(c). Mod(E) is a preferred class of models of T, iff Th(E) is an extension of T.

Proof. Let E be an extension of T. Then E has a model m. It is not hard to see, that there is no other m', such that $m' \models T$, and $D(m) \subset D(m')$, because, otherwise, there is $\varphi \in D$, such that $E, \varphi \not\vdash \bot$, and $\varphi \notin E$, which means, that E is not a maximal consistent set of the form $Th(T \cup Q)$, where $Q \subseteq D$, i.e. E is not an extension. So, m is a maximal model of T.

$e(m,T) = \{m' \mid D(m')=D(m), \text{ and } m'\models T\} = Mod(E) \in Q$.

So, m is a preferred model of T, and Mod(E) is a preferred class of models of T.

Let m be a preferred model of T. Then $E'=Th(T \cup D(m))$ is a maximal set for T, because, otherwise, there is $\varphi\in D\setminus E'$, such that $E',\varphi\nvdash\bot$, so, there is m', such that $m'\models E'$, $m'\models\varphi$, $m<m'$, which means, that m is not a preferred model of T.

$$Mod(E') = \{m' \mid D(m')=D(m),\ m'\models T\} = e(m,T) \in Q,$$

so, $E'\in C$, consequently, E' is an extension of T.

Let Mod(E) be a preferred class of models of T. Then there is m - a preferred model of T, such that $Mod(E)=e(m,T)$, $Th(T\cup D(m))$ is an extension of T, and $Mod(T\cup D(m))=e(m,T)$. So, $Th(E)=Th(T\cup D(m))$, i.e. Th(E) is an extension of T. ∎

Introduce abbreviations:

(a). $T\models^1\varphi$ iff either there is a preferred class of models of T, such that φ is true in all models of this class, or T has no preferred models;

(b). $T\models^2\varphi$ iff φ is true in all preferred models of T.

Theorem 3. (Completeness and soundness)
Let $T\subseteq L$, $T\nvdash\bot$, $\varphi\in L$. Then: $T\vdash^1\varphi$ iff $T\models^1\varphi$ and $T\vdash^2\varphi$ iff $T\models^2\varphi$.

This theorem is an obvious consequence of the Theorem 4.

In the next sections of the paper I consider a few examples of abductive systems, and their correspondence with some other approaches to Non-Monotonic reasoning.

3. Non-Monotonic Modal Logic

Consider an abductive system $U_{Ax} = \langle L;\vdash_{Ax};D;C\rangle$, where

1) L is a propositional modal language;

2) $\vdash_{Ax}$ is defined by the following deductive system:

(1°). all tautologies;

(Ax). any set of axioms;

$$(MP): \frac{\varphi,\ \varphi\supset\psi}{\psi}; \qquad (Nec): \frac{\varphi}{\Box\varphi} \ ;$$

3) $D = \{\neg\Box\varphi \mid \varphi \in L\}$;

4) $C = \{E \mid E\subseteq L, \text{ and } \Box\varphi\supset\varphi,\ \Box\varphi\supset\Box\Box\varphi,\ \neg\Box\varphi\supset\Box\neg\Box\varphi \in E \text{ for any } \varphi\in L\}$.

Theorem 4. Let $T,E \subseteq L$. Then E is an extension of T in U_{Ax} iff

$$E = Th_{Ax}(T \cup \{\neg\Box\varphi \mid \varphi \notin E\}).$$

Proof. a). Let $E = Th_{Ax}(T \cup \{\neg\Box\varphi \mid \varphi \notin E\})$. Then E is a maximal set for T in U_{Ax}, because, otherwise, there is $\varphi \in L$, such that $E, \neg\Box\varphi \nvdash \bot$ and $\neg\Box\varphi \notin E$, so $\varphi \in E$, $\Box\varphi \in E$., which leads to contradiction with $E, \neg\Box\varphi \nvdash \bot$.

If $\Box\varphi \in E$, then $\neg\Box\varphi \notin E$, $\varphi \in E$, $\Box\Box\varphi \in E$, so, $\Box\varphi \supset \varphi \in E$, and $\Box\varphi \supset \Box\Box\varphi \in E$.

If $\Box\varphi \notin E$, then $\varphi \notin E$, $\neg\Box\varphi \in E$, so, $\Box\varphi \supset \varphi \in E$, and $\Box\varphi \supset \Box\Box\varphi \in E$.

If $\neg\Box\varphi \in E$, then $\Box\neg\Box\varphi \in E$, so, $\neg\Box\varphi \supset \Box\neg\Box\varphi \in E$.

If $\neg\Box\varphi \notin E$, then $\varphi \in E$, $\Box\varphi \in E$, so, $\neg\Box\varphi \supset \Box\neg\Box\varphi \in E$.

Thus, E is a maximal set for T in U_{Ax}, and $E \in C$, so, E is an extension of T in U_{Ax}.

b). Let E be an extension of T in U_{Ax}. Then

(1). $E = Th_{Ax}(T \cup \{\neg\Box\varphi \mid E, \neg\Box\varphi \nvdash \bot\})$

(2). for any $\varphi \in L$: $\Box\varphi \supset \varphi$, $\Box\varphi \supset \Box\Box\varphi$, $\neg\Box\varphi \supset \Box\neg\Box\varphi \in E$.

It is not hard to see, that, the second condition (2) implies the equivalence of $\varphi \notin E$ and $E, \neg\Box\varphi \nvdash \bot$ for any $\varphi \in L$. So,

$$E = Th_{Ax}(T \cup \{\neg\Box\varphi \mid \varphi \notin E\}). \blacksquare$$

Let Ax be defined by some axiom schemata from the following list:

(K) $\Box(\varphi \supset \psi) \supset (\Box\varphi \supset \Box\psi)$

(T) $\Box\varphi \supset \varphi$

(4) $\Box\varphi \supset \Box\Box\varphi$

(5) $\neg\Box\varphi \supset \Box\neg\Box\varphi$

In this case U_{Ax} becomes equivalent to McDermott's propositional Non-Monotonic Modal Logic (NML), where the definition of extension (see [7],[8]) has the following form:

E is an extension of T in NML

iff

$$E = Th_{Ax}(T \cup \{\neg\Box\varphi \mid \varphi \notin E\}).$$

In [8] it is shown, that Moore's Autoepistemic Logic [5] is equivalent to NML, based on the modal system K45 = ⟨ 1,K,4,5, MP, Nec ⟩. So, Autoepistemic Logic also can be expressed in terms of U_{Ax}.

Consider the semantics for U_{Ax}. A *Kripke-structure* for U_{Ax} is a triple $G = \langle W,R,V \rangle$, where W is a nonempty set of worlds ; R is a relation on W; and for each $u \in W$, $V(u)$ is a set of propositional variables which are said to be true in the world u. The forcing relation $\langle u,G \rangle \Vdash \varphi$ (the formula φ is true in the world u of the structure G) is defined inductively as follows:

$\langle u,G \rangle \Vdash p$ iff $p \in V(u)$; $\langle u,G \rangle \Vdash \varphi \supset \psi$ iff $\langle u,G \rangle \Vdash \psi$ or not $\langle u,G \rangle \Vdash \varphi$; $\langle u,G \rangle \Vdash \neg\varphi$ iff not $\langle u,G \rangle \Vdash \varphi$; $\langle u,G \rangle \Vdash \Box\varphi$ iff for each v with uRv, $\langle v,G \rangle \Vdash \varphi$.

By definition, $G \models \varphi$, iff for each $u \in W$, $\langle u,G\rangle \models \varphi$.

Let

1). M_{Ax} be the set of all Kripke-structures G, such that $G \models Ax$;

2). $\models_{Ax} \subseteq M_{Ax} \times L$, such that $G \models_{Ax} \varphi$ iff $G \models \varphi$;

3). $\ll \subseteq M \times M$, such that: $G_1 \ll G_2$ iff

$$\{\neg\Box\varphi \mid \varphi \in L,\ G_1 \models \neg\Box\varphi\} \subseteq \{\neg\Box\varphi \mid \varphi \in L,\ G_2 \models \neg\Box\varphi\};$$

4). $Q = \{H \mid H \subseteq M$, and $H \models (\Box\varphi \supset \varphi) \& (\Box\varphi \supset \Box\Box\varphi) \& (\neg\Box\varphi \supset \Box\neg\Box\varphi)$ for any $\varphi \in L\}$.

It is easy to show, that if Ax includes all instances of the axiom scheme (K): $\Box(\varphi \supset \psi) \supset (\Box\varphi \supset \psi)$, then $S_{Ax} = \langle M_{Ax}; \models_{Ax}; \ll; Q\rangle$ is a formal semantics for $\cup_{Ax}$.

If Ax corresponds to one of modal systems, used for McDermott's propositional Non-Monotonic Modal Logic, then S_{Ax} can be considered as a formal semantics for this logic.

4. Logic of Knowledge and Beliefs (KBL)

Consider an abductive system $KBL = \langle L; \vdash; D; C\rangle$, where

1) L is a bimodal propositional language with modalities K and B ($K\varphi$ means "I know φ", $B\varphi$ means "I belief φ");

2) $\models$ is defined by the following deductive system:

1°. All tautologies

2°. $K(\varphi \supset \psi) \supset (K\varphi \supset K\psi)$

3°. $B(\varphi \supset \psi) \supset (B\varphi \supset B\psi)$

4°. $K\varphi \supset KK\varphi$

5°. $K\varphi \supset \varphi$

6°. $B\varphi \supset KB\varphi$

7°. $K\varphi \supset B\varphi$

8°. $B\neg\varphi \supset \neg B\varphi$

$$\text{MP:} \frac{\varphi, \varphi \supset \psi}{\psi} \qquad \text{Nec}^{*}: \frac{\vdash \varphi}{\vdash K\varphi};$$

3) $D = \{K\neg K\varphi \mid \varphi \in L\} \cup \{K\neg B\varphi \mid \varphi \in L\}$;

4) $C = \{E \mid E \subseteq L$, and $\neg K\varphi \supset K\neg K\varphi$, $\neg B\varphi \supset K\neg B\varphi \in E$ for any $\varphi \in L\}$.

<u>Theorem 5</u>. Let $E,T \subseteq L$. Then E is an extension of T in KBL, iff

$$E = Th(T \cup \{K\neg K\varphi \mid K\varphi \notin E\} \cup \{K\neg B\varphi \mid B\varphi \notin E\}).$$

The proof of this theorem is completely similar to the proof of the Theorem 4.

It is easy to show, that if T includes all formulas of the form $B\varphi \supset K\varphi$, where $\varphi \in L$, then KBL becomes equivalent to Mcdermott's NML, based on the propositional modal system S4=⟨1,K,T,4,MP,Nec⟩.

A *Kripke-structure* for KBL is a guadruple $\langle W,R_K,R_B,V\rangle$, where

1) W is a nonempty set of worlds;

2) R_K is a reflexive and transitive relation on W;

3) R_B is a relation on W, such that:

  (a). $R_B \subseteq R_K$;

  (b). for any $u \in W$ there is $v \in W$, such that uR_Bv;

  (c). for any $u,v,t \in W$, uR_Kv and vR_Bt imply uR_Bt;

4) for each $u \in W$, V(u) is a set of propositional variables.

Consider a guadruple $S_{KBL} = \langle M,\models,<<,Q\rangle$, where

1) M is the set of all pairs ⟨u,G⟩, where G is a Kripke structure for KBL, and u is a world in G;

2) $\models$ is the forcing relation, defined inductively as follows:

  $\langle u,G\rangle \models p$ iff $p \in V(u)$;

  $\langle u,G\rangle \models \neg\varphi$ iff not $\langle u,G\rangle \models \varphi$;

  $\langle u,G\rangle \models \varphi\supset\psi$ iff $\langle u,G\rangle \models \neg\varphi$ or $\langle u,G\rangle \models \psi$;

  $\langle u,G\rangle \models K\varphi$ iff for each v with uR_kv, $\langle v,G\rangle \models \varphi$;

  $\langle u,G\rangle \models B_\varphi$ iff for each v with uR_Bv, $\langle v,G\rangle \models \varphi$;

3) $>> \subseteq M \times M$, such that

$$m_1 << m_2 \text{ iff } D(m_1) \subseteq D(m_2);$$

4) $Q = \{H \mid H\subseteq M, \text{ and } H\models (\neg K\varphi\supset K\neg K\varphi)\&(\neg B\varphi\supset K\neg B\varphi) \text{ for any } \varphi\in L\}$.

It is not hard to see that the quadruple S_{KBL} is a formal semantics for KBL.

The logic KBL can be considered as a formalization of reasoning of a person about his own knowledge and beliefs. Each interpretation m in S_{KBL} defines a world outlook of a person, i.e. his knowledge and beliefs. One interpretation is more preferable, than another one, if it has less ungrounded knowledge and beliefs. Following the Definition 4, an interpretation m is a *preferred model* of a set of formulas T, iff

1). m is a model of T with the minimal knowledge and beliefs;

2). all formulas of the form $\neg K\varphi\supset K\neg K\varphi$ and $\neg B\varphi\supset K\neg B\varphi$ are true in any model of T, which is equivalent to m (has the same knowledge and beliefs).

5. Bimodal Non-Monotonic Logic (BNL)

Bimodal Non-Monotonic Logic (BNL) is the following abductive system BNL = $\langle L,\vdash,D,G\rangle$, where

1) L is a propositional bimodal language with modalities A and L;

2) $\vdash$ is defined by the following deductive system:

1°. All tautologies

2°. $L\varphi \supset A\varphi$

3°. $L(\varphi\supset\psi) \supset (L\varphi\supset L\psi)$

4°. $A(\varphi\supset\psi) \supset (A\varphi\supset A\psi)$

$$\text{MP:}\frac{\varphi, \varphi \supset \psi}{\psi} \qquad \text{Nec}^{*}: \frac{\vdash \varphi}{\vdash L\varphi}$$

3) $D = \{\neg L\varphi \mid \varphi\in L_o\} \cup \{\neg A\varphi \mid \varphi\in L_o\}$, where L_o is the set of all formulas without modalities;

4) $C = \{E \mid E\subseteq L$, and $L\varphi\equiv A\varphi \in E$ for any $\varphi \in L_o\}$.

A *Kripke structure* for BNL is a quadruple $G = \langle W;R_L;R_A;V\rangle$, where W is a nonempty set of worlds, $R_L,R_A \subseteq W\times W$,and $R_A \subseteq R_L$, and for each $u\in W$, V(u) is a set of propositional variables.

An *interpretation* is a pair $\langle u,G\rangle$, where $u\in W$.

Let $S_{BNL} = \langle M,\models,<<,Q\rangle$, where

1) M is the set of all interpretations;

2) $\models$ is the forcing relation, defined similarly to KBL;

3) $<< \subseteq M \times M$, such that $m_1<<m_2$ iff $D(m_1) \subseteq D(m_2)$;

4) $Q = \{H \mid H\subseteq M$, and $H\models L\varphi\equiv A\varphi$ for any $\varphi\in L_o\}$.

It is easy to show, that:

(a). S_{BNL} is a formal semantics for BNL;

(b). An interpretation m is a preferred model of a set of formulas T in S_{BNL}, iff

1). $m \models T$;

2). there is no other m', such that $m' \models T$ and $m < m'$;

3). $m \models L\varphi\equiv A\varphi$ for any $\varphi \in L_o$.

Consider the correspondence of BNL with the Logic of Grounded Knowledge [3].

The formal language L_{GK} for the Logic of Grounded Knowledge (GK) is the same as for BNL.

A *Kripke structure* for GK is a quadruple $G=\langle W,R_L,R_A,V\rangle$, where W is a nonempty set, $R_L,R_A \in W \times W$, and for each $u \in W$, V(u) is a set of propositional variables.

An *interpretation* is a pair $\langle u,G\rangle$, where $u \in W$. The forcing relation $\langle u,G\rangle \models \varphi$ is defined similarly to KBL.

The preference relation $<$ on interpretations is defined as follows:

$$m_1<m_2 \quad \text{iff} \quad A(m_1) = A(m_2), \text{ and } L(m_2) \subset L(m_1), \text{ where}$$
$$A(m)=\{\varphi \mid m\models A\varphi,\ \varphi\in L_0\}, \text{ and } L(m)=\{\varphi \mid m\models L\varphi,\ \varphi\in L_0\}.$$

An interpretation m is called a *preferred* model of $T \subseteq L_{GK}$ in GK iff

1) $m \models T$;

2) there is no other m'. Such that $m' \models T$ and $m < m'$;

3) $L(m) = A(m)$.

Obviously, any preferred model of T in S_{BNL} is a preferred model of T in GK.

Theorem 6. Let $T \subseteq L_{GK}$, and any $\varphi \in T$ has the form: $L\varphi_1 \& \ldots \& L\varphi_n \ \& \neg A\psi_1 \ldots \neg A\psi_m \supset L\chi$, where $\varphi_1 \ldots \varphi_n, \psi_1 \ldots \psi_m, \chi \in L_0$. Then any preferred model of T in GK is a preferred model of T in S_{BNL}.

Proof. Let m be a preferred model of T in GK, but not in S_{BNL}. This means, that there is m', such that $m' \models T$, $L(m') \subseteq A(m')$, and $D(m) \subseteq D(m')$, where $D(m) = \{\neg A\varphi | \varphi \in L_o\} \cup \{\neg L\varphi | \varphi \in L_o\}$. Consider three possible variants.

1). $A(m') \subset A(m)$, $L(m') = L(m)$. This is impossible, because $L(m') \subseteq A(m')$.

2). $A(m') = A(m)$, $L(m') \subset L(m)$. In this case m is not a preferred model of T in GK, which is wrong.

3). $A(m') \subset A(m)$, $L(m') \subset L(m)$. In this case (because of the special form of the formulas from T) there is m'', such that $L(m'') \subset L(m)$, and $A(m'') = A(m)$, so m is not a preferred model of T in GK, which is wrong.

■

Let $BNL^* = \langle L, \vdash, D^*, C \rangle$, where $D^* = D \cup \{A\varphi | \varphi \in L_o\}$, and $S^*_{BNL} = \langle M, \models, <<^*, Q \rangle$, where $<<^*$ corresponds to D^*, i.e. for any $m_1, m_2 \in M$, $m_1 <<^*_2 m$ iff $D^*(m_1) \subseteq D^*(m_2)$.

Theorem 7. Let $m \in M$, $T \subseteq L$. Then m is a preferred model of T in GK, iff m is a preferred model of T in S^*_{BNL}.

Proof. Consider S^*_{BNL}. $m_1 < m_2$ in S^*_{BNL}, iff $D^*(m_1) \subset D^*(m_2)$, where $D^*(m) = \{\neg L\varphi | \varphi \in L_o\} \cup \{\neg A\varphi | \varphi \in L_o\} \cup \{A\varphi | \varphi \in L_o\}$. So, $m_1 < m_2$ in S^*_{BNL}, iff $L(m_2) \subset L(m_1)$, and $A(m_2) = A(m_1)$. Thus, m is a preferred model of T in S^*_{BNL}, iff

(1). m is a maximal model of T in S^*_{BNL}, and

(2). $L(m) = A(m)$.

Let m be a preferred model of T in S^*_{BNL}, but not in GK. Then there is m', such that $L(m') \subset L(m)$, $A(m') = A(m)$. $m \models L\varphi \supset A\varphi$ for any $\varphi \in L_o$, so, $m' \models L\varphi \supset A\varphi$ for any $\varphi \in L_o$, so, m' is an interpretation in S^*_{BNL}, so, m is not a preferred model of T in S^*_{BNL}, which is wrong.

Let m be a preferred model of T in GK, but not in S^*_{BNL}. Then there is $m' \in M$, such that $L(m') \subset L(m)$, and $A(m')=A(m)$, so, m is not a preferred model of T in GK. ■

Thus, BNL^* can be considered, as a proof theoretical reformulation of the logic GK. Obviously, any extension of T in BNL is an extension of T in BNL^*. The advantage of BNL is, that it has less ungrounded extensions, than BNL^*. For example, $T=\{Ap \supset Lp\}$ has unique extension $E_1=Th(\{\neg Ap, \neg Lp\})$ in BNL. But in BNL^* T has the second extension $E_2=Th(\{Ap,Lp\})$. So, $T \vdash^1 Lp$ in BNL^*, which is not good for some applications.

Consider the correspondence of BNL with Default Logic [5]. A propositional *default theory* is a pair <T,D>, where T is a set of propositional sentences, and D is a set of expressions of the form $\varphi...\varphi_n:\psi...\psi_m \,/\, \chi$, called *defaults*.
I refer to [6] for the definition of an extension for a default theory. It is shown in [3] that propositional Default Logic can be expressed in terms of GK, using the following translation: for any default theory <T,D>, $f(T,D)=$
$\{\varphi \mid \varphi \in T\} \cup \{L\varphi_1 \& ... \& L\varphi_n \& \neg A \neg \psi_1 \& ... \& \neg A \neg \psi_m \supset L\chi \mid \varphi_1 ... \varphi_n : \psi_1 ... \psi_m / \chi \in D\}$.

<u>Theorem 8.</u> (Shoham,Lin, [3]). A consistent set of propositional sentences E is an extension for a default theory <T,D>, iff there is m - a preferred model of f(T,D) in GK, such that E = L(m).

<u>Theorem 9.</u> A consistent set of propositional sentences E is an extension for a default theory <T,D> iff there is E'- an extension of f(T,D) in BNL, such that $E = \{\varphi \mid L\varphi \in E', \varphi \in L_o\}$.

<u>Proof.</u> This theorem is an obvious consequence of the Theorem 6 the Theorem 8.■

The Stable Model Semantics [1] for the Propositional Logic Programming also can be formulated in terms of BNL. Following [1], a *logic program* P is a set of clauses of the form $q \leftarrow p_1...p_m, q_1...q_n$, where q, $p_1...p_m$, $r_1...r_n$ are atoms (propositional variables).
I refer to [1] for the definition of a stable model of such a program. For any program P define the translation f(P):

$$f(P) = \{Lp_1 \& ... \& Lp_m \& \neg Ar_1 \& ... \& \neg Ar_n \supset Lq \mid q \leftarrow p_1...p_m, \neg r_1 ... \neg r_n \in P\}.$$

<u>Theorem 10.</u> A set of atoms E is a stable model of P, iff there is

E' - an extension of f(P) in BNL, such that

$$E = \{p \mid Lp \in E', p - \text{atom}\}.$$

In [4] it is shown, that Stable Model Semantics for Logic Programming can be formulated in terms of slightly generalized Default Logic (Proposition 3.2, [4]). The Theorem 10 is a straight consequence of the Proposition 3.2 (see [4]), and the Theorem 9.

REFERENCES

1. Gelfond M., Lifschitz V., . Stable semantics for Logic Programs. In: Proceedings of 5th International Symposium Conference on Logic Programming, Seattle, 1988.
2. Kakas A.C., Mancarella P., On the relation between truth maintenance and abduction . In: Preprints of the Third International Workshop on Nonmonotonic Reasoning, South lake Tahoe, California, 1990.
3. Lin F.,Shoham Y., Epistemic Semantics for Fixed-Point Non-Monotonic Logics. In: R.Parikh, ed., Proc. of the Third Conference on Theoretical Aspects of Reasoning About Knowledge. San Mateo: Morgan Kaufmann Publishers,1990.
4. Marek W., Nerode A., Remmel J., A Theory of Nonmonotonic Rule Systems. In: Proceedings of the Fifth Annual IEEE Symposium on Logic in Computer Science, Philadelphia, 1990.
5. Moore R.C., Semantical considerations on nonmonotonic logic., Artificial Intelligence, 25(1985),75-94.
6. Reiter R., A logic for default reasoning, Artificial Intelligence, 13(1980),81-132.
7. McDermott D., Nonmonotonic logic II : Nonmonotonic modal theories, Journal of the ACM, 29,33-57.
8. Shvartz G.F., Autoepistemic modal logics. In: R.Parikh, ed. Proc. of the Third Conference on Theoretical Aspects of Reasoning About Knowledge, San Mateo, Morgan Kaufmann Publishers, 1990.

PROPERTIES OF ALGORITHMIC OPERATORS

Vladimir B. Borshchev
All-Union Institute of Scientific and Technical Information of Academy of Sciences of the USSR
Usiyevicha ul 20a, Moscow, 125219, USSR

Abstract

We consider operators working on lattices of interpretations of logic programs. Operators defined by programs are called algorithmic. Some properties of algorithmic operators, such as monotonicity and compactness, are used to define fix-point semantics. But what are necessary and sufficient conditions for such operators? The main result of this paper may be formulated, roughly speaking, as follows: an operator is algorithmic, iff its result on a given interpretation is the sum of its results on finite and restricted parts of this interpretation, i.e., is in essence defined by a certain algebra. This condition, which we called as n,k-truncateness of an operator, may be divided into several simpler conditions dealing with the generalization of the notion of monotonicity and with the "breadth" n (the maximum number of terms) and the "height" k (the maximum height of terms) of subsystems which form the result of the operator.

0. Introduction

Operators working on lattices of interpretations are the main notion of fix-point semantics of logic programs. Operators defined by programs are called algorithmic. Some properties of algorithmic operators, such as monotonicity and compactness, are used to define this semantics. But what are necessary and sufficient conditions for such operators? The main result of this paper may be formulated, roughly speaking, as follows: an operator is algorithmic, iff its result on a given interpretation is the sum of its results on finite and restricted parts of this interpretation, i.e., is in essence defined by a certain algebra.

This work has a long history. At first this result was

established for the so called neighbourhood version of logic programming [see Borchchev and Khomyakov 1972]. But in this very simple version we did not use compound terms. For other versions, the corresponding results were published in [Borchchev and Khomyakov 1976, Borchchev 1986], but there were some inaccuracies in definitions and proofs.

Here this result is formulated for the "homogeneous" version that does not distinguish between predicate and functional symbols, hence identifying terms and atoms, as in PROLOG-II [Colmerauer et al 1983].

1. Homogeneous version of logic programming

1.1. Notation and terminology. Let Var be a set of variables and Σ be a functional signature, i.e. a set of functional symbols (constants are defined as functional symbols of arity zero). As usual, constants and variables are terms and if f is a n-ary functional symbol and $t_1,\dots,t_n$ are terms, then $f(t_1,\dots,t_n)$ is a term. A term is ground if no variables occur in it. Denote by $F_\Sigma(Var)$ the set of all terms and by I_Σ the set of all ground terms. It is easy to see that $F_\Sigma(Var)$ is a free Σ-algebra and $I_\Sigma \subset F_\Sigma(Var)$ is an initial Σ-algebra. As usual, a substitution is a mapping $\theta\colon Var \dashrightarrow F_\Sigma(Var)$, which is naturally extended to the homomorphism $\theta\colon F_\Sigma(Var) \dashrightarrow F_\Sigma(Var)$.

On the set $F_\Sigma(Var)$ of terms the relation $<$ of strict partial order ("to be a proper subterm)" is defined in the usual way:

1) if $t = f(s_1,\dots,s_l)$, then $s_i < t$ for $i = 1,\dots,l$;
2) if $r < s$ and $s < t$, then $r < t$ for every r, s and t;

We write $s \leqslant t$, if $s < t$ or $s = t$.

Definitions. A *homogeneous system (h-system*, for brevity) is an arbitrary set of terms from $F_\Sigma(Var)$.

For a term t put $Term(t) = \{s \mid s \leqslant t\}$ and for an h-system H put $Term(H) = \bigcup Term(t),\ t \in H$.

Let G and H be h-systems and θ be a substitution such that $\theta G \subseteq H$. Then it is natural to say that θ is a homomorphism from G to H and to use the standard notation $\theta\colon G \dashrightarrow H$.

As in the "classical" version (see, for example, [Lloyd 1984]), a *program* is a finite set of *rules*. A rule is a pair $\langle t_0, B\rangle$, where $t_0 = f(t_1,\dots,t_n)$ is a term from $F_\Sigma(Var)$ and $B = \{s_1,\dots,s_m\}$ is a finite h-system, possibly empty. Terms $t_0,s_1,\dots,s_m$ in the rule cannot be variables. t_0 is called the head and B the body of the rule and the standard notation t_0 <-- $s_1,\ldots,s_m$ is used. We say that this rule defines the symbol f, the outermost functional symbol of the term t_0.

So, homogeneous programs differ from classical ones by the fact that they use symbols of the same signature for the construction of objects, as well as for descriptions of connections between the objects. Technically, only terms are used in the rules (there are no atoms) and the outermost symbols of these terms may occur inside of these and other terms, as, for example, in the rule $f(g(X,Y))$ <-- $g(Y,f(X))$.

An *interpretation* of a program is an arbitrary h-system $Int \subseteq I_\Sigma$. An interpretation Int is called a *model* of a program P, if for any ground term t the following condition holds:

$t \in Int$ <=> there exist a rule $\langle t_0,B\rangle \in P$ and a homomorphism θ: B --> Int, such that $\theta t_0 = t$.

1.2. ***P*-algebras, operators and their fixed points.** Every rule $r = \langle t_0,B\rangle$, where $B = \{s_1,\ldots,s_m\}$, defines on $F_\Sigma(Var)$, and also on I_Σ, a partial m-ary operation, which we also denote by r:

r is defined on terms $t_1,\ldots,t_m$, if there exists a homomorphism θ: B --> $F_\Sigma(Var)$ such that $\theta s_1 = t_1,\ldots,\theta s_m = t_m$; in this case the term θt_0 is the result of the operation r (this operation, generally speaking, is not single-valued because t_0 may contain variables not occuring in terms of B).

In this way the program P defines, on the free Σ-algebra $F_\Sigma(Var)$, new operations marked by symbols of rules from P, so we have an algebra in the signature P or a P-algebra with the same carrier $F_\Sigma(Var)$.

The class of h-systems, the subsets of $F_\Sigma(Var)$, and the class of interpretations, the subsets of I_Σ, are complete lattices with the operations $\cup$, $\cap$ and the order $\subseteq$. We will consider operators on these lattices. An *operator* is a mapping from subsets of $F_\Sigma(Var)$ to subsets of $F_\Sigma(Var)$ (or from subsets of I_Σ to subsets of I_Σ). An operator is *monotone* if $A \subseteq B$ ==> $J(A) \subseteq J(B)$. An operator is *compact* if $t \in J(H)$ ==> $t \in J(F)$ for some finite $F \subseteq H$.

The P-algebra defines the following operator J_P on these lattices. Let $H \subseteq F_\Sigma(Var)$, then

$J_P(H) = \{t \mid$ there exist a rule $r = \langle t_0, B\rangle$, $r \in P$ and a homomorphism $\theta: B \to H$ such that $\theta t_0 = t$, i.e. the operation r is defined on $\theta B \subseteq H$ and $r(\theta B) = t\}$.

On the interpretations, the subsets of I_Σ, the operator J_P is defined in the same way. But in this case we must only consider substitutions of the type $\theta: Var \to I_\Sigma$.

As for ordinary logic programs, the operator J_P is monotone and compact. Its fixed points on the lattice of interpretations are the models of the program P. Note that fixed points of J_P are the open subalgebras of our P-algebra.

2. Necessary and sufficient conditions for algorithmic operators

An operator J is called *algorithmic*, if there exists a program P such that $J = J_P$. We are interested in necessary and sufficient conditions for an operator on the lattices of h-systems to be algorithmic. A logic program is restricted in size: its number of rules is by hypothesis finite, the number of terms in the body of a rule is finite, and the maximum "heights" of the terms occuring in a rule is bounded. For these reasons, as we show below, the result of an operator defined by some program for any h-system is the sum of the results of this operator for restricted subsystems of this system. On the other hand, such local definiteness is sufficient for an operator to be algorithmic. To explicate local properties of operators we will need some definitions.

First of all let us define the notion of *height* $\|t\|$ of a term t:

I) if t is a constant or a variable, then $\|t\| = 1$;

2) if $t = f(s_1, \ldots, s_n)$, $\|s_i\| = k$ for some i, $1 \leqslant i \leqslant n$, and for every j, $1 \leqslant j \leqslant n$, we have $\|s_j\| \leqslant k$, then $\|t\| = k + 1$.

We say that an h-system H has height k ($\|H\| = k$ in writing), if there exists a term $t \in H$ such that $\|t\| = k$, and for every term $t' \in H$ we have $\|t'\| \leqslant k$.

Denote by $Var(t)$ the set of variables occuring in a term t and given a h-system H put $Var(H) = \cup Var(t)$, $t \in H$.

Let us fix the natural numbers n and k. We say that the h-system A is an *n,k-system*, if $|A| \leqslant n$ and $\|A\| \leqslant k$. We say also that an operator J is *n,k-truncated (n,k-resticted)*, if for every h-system H and every term $t \in F_\Sigma(Var)$ we have

$t \in J(H)$ <==> there exist a n,k-system A, a term $s \in J(A)$ and a homomorphism $\theta: A \dashrightarrow H$, such that $\theta s = t$.

We have already noted that for a rule of program whose head contains variables which do not occur in the body, the corresponding operation is multi-valued. So the result of an operator defined by such a program even for a finite h-system contains a infinite set of terms. But this infinity is regular: the head of every such rule is like a matrix, forming terms contained in the set of results of the operator.

An operator J is called *regular* if for any finite h-system H there exists a finite h-system $C \subseteq J(H)$ such that

$t \in J(H)$ <==> there exist a term $s \in C$ and a substitution λ, which is identical for any variable $X \in Var(H)$, such that $\lambda s = t$.

The h-system C is said to be the matrix realizing J on H.

Theorem 1. An operator J is algorithmic iff it is regular and n,k-truncated for some n and k.

Proof. First of all we show that for every program P the operator J_P is n,k-truncated and regular.

n,k-truncateness. For every rule $r = \langle t_0, B\rangle$ from P we have $t_0 \in J_P(B)$. Put $n_r = |B|$, $k_r = \|B\|$, and also $n_P = \max n_r$, $k_P = \max k_r$, $r \in P$. It is easy to see that J_P is n_P,k_P-truncated.

Regularity. For every rule $r = \langle t_0, B\rangle$ put $Varg(r) = Var(t_0) - Var(B)$, $m_r = |Varg(r)|$ and $m_P = \max m_r$, $r \in P$. Let H be a finite h-system. Fix a set of variables $Varg(P,H) \subseteq Var$ such that $|Varg(P,H)| = m_P$ and $Varg(P,H) \cap Var(H) = \emptyset$. Consider the h-system

$C = \{t \mid$ there exist a rule $r = \langle t_0, B\rangle \in P$ and a homomorphism $\theta: B \dashrightarrow H$, such that $t = \theta t_0$ and $\theta X \in Varg(P,H)$ for every variable $X \in Varg(r)$, and if $X,Y \in Varg(r)$ and $X \neq Y$, then $\theta X \neq \theta Y\}$.

It is easy to see that C is a finite matrix, realizing J on H.

Conversely, let J be an n,k-truncated regular operator. We construct the program P in the following way. Consider the set $T(n,k)$, containing a representative from every class of isomorphic n,k-systems. Obviously, this set is finite. Let $B \in T(n,k)$ and C be the matrix realizing J on B. For every $t \in C$ consider the rule $\langle t,B\rangle$. Denote by P the set of all such rules.

Let us show that $J_P = J$. Indeed, suppose that for an h-system H we have $t \in J_P(H)$. Then there exist a rule $<t_0, B> \in P$ and a substitution θ, such that $\theta B \subseteq H$ and $\theta t_0 = t$. But $t_0 \in J(B)$, therefore by definition of n,k-truncateness of operator, we have $\theta t_0 \in J(H)$. Thus $J_P(H) \subseteq J(H)$.

Now suppose that $t \in J(H)$. From the n,k-truncateness of J it follows that there exist n,k-system B, a term $t_0 \in J(B)$ and a substitution θ such that $\theta B \subseteq H$ and $\theta t_0 = t$. But then for some h-system $B' \in T(n,k)$ and some term $t_0' \in J(B)$ there exists an isomorphism δ such that $\delta B' = B$ and $\delta t_0' = t_0$. Let C be the matrix realizing J on B'. Then for some term $t_0'' \in C$ we have $\lambda t_0'' = t_0'$ and $\theta\delta\lambda t_0'' = t$, where λ is a substitution identical on $Var(B')$. We have $\theta\delta\lambda B' = \theta\delta B' \subseteq H$ and therefore $t \in J_P(H)$. Thus, $J(H) \subseteq J_P(H)$.▌

The property of n,k-truncateness is a strengthening of the property of compactness. In essence, the Theorem 1 claims that an operator is algorithmic when its values are defined by some algebra and the operations of this algebra are identical for terms which have identical "tops" up to a given height. The condition of regularity says that if some operation of this algebra is multy-valued, its results are "reprinted" by means of some matrix.

The condition of n,k-truncateness of an operator may be divided into several simpler conditions dealing with the "breadth" (n) and the height (k) of subsystems which form the result of the operator and with the generalization of the notion of monotonicity.

We say that an operator J is *n-truncated in breadth* if there exists a natural number n such that for every h-system H we have

$$J(H) = \cup J(A),\ A \subseteq H \text{ and } |A| \leqslant n.$$

An operator J is said to be *uniform* if for all h-systems A and H for which there exists a homomorphism θ: $A \dashrightarrow H$ we have $\theta J(A) \subseteq J(H)$.

It is easy to see that the uniformity of an operator implies its monotonicity. It is clear also that for isomorphic systems the results of such operator are isomorphic.

Proposition 2. Every n,k-truncated operator is n-truncated in breadth and uniform.

The proof follows immediately from the corresponding definitions.▌

In order to speak of the truncateness in height, we will need some standard notions concerning h-systems (such as congruence and factor system), as well as special ones (k-top).

For an arbitrary h-system H we will consider equivalence relations on the set $Term(H)$ and will denote these relations by the symbol $\approx$, sometimes indexed. We will be interested in the relations for which the following condition holds:

if $t = f(s_1,\dots,s_m)$ and $t' = g(s'_1,\dots,s'_n)$, $m,n \geqslant 0$, then
$t \approx t' \iff f = g,\ m = n$ и $s_1 \approx s'_1,\dots,\ s_n \approx s'_n$ (1)

Denote by $Term(H,\approx)$ the set of equivalence classes for the relation $\approx$. The relation $<$ on the set $Term(H)$ induces the relation $<$ on the set $Term(H,\approx)$:

let $T,S \in Term(H,\approx)$, then
$T < S \iff$ there exist terms $t,s \in Term(H)$, such that $t \in T$, $s \in S$ and $t < s$.

It is easy to see that the relation $<$ on $Term(H,\approx)$ is not necessarily a partial order. For example, if for the term $f(f(f(X)))$ we put $f(f(f(X))) \approx f(X)$ and $f(f(X)) \approx X$ and denote the corresponding equivalence classes by T and S, we will have a "cycle": $T < S$ and $S < T$.

An equivalence relation $\approx$ on the set $Term(H)$ of the h-system H is called a *congruence* on H, if the condition *(1)* holds and the relation $<$ on $Term(H,\approx)$ is a strict partial order.

Let $\approx$ be a congruence on the h-system H. We will construct the h-system $H/\approx$ and the homomorphism $[\approx]$: $H \to H/\approx$. First of all we construct the set $Term(H/\approx)$, assigning to every equivalence class T of terms from $Term(H)$ the term $\mathbf{t}$ from $Term(H/\approx)$ and putting $[\approx](t_i) = \mathbf{t}$ for every $t_i \in T$. We will use induction over the height of the terms:

1) If all terms from T are variables, then $\mathbf{t}$ is a variable that differs from all variables corresponding to the other equivalence classes of terms from $Term(H)$.

2) If T contains a constant a, then $\mathbf{t} = a$.

3) If T contains a term $f(s_1,\dots,s_m)$, $m \geqslant 1$, we assign to this class a term $\mathbf{t} = f(\mathbf{s}_1,\dots,\mathbf{s}_m)$, where $\mathbf{s}_1,\dots,\mathbf{s}_m$ are terms, assined to the classes, containing terms $s_1,\dots,s_m$ respectively (i.e.. $[\approx](s_i) = \mathbf{s}_i$). It follows from condition *(1)* that our construction is well-defined.

We define $H/\approx$ in the following way:.
$H/\approx = \{\mathbf{t} \in Term(H/\approx) \mid$ there exists a term $t \in H$ such that $[\approx](t) = \mathbf{t}\}$.

The h-system $H/\approx$ is said to be the *factor system* of the h-system H by the congruence $\approx$. It is easy to see that our construction defines $H/\approx$ up to isomorphism.

As usual, every homomorphism θ: A --> H defines a congruence $\underset{\theta}{\approx}$ on A: for $s,t \in Term(A)$ we have

$$s \underset{\theta}{\approx} t \iff \theta s = \theta t.$$

It is easy to see that the congruence defined by the homomorphism $[\approx]$: H --> $H/\approx$ coincides with the congruence $\approx$.

The monomorphism $\theta/\approx$: $A/\underset{\theta}{\approx}$ --> H is defined in a natural way: if a term $\mathbf{t} \in A/\underset{\theta}{\approx}$ is assigned to an equivalence class T of terms from $Term(A)$, then $\theta/\approx(\mathbf{t}) = \theta t$, where t is some term from T.

Thus the following diagram is commutative:

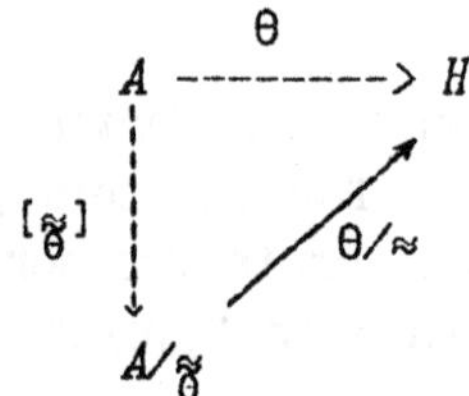

For h-systems which are singletons the notation given above can be simplified. If $A = \{s\}$ and $H = \{t\}$, then for the homomorphism θ: A --> H we will use the notation θ: s --> t. This homomorphism is called a *top* of the term t. The only term of the factor system $A/\underset{\theta}{\approx}$ will be denoted by $s/\underset{\theta}{\approx}$.

We will need the notion of a *free term*:

1) every variable $X \in Var$ is a free term;

2) every ground term is a free term;

3) if $s_1,\dots,s_m$ are free terms, $Var(s_i) \cap Var(s_j) = \emptyset$ for $i \neq j$ and f is a m-ary functional symbol, then $f(s_1,\dots,s_m)$ is a free term.

A top θ: s --> t is called a *free top* of the term t, if s is a free term. A free top θ: s --> t is called a *free k-top* of the term t, if:

I) $\|s\| = k$ when $\|t\| \geq k$ and $\|s\| = \|t\|$ when $\|t\| \leq k$;

2) for every free top λ: s' --> t such that $\|s'\| \leq k$, there exists a free top δ: s' --> s such that $\theta\delta = \lambda$.

If θ: s --> t is a free k-top of the term t, then the homomorphism $\theta/\approx$: $s/\approx_\theta$ --> t is said to be a *k-top* of the term t.

Note that the height $\|s/\approx_\theta\|$ of the term $s/\approx_\theta$ in the k-top $\theta/\approx$: $s/\approx_\theta$ --> t can exceed k. For example, the free 3-top and 3-top for the term $t = h(f(f(f(a))),f(f(a)))$ have the following form

θ: $h(f(X),f(Y))$ --> t, where $\theta X = f(f(a))$ and $\theta Y = f(a)$;

$\theta/\approx$: $h(f(f(Z)),f(Z))$ --> t, where $\theta Z = f(a)$.

Let H be an arbitrary h-system and F be an h-system such that all its terms are free. A homomorphism δ: F --> H is said to be a free k-top of h-system H, if

I) $\delta F = H$;

2) for every terms $s \in F$ and $t \in H$ such that $\delta s = t$, the top δ: s --> t is a free k-top.

The monomorphism $\delta/\approx$: $F/\approx_\delta$ --> H defined by the homomorphism δ, is said to be a k-top of the h-system H.

And as for the height of k-top of term, the height $\|F/\approx_\delta\|$ of the h-system $F/\approx_\delta$ in the k-top of h-system H can exceed k.

Proposition 3. Let H be a h-system and δ: A --> H be its k-top. Then

1) For every h-system B such that

a) $\|B\| \leqslant k$ and

b) there exists a homomorphism θ: B --> H,

we can define a homomorphism λ: B --> A such that $\delta\lambda t = \theta t$ for every $t \in B$.

2) Let C be an h-system, β: C --> H be a homomorphism and for every h-system B, such that the conditions a) and b) holds, there exists a homomorphism ε: B --> C such that $\beta\varepsilon t = \theta t$ for every $t \in B$. Then there exist a homomorphism μ: A --> C such that $\beta\mu s = \delta s$ for every term $s \in A$ (see the diagram below).

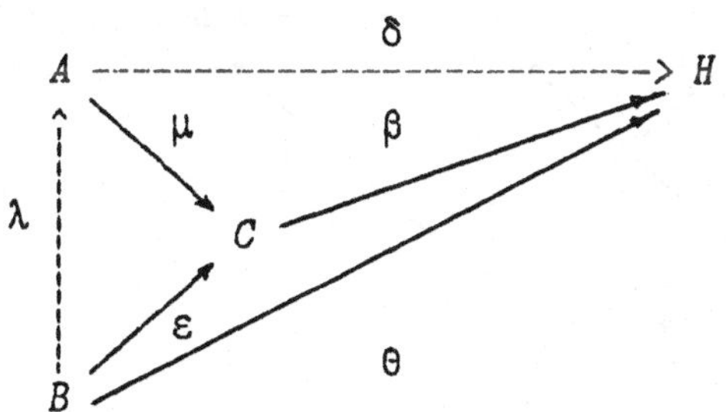

Proof.

1) Let α: F --> B be a free k-top of h-system B and η: G --> A be a free k-top of h-system A. Therefore $\delta\eta$: G --> H is a free k-top of h-system H (see the diagram below). Then for all terms $r \in F$, $s \in B$ and $t \in H$ such that $\alpha r = s$ and $\theta s = t$, the homomorphism $\theta\alpha$: r --> t is a free top of the term t. By definition of free k-top, every free k-top of the term t whose height does not exceed k, is a free top of a free k-top of the term t. Thus there exists a homomorphism γ: F --> G, such that for every term $r \in F$ we have

$$\delta\eta\gamma r \quad = \quad \theta\alpha r \qquad (1)$$

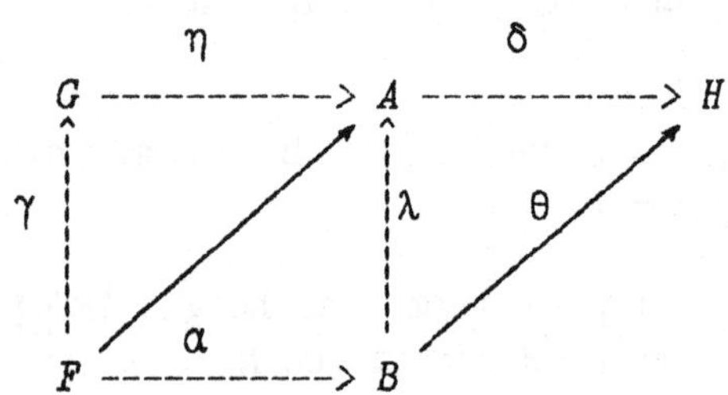

Now we define a homomorphism λ: B --> A, so that $\lambda\alpha r = \eta\gamma r$ for every $r \in F$. For every variable $X \in Var(B)$ there exists a variable $Y \in Var(F)$ such that $\alpha Y = X$, so put $\lambda X = \eta\gamma Y$. It is easy to see since δ is a monomorphism and condition *(1)* holds that λ is well defined and λ is the required homomorphism.

Let C be h-system and β: C --> H be a homomorphism such that the conditions of the Proposition 3 hold. Let η: G --> A be as above: a free k-top of the h-system A. We have $\|r\| \leqslant k$ for every $r \in G$, so, by the conditions of Proposition 3, there exists a homomorphism ε:G --> C such that $\beta\varepsilon r = \delta\eta r$ for every $r \in G$ (see the diagram below).

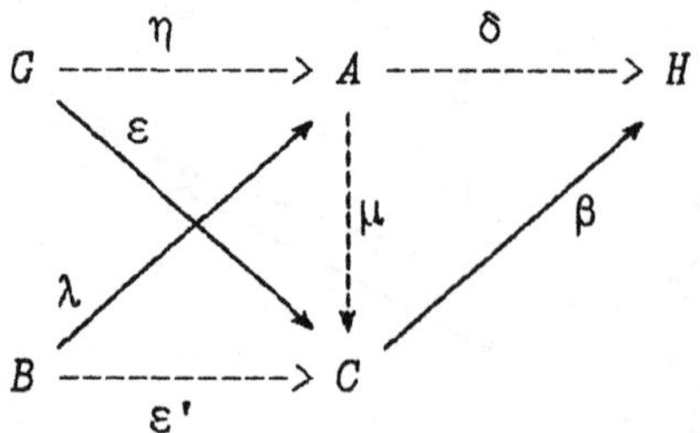

It is easy to see that if we have $s_1 \underset{\varepsilon}{\approx} s_2$, then $s_1 \underset{\eta}{\approx} s_2$ for some terms $s_1,s_2 \in Term(G)$, since δ is a monomorphism. Below we show, that the converse is also true, so the equvalences $\underset{\varepsilon}{\approx}$ and $\underset{\eta}{\approx}$ coincide. Let us define the homomorphism μ: A --> C: put $\mu X = t$ for every variable $X \in Var(A)$, if $\varepsilon Y = t$ for some variable $Y \in Var(G)$ such that $\eta Y = X$. It is easy to see that this definition of homomorphism μ is correct and the necessary conditions hold.

Now let us show that the equivalences $\underset{\varepsilon}{\approx}$ and $\underset{\eta}{\approx}$ on G always coincide. Suppose the converse, i.e. that for some terms $s_1,s_2 \in Term(G)$ which are subterms of terms $r_1,r_2 \in G$ correspondingly, we have $s_1 \underset{\eta}{\approx} s_2$ and at the same time, $\varepsilon s_1 \neq \varepsilon s_2$. Note that s_1 and s_2 can be are subterms of the same term, i.e. r_1 and r_2 are not necessarily distinct. Without loss of generality we can assume that the terms s_1 and s_2 are maximal, since for all terms s_1' and s_2' such that $r_1 \geqslant s_1' \geqslant s_1$ and $r_2 \geqslant s_2' \geqslant s_2$, we have $\eta s_1' \neq \eta s_2'$.

Take the h-system $B = \{r_1',r_2'\}$, where terms r_1' and r_2' are made from the terms r_1 and r_2 by the replacement of their subterms s_1 and s_2 by the variable X. Obviously, $\|B\| \leqslant k$. Define a homomorphism λ: B --> A as follows: $\lambda X = \eta s_1 = \eta s_2$ и $\lambda Y = \eta Y$ for every variable $Y \neq$ $\neq X$. Thus the homomorphism $\delta\lambda$: B --> H is defined. But it is easy to see that there does not exists a homomorphism ε': B --> C such that $\beta\varepsilon' r = \delta\lambda r$ for every term $r \in B$. So equivalences $\underset{\varepsilon}{\approx}$ and $\underset{\eta}{\approx}$ have to coincide.∎

Proposition 3 states that if δ: A --> H is a k-top, then any h-system B such that $\|B\| \leqslant k$ and can be embedded in H, can be embedded in A.Moreover, A is a minimal system such that this condition holds.

An operator J is said to be *k-truncated in height*, if for any h-system H and any term $t \in J(H)$ there exists k-top δ: A --> H and a term $s \in J(A)$ such that $\delta s = t$.

Proposition 4. Let J be n,k-truncated operator. Then J is k-truncated in height.

Proof. Let J be n,k-truncated operator, h be an arbitrary h-system, $t \in J(H)$ and δ:A --> H be a k-top. It follows from the n,k-truncateness of J that there exist an n,k-system B, term $r \in J(B)$ and the homomorphism θ: B --> H such that $\theta r = t$. On the other hand statement 1) of Proposition 3 implies that there exists a homomorphism λ: B --> A such that $\delta\lambda X = \theta X$ for every variable $X \in Var(B)$. But for the variables from $Varg(r) = Var(r) - Var(B)$ the substitutions $\delta\lambda$ и θ not compatible, i.e. for a variable $Y \in Varg(r)$ the equality $\delta\lambda Y = \theta Y$ does not always hold. For this reason we define the substitutions λ' and δ', for which this equality will always hold, with the help of substitutions λ and δ:

1) for every variables $X \in Var(B)$ and $Z \in Var(A)$ put $\lambda'X = \lambda X$ and $\delta'Z = \delta Z$;

2) if for $Y \in Varg(r)$ there exists the term $\delta^{-1}\theta Y$, then put $\lambda'Y = \delta^{-1}\theta Y$ (and, so $\delta\lambda Y = \theta Y$);

3) if for $Y \in Varg(r)$ the term $\delta^{-1}\theta Y$ does not exists , then choose a variable $U \in Var$ such that $U \notin Var(A)$ and $U \notin Var(\lambda'W)$ for every variable $W \in Varg(r)$, $W \neq Y$ and put $\lambda'Y = U$ and $\delta'U = \theta Y$.

It is easy to see that the monomorphism $\delta': A \dashrightarrow H$ defined in this way is k-top and for the term $s = \lambda'r$ we have $\delta's = t$. ▌

The meaning of Proposition 4 is that the result of n,k-truncated operator for h-system h is defined by its k-top.

Proposition 5. Let J be an operator, uniform, n-truncated in breadth and k-truncated in height for some n and k. Then J is $n.k'$-truncated operator for some $k' \geqslant k$.

Proof. Let $\delta: A \dashrightarrow H_n$ be k-top of the h-system H_n, $|H_n| \leqslant n$ and $k_A = \|A\|$. It is easy to see that for a fixed signature Σ there exists a natural number k' such that $k_A \leqslant k'$ for all such k-tops. Now let us show that the operator J is n,k'-truncated.

Really, let $t \in J(H)$ for some h-system H. Since J is n-resticted in breadth, then $t \in J(H_n)$ for some h-system $H_n \subseteq H$, $|H_n| \leqslant n$. And since J is k-truncated by height, there exist k-top $\delta: A \dashrightarrow H_n$ and a term $s \in J(A)$ such that $\delta s = t$. Since $\|A\| \leqslant k'$, then A is n,k'-system. On the other hand, the uniformity of J implies that if $r \in J(B)$, where B is n,k'-system, and there exists a homomorphism $\theta: B \dashrightarrow H$, then $\theta r \in J(H)$.▌

Thus Propsitions 2 - 5 enable us reformulate the iff-conditions for algorithmic operators:

Theorem 6. An operator J is algorithmic iff it is regular, uniform, n-truncated in breadth and k-truncated in height for some n and k.▌

3. Conclusion

The real point of this paper is a *restricted diversity* of

algorithmic systems and corresponding operators. First of all, model-theoretic semantics demonstrates the "local well-formedness" principle: models of logic programs can be defined by means of a finite set of patterns. Really, in such a model every term (or every atom in the classical case) has a proper neighbourhood which is an image of a body of some rule modified by some substitution.

This local well-formedness principle corresponds to local properties of algorithms discussed by Turing [1936] and Kolmogorov-Uspenski [1957]. Every algorithm at every step of its work makes a local transformation of its data. The scope of these transformations is limited, so there exists only a limited number of such transformations. Thus logic programming reformulates this dynamic locality principle of Turing and Kolmogorov-Uspenski in static form.

Theorem 1 is another form of the local well-formedness principle. Theorem 6 divides this property into several simpler conditions.

References

Borshchev, V.B. and Khomyakov, M.V. (1972) *Schemes for Functions and Relations*. Workshop on Automatic Processing of Texts of Natural Languages. Yerevan, (The same text in Investigations in formalized languages and nonclassical logics, Moscow, Nauka 1974, 23-49).(in Russian).

Borshchev, V.B. and Khomyakov, M.V. (1976) *Club systems*. Nauchno-Tekhnicheskaya Informatsia, Seriya 2, 8, 3-6 (in Russian).

Borshchev, V.B. (1986) *Logic programming*. Tekhnicheskaya kibernetika, 2, 89-109, (in Russian).

Colmerauer A., Kanoui H., van Caneghem M. (1983) *Prolog, bases theoriques et developpements actuels*. Technique et Science Informatiques, 4. 277-311.

Kolmogorov, A.N. and Uspenski, V.A. (1957) *To the Definition of Algorithm*. Uspekhi Matematicheskikh Nauk., 13, 4 3-28, (in Russian).

Lloyd J.W. (1984) *Foundations of logic programming*. Springer Verlag, 124p.

Turing, A.M. (1936) *On Computable Numbers, with an Application to the Entsheidungproblem*. Proc. London Math. Soc. Ser. 2, 42, 3-4 230-265.

DEEP LOGIC PROGRAM TRANSFORMATION USING ABSTRACT INTERPRETATION

Dmitri Yu. Boulanger

Institute for Informatics Problems
USSR Academy of Science
30/6 Vavilov str., Moscow 117900, USSR
blng@sms.ccas.msk.su

Abstract This paper presents a procedure for deep transformation of logic programs, which is based upon partial evaluation of source logic programs and includes unfolding, term rewriting and deriving new predicates. Controlling of these operations is based upon abstract OLDT interpretation, which is used to produce transformation guide for a corresponding residual code generator. The abstract interpretation schema includes the rules for abstract domains construction, which can be tuned for a particular application. The algorithms are applicable for deep transformations, which occur when compiling meta-interpreters.

1. INTRODUCTION

Transformation of logic programs by partial evaluation has attracted great interest in recent years. Partial evaluation was introduced as an optimization technique in [FUT77]. Partial evaluation consists of deriving a specialized version of a program (residual program) wrt some known input data. In the context of logic programming partial evaluation has been practically fruitful when applied to meta-programs. In order to obtain efficient residual code deep transformations of meta-programs are required. There are two important issues in this respect.

Transformation of the source program can be done by unfolding. During this process the current atom is either unfolded or not unfolded. Controlling the unfolding process is an important task. In our approach this problem is solved using an abstract interpretation technique [ABH87]. The abstract interpretation is applied to the abstract model of the source program, where only properties of interest from the source program are included. Using different models we can generate different variants of the residual program. Abstract interpretation mechanism uses special variant of abstract OLDT resolution.

The second problem is code size of residual program. Our algorithms generate special data structures, which can be used to keep residual code to a reasonable size.

The abstract interpretation algorithm derives two kinds of elementary transformations, which have to be executed by code generator. The first one is a sequence of atoms of source program, which have to be unfolded. The second one is a special data structure, which is used to derive new predicates by renaming and rewriting. These two are mutually dependent and can be controlled by parameters of the abstraction schema used. The parameters of the abstraction schema are used to tune the transformation process for a particular application.

The outline of the paper is as follows. In the next section the basic concepts of logic programming and OLDT resolution are given. The third section introduces our abstract interpretation schema. In the fourth section the algorithm to derive residual code structure from abstract OLDT tree is given, while the fifth section contains a description of the code generator. The last part of the paper is a brief consideration of source program and residual program equivalence and some practical experience.

2. PRELIMINARIES

In this section we recall necessary notions of foundations of logic programming and give a description of OLDT tree structure, which will be used as a basis for deriving logic program transformation algorithms. We also define special partial orderings to be used for abstract interpretation schema construction.

2.1. Terms, Expressions and Logic Programs

Let Σ be *an alphabet*, in which all programs and goals are written. We will assume that

$$\Sigma = \text{FUNC} \cup \text{VAR}, \quad \text{FUNC} \cap \text{VAR} = \emptyset$$

where FUNC is a finite set of functors and VAR is a infinite set of variables. For any functor $f \in$ FUNC its arity is given. Constants are functors with arity 0. Then

$$\text{TERM} = \text{TERM}(\Sigma)$$

will denote the set of all finite structures composed of functors and variables from the alphabet Σ. [LLO87] is the default reference for any concepts not explained herein. The alphabet Σ will be considered a parameter for the terms construction.

Idempotent substitutions is a set of almost everywhere identity functions [EDE85]:

$$\text{SUBST} = \{\ \theta \in \text{VAR} \to \text{TERM}\ \mid\ \text{dom}(\theta) = \{x \in \text{VAR} \mid x \neq \theta(x)\} \text{ is finite} \wedge \theta = \theta\theta\} \cup \{\text{fail}\}.$$

If a substitution has the form $\{X_1/Y_1, X_2/Y_2, \ldots, X_n/Y_n\}$, where all the Y_i are new and distinct variables, it is called a *renaming substitution*. We will use upper case characters for logic variables.

Let EXPR denote *set of expressions*, which can be constructed from terms and special function symbols not occurring in Σ. These symbols will be defined later and will be used to compose clauses, goals and other necessary expressions. The set of all terms TERM can also be considered a parameter for the expressions construction, i.e.

$$\text{EXPR} = \text{EXPR(TERM)}, \quad \text{TERM} \subseteq \text{EXPR}.$$

For any substitution $\theta \in$ SUBST and an expression $E \in$ EXPR, $E\theta \in$ EXPR denotes an expression, which is obtained from E by simultaneously applying substitution θ to all variables occurring in E. The variables occurring in $E \in$ EXPR we will denote by var(E). Note that no substitution can replace a variable by an expression, which is not a term. We will assume that $E\theta$ = fail if θ = fail for any expression $E \in$ EXPR, where fail $\notin \Sigma$, fail $\in$ EXPR.

It can be shown that (EXPR, «) is *a set equipped with preoder*, where « is a relation on EXPR such that $E_1 \ll E_2$, $E_1,E_2 \in$ EXPR if there is a substitution θ for which $E_1 = E_2\theta$. From this preoder *an equivalence relation* $\simeq$ on EXPR can be defined as follows:

$$E_1 \simeq E_2 \text{ iff } E_1 \ll E_2 \wedge E_2 \ll E_1, \quad E_1,E_2 \in \text{EXPR}.$$

This equivalence relation embodies the notion of *variance*. We will denote by $\text{EXPR}_{\simeq}$ the quotient set with respect to variance of EXPR over variables from Σ. Then $(\text{EXPR}_{\simeq},\leq)$ is *a set equipped with partial ordering*, which is induced by preordering relation «. Elements of a set $\text{EXPR}_{\simeq}$ represent equivalence classes of expressions. If $\mathbb{E} \subseteq$ EXPR, then $\mathbb{E}_{\simeq}$ denotes the set of corresponding equivalence classes and if $E \in$ EXPR, then $E_{\simeq}$ denotes equivalence class for E. The same technique to factor out variance and usual renaming of logic variables is used in [KER90].

Equivalence class for any expression $E \in$ EXPR can be represented by *the canonical form of the expression* E, which is obtained by replacing all variables var(E) by canonical variables from a set

$$\mathbb{Z} = \{Z_1,Z_2,\ldots Z_n,\ldots\}, \quad \mathbb{Z} \cap \text{VAR} = \emptyset.$$

The canonical form of an expression E is obtained by replacing the first variable, occurring in E by Z_1, the second by Z_2 and so on [GCS88].

Example 2.1. If f,g $\in$ FUNC and X,Y $\in$ VAR, then
$\{f(X,Y,X),\ f(X,X,Y)\}_{\simeq} = \{f(Z_1,Z_2,Z_1),\ f(Z_1,Z_1,Z_2)\}$
$\{g(X,Y),\ g(Y,X)\}_{\simeq} = \{g(Z_1,Z_2)\}$
$X_{\simeq} = Y_{\simeq} = Z_1$
$\text{VAR}_{\simeq} = \{Z_1\}$ ■

A set of atoms ATOM $\subseteq$ EXPR is a set of terms, i.e. ATOM = TERM. Note that we accept atoms, which can be variables. We will not distinguish predicates and functional symbols. Other syntactic elements of logic programs can be defined using atoms as follows:

$$\text{BODY} = \{B \subseteq \text{EXPR} \mid B \equiv A_1,A_2,\ldots,A_n,\ \ n\geq 1,\quad A_i \in \text{ATOM}\},$$
$$\text{CLAUSE} = \{C \subseteq \text{EXPR} \mid C \equiv H\text{:-}B,\ H \in \text{ATOM},\ B \in \text{BODY} \vee B \equiv \text{true}\}$$

where symbols ","/2, ":-"/2 and "true"/0 do not belong to the alphabet Σ. A clause of the form H :- true is called *a unit clause*.

In what follows we will consider only queried logic programs. *A queried logic program* is a set of clauses and a unique *initial goal clause* of the following form:

$$\varepsilon(V_1,V_2,\ldots,V_n) \text{ :- } B, \quad B \in \text{BODY}, \ \{V_1,V_2,\ldots,V_n\} = \text{var}(B), \ n \geq 0$$

where functors $\{\varepsilon/n \mid n=0,1,\ldots\}$ do not belong to the alphabet Σ. A goal clause and its head are considered to be expressions. *A set of all queried logic programs* we will denote by PROG.

A computed answer substitution $\theta \in$ SUBST for a goal of queried logic program P is obtained by restricting the composition of $\theta_1 \circ \theta_2 \circ \ldots \circ \theta_n$ to the variables of the head of a goal clause, where $\theta_1,\theta_2,\ldots,\theta_n$ is the sequence of mgu's used in SLD-refutation of program P for its initial goal. Every instantiation of an answer substitution is a *correct answer substitution*. CAS(P) denotes the set of all correct answer substitutions of queried logic program P.

Definition 2.2. (*CAS-equivalence*)
Let $P_1,P_2 \in$ PROG. P_1 and P_2 are said to be CAS-equivalent iff CAS(P_1) is identical to CAS(P_2). ∎

Any program P ∈ PROG is a set of expressions. Thus we can define a *canonical form of the program* P as a set of corresponding canonical clauses obtained from the original clauses of P. $\text{PROG}_{\cong}$ denotes *the set of all canonical queried logic programs*. If two logic programs have identical canonical forms, then they are CAS-equivalent.

Definition 2.3. (*Partial order for the canonical logic programs*)
Let $P_1,P_2 \in \text{PROG}_{\cong}$. Then $P_1 \leq P_2$ iff for any clause $C_1 \in P_1$ there is a clause $C_2 \in P_2$ such that $C_1 \leq C_2$, where canonical clauses C_1 and C_2 are considered canonical expressions with defined above partial order. ∎

Example 2.4. Let FUNC = {a,b,1,2}, Σ = FUNC ∪ VAR. Then $P_1 \leq P_2 \leq P_3$, where

P_1 = { $\varepsilon(Z_1)$:- $a(Z_1)$, a(1) :- true, a(2) :- true, b :- $a(Z_1)$ }
P_2 = { $\varepsilon(Z_1)$:- $a(Z_1)$, $a(Z_1)$:- true, b :- $a(Z_1)$ }
P_3 = { $\varepsilon(Z_1)$:- Z_1, Z_1 :- true, Z_1 :- Z_2 } ∎

2.2. OLDT Tree Structure

OLD resolution with tabulation (or OLDT) was discovered independently by several authors [TAS86,SGG86,VIE89]. OLDT resolution will be used as a basis of operational semantics of logic programs PROG. We define *the operational semantics* I of logic programs as a mapping from canonical logic programs to computational OLDT trees as follows:

$$\text{I: } \text{PROG}_{\cong} \longrightarrow \text{OLDT\_TREE}$$

where OLDT_TREE is a *set of canonical OLDT trees*. Detailed description of the algorithms, which can be used to construct OLDT tree structure can be found in [TAS86,VIE89,PLU90]. Herein we describe only basic properties of OLDT tree structure, which will be used in the next sections.

Let NODE = {0,1,2,...} denotes a *set of the nodes of OLDT tree* T and node 0 represents root of the tree. Any node i of the tree T has a *state* denoted by st i ∈ $\text{STATE}_{\simeq}$, where STATE ⊆ EXPR is a set of all states and

$$\text{STATE} = \{\varepsilon/n \mid n=0,1,\ldots\} \cup \{\text{true}\} \cup \{\text{fail}\} \cup \text{BODY},\ \text{true},\text{fail},\varepsilon/n \notin \Sigma.$$

We have to equip the set of canonical states of OLDT tree with special partial order to be used later for the abstract interpretation schema construction.

Definition 2.5. (*Partial order for canonical states of OLDT tree*)
Partial order ≤ for canonical states is the following extension of the ordering relation for canonical expressions: fail ≤ S, S ∈ $\text{STATE}_{\simeq}$, S ≠ ε/n, n=0,1,2,... (Figure 2.1) ∎

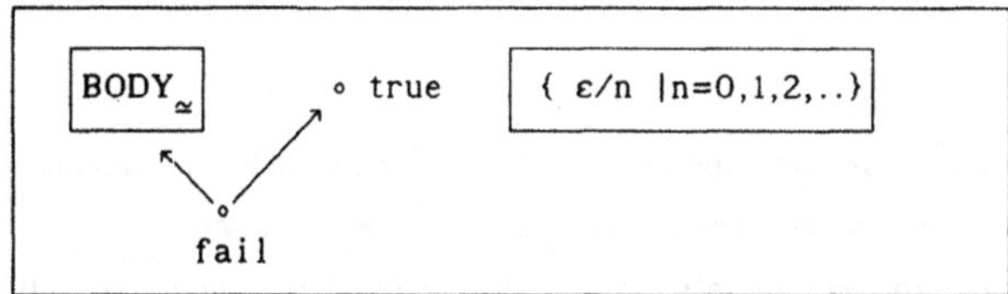

Figure 2.1. Partial order for canonical states.

The root of tree T constructed for logic program P has a state st 0 = H, where H is a head of the goal clause of P. *Arcs* of an OLDT tree are directed away from the root. Only *leaf nodes* of OLDT tree can have *true* or *fail* state. Let i be a node of an OLDT tree. Then

- *pred i* denotes a set of predecessor nodes of node i, i ∉ pred i;
- *succ i* denotes a set of successor nodes of node i, i ∉ succ i;
- *call i* denotes a canonical form of the first atom of the the state of node i or is equal to true (if st i = true) or fail (if st i = fail) and call 0 = ε.

Each OLDT tree arc has a *label*. If node i is not a root, then arc i denotes incoming arc label for this node. The arc labels can be from the set $\text{CLAUSE}_{\simeq}$ or from the set XRULE ⊆ $\text{ATOM}_{\simeq} \times \text{ATOM}_{\simeq}$. A set of x-rules XRULE is a *set of lemmas* [VIE89] generated during OLDT tree construction and can be defined as follows. Let i and j be OLDT tree nodes such that i ∈ pred j, i ≠ 0, st i ∉ {true,fail}, st j ≠ fail. Then we can define predicates *noexit i j* and function *answ i j* as follows:

- noexit i j iff $\quad$ st i = $(A_1,A_2\ldots,A_n)_{\simeq} \wedge$
  not ($\exists \tilde{j} \in$ (succ i ∩ pred j).st $\tilde{j}$ = $((A_2,\ldots,A_n)\theta)_{\simeq}$);

- answ i j = $(A_1\theta)_{\simeq}$ iff $\quad$ noexit i j ∧ $\quad$ st i = $(A_1,A_2\ldots,A_n)_{\simeq} \wedge$
  (st j = $((A_2,\ldots,A_n)\theta)$ ∨ st j = (true.θ))

where $\theta \in$ Subst, $A_1,\dots,A_n \in$ Atom, $n\geq 1$. Now we can define *the set of x-rules* as

$$XRULE = \{<A_1,A_2> \mid \exists i,j \in NODE.\ call\ i = A_1 \wedge answ\ i\ j = A_2\}$$

i.e. each x-rule has a form $<A_{\cong},(A\sigma)_{\cong}>$, $A \in$ ATOM, $\sigma \in$ SUBST and corresponds to the unit clause $A\sigma$:-true, where substitution σ is a computed answer substitution for the goal A. A path from node i to node j, which was used for x-rule generation is called *a proof* for the node i.

OLDT tree nodes can also have *labels* from a set

$$\{\pm,\ repeat(j) \mid j \in NODE\}.$$

If node i has a label, then *lab i* denotes its label. The node labels have to satisfy the following conditions:

- lab j = repeat(i) iff call i = call j and i<j, i,j $\in$ NODE;
- lab i = $\pm$ iff $\exists$j $\in$ NODE. lab j = repeat(i), i<j, i $\in$ NODE.

Now we can define *the computation rule of OLDT resolution*

$$xres:\ STATE_{\cong} \times \left(CLAUSE_{\cong} \cup XRULE\right) \rightarrow STATE_{\cong}$$

as follows:

$$xres(S,R) = \begin{cases} res(S,R), & \text{if lab } i \neq \text{repeat and } R \in CLAUSE \\ \rho(S,R), & \text{if lab } i = \text{repeat and } R \in XRULE \end{cases}$$

where st i = S, res: $STATE_{\cong} \times CLAUSE_{\cong} \rightarrow STATE_{\cong}$ is a *standard resolution rule* for resolving nodes with non repeated calls and ρ: $STATE_{\cong} \times XRULE \rightarrow STATE_{\cong}$ is a *special resolution* rule for resolving only nodes with repeated calls. Standard resolution rule is not defined for true or fail nodes and

$$res((A_1,\dots,A_n)_{\cong},(H{:-}B)_{\cong}) = \begin{cases} ((B\mu,A_2,\dots,A_n)\theta)_{\cong}, & \text{if } B\neq true,\ n\geq 1 \\ ((A_2,\dots,A_n)\theta)_{\cong}, & \text{if } B=true,\ n\geq 2 \\ (true.\theta)_{\cong}, & \text{if } B=true,\ n=1, \end{cases}$$

where θ = mgu(A_1,Hμ) (if mgu does not exist, then θ = fail) and μ is a renaming substitution. The computation rule

$$\rho((A_1,\dots,A_n)_{\cong},<A_{\cong},(A\sigma)_{\cong}>) = \begin{cases} ((A_2,\dots,A_n).\theta)_{\cong}, & \text{if } n\geq 2 \\ (true.\theta)_{\cong}, & \text{if } n=1, \end{cases}$$

where μ is a renaming substitution and θ = mgu(A_1,$A\sigma\mu$), can be applied only if $A_{\cong} = A_{1\cong}$.

The basic property of any OLDT tree structure is the following: there is an arc labeled by R from a node with state S_1 to a node with a state S_2 iff

$$xres(S_1,R) = S_2.$$

Note that if node i is a leaf of OLDT tree, then st i ∈ {true,fail} or lab i = repeat(j) for some node j. The latter occurs only for looping programs.

Figure 3.2 shows an example of OLDT tree structure.

OLDT refutation is a path from the root to a true node. Any computed answer substitution has a corresponding OLDT refutation. Note that any OLDT refutation from program P is an SLD-refutation from P1, where P1 is P plus unit clauses, which correspond to generated x-rules. Thus *soundness* of OLDT is implied by soundness of SLD. A *completeness* proof of OLDT resolution has been given by Tamaki and Sato in [TAS86]. They have also shown that OLDT always terminates and gives a finite set of solutions for programs with finite Herbrand models. Powerful sufficient conditions for *finiteness* of OLDT resolution have been suggested in [SEI88,PLU90].

3. ABSTRACT INTERPRETATION OF LOGIC PROGRAMS USING OLDT RESOLUTION

In this section we introduce an abstract interpretation technique, which correctly mimics the concrete one discussed above. Our abstract interpretation schema involves the rules for abstract domains construction and the modifications of standard OLDT resolution algorithms and can be regarded as an extension of the one suggested in [GCS88]. An excellent tutorial [BRS88] contains all the basic notions of the abstract interpretation schema, which is described below.

3.1. Abstract Domains, Concretization and Abstraction

The general rules for the abstract domains construction are the following. Let us suppose that $\mathbb{D} \subseteq \mathbb{S}$, where $(\mathbb{S},\le)$ is a set equipped with a partial order. The subset $\mathbb{D}$ is considered a set of *descriptors* of elements from $\mathbb{S}$ (*abstract domain*). Then we can define the following mappings [COU77]:

- γ: $\mathbb{D} \longrightarrow 2^{\mathbb{S}}$, *the concretization function*;
- α: $2^{\mathbb{S}} \longrightarrow \mathbb{D}$, *the abstraction function*

Following [BRS88,MAS89,MAS91] we define concretization function as

$$\gamma\ d = \{s \in \mathbb{S}\ |s \le d\},\ d \in \mathbb{D}.$$

It is clear that γ is *monotonic*, i.e. if $d_1 \le d_2$, $d_1,d_2 \in \mathbb{D}$, then $\gamma\ d_1 \subseteq \gamma\ d_2$. For any subset $S \subseteq \mathbb{S}$ we can define

$$\Gamma(S) = \{d \in \mathbb{D} | S \subseteq \gamma d\}.$$

In what follows we will assume that the abstraction function α is any function for which the following holds:

1. if $\Gamma(S) \neq \emptyset$, then $\alpha S \in \Gamma(S)$, and αS is undefined otherwise.
2. If $S_1 \subseteq S_2$ and $\Gamma(S_1) \neq \emptyset$, $\Gamma(S_2) \neq \emptyset$, then $\alpha\ S_1 \leq \alpha\ S_2$.

Note that if the set of descriptors $\mathbb{D}$ satisfies the following condition

$$\forall s \in \mathbf{S}\ \exists \tau \in \mathbb{D}.\ \tau \geq s$$

and the set $S \subseteq \mathbf{S}$ contains exactly one element, then $\Gamma(S) \neq \emptyset$ and thus the abstraction function can be defined. Above condition holds for all abstract domains, which will be introduced in this section. In what follows we use the abstraction functions only for singleton sets.

Proposition 3.1. The functions α and γ have the following properties:

1. $\alpha(\gamma d) \geq d,\ d \in \mathbb{D}$
2. $\gamma(\alpha S) \supseteq S,\ S \subseteq \mathbf{S}$ ■

We do not restrict the concretization and the abstraction relations to the "best" ones for which $\alpha(\gamma d) = d,\ d \in \mathbb{D}$ holds [BRS88,MAS89,MAS91].

3.2. Abstract Terms, Expressions and Logic Programs

The particular mechanism for the construction of abstract domains is based upon considering subsets of the concrete domain of canonical expressions rather than the more usual "abstract substitutions" [ABH87,GCS88]. Our abstract interpretation technique is based upon interpretation of abstract logic programs. *Abstract logic programs* are constructed from abstract terms. *Abstract terms* (or term descriptors) are defined as any subset $\mathbb{D} \subseteq$ TERM, which contains infinite set of variables, i.e. VAR $\subseteq \mathbb{D}$. *Abstract expressions* $EXPR_{\mathbb{D}}$ are constructed from abstract terms $\mathbb{D}$ as it was explained in Section 2. Abstract terms are considered a parameter for abstract expressions construction, i.e. $Expr_{\mathbb{D}} = Expr(\mathbb{D})$. Other syntactic elements of abstract programs can be defined as follows:

$$ATOM_{\mathbb{D}} = \mathbb{D}$$
$$CLAUSE_{\mathbb{D}} = EXPR_{\mathbb{D}} \cap CLAUSE$$
$$STATE_{\mathbb{D}} = EXPR_{\mathbb{D}} \cap STATE$$
$$PROG_{\mathbb{D}} = 2^{(Expr_{\mathbb{D}})} \cap PROG\ .$$

Note that for any set of abstract terms $\mathbb{D}$

$$EXPR_{\mathbb{D}} = VAR \cup \{\varepsilon/n \mid n=0,1,\ldots\} \cup \{true\} \cup \{fail\} \cup \ldots$$

and *abstract logic programs* $PROG_{\mathbb{D}}$ are logic programs constructed using subset of terms $\mathbb{D}$.

Abstract terms $\mathbb{D}$ can be expressed in different ways. There are two simple examples:

- define the alphabet $\Sigma^{\sim} \subseteq \Sigma$ such that $\Sigma^{\sim} = FUNC^{\sim} \cup VAR$, $FUNC^{\sim} \subseteq FUNC$. Then

$$\mathbb{D} = TERM(\Sigma^{\sim}).$$

- define subset $FUNC^{\sim} \subseteq FUNC$. Then $t \in \mathbb{D}$ iff t does not contain any subterm with main functor $f \in FUNC^{\sim}$, whose norm $|t|$ more than given limit. For a subset $T \subseteq TERM$ *a norm* is a mapping $|...|: T \rightarrow Integer$ [PLU90].

The abstract interpretation schema what follows involves three abstract domains $\mathbb{D}_0, \mathbb{D}_1, \mathbb{D}_2 \subseteq TERM$ used as descriptors of the terms. The set of abstract logic programs $PROG_{\mathbb{D}_0}$, the set of abstract states of OLDT tree $STATE_{\mathbb{D}_1}$ and the set of abstract answers $ATOM_{\mathbb{D}_2}$ of the proofs of OLDT tree are constructed as explained above using these abstract domains, respectively. For this purpose we need three pairs of the concretization and abstraction functions constructed following the general schema above:

1. $\gamma_{PROG} : (PROG_{\mathbb{D}_0})_{\simeq} \longrightarrow 2^{Prog_{\simeq}}$, $\quad \alpha_{PROG} : 2^{Prog_{\simeq}} \longrightarrow (PROG_{\mathbb{D}_0})_{\simeq}$
2. $\gamma_{STATE} : (STATE_{\mathbb{D}_1})_{\simeq} \longrightarrow 2^{STATE_{\simeq}}$, $\quad \alpha_{STATE} : 2^{STATE_{\simeq}} \longrightarrow (STATE_{\mathbb{D}_1})_{\simeq}$
3. $\gamma_{ANSW} : (ATOM_{\mathbb{D}_2})_{\simeq} \longrightarrow 2^{ATOM_{\simeq}}$, $\quad \alpha_{ANSW} : 2^{ATOM_{\simeq}} \longrightarrow (ATOM_{\mathbb{D}_2})_{\simeq}$

where all the necessary partial orderings were defined in Section 2. In what follows we will use the following restriction for the abstraction functions: $\alpha_{STATE}\{fail\} = fail$. Then the following proposition holds.

Proposition 3.2. The abstraction function α_{STATE} and the concretization function γ_{STATE} have the following properties:

$$\gamma_{STATE}\ \varepsilon(...) = \{\varepsilon(...)\}, \qquad \alpha_{STATE}\ \{\varepsilon(...)\} = \varepsilon(...),$$
$$\gamma_{STATE}\ true = \{true, fail\} \qquad \alpha_{STATE}\ \{true\} = true,$$
$$\gamma_{STATE}\ fail = \{fail\}, \qquad \alpha_{STATE}\ \{fail\} = fail \ ■$$

Example 3.3. Consider the logic programs P_1, P_2, P_3 and the alphabet Σ from Example 2.4. If $\mathbb{D}_0 = VAR$, then $P_1, P_2 \in \gamma_{PROG}\ P_3$, but if $\mathbb{D}_0 = TERM(\Sigma^{\sim})$, $\Sigma^{\sim} = FUNC^{\sim} \cup VAR$, $FUNC^{\sim} = FUNC \setminus \{1,2\}$, then $P_1 \in \gamma_{PROG}\ P_2$ and $P_1 \in \gamma_{PROG}\ P_3$ ■

3.3. Abstract Interpretation of Logic Programs

Abstract interpretation of canonical logic programs is a mapping

$$A: (PROG_{\mathbb{D}_0})_{\simeq} \longrightarrow OLDT\_TREE_{\mathbb{D}_1,\mathbb{D}_2}$$

where $PROG_{\mathbb{D}_0}$ is the set of abstract logic logic programs and $OLDT\_TREE_{\mathbb{D}_1,\mathbb{D}_2}$ denotes *the set of abstract OLDT trees* whose states from the set $STATE_{\mathbb{D}_1}$ and x-rules from the set

$(\mathrm{ATOM}_{\mathbb{D}1})_{\simeq} \times (\mathrm{ATOM}_{\mathbb{D}2})_{\simeq}$. The abstract OLDT tree structure can be generated using known algorithms [TAS86,VIE89,PLU90] with only slight modifications as follows.

The OLDT resolution algorithms are based upon two important mechanisms. The first one is for resolving state of the current node (*top down mechanism*) and the second one for collecting x-rules using generated proofs (*bottom up mechanism*). Our abstract interpretation schema uses top down resolution of the following form

$$\mathrm{xres}_{\mathbb{D}1}(S,R) = \alpha_{\mathrm{STATE}}\,\{\mathrm{xres}(S,R)\}, \quad S \in (\mathrm{STATE}_{\mathbb{D}1})_{\simeq}, \quad R \in (\mathrm{CLAUSE}_{\mathbb{D}0})_{\simeq} \cup \mathrm{XRULE}_{\mathbb{D}1,\mathbb{D}2}$$

where

$$\mathrm{XRULE}_{\mathbb{D}1,\mathbb{D}2} \subseteq (\mathrm{ATOM}_{\mathbb{D}1})_{\simeq} \times (\mathrm{ATOM}_{\mathbb{D}2})_{\simeq}.$$

The bottom up mechanism uses the following abstract answers obtained from the proofs, which have been generated by top down mechanism:

$$\mathrm{answ}_{\mathbb{D}2}\ i\ j = \alpha_{\mathrm{ANSW}}\,\{\mathrm{answ}\ i\ j\},\ i,j \in \mathrm{NODE}$$

where NODE denotes the set of nodes of abstract OLDT tree structure.

The abstract terms $\mathbb{D}_0$ are used to describe properties of interest of the source logic program while the others can be used to control structure of the residual code, which can be obtained using abstract OLDT tree (Sections 4 and 5).

```
g :  ε(L,X) :- ll(L,[a],X).

ll:  ll(L1,L2,X) :- app(L1,L2,L), l(L,X).

a1:  app([],L,L) :- true.
a2:  app([H|T],L,[H|R]) :- app(T,L,R).

l1:  l([],[]) :- true.
l2:  l([H|R],[1|X]) :- l(R,X).
```

Figure 3.1. List length logic program

Example 3.4. Consider the logic program, which is shown in Figure 3.1. It computes the length of two lists and has a partially instantiated goal. Figure 3.2 shows the abstract OLDT tree structure constructed for this program above using the following parameters of the abstraction schema:

$$\mathbb{D}_0 = \mathbb{D}_1 = \mathrm{TERM}(\Sigma),\ \mathbb{D}_2 = \mathrm{VAR}$$

where Σ is the alphabet of the program. Note that in this example abstract interpretation differs from the concrete one in the following: any collected x-rule has the form <A,Z1>, $A \in \mathrm{ATOM}_{\simeq}$, $Z_1 \in \mathbb{Z}$, that is, in fact, all x-rules were suppressed. Clearly that if the abstract interpretation schema uses $\mathbb{D}_2$ = VAR, then not more than one x-rule can be collected for any node ∎

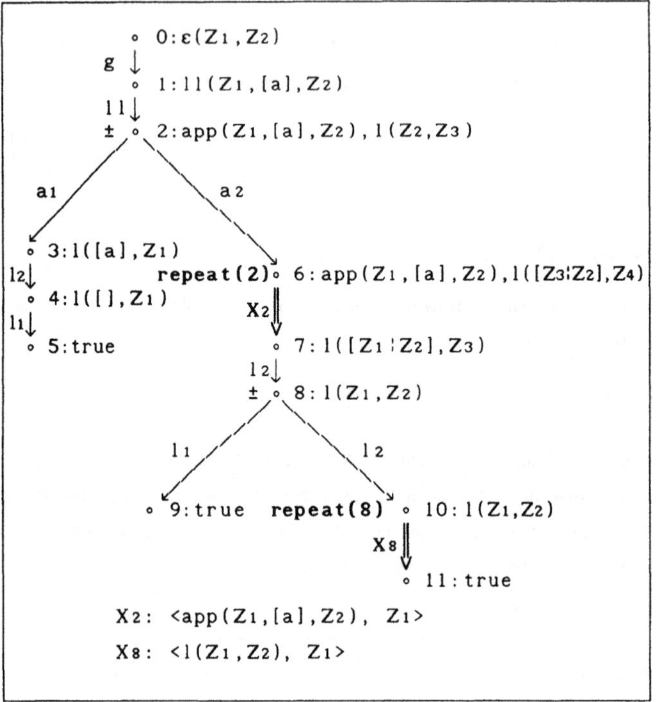

Figure 3.2. Abstract OLDT tree structure for the list length program

4. DERIVING RESIDUAL LOGIC PROGRAM STRUCTURE FROM ABSTRACT OLDT TREE

In this section we give the algorithms to derive from OLDT tree special data structures to be used as a transformation guide for residual code generation, which is introduced in the next section. These algorithms generate sequences of atoms to be unfolded and skeletons for new predicates, which are used to assemble the residual code from the unfolded sequences of atoms.

4.1. Abstract OLDT Tree Transformation

An abstract OLDT tree structure as introduced above contains some branches and subtrees, which are redundant for deriving properties of residual code. Thus we need the transformation algorithm of OLDT tree structure into the one, which exactly represents residual code structure. We denote the set of this new trees by TS_TREE.

Before going into details we have to give important definitions. Let suppose that OLDT tree T and the set of its nodes NODE are given.

Definition 4.1. (*Continuous path of OLDT tree*)
A path in OLDT tree from the node i to the node j, i,j ∈ NODE, i ∈ pred j is called continuous iff it does not contain arcs labeled by x-rules and any *internal node* k (k ∈ pred i ∩ succ j) of this path has no label ∎

Definition 4.2. (*Maximal continuous path of OLDT tree*)
The continuous path from node i to node j is a maximal one if the following holds:

1. pred i = ∅ ∨ arc i ∈ XRULE or node i has some label
2. st j ∈ {true,fail} or node j has some label ∎

A maximal continuous path can contain degenerated nodes defined as follows. Suppose that there is a maximal continuous path from node i to node j. We can enumerate these nodes as i,i+1,...,j-1,j. Then for any node of the path can be computed a *weight* w as follows:

$$
\begin{aligned}
w(i\ \ ) &= 0 & & \\
w(i+1) &= \text{length}(c_1), & c_1 &= \text{arc}(i+1) \\
w(i+2) &= \text{length}(c_2) + w(i+1)-1, & c_2 &= \text{arc}(i+2) \\
&\dots & & \\
w(j\ \ ) &= \text{length}(c_n) + w(j-1)-1, & c_n &= \text{arc}(j\ \ )
\end{aligned}
$$

where length(c) denotes number of atoms in the body of clause c.

Definition 4.3. (*Degenerated node of OLDT tree*)
The internal node k of maximal continuous path of OLDT tree is called a degenerated node iff w(k) = 0 ∎

Note that for any degenerated node j there is a path from node i to node j, which is the proof for node i.

Now we can describe transformation of the OLDT tree structure into TS_TREE one. The transformation algorithm involves three macro steps:

1. *Deleting superfluous branches and subtrees.*

   - delete all arcs and nodes, which do not belong to some path from the the OLDT tree root to the leaf node with a true state;

   - delete any leaf node and its incoming arc, which has a true state and incoming arc labeled by x-rule;

   - delete any subtree excluding its root i ∈ NODE for which the following holds:

     lab i = repeat(j), i ∈ pred j, st i = st j, noexit i j, i,j ∈ NODE.

2. *Assigning new labels to the nodes of OLDT tree.*

For the nodes of TS_TREE structure we extend the set of labels by two new labels "+" and "-". These have to be assigned to the following nodes:

- the root of OLDT tree receives "+" label;
- all degenerated nodes receive "-" labels;
- any node with incoming arc labeled by x-rule (*x-node*), which has not been given any label yet receives "-" label.

Note that after this *obligatory label assignment* has been done, any non leaf node without a label can be given any label from the set {±,+,-}. Nothing from what follows will be incorrect.

3. *Reducing arcs of OLDT tree.*

- generate the set of sequences of the arc labels, which corresponds to the set of maximal continuous paths of the modified OLDT tree obtained after steps 1 and 2, that is the maximal continuous paths have to be generated taking into account all the node labels, which have been assigned to the nodes. Let us denote this set by TS; it will be called *the set of transformation sequences.*
- delete all the internal nodes of maximal continuous paths of the modified OLDT tree together with their incoming and outcoming arcs;
- any pair of nodes (i,j) from the rest part of OLDT tree, where i was the first node and j was the last node of some maximal continuous path, connect with an arc from i to j labeled by corresponding sequence from TS.

For the TS_TREE structure we will use all the notions introduced above for OLDT tree except states. Instead of the states we will use only *calls*, call(i), i ∈ NODE, where NODE denotes *the set of TS tree nodes*. For the nodes from NODE we preserve the numbering inherited from the OLDT tree. The set of transformation sequences TS will be used as a guide for the source program transformation by unfolding.

Example 4.4. Consider abstract OLDT tree shown in Figure 3.2. It has two degenerated nodes {3,7}. The TS tree can be obtained by deleting the node 11 and replacing the path [0,1,2] by the single arc 0 ⟶ 2 labeled by <g,l1>, the path [3,4,5] by the arc 3 ⟶ 1 labeled by <l2,l1>. The corresponding set of transformation sequences is

TS = { <g,l1>, <a1>, <a2>, <l2,l1>, <l2>, <l1> } ■

4.2. Deriving Predicates of Residual Code

The TS_TREE structure defines predicate skeletons of the residual code. We need several definitions.

Definition 4.5. (*Clause skeletons*)

The set of clause skeletons CSKEL ⊆ NODE × TS is defined as follows: cskel = <i,seq> ∈ CSKEL iff i ∈ NODE, seq ∈ TS and there is an arc in TS tree labeled by seq and outcoming from node i and incoming to some node j ∈ NODE. We will use the following denotations: h(cskel) = i, t(cskel) = j, w(cskel) = w(j), where w(j) was computed for the node j of the OLDT tree, which has been obtained after steps 1 and 2 of the transformation algorithm described above ∎

The weight of the clause skeleton w(cskel) gives the number of atoms in the body of corresponding residual clause. Thus we can define the atoms of residual code.

Definition 4.6. (*Atom skeletons*).

The set of atom skeletons ASKEL ⊆ CSKEL × Integer is defined as follows:

$$ASKEL = \{<cskel,pos> \mid cskel \in CSKEL,\ 0 \le pos \le w(cskel)\}$$

where pos denotes the atom number in the clause (pos = 0 corresponds to the head) ∎

Now we can define *the set of predicate nodes* PNODE ⊆ NODE as follows:

$$PNODE = \{i \in NODE \mid \exists cskel = <i,seq> \in CSKEL\}.$$

Clearly that if i,j ∈ PNODE and i ≠ j, then call i ≠ call j and lab i ∈ {+,-,±} for any i ∈ PNODE, that is each predicate node has a unique call pattern. Every node i ∈ PNODE defines a predicate of residual code, that is it can be defined the mapping

$$cpoint: PNODE \longrightarrow ASKEL$$

which associates predicate nodes and corresponding *invocation points* from the clause bodies by means of the following algorithm:

$$cpoint\ i = \begin{cases} deref(<cskel,1>) & \text{iff } \exists cskel \in CSKEL.\ t(cskel) = i \\ deref((cpoint\ j)+1) & \text{iff arc } i \text{ is labeled by x-rule} \end{cases}$$

where j ∈ NODE is a direct predecessor of i ∈ PNODE and formally

$$<cskel,pos> + offs = <cskel,(pos+offs)>,\ offs \in Integer.$$

The dereferencing of an invocation point, which corresponds to an atom of some clause, can be done as follows:

$$deref\ <cskel,pos> = \begin{cases} <cskel,pos>) & \text{iff } 0 < pos \le w(cskel) \\ deref(\ cpoint(h(cskel)) + (pos-w(cskel)\ ) & \text{iff } pos>w(cskel) \end{cases}$$

Proposition 4.7. For any atom skeleton <cskel,pos> ∈ ASKEL, pos>0 (invocation point) there is at least one predicate node i ∈ PNODE such that cpoint i = <cskel,pos> ∎

This proposition gives the possibility to define correctly the mapping

$$\text{pcalls: ASKEL} \longrightarrow 2^{\text{PNODE}}$$

which relates atom skeletons and predicate nodes. Note that an atom can be related to several predicate nodes. An atom skeleton is called *degenerated* one if it has more than one related predicate node. The algorithm is the following:

$$\text{pcalls } \langle \text{cskel,pos} \rangle = \begin{cases} \{ i \in \text{PNODE} \mid \text{cpoint } i = \langle \text{cskel,pos} \rangle \}, \text{if pos}>0 \\ \{ i \}, \ \text{cskel} = \langle i,\text{seq} \rangle, \ i \in \text{PNODE}, \quad \text{if pos}=0 \end{cases}$$

The abstract residual "pseudo" program can be constructed using the mapping *pcalls* as follows: any clause skeleton should be transformed to a *"pseudo" clause* |cskel| of the form

$$|\text{cskel}| = \begin{cases} N_0 \text{ :- } N_1, N_2, \ldots, N_m \text{ if } w(\text{cskel}) = m \geq 1 \\ N_0 \text{ :- true} \qquad \text{if } w(\text{cskel}) = 0 \end{cases}$$

where

$$N_k = \begin{cases} i \qquad \text{if pcalls } \langle \text{cskel},k \rangle = \{i\} \\ [i_1; i_2; \ldots i_k] \text{ if pcalls } \langle \text{cskel},k \rangle = \{i_1, i_2, \ldots i_k\} \end{cases}$$

$k = 0,1,\ldots,m.$

Note that that, in fact, this algorithm derives an AND-OR tree [BRS88] from OLDT one.

4.3. Recursive and Non Recursive Predicates

Any predicate node has a label from the set {±,+,-}. The intended interpretation of these labels is the following: predicates with "+" label can not be unfolded during residual code generation while the ones labeled by "-" label can be. The label "±" denotes the predicate, which are not known to be unfolded or not. The following algorithm is capable to classify the "±" predicates to the "+" or "-" ones.

The abstract residual pseudo program can be considered a set of clauses composed only from predicate names. Atom of the form $[i_1;i_2;\ldots i_n]$ have to be interpreted as *a group of alternatives* and will be called *a special group*. Keeping this comments in mind *the dependency graph* DG can be constructed for the pseudo program.

For two predicate nodes, we say *i depends on j*, written $i \xrightarrow{\pi} j$, if j occurs in the body of some of the clause with head i. The relation "$\xrightarrow{\pi}$" defines predicate dependency graph DG ⊆ PNODE × PNODE for residual pseudo program. Any vertex of DG has a label from the set {±,+,-}.

Recall that in *directed graph* G(V,E) a *strongly connected component* (SCC) is a subgraph $G_1(V_1,E_1)$, $V_1 \subseteq V$, $E_1 \subseteq E$ such that two vertices $v,w \in V_1$ iff there is a path from v to w and vice versa. *A feedback vertex* of a directed graph G is a vertex contained in any cycle of G. An algorithm to find all maximal strongly connected components has been given in [AHU74] and a linear-time algorithm for finding all feedback vertices can be found in [GAT78].

We need two transformations del^+, del^-: DG $\longrightarrow$ DG. The algorithm for del^- is the following:

1) delete all vertices of the dependency graph labeled by "-" label together with all their incoming and outcoming edges;

2) add directed edges instead of deleted ones to recreate all broken after step 1 directed paths of the rest part of the dependency graph.

The algorithm for del^+ includes only the first step of the above one to be executed for "+" labels.

Now we can present the algorithm for replacing "±" labels by "+" or "-" ones. Let us suppose that predicate dependency graph G_0 for the pseudo program is given. Then the algorithm includes the following steps:

1) if $G_0 \neq \emptyset$, then generate $G_1 = del^-(del^+(G_0))$; stop otherwise.

2) find all vertices of G_1, which do not belong to any maximal SCC of G_1; assign to these vertices "-" labels.

3) generate $G_2 = del^-(G_1)$.

4) generate the set of all maximal SCC of G_2. For every maximal SCC $\mathbb{M}$ find some SCC $M_1 \subseteq \mathbb{M}$, which has at least one feedback vertex (such subcomponent always exists), choose one of them, assign "+" label to the chosen feedback vertex and "-" label to the all vertices of the rest part of subcomponent M_1, let $G_0 = M_1$ and go to step 1.

This algorithm is an extension of that suggested in [PLU90].

Let us make several important remarks about the set of all solutions, which can be generated by the algorithm. Any generated solution L is a tuple $\langle lab_0, lab_1, \ldots lab_n \rangle$ containing only "+" and "-" labels, where $\{0,1,\ldots,n\}$ denotes the predicates of the pseudo program. Let denote the set of all tuples, which can be generated by $\mathbb{L}$. Then $L \leq \tilde{L}$, $L, \tilde{L} \in \mathbb{L}$ iff $lab_i \leq \tilde{lab_i}$ for all predicates i, where "-" label is less than "+" label. Let us denote by $\mathbb{L}_{min} \subseteq \mathbb{L}$ *the set of all minimal solutions*:

$$\mathbb{L}_{min} = \{L \in \mathbb{L} \mid \text{iff does not exist } \tilde{L} \in \mathbb{L} \text{ such that } \tilde{L} \leq L\}$$

Clearly that any solution, which can be generated by the algorithm above, can be obtained from some element of $\mathbb{L}_{min}$ by replacing some "-" labels by "+" ones.

For special cases among the tuples from $\mathbb{L}_{min}$ we can choose the best ones. Let us suppose that at the first arrival to the step 4 of the algorithm above one of the maximal SCC, say $\mathbb{M}$, has at least one feedback vertex. Then the best solution is to assign "+" labels only to these feedback vertices because only these label tuples can eliminate *mutually recursive predicates* from residual code [PLU90]. Otherwise the maximal SCC $\mathbb{M}$ will generate mutually recursive predicates. We believe that in this context mutually recursive predicates always not better than linear recursive ones.

Example 4.8. Figure below shows some pseudo program and the corresponding dependency graph with initial set of labels. $\mathbb{L}_{min}$ contains only two solutions, which are shown below dependency graph. First solution was generated using the feedback node 3 for SCC = {1,2,3} as a "+" node, while the second one the feedback node 1 for SCC = {1,3}, which is not optimal. Two variants of unfolded source program are shown in the figure.

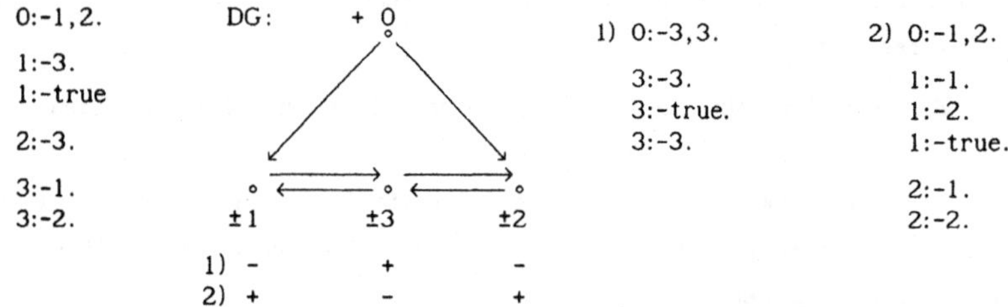

■

Example 4.9. The set of clause skeletons CSKEL and the residual pseudo program for the abstract OLDT tree shown in Figure 3.2 are the following:

```
<0,<g,l1>>        +    0:-2,[3;7].

<2,<a1>>          +    2:-true.
<2,<a2>>               2:-2.

<8,<l1>>          +    8:-true.
<8,<l2>>               8:-8.

<3,<l2,l1>        -    3:-true.

<7,<l2>>          -    7:-8
```

■

5. RESIDUAL CODE GENERATION

In this section we give the algorithms for the residual code generation using the abstract residual code structure. There are three important operations: abstract residual code structure generation and its concretization, unfolding and term rewriting.

5.1. Generation of Abstract Residual Code Structure

Let us suppose that abstract residual pseudo program together with some label assignment L $\in$ $\mathbb{L}$ and the set of clause skeletons CSKEL are given. Then the abstract residual code structure can be generated by unfolding the heads and the bodies of every pseudo clause |cskel|, cskel ∈ CSKEL. The head unfolding is the following transformation:

```
i :- B
  ⇓
i :- c1,c2,...,cm,B
```

where cskel = <i,seq>, seq = $<c_1,c_2,\ldots,c_m>$ ∈ TS, |cskel| = i:-B, i ∈ PNODE. The body unfolding have to be done only after all the heads have been unfolded. There are three variants of substitution for i ∈ PNODE in the body:

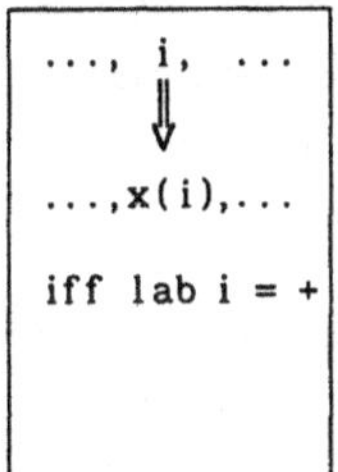

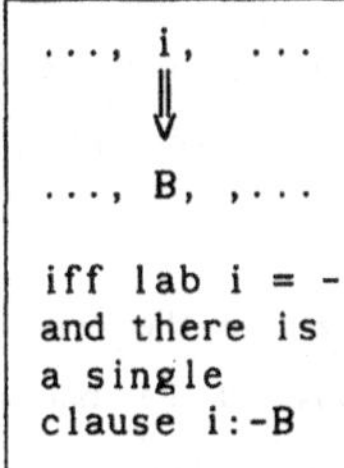

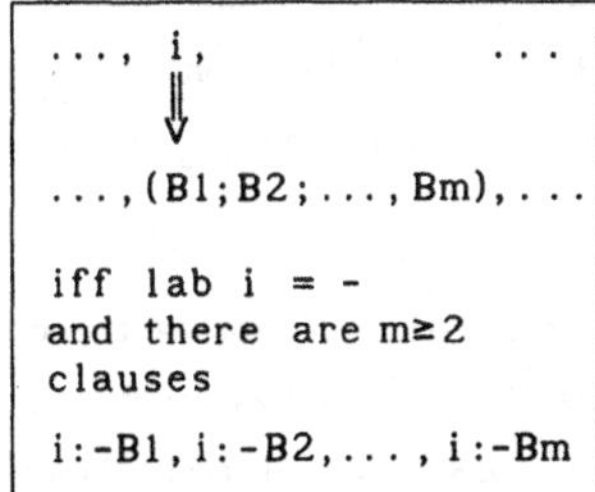

These transformations have to be continued until all the bodies will contain only constructs x(j), lab j = +, j ∈ PNODE and the arc labels of OLDT tree. Some of them can be collected in groups of alternatives. As a final step replace also all the heads i by x(i) and delete all the clauses with the head x(j), lab j = -, j ∈ PNODE. Note that any label assignment generated by the algorithm described in section 4 ensure termination of these transformations.

Example 5.1. After simple optimization we can obtain the following abstract residual code structure, which corresponds to Example 4.9:

```
x(0) :- g, l1, x(2), l2, [ l1; x(8) ]

x(2) :- a1.
x(2) :- a2, x(2).

x(8) :- l1.
x(8) :- l2, x(8).
```

■

Note that the clause bodies of the abstract residual code structure can contain two types of groups: *normal groups* of the form $(i_1;i_2;\ldots i_m)$ and *special groups* of the form $[j_1;j_2;\ldots j_n]$. Both of them should be considered a list of alternatives if we are interested in CAS-equivalence of the source and residual logic programs. The special groups are important to preserve equivalence of Prolog programs (see Example 4.5).

5.2. Concretization of Unfolded Residual Code Structure

The abstract residual code structure is constructed using transformation sequences seq $\in$ TS, which contain clauses of the abstract logic program. The following algorithm performs "translation" of the abstract residual code structure into the concrete one. This transformation is called *a concretization in the context of a source program.*

Let $P = \{c_1,c_2,\ldots,c_m\}$, $c_i \in CLAUSE_{\simeq}$, $i=1,2,\ldots,m$, $m\geq1$ is a canonical source logic program and $(P_{\mathbb{D}0})_{\simeq} = \{d_1,d_2,\ldots,d_n\}$, $d_i \in (CLAUSE_{\mathbb{D}0})_{\simeq}$, $i=1,2,\ldots,n$, $1\leq n\leq m$ is a canonical abstract logic program, which has been used for OLDT tree construction. Then the abstract residual code structure refers only abstract clauses d_i, $i=1,2,\ldots,m$.

Let

$$\gamma\colon (CLAUSE_{\mathbb{D}0})_{\simeq} \longrightarrow 2^{CLAUSE_{\simeq}}$$

be the concretization function, which is defined following the general schema described in Section 3.1. Using this concretization function we can define the concretization of abstract clause $d \in (CLAUSE_{\mathbb{D}0})_{\simeq}$ *in the context of source logic program* P as follows:

$$\gamma_P\ d = \gamma\ d\ \cap P.$$

Now the translation of abstract code structure into the concrete one can be done very easy: replace in the bodies each abstract clause d by normal group $(c_1;c_2;\ldots c_m)$, if $\gamma_P\ d = \{c_1,c_2,\ldots,c_m\}$, $m\geq2$; if $m=1$, then replace abstract clause d by c, $\gamma_P\ d = \{c\}$.

The concretization algorithm can be modified to preserve the number and the order of solutions of source program if abstract OLDT tree has been constructed using Prolog-like strategy for ordered canonical abstract program with fixed order of clauses, i.e. for a "pure" Prolog program.

5.3. Code Generation Using Concrete Residual Code Structure

Before unfolding and term rewriting the code structure can be optimized in different ways. This optimization can be done keeping in mind the balance between efficiency and reasonable residual code size. An optimization depends on the mechanism to be used as interpreter for residual code. The are two basic variants: the top down based upon **WAM** [WAR83] and the bottom up [CGT90]. The bottom up technique is used for deductive database systems. During this optimization groups can be unfolded or reconstructed.

Code generation includes two types of operations: standard unfolding and term rewriting. Unfolding of the body of the clause structure can be done using well known technique [SAH91]. Herein we describe in details only term rewriting algorithm, which can be regarded as an opposite to the one suggested in [GAB90].

The term rewriting can be done as follows. Let us suppose that clause F is a result of unfolding some clause structure. Let an atom A of F corresponds to some construct x(i) in the clause structure. Then the atom have to be rewritten if $i\neq0$. The atom rewriting includes three steps:

1) find some term $t \in TERM(\Sigma)$, such that

$$\text{call } i = t_{\alpha},\ var(t) \cap var(F) = \emptyset,\ var(t) = \{V_1,V_2,\ldots V_n\} \subseteq VAR,\ n \geq 0$$

2) generate new atom

$$A^{\sim} = \begin{cases} \text{"p\_i"}(V_1,V_2,\ldots V_n) & \text{iff call } i = p(\ldots) \\ \text{"p\_i"} & \text{iff call } i = p \\ V_1 & \text{iff call } i \text{ is a canonical variable} \end{cases}$$

3) obtain the clause $F^{\sim}$ form F by replacing the atom A by the atom $A^{\sim}$ and generate the new clause $F_{new} = F^{\sim}.\theta$, where $\theta = mgu(A,t)$. Replace clause F by F_{new}. Note that if $mgu(A,t) = fail$, then new clause F_{new} is considered to be empty.

The renaming of atoms have to be continued until all atoms of the residual program (except ε-headers of the goal clauses) will be renamed. Note that all the atoms, which correspond to the nodes of TS tree with non-variable calls, will be given a new name, while for the others their original names will be preserved.

Example 4.5. The residual code for the list length program obtained from the the abstract residual code structure from Example 5.1 is the following:

```
ε(L,X) :- X=[1|Y], app_2(L,[_|T]), [ T=[], Y=[]; l_8(T,Y) ].

app_2([],[a]).
app_2([H|T],[H|R]) :- app_2(T,R).

l_8([],[]).
l_8([H|T],[1|X]) :- l_8(T,X).
```

This program is CAS-equivalent to the original one. Completely unfolded variant of the goal clause has the following form:

```
ε(L,X) :- X=[1], app_2(L,[_]).
ε(L,X) :- X=[1|Y], app_2(L,[_|T]), l_8(T,Y).
```

If we consider this code as a "pure" Prolog program, then this variant of residual code has two solutions of the form ε([],[1]), while the source program has only one solution of this form. This effect has come from unfolded special group. The following variant of the goal clause preserves the order and number of solutions of the source program:

```
ε(L,X) :- X=[1|Y], app_2(L,[_|T]), ( T=[],Y=[], /; l_8(T,Y) ).
```

where "/" is a variant of cut built-in predicate, which cuts only alternatives of the current group. Note that there is another CAS-equivalent variant for the goal clauses

```
ε(L,X) :- L=[],   X=[1].
ε(L,X) :-         X=[1|Y], app_2(L,[_|T]), l_8(T,Y).
```

which does not preserve the number and the order of solutions of the source program ∎.

6. EQUIVALENCE OF SOURCE AND RESIDUAL CODE

It is reasonable to take into consideration two kinds of equivalence: CAS-equivalence and extended equivalence. *The extended CAS-equivalence* implies preserving the order and the number of solutions for ε-predicate of the source program when source program executed using Prolog-like computation rule. The extended CAS-equivalence is very important for applications such that [SAH91].

The transformation procedure presented in the paper ensures the CAS-equivalence of the source and the residual programs. Herein we will not give complete proof of this proposition, but only remark that the CAS-equivalence follows from the correctness [ABH87,BRS88] of the abstract interpretation schema described in Section 3. The proof for the latter can be obtained as an extension of that suggested in [GCS88] and observation that the abstract calls of the TS tree correctly represent the concrete ones, which can occur during execution of the source program.

The extended CAS-equivalence is an open problem (see Example 5.2) because it is not clear how to use the built-in predicate "/" in more complex cases when mutual recursion occurs. On the other hand if the abstract residual code structure does not contain special groups, then the transformation algorithm ensures the extended CAS-equivalence.

An other important question is the finiteness of the TS_TREE structure. It follows from the description of the abstract interpretation schema, that for any program can be found the parameters $\mathbb{D}_0$, $\mathbb{D}_1$ and $\mathbb{D}_2$, which ensure the termination. But the problem is to find the ones, which produce the most deep partial evaluation of the source program. As a rule, it is uneasy to prove finiteness of TS tree for such abstractions. We believe that the approach for deriving finite proofs of logic programs suggested in [PLU90] can be applied for this purpose.

7. APPLICATIONS

One of the intendant applications of the transformation procedure presented above is the deep transformation of logic programs. This problem occurs when compiling meta-interpreters into the flat code [NEU86,NEU90]. The transformation technique has been tested for several popular meta-interpreters, which were given to the author by Gustav Neumann, Vienna University of Economics and Business Administration. The main target was to perform optimization of object level program using meta-interpreter as an operational semantics rather than to delete only meta-interpretation level [NEU90]. It was necessary to design special abstraction schema for each particular meta-interpreter using the different sets $\mathbb{D}_0$, $\mathbb{D}_1$, and $\mathbb{D}_2$.

The most serious example was to implement the optimization of deductive database queries, which is implied by bottom up meta-interpreter suggested in [BRY89]. During these experiments was discovered the great importance of optimal label classification algorithm, which is given in Section 4.

8. CONCLUSION

In this paper we have presented a sketch of the transformation technique, which can be used for deep logic program transformation using unfolding, term rewriting and deriving new predicates. Abstract interpretation is used to control these operations. Abstract interpretation schema is based on that suggested in [GCS88]. We have extended it by careful investigation of abstract OLDT tree structure, what has given the possibility to derive the detailed structure of residual program. We have also used more flexible abstract domain construction algorithms, which are similar to those presented in [KER90].

The resulting basic transformations is similar to those suggested in [BEL90,GAB90,SAH91], but from our point of view the presented approach is more integrated and flexible because of the possibility to control the residual code quality using high level concepts and to force the partial evaluator to penetrate very deeply into nested recursions.

Acknowledgments I would like to thank Gustav Neumann, who have sent me his very interesting report and a diskette with a set of popular meta-interpreters, which were used as the main tests for the algorithms. I would like also to express gratitude to Bart Demoen for his comments on draft of this paper.

REFERENCES

[ABH87] *Abstract Interpretation of Declarative Languages*, ed. Abramsky, S., Hankin, C., Ellis Horwood Ltd., New York, 1987.

[BEL90] Benkerimi, K., Lloyd, J., *A Partial Evaluation Procedure for Logic Programs*, Proc. 1990 North American Conf. on Logic Programming, 1990, 343-358.

[AHU74] Aho, A., Hopcroft, J., Ullman, J., *The design and Analysis of Computer Algorithms*, Addison-Wesly, Reading, Mass., 1974.

[BRS88] Bruynooghe, M., De Schreye, D., *Tutorial Notes for: Abstract Interpretation in Logic Programming, Part I*, K.U. Leuven, Dept. of Computer Science, Leuven, Belgium, 1988.

[BRY89] Bry, F., *Query Evaluation in Recursive Databases: Bottom up and Top down Reconciled*, Data & Knowledge Engineering, 1989, Vol.5, 289-312.

[CGT90] Ceri, S., Gottlob, G., Tanca, L., *Logic Programming and Databases*, Springer, Berlin, 1990.

[COU77] Cousot, P., Cousot R., *Abstract Interpretation: A Unified Lattice Model for Static analysis of Programs by Construction of Approximation of Fixpoints*, Proc. 4-th ACM POPL Symp., 1977, 238-252.

[EDE85] Eder, E., *Properties of Substitutions and Unification*, Journal of Symbolic Computation, 1985, Vol.1, No.6, 31-46.

[GAB90] Gallaher, J., Bruynooghe, M., *Some Low Level Transformations for Logic Programs*, Proc. 2-nd Workshop on Meta-Programming in Logic, Leuven, Belgium, 1990, 229-244.

[GAT78] Garey, M., Tarjan, R., *A Linear-Time Algorithm for Finding all Feedback Vertices*, Information Processing Letters, 1978, Vol.7, 274-276.

[GCS88] Gallaher, J., Codish, M., Shapiro., E. *Specialization of Prolog and FCP Programs Using Abstract Interpretation*, New Generation Computing, 1988, Vol.6, Nos.2-3, 159-186.

[FUT71] Futamyra, Y., *Partial Evaluation of Computation Process - an Approach to a Compiler-Compiler*, Systems, Computers, Control, 1971, Vol.25, 45-50.

[KER90] Kemp, R., Ringwood, G., *An Algebraic Framework for Abstract Interpretation of Definite Programs*, Proc. 1990 North American Conf. on Logic Programming, 1990, 343-358.

[LLO87] Lloyd, J., *Foundations of Logic Programming*, Springer, Berlin, 1987.

[MAS89] Mariott, K., Sondergaard, H., *Semantic-Based Data Flow Analysis of Logic Programs*, Information Processing 89, ed. Ritter, G., North-Holland, IFIP, 1989.

[MAS91] Mariott, K., Sondergaard, H., *Bottom-up Data Flow Analysis of Normal Logic Programs*, Journal of Logic Programming, 1991.

[NEU86] Neumann, G., *Meta-Interpreter Directed Compilation of Logic Programs to Prolog*, IBM T.J.Watson Research Center, RC 12113(#54357), New York, 1986.

[NEU90] Neumann, G., *Transforming Interpreters into Compilers by goal classification*, Proc. 2-nd Workshop on Meta-Programming in Logic, Leuven, Belgium, 1990, 205-217.

[PLU90] Plumer, L., *Terminating Proofs for Logic Programs*, Lecture Notes In AI, 446, Springer, Berlin, 1990.

[TAS86] Tamaki, H., Sato, T., *OLD resolution with tabulation*, Proc. 3-rd Int. Conf.on Logic Programming, London, 1896, 84-98.

[SAH90] Sahlin, D., *An Automatic Partial Evaluator for Full Prolog*, SICS Dissertation Series 04, Stockholm, Sweden, 1991.

[SEI88] Seki, H., Itoh, H., *A Query Evaluation Method for Stratified Programs under Extended CWA*, Proc. 5-th Int. Conf. and Symp. on Logic Programming, Seatle, 1988, 195-211.

[SGG86] Smith, D., Genesereth, M., Ginsburg, M., *Controlling Recursive Inference*, Artificial Intelligence, 1987, Vol.30, No.3, 343-389.

[VIE89] Vieille, L., *Recursive Query Processing: The Power of Logic*, Theoretical Computer Science, 1989, Vol.69, No.1, 1-53.

[WAR83] Warren, D., *An Abstract Prolog Instruction Set*, Technical Note 309, SRI International, Menlo Park, CA, Oct.1983

Objects in a Logic Programming Framework *

Antonio Brogi[1], Evelina Lamma[2], Paola Mello[2]

[1] Dipartimento di Informatica, Università di Pisa
Corso Italia 40, 56125 Pisa, Italy
brogi@dipisa.di.unipi.it

[2] DEIS, Università di Bologna
Viale Risorgimento 2, 40136 Bologna, Italy
{evelina,paola}@deis33.cineca.it

Abstract

Some basic notions of object-oriented programming such as objects, messages and inheritance are provided of a clean definition according to the logic programming paradigm. Objects are represented by logic theories, while inheritance mechanisms are expressed through meta-level axioms. The sending of a message is interpreted as a request for the proof of a formula. The main contribution of the paper is to provide a clean semantic characterization for these notions defining both a proof-theory and a compositional model-theory.

1 Introduction

The procedural interpretation of logic, first proposed in [18], has led to the use of logic as a programming language. The main features of logic programming are the expressive power of logic, together with the sound and clean semantics of the corresponding computational model. Logic programming languages, however, lack of some characteristics which are fundamental in large applications and that are typical of object-oriented features, such as modularity, information hiding, inheritance and message-passing.

Following these considerations, several ways of combining object-oriented and logic programming have been proposed so far (e.g. [1,8,9,10,15]). However, the focus of the above mentioned works is more on the integration of the two paradigms rather than on the semantic characterization of the resulting language.

*Work partially supported by CNR "Progetto Finalizzato Sistemi Informatici e Calcolo Parallelo".

In this paper an interpretation of classical object-oriented concepts within a logic-based framework is studied. Instead of presenting yet another kind of integration of object-oriented and logic programming, the proposed logic-based interpretation is intended to give a semantic characterization of some notions of the object-oriented paradigm. Roughly, an object is represented by an extended logic theory, while messages are requests to other objects for proving a formula. Inheritance links are expressed by meta-level axioms.

From the semantic point of view, the Closed World Assumption (CWA for short) [17] is usually adopted to mean that a theory has (or is supposed to have) complete knowledge on some domain. When integrating the logic and object-oriented paradigm, the CWA seems to be no more a justified inference rule. Generally speaking, it is not assumed that an object has complete knowledge, since it can refer both explicitly (by sending messages) and implicitly (by inheritance mechanisms) to other objects. Correspondingly, logic theories have to be interpreted as *open* theories [3] instead of theories under the CWA.

After introducing a proof-theory through a set of inference rules, a model-theory is defined in terms of Herbrand models. Each object - interpreted as an *open* logic theory - is denoted by a set of Herbrand models [3] which capture the possible interactions with the external world (i.e. the other objects). The semantics of a program (i.e. a set of objects) is obtained from the semantics of the separate objects in a compositional way. At the end of the composition process, when all the interactions among objects have been taken into account, the semantics of an object is denoted by a unique Herbrand model, as in standard logic programming [18].

2 Objects in Logic Programming

Although object-oriented programming has a long history, there is no generally accepted definition or formal semantics for this paradigm. Nevertheless, as pointed out in [11], many properties are associated with object-oriented programming, such as: "Everything is an object with methods", "All activities are expressed by message passing and method invocation" or "Inheritance is the main structuring mechanism".
These properties will be discussed, in the following, from a more formal point of view, based on a logic programming semantics. We will refer to an extended logic language derived from the Class Template Language [13] and encompass only those aspects of object-oriented programming that are directly related to inheritance, message invocation and method determination.

2.1 Logic Objects

The first problem to be faced with is how to embed the concept of object in a logic programming language. In this setting, an object can be viewed either as a (structured) term [1] or as a logic theory [5,10,13,15]. In this paper we follow the second approach and we consider an object as a set of clauses, univocally referred to by a constant, and a method as a clause. This choice leads to two important features, inherited from logic programming. Input and output parameters of methods are not statically fixed, but dynamically determined by using the unification mechanism. Moreover, the same method can have multiple definitions in the same object. We will refer to this form of non-determinism as *intra-object non-determinism.*

Example 2.1 *Let us consider the following program, where different logic objects are defined as a set of (extended) clauses and connected in hierarchy.*

```
object(root)
% empty

object(animal)
likes(eating) ←
mode(walk) ←
mode(run) ← self:legs(2)
move(itself) ← self:mode(walk)

object(bird)
legs(2) ←
mode(fly) ←

object(person)
legs(2) ←
likes(women) ← self:sex(male)
likes(jogging) ← self:mode(run)
likes(birds) ← bird:mode(fly)

object(peter)
sex(male) ←
likes(something)  ← super:likes(X)
```

To specify the following is_a *hierarchy among the objects:*

```
                          is-a    bird
        is-a            ↗
root  ------→  animal
                        ↘
                          is-a   person --is-a--→ peter
```

meta-level axioms of the form bird is_a animal *are defined.* □

2.2 Messages

Objects communicate each other by means of messages. Following [5,15], each message is interpreted as a request to another object for the proof of a goal. A message has the following form: *Object* : *Goal*, where *Goal* is to be proved by the receiver *Object*. Such a message-passing primitive can be interpreted as trying to prove some formula by using the set of clauses associated with *Object*. This corresponds to a meta-level call of the kind *demo*(*Object*, *Goal*) (see [5]). Notice that *Goal* can contain variables so that unification provides a bi-directional communication mechanism.

2.3 Inheritance

Inheritance relationships in object-oriented systems are used to achieve non-replication of behaviour. A class can inherit part of its instance specification from more general classes (called superclasses).

From a logic programming point of view, several authors have proposed the interpretation of inheritance mechanisms as a combination of logic theories [4,13,14,16]. Superclass relations are expressed through meta-level axioms of the form O_i *is_a* O_j. An empty theory *root* denotes the root of the hierarchy. The execution of a message of the kind $O : G$ corresponds to the proof of G in the theory obtained by composing O with all its superclasses (until *root* is reached). By means of inheritance, each object can be interpreted as an incremental modification of an existing object. Inheritance introduces a second form of non-determinism (referred to as ***inter-object non-determinism***) since different clauses belonging to different objects in an inheritance chain can possibly unify with a formula G. $self$ references allow us to obtain the same conceptual effect as direct modification. By analogy with traditional object-oriented systems, in logic objects the label $self$ explicitly refer to the overall set of clauses. While we explicitly deal with this kind of messages, other proposals adopts the concept of parametric unit [16], or *lazy binding* [14] to implement the *self* behaviour.
Let us consider example 2.1. It defines an inheritance tree where inter-object non-determinism is present, due to the fact that *mode* and *likes* methods have multiple definitions in different objects (i.e. *animal* and *bird* with respect to *mode*, and *animal*, *person* and *peter* with respect to *likes* message).

3 An Operational Semantics for Logic Objects

In this section an operational semantics for logic objects and inheritance is defined in terms of inference rules by suitably extending the operational semantics of pure logic programming.

Definition 3.1 *Given a program P, let $O_i, O_j, Self$ denote objects of P, A, A' atoms, g a message, G a (conjunction) of messages, and $\vartheta, \gamma, \epsilon$ substitutions. A formula $F\vartheta$ is derivable from a program P ($P \vdash F\theta$) if there exists a proof for root root $\vdash_\vartheta F$. A proof for $O\ O' \vdash_\vartheta F$ is a tree such that:*

- *The root node is labelled by $O\ O' \vdash_\vartheta F$*
- *All the leaves are labelled by the empty formula (true)*
- *The internal nodes are derived by using the following inference rules*

$$(1)\quad \frac{}{O_i\ Self \vdash_\epsilon\ true}$$

$$(2)\quad \frac{O_i\ Self \vdash_\vartheta\ g\ and\ O_i\ Self \vdash_\gamma (G)\vartheta}{O_i\ Self \vdash_{\vartheta\gamma}\ (g,G)}$$

$$(3)\quad \frac{A' \leftarrow\ G\ is\ a\ clause\ of\ O_i\ and\ \vartheta = mgu(A,A')\ and\ O_i\ Self \vdash_\gamma (G)\vartheta}{O_i\ Self \vdash_{\vartheta\gamma}\ A}$$

$$(4)\quad \frac{O_i\ is\_a\ O_j\ and\ O_j\ Self \vdash_\gamma A}{O_i\ Self \vdash_\gamma\ A}$$

$$(5)\quad \frac{O_j\ O_j \vdash_\gamma A}{O_j\ Self \vdash_\gamma\ O_j : A}$$

$$(6)\quad \frac{Self\ Self \vdash_\gamma A}{O_i\ Self \vdash_\gamma\ self : A}$$

□

Comments to these inference rules are to some extent inspired by the informal, canonical operational semantics of the method lookup algorithm of Smalltalk ([11], pp. 61-64), one of the most widely used operational semantics for inheritance of object-oriented languages. Intuitively, as it happens in the function *lookup*, at each step of kind: $O_i\ Self \vdash G$ the original object $Self$ is kept, which received the message (formula) G to be proved and the object O_i in which the method (clause) is sought (or found). The true formula is always proved (rule 1), while a conjunction of formulae is proved if its parts are separately proved (rule 2). When an atomic formula A has to be proved, unifying clauses are sought in the object O_i (rule 3), or in the superclasses of O_i. $Self$ and G do not change, while O_i is non-deterministically bound to each of its ancestors (rule 4). When a message is sent to an object O_j, the self-reference is bound to O_j, and the search for unifying clauses is forced in O_j or its superclasses (rule 5). When a message is sent to $self$, unifying clauses are searched in the object $Self$ or in its superclasses, regardless of the object O_i (rule 6).

Notice that the inference rules specified above implicitly express two forms of non-determinism, lacking in procedural object-oriented systems, i.e. intra-object non-determinism (rule 3), and inter-object non-determinism (rules 3 and 4).

Example 3.2 *Let us consider the objects of example 2.1 and the message person : likes(jogging). A proof for this message is reported below (for the sake of simplicity substitutions are omitted).*

root	*root*	$\vdash$ *person* : *likes(jogging)*	*by rule* (5)
person	*person*	$\vdash$ *likes(jogging)*	*by rule* (3)
person	*person*	$\vdash$ *self* : *mode(run)*	*by rule* (6)
person	*person*	$\vdash$ *mode(run)*	*by rule* (4)
animal	*person*	$\vdash$ *mode(run)*	*by rule* (3)
animal	*person*	$\vdash$ *self* : *legs*(2)	*by rule* (6)
person	*person*	$\vdash$ *legs*(2)	*by rule* (3)
person	*person*	$\vdash$ *true*	*by rule* (1)

□

4 A Declarative Semantics for Logic Objects

This section presents a declarative semantics for logic objects and inheritance obtained by suitably extending the declarative semantics of pure logic programming [18]. In this sense, the approach here followed is different from the one adopted in [16], where the declarative semantics is defined indirectly by showing how to translate inheritance into systems of logic programs and show the equivalence between the respective operational semantics. Since distinguishing properties of object-oriented programming are modularity, re-usability and incrementality, a satisfactory declarative semantics should respect these properties and separately model each object as a single "open" component.

The semantics of a program, i.e. a set of objects, is here obtained from the semantics of the separate objects in a compositional way. First, each object - interpreted as an ***open*** logic theory - is denoted by a set of Herbrand models which capture the possible interactions with the other objects. Then the hierarchies among objects are taken into account by suitably composing the models of the objects belonging to the hierarchies themselves. Finally, pending messages to other objects are solved again by composing the obtained models. Notably, at the end of the process, each object is denoted by a unique Herbrand model which corresponds to the set of formulae which can be proved in the object.

4.1 Logic Objects as Admissible Herbrand Models

Objects may refer to other objects both implicitly (through inheritance links) and explicitly (through messages). From the semantic point of view, the Closed World Assumption [17] does not adequately model this aspect of incompleteness of knowledge, and correspondingly each object cannot be separately denoted by its least Herbrand model. To cope with this problem, objects are interpreted as "open" entities, and correspondingly are denoted by a set of Herbrand models. Roughly, each model is supported by the assumption of a set of hypotheses which may hold because of the interaction with the other objects.

In the following, let P be a program, i.e. a set of objects. The Herbrand base of an object O, B_O, is built up starting from the Herbrand universe of the program, H_P, and the predicate symbols occurring in O. The Herbrand universe H_P is the standard one. The names of the objects are not included in H_P. We first define the set of hypotheses of an object.

Definition 4.1 *Given an object O, let $Msg(O)$ be the set of messages occurring in ground(O).* □

Then the notion of *admissible Herbrand models* is introduced, according to [3].

Definition 4.2 *Given an object O, let $H \subseteq Msg(O)$ and $M \subseteq B_O$. M is an* admissible Herbrand model *(AHM) for O under hypotheses H iff M is the least Herbrand model of the logic program π_O, obtained from ground(O) by:*

- *Deleting each message which belongs to H*
- *Replacing each message $self : A$ which does not belong to H with A*
- *Deleting each clause containing a message of kind: $Obj : A$ in its body, if the message does not belong to H.*

□

The notations M *if* H and M^H will be used as shorthands for *M is an admissible Herbrand model under the hypotheses H.* The set of admissible Herbrand models of an object O will be denoted by $AHM(O)$. Actually, each admissible Herbrand model, M^H, of a ground object O can be built up by setting a set of messages, H, occurring in the clause bodies of O, and then reducing O to a standard logic programming theory, π_O, according to definition 4.2. M^H is an admissible Herbrand model of O iff M is the least Herbrand model of π_O.
The admissible Herbrand model corresponding to the empty set of hypotheses represents formulae that are true independently of beliefs on the external world, i.e. the model of the object under CWA. Let us note that object *root* is associated with the set $\{\{\}^{\{\}}\}$.

Example 4.3 *For the object animal of the program in example 2.1, the set of messages occurring in clause bodies is:*
$Msg(animal) = \{\ self : legs(2),\ self : mode(walk)\ \}$.
As a consequence we have 2^2 sets of messages for animal. The denotation of the open object animal is given by the following admissible Herbrand models, $AHM(animal)$=
{likes(eating), mode(walk), move(itself)} if {},
{likes(eating), mode(walk), move(itself), mode(run)} if {self:legs(2)},
{likes(eating), mode(walk), move(itself)} if {self:mode(walk)},
{likes(eating), mode(walk), move(itself), mode(run)} if {self:legs(2), self:mode(walk)}
□

4.2 Logic Objects in Hierarchies

So far, we have characterized the semantics of each object O of a program P by the set of its admissible Herbrand models. In this section, it is shown how to determine, in a compositional way, the semantics of each object O when taking into account the hierarchy to which O belongs. Since each object is denoted by the set of its admissible Herbrand models, the object composition along one hierarchy corresponds to a function working on sets of admissible models.
In [3], a composition operator τ (monotonic and continuous) is introduced which works on a set of (admissible) Herbrand models and maps Herbrand interpretations into Herbrand interpretations. Here its definition is recalled.

Definition 4.4 *Let S be a set of admissible Herbrand models, the composition operator τ mapping Herbrand interpretations into Herbrand interpretations is defined as follows:*
$\tau_S(I) = (\bigcup \{M \mid M^H \in S \text{ and } H \subseteq I\}) \cup I$ □

The τ operator can be conveniently applied also to the case of logic objects connected into hierarchies. Let the hierarchy of O (denoted as $Anc(O)$) be the set of objects related to O by the transitive closure of the *is_a* relationship (i.e. the ancestors of O). When an object is connected in hierarchy, *self* messages can be completely solved taking into account the hierarchy, whereas explicit messages to external objects remain unsolved.

Definition 4.5 *Given an object O, the set of its external messages is defined as:*
$Ext\_Msg(O) = \{Obj : A \in Msg(O) \mid Obj \neq self\}$. □

We can now state the compositionality result for objects connected in hierarchy by using the admissible model semantics and the τ operator.

Definition 4.6 *Given an object O_0, let $Anc(O_0) = \{O_1, \ldots, O_n\}$.*
Let $H \subseteq \bigcup_{i=0,\ldots,n} Ext\_Msg(O_i)$. Let $\overline{AHM(O_i)}$ be the set of admissible Herbrand models obtained from $AHM(O_i)$ by replacing in each admissible Herbrand model $M_i^{H_i} \in AHM(O_i)$ each hypothesis $\in H_i$ of the kind $self : A$ with A and by deleting each hypothesis $\in H_i$ of the kind $obj : A$ if $obj : A \in H$.
M is an admissible Herbrand model of O connected in hierarchy under the set of hypotheses H if and only if: $M = \tau_{\overline{AHM(O_0)} \cup \overline{AHM(O_1)} \ldots \cup \overline{AHM(O_n)}} \uparrow \omega(\{\})$. □

An admissible Herbrand model M^H of an object O hierarchically connected is obtained by assuming a set of external messages and by solving all *self* messages with respect to O and its ancestors through the least fixpoint of the τ operator.

Notation: Let $HAHM(O)$ be the denotation of the object O when connected in hierarchy, i.e. the set of admissible Herbrand models obtained according to definition 4.6.

Example 4.7 *Let us consider the program of example 2.1. The $HAHM(person)$, obtained by composing the admissible Herbrand models of objects person and its ancestor animal is:*
{legs(2), likes(eating), mode(walk), mode(run), move(itself), likes(jogging) } if {},
{legs(2), likes(eating), mode(walk), mode(run), move(itself), likes(jogging), likes(birds)} if {bird : mode(fly) } □

4.3 Denoting the Program

Our aim is to give a denotation of each object O of a program P in terms of a unique *Herbrand model*, as it happens in standard logic programming semantics. This canonical Herbrand model denoting O (referred to as M_O in the following) is obtained starting from the $HAHM$s of all the objects of the program by solving also the hypotheses corresponding to *external messages*. From the declarative point of view, M_O represents the set of (ground) atoms A such that $O : A$ logically follows from P. From the operational point of view, M_O represents the set of (ground) atoms A such that the message A sent to O (i.e. $O : A$) succeeds in P.
The declarative semantics of a program P is defined as the ordered collection of the canonical Herbrand models of its composing objects. More formally, let P be composed by the objects $O_1, \ldots, O_n$, and I be an ordered collection of Herbrand interpretations $\langle I_1, \ldots, I_n \rangle$ (where $I_i \subseteq B_P$). We will call I in the following a *program interpretation.* An ordering relationship on program interpretations can be defined as follows.

Definition 4.8 *Given a program P, let $I = \langle I_1, \ldots, I_n \rangle$, $J = \langle J_1, \ldots, J_n \rangle$ be two program interpretations.*
$I \preceq J$ iff $I_i \subseteq J_i \; \forall\, i = 1, \ldots, n$
The meet ($\sqcup$) operator is defined as:
$I \sqcup J = \langle I_1 \cup J_1, \ldots, I_n \cup J_n \rangle$, where $\cup$ is the usual set union operator.
The join ($\sqcap$) operator is defined as:
$I \sqcap J = \langle I_1 \cap J_1, \ldots, I_n \cap J_n \rangle$, where $\cup$ is the usual set intersection operator. □

The domain of program interpretations, $\Im$, is a complete lattice, where $\bot = \langle \{\}, \ldots, \{\} \rangle$ is obtained as $\sqcap \Im$ and $\top = \langle B_P, \ldots, B_P \rangle$ is obtained as $\sqcup \Im$.

In the sequel we introduce a new composition operator ϕ as a natural extension of the one introduced in definition 4.4 for a single object. ϕ still works on a set of (admissible) Herbrand models and maps program interpretations into program interpretations.

Definition 4.9 *The composition operator $\phi : \Im \mapsto \Im$ is defined as follows:*
$\phi_S(\langle I_1, \ldots, I_n \rangle) = \bigsqcup \{\langle I_1, \ldots, I_i \cup M, \ldots, I_n \rangle \mid M^H \in HAHM(O_i)$ and $H = \{\}$ or $(H = \{O_j : A^j, \ldots, O_k : A^k\}$ and $A^j \in I_j$, and $\ldots$, and $A^k \in I_k)\}$
where S is the ordered collection of the denotation of P objects when connected in hierarchy, i.e.: $S = \langle HAHM(O_1), \ldots, HAHM(O_n) \rangle$. □

In practice, at each step the ϕ operator solves external messages of the kind $O_i : A^i$ if A^i belongs to the interpretation corresponding to object O_i.
The ϕ operator is monotonic and continuous and its least fixpoint ($\phi_S \uparrow \omega(\bot)$) corresponds to the denotation of the program P, i.e.: $lfp(\phi_S) = \langle M_{O_1}, \ldots, M_{O_n} \rangle$.

Example 4.10 *Let us consider a program P composed of the objects animal, person, and bird defined in example 2.1. The declarative denotation of P is given by :*
$\langle M_{animal}, M_{person}, M_{bird} \rangle$, *where* :
$M_{animal} = \{likes(eating), mode(walk), move(itself)\}$
$M_{person} = \{legs(2), likes(eating), mode(walk), mode(run), move(itself),$
$likes(jogging), likes(birds)\}$
$M_{bird} = \{likes(eating), mode(walk), move(itself), mode(run), mode(fly)\}$ □

5 Extensions and Related Works

As such, many important aspects of object-oriented programming have been omitted, including instances, instance variables and non-monotonic inheritance. In the following, we briefly discuss these issues.

Classes and Instances
In our framework, a logic object is conceived simply as an operation descriptor, like a module. We lack of instances with a private state that can be repeatedly changed. State change is still a problem in logic programming since multiple assignments of values on logic variables cannot be performed. To overcome this problem, some proposals simulates the assignment statement in a logic framework by using the built-in predicates *assert* and *retract* of Prolog [9]. However, this approach is not satisfactory, since it lacks of a declarative characterization. One solution is proposed in [6], where *intensional variables* are introduced to keep trace of state changes without side effects. Another satisfactory, even if limited solution, could be as in [13] to represent classes as *generic objects* with variables specified in their names. Such variables perform some of the same functions as instance variables in conventional object-oriented programming (although they do not change state). State change can be simulated by creating new instances. From the declarative point of view, as pointed out in [13], variables in the class name stands for infinitely many ground instantiations of the class, determined by substituting terms of the Herbrand universe to the specified variables. In other proposals, supported by a concurrent logic programming language, state change is simulated by unification and recursion (see [8,15]). In [2] and [7] multi-headed clauses are used to implement objects with state change. In [12], a goal-continuation passing style of programming is used to model the state of objects.

Non-Monotonic Inheritance
Differently of [16], the form of inheritance introduced in our framework is monotonic since all possible definitions of a method in the hierarchy are taken into account (inter-object non-determinism). From the declarative point of view, the canonical Herbrand

model of an object is always a superset of those of its superclasses. However, in particular applications the introduction of default, exception concepts and incremental modification can be useful to take into account only the solution found in the most specific class. Dealing with non-monotonicity implies a deep revision of our semantics, with reference to both the operational and declarative point of view. We think that the best way to deal with non-monotonicity should be by introducing different forms of negation and constraints.

6 Conclusions

While logic programming is based on a well-founded formalization, object-oriented programming is often defined by specific programming language constructs and lacks of a clear formalization. It is then of interest from both the object-oriented and the logic programming point of view, to study the combination of the two paradigms, mainly for the impact on the semantics. This paper focuses, rather than on the description of a new proposal of integration between object-oriented and logic programming and a specific language to express them, on this integration from the semantic point of view. This leads to the possibility of giving a clear operational and a compositional declarative semantics, obtained as extension of the ones given for standard logic programming, for concepts typical of the object-oriented paradigm such as classes, messages, self-references and inheritance, when introduced in a logic programming setting.

References

[1] H. Ait-Kaci and R. Nasr. Login: a logic programming language with built-in inheritance. *Journal of Logic Programming*, 3:182–215, 1986.

[2] J.M. Andreoli and R. Pareschi. Linear objects: logical processes with built-in inheritance. In D.H.D. Warren and P. Szeredi, editors, *Proc. Seventh International Conference on Logic Programming*, pages 495–510. The MIT Press, 1990.

[3] A. Brogi, E. Lamma, and P. Mello. Open Logic Theories. In *Proceedings of Second Workshop on Extensions of Logic Programming*. Kista, January 1991.

[4] A. Brogi, P. Mancarella, D. Pedreschi, and F. Turini. Composition Operators for Logic Theories. In J.W. Lloyd, editor, *Computational Logic, Symposium Proceedings*, pages 117–134. Springer-Verlag, Brussels, November 1990.

[5] A. Brogi and F. Turini. Metalogic for Knowledge Representation. In J.A. Allen, R. Fikes, and E. Sandewall, editors, *Principles of Knowledge Representation and Reasoning: Proceedings of the Second International Conference*, pages 100–106, Cambridge, CA, 1991. Morgan Kaufmann.

[6] W. Chen and D.H. Warren. Objects as intensions. In R. A. Kowalski and K. A. Bowen, editors, *Proc. Fifth International Conference on Logic Programming*. The MIT Press, 1988.

[7] J.S. Conery. Logical objects. In R. A. Kowalski and K. A. Bowen, editors, *Proc. Fifth International Conference on Logic Programming*, pages 420–434. The MIT Press, 1988.

[8] K. Furukawa et al. Mandala: A logic based knowledge programming system. In *Proc. Int'l Conf. FGCS-84*, Tokyo (J), 1984.

[9] K. Fukunaga and S. Hirose. An experience with a Prolog-based object-oriented language. In *Proceedings of OOPSLA-86*. ACM Press, Portland (Oregon), 1986.

[10] H. Gallaire. Merging objects and logic programming: Relational semantics. In *AAAI-86 Conference Proceedings*, 1986.

[11] A. Goldberg and D. Robson. *Smalltalk-80, The Language and its Implementation*. Addison Wesley, 1983.

[12] J. Hodas and D. Miller. Representing objects in a logic programming language with scoping constructs. In D.H.D. Warren and P. Szeredi, editors, *Proc. Seventh International Conference on Logic Programming*. The MIT Press, 1990.

[13] F.G. McCabe. *Logic and Objects*. PhD thesis, University of London, November 1988.

[14] P. Mello. Inheritance as combination of Horn clause theories. In D. Lenzerini, D. Nardi, and M. Simi, editors, *Inheritance Hierarchies in Knowledge Representation*. J. Wiley and Sons, 1991.

[15] P. Mello and A. Natali. Objects as Communicating Prolog Units. In *Proceedings of ECOOP 87*, number 276 in LNCS. Springer-Verlag, 1987.

[16] L. Monteiro and A. Porto. A transformational view of inheritance in logic programming. In D.H.D. Warren and P. Szeredi, editors, *Proc. Seventh International Conference on Logic Programming*. The MIT Press, 1990.

[17] R. Reiter. On closed world data bases. In H. Gallaire and J.Minker, editors, *Logic and Data Bases*. Plenum, 1978.

[18] M.H. van Emden and R.A. Kowalski. The semantics of predicate logic as a programming language. *Journal of the ACM*, 23(4):733–742, 1976.

Integrity Verification in Knowledge Bases

François Bry[1], Rainer Manthey[1], and Bern Martens[2]

[1]: *ECRC, Arabellastraße 17, D - 8000 München 81, West-Germany*
[2]: *Department of Computer Science, Katholieke Universiteit Leuven, Celestijnenlaan 200A, B - 3030 Heverlee, Belgium*

fb@ecrc.de *rainer@ecrc.de* *bern@cs.kuleuven.ac.be*

ABSTRACT *In order to faithfully describe real-life applications, knowledge bases have to manage general integrity constraints. In this article, we analyse methods for an efficient verification of integrity constraints in updated knowledge bases. These methods rely on the satisfaction of the integrity constraints before the update for simplifying their evaluation in the updated knowledge base. During the last few years, an increasing amount of publications has been devoted to various aspects of this problem. Since they use distinct formalisms and different terminologies, they are difficult to compare. Moreover, it is often complex to recognize commonalities and to find out whether techniques described in different articles are in principle different. A first part of this report aims at giving a comprehensive state-of-the-art in integrity verification. It describes integrity constraint verification techniques in a common formalism. A second part of this report is devoted to comparing several proposals. The differences and similarities between various methods are investigated.*

1 Introduction

Knowledge bases extend databases by managing general, nonfactual information. Nonfactual knowledge is represented by means of deduction rules and integrity constraints [GMN84, Rei84, Ull88]. Deduction rules – called, in some cases, views – permit to derive definite knowledge from stored data. Typical examples of deduction rules define flight connections in terms of direct flights, and subordination of employees to managers from membership to and leadership of departments. In contrast, integrity constraints can be used for expressing indefinite and negative information. An integrity constraint can for example claim the existence of a manager in each department without choosing a definite person. Another constraint may require a direct flight on every day from Munich to Paris-Charles-de-Gaule or to Paris-Orly, without imposing the choice of one of the two airports. An integrity constraint can also forbid that someone manages two distinct departments. Restricted kinds of integrity constraints, called dependencies, have been often considered expressive enough for database applications [FV83]. However, as dependencies can neither express complex relationships, nor existence and disjunctive properties, more general integrity constraints are often needed for knowledge base applications.

Some integrity constraints express properties that must be fullfilled in each of the states resulting from updating a knowledge base. They are called 'state' or 'static constraints'. Other constraints refer to both the current and the updated knowledge base. They are called 'transition' or 'dynamic constraints'. Assuming that the current state satisfies the static constraints, those can be viewed as post-conditions for updates, i.e., as a simple form of dynamic constraints. As noted in [Ull88], static integrity constraints can be expressed using the knowledge base query language as yes/no queries, i.e., formally, as closed formulas [GMN84, Rei84]. For expressing dynamic constraints, however, a slight extension of the query language is needed: Statements referring to the current state must be distinguished from statements related to the updated knowledge base.

If a knowledge base is compelled to satisfy static integrity constraints, these constraints have to be verified after each update. This can be achieved by handling the static integrity constraints as post-conditions and simply re-evaluating them in the new state. This naive approach is however often inefficient, for most updates affect only part of the data and therefore only part of the static integrity constraints. Methods have been proposed that profit from the satisfaction of the static integrity constraints in the state prior to the update for simplifying their evaluation in the updated knowledge base. Consider for example a static integrity constraint requiring that all employees speak English. If it is satisfied in the current state, it is sufficient to check that every new employee speaks English for ensuring the desired property in all states. With more general integrity constraints, the definition and the computation of such an update-dependent necessary and sufficient condition is often less trivial. This article is devoted to analyzing methods that systematically generate such conditions from updates and general static integrity constraints.

These methods are presented in the literature from various points of view. In particular, some authors describe the simplification of a static constraint with respect to an update as the specialization of a logic programm – the static constraint – with respect to some input data – the update. The same problem has also been regarded as the generation of a dynamic constraint from a static one. Indeed, simplified forms of static constraints on the one hand impose conditions on the updated state, and on the other hand refer, through the update definition, to the knowledge base state prior to the update.

Processing static integrity constraints has been investigated since declarative query languages are considered, i.e., roughly since relational databases are studied. Early work on integrity verification has investigated whether simple updates could affect some static integrity constraints. The specialization of static integrity constraints with respect to updates has been proposed first by Nicolas in [Nic79] and by Blaustein in [Bla81]. These two seminal papers considered relational databases, i.e., knowledge bases without deduction rules. The necessary extensions needed in the presence of deduction rules have been investigated by several authors from various points of view during the mid-eighties. Since then, a considerable number of articles on integrity verification has been published. Over the last three years, more than fifteen approaches have been described. Since different terminologies and formalisms are used, it is rather difficult to compare the various proposals and to recognize whether methods from different authors significantly differ. This article aims at giving a comprehensive state-of-the-art in integrity verification.

With the aim of overcoming differences in form not reflecting differences of principles, we first describe the techniques involved in integrity constraint verification in a common formalism. This formalization gives rise to recognizing strong similarities between two aspects of integrity verification – the specialization of constraints and the computation of resulting changes. We show that, in fact, computing specialized constraints is a form of partial computation of resulting changes. Then, we consider several proposals. Relying on the formalization we introduced, we analyse their commonalities and differences.

This article consists of five sections, the first of which is this introduction. In Section 2, we present background notions. We propose a formalization of deduction rules, updates, and integrity constraints. Section 3 is devoted to describing the principles of integrity verification. We first consider specializing constraints with respect to updates, then the propagation of updates through deduction rules. Finally, we give a unified view of these two issues. Section 4 is an overview of the literature on integrity verification. We summarize the article and we indicate directions for further research in Section 5.

2 Knowledge Bases: A Formalization

Various languages can be adopted for expressing knowledge, but stylistic differences do not necessarily reflect differences of principles. A same deduction rule for example can be expressed as an SQL view or à la PROLOG – even recursive rules are not forbidden by the definition of SQL [Dat85], although they are rejected by the current implementations. The choice of either style does not impose any evaluation paradigm, though most current systems associate with each style given evaluation techniques. In this section, we introduce a formalism for describing the components of knowledge bases that are relevant to integrity verification: Deduction rules, updates, and integrity constraints. We emphazise that the formalism described here is not proposed as a language for knowledge base designers or users: Styles closer to natural or programming languages could be more convenient for users that are not familiar with mathematical notations.

We think that the formalism proposed below permits to describe large classes of systems. In particular, it can be used for formalizing the conventional update languages of databases as well as updating facilities more recently considered in logic programming, deductive databases, and knowledge bases.

We rely on the nowadays classical formalization of relational and deductive databases in logic [GMN84, Rei84, Ull88]. A knowledge base consists of finitely many relations. A relation is a set of tuples. Relations are not necessarily in first normal form, i.e., tuples may contain nested terms. Relation names are interpreted in logic as predicate symbols, and term constructors occurring in tuples as function symbols. Thus, a knowledge base KB defines a finite first-order logic language $\mathcal{L}$(KB). The variables of this language range over the components of tuples: This is a domain calculus. Though no more expressive than tuple calculi, domain calculi are more convenient for a simple description of integrity constraint specialization and computation of resulting changes. However, a tuple calculus may be more convenient in other contexts. We recall that information which is not provable is considered proven false *(closed world assumption).*

It is often convenient to strongly restrict the syntax of deduction rules, in order to focus on the core of some issue. In particular, many studies on query evaluation consider DATALOG [Ull88] or DATALOG^{NOT} [AV88] rules, for investigating respectively the evaluation of recursive rules, or the processing of negation. Since integrity constraints are in general expressed in a more permissive syntax, it is sensitive to allow, in a study on integrity verification, a free syntax for deduction rules as well. We shall therefore not explicitly mention syntactical restrictions imposed by a query evaluator such as, for example, the exclusion of recursive predicate definitions in deduction rules, of explicit universal quantifications, or of function symbols. We assume to have a query evaluator at our disposal and we assume that deduction rules and integrity constraints fulfill the syntactical restrictions that are necessary for the soundness, completeness, and exhaustivity of the considered evaluator.

Integrity verification however does require a syntactical restriction, namely, that deduction rules and integrity constraints are 'range-restricted'. We first recall the definition of deduction rules and static integrity constraints. Then, we define ranges and range-restriction. We introduce a formalism for expressing general updates in terms of deduction rules. Finally, we give a definition of dynamic constraints.

2.1 Deduction Rules and Static Integrity Constraints

A deduction rule is an implicative formula which defines – part or all – of a relation.

Definition 1:
A rule defining an n-ary relation p is denoted by $p(t_1, ..., t_n) \leftarrow Q$ where Q is a well-formed formula [Smu88] – possibly containing quantifiers –, and where the t_is are terms. The atom $p(t_1, ..., t_n)$ is called the *head* of the rule, the formula Q its *body*.

Formally, the rule $p(t_1, ..., t_n) \leftarrow Q$ denotes the formula $\forall x_1 ... \forall x_k\ [Q \Rightarrow p(t_1, ..., t_n)]$ where $x_1, ..., x_k$ are the variables occurring freely in $p(t_1, ..., t_n)$ or Q.

Intuitively, the evaluation of Q over the considered knowledge base KB returns values for the x_is. These instantiations affect those terms t_i in which some x_js occur and define p-tuples. Q can be viewed as a query with target list $(x_1, ..., x_k)$. More formally, if σ is an instantiation of the x_js such that $Q\sigma$ holds in KB – noted, $KB \vdash Q\sigma$ – then the instantiated term $p(t_1, ..., t_n)\sigma$ belongs to the extension of p – $KB \vdash p(t_1, ..., t_n)\sigma$.

Definition 2:
A static integrity constraint is a closed well-formed formula [Smu88] in the language of the knowledge base, i.e., a formula the variables of which are all quantified.

It is worth noting that some knowledge can be represented either by means of deduction rules, or as integrity constraints. This is the case, for example, for the statement:

$$\forall x\ [\ employee(x) \Rightarrow speaks(x, English)\ ]$$

Other statements, however, can only be used as integrity constraints, like, for example:

$$\forall x\ [\ employee(x) \Rightarrow (speaks(x, English) \lor speaks(x, German))\ ]$$
$$\forall x\ [\ project(x) \Rightarrow (\exists y\ employee(y) \land leads(y, x))\ ]$$

More generally, in the current state-of-the-art in query evaluation, deduction rules cannot serve for defining existential and disjunctive statements. Such properties must be expressed by means of integrity constraints.

For the sake of simplicity, we shall restrict ourselves to the connectives $\neg$, $\land$, $\lor$. We assume that $\Rightarrow$, $\Leftarrow$, and $\Leftrightarrow$ are expressed through $\neg$, $\land$, and $\lor$ according to the well-known equivalences: $(F \Rightarrow G) \Leftrightarrow (\neg F \lor G)$, $(F \Leftarrow G) \Leftrightarrow (F \lor \neg G)$, and $(F \Leftrightarrow G) \Leftrightarrow [(\neg F \lor G) \land (F \lor \neg G)]$. For readibility reasons, however, we shall permit an explicit use of $\Rightarrow$ in combination with universal quantifiers. Consider for example the following universal integrity constraint:

$$\forall x \; [\; employee(x) \Rightarrow speaks(x, English) \;]$$

Its meaning is rather clear: It requires to check that every employee speaks English. This meaning is somehow hidden in the equivalent form:

$$\forall x \; [\; \neg \; employee(x) \vee speaks(x, English) \;]$$

2.2 Range-Restriction

In order to ensure that deduction rules define ground tuples – i.e., tuples without unbound variables – the bodies of rules have to provide ranges for the free variables. Before formally defining ranges, we illustrate this notion with examples. The variables x and y in the rule $p(x, y) \leftarrow q(x) \vee r(y)$ are not both bound by an evaluation of the body $q(x) \vee r(y)$: This body is not a range for x and y. In contrast, evaluating the body of the rule $s(x, y) \leftarrow t(x) \wedge u(y) \wedge (q(x) \vee r(y))$ binds all variables. The following definition considers the various possible combinations of logical connectives that yield expressions – called 'ranges' – whose evaluations bind a set of variables.

Definition 3:
A *range* R for variables $x_1, ..., x_n$ is inductively defined as follows:

1. $p(t_1, ..., t_k)$ is a range for $x_1, ..., x_n$ if p is a k-ary predicate symbol and if the t_is are terms such that each variable x_j occurs in at least one of them.
2. $R_1 \wedge R_2$ is a range for $x_1, ..., x_n$ if R_1 is a range for $y_1, ..., y_k$, if R_2 is a range for $z_1, ..., z_h$ and if $\{y_1, ..., y_k\} \cup \{z_1, ..., z_h\} = \{x_1, ..., x_n\}$.
3. $R_1 \vee R_2$ is a range for $x_1, ..., x_n$ if R_1 and R_2 are both ranges for $x_1, ..., x_n$.
4. $R \wedge F$ is a range for $x_1, ..., x_n$ if R is a range for $x_1, ..., x_n$ and F is a formula with free variables in $\{x_1, ..., x_n\}$.
5. $\exists y_1...y_p$ R is a range for $x_1, ..., x_n$ if R is a range for $y_1, ..., y_p, x_1, ..., x_n$.

Integrity verification does not only require that the body of a deduction rule is a range for its free variables. Similar conditions, formalized by the following definition, must as well be imposed on the quantified variables that occur in bodies of deduction rules and in static integrity constraints.

These conditions are very similar to typed quantifications. Typed quantifications assign types to each quantified variable. If types are represented by unary predicates t_i, a typed quantification has one of the forms $\exists x \; t_i(x) \wedge F$ or $\forall x \; t_i(x) \Rightarrow F$. The definition of range-restricted formulas is based on the same idea, but allows to 'type' variables with ranges. For legibility reasons, we denote multiple quantifications $\exists x_1...\exists x_n$ and $\forall x_1...\forall x_n$ by $\exists x_1...x_n$ and $\forall x_1...x_n$, respectively.

Definition 4:
A closed formula F is *range-restricted* if all the quantified subformulas SF of F have one of the following forms:

- $\exists x_1...x_n \; R$
- $\exists x_1...x_n \; R \wedge G$
- $\forall x_1...x_n \; \neg R$

- $\forall x_1 ... x_n \; R \Rightarrow G$

  where R is a range for x_1, ..., x_n and G is a formula with all free variables in $\{x_1, ..., x_n\}$.

  A deduction rule $p(t_1, ..., t_n) \leftarrow Q$ is *range-restricted* if the closed formula $\forall x_1 ... x_k [Q \Rightarrow p(t_1, ..., t_n)]$ is range-restricted, where x_1, ..., x_k are the free variables of $[Q \Rightarrow p(t_1, ..., t_n)]$.

  An open formula F with free variables x_1, ..., x_k is *range-restricted* if its existential closure $\exists x_1 ... x_k$ F is range-restricted.

The notion 'range-restriction' has been defined first for formulas in clausal form in [Nic79]. The more general definition given above, as well as Definition 3, was given in [Bry89b]

Range-restricted formulas are definite [Kuh67] and domain independent [Fag80]. These two equivalent properties characterize the formulas whose evaluation is not affected by updates that have no effects on the extensions of the relations mentioned in the formulas. Although some domain independent formulas are not range-restricted, it is necessary in practice to consider that class of formulas since definiteness and domain independence are undecidable [Di 69]. Informally, the proof given in [Di 69] makes use of the fact that there is no decision procedure for recognizing tautologies in first-order logic. Tautologies are definite, indeed, for they are not affected by any update since they always evaluate to true.

In contrast with definiteness and domain independence, range-restriction is defined syntactically. It is therefore decidable. It is worth noting that all proposed query languages restrict themselves to range-restricted queries. Range-restricted deduction rules and integrity constraints are assumed in most studies on integrity verification. Although this property is not mentioned in other studies, the methods they describe are correct only if rules and constraints are range restricted. In more theoretical studies, other decidable classes of domain independent formulas have been proposed, e.g., the various classes of allowed formulas [LT86, VGT87, She88], and the formulas a bit misleadingly called 'definite' in [Ull88]. For each closed formula in one of these classes, an equivalent range-restricted formula can be constructed by applying equivalence preserving syntactical transformations [Bry89a].

2.3 Rule-Based Specification of Updates

In this section, we describe a formalism for defining set updates. Many studies on integrity verification mention only single tuple updates – i.e., the insertion or the removal of a given tuple. The reason for this apparent restriction is that the techniques involved in integrity verification can be extended without difficulties from single tuple updates to set updates. We first define set updates. Then, we show that more general updates – such as intensional and sequential updates – can be expressed in terms of deduction rules. Thus, an integrity verification method that takes deduction rules and set updates into account is in fact also applicable to more general updates.

Set updates are necessary since the modifications of a knowledge base that are meaningful from the application viewpoint, often affect more than one relation. If, for example, an employee leaves a company, one may have to remove a first tuple from a relation listing the affectation of employees to projects, and a second tuple from another relation with administrative data (address, salary, etc.). An update may affect deduction rules as well as factual data. Since relations are the usual way to represent sets in knowledge bases, we shall rely on two relations 'to-be-removed' and 'to-be-inserted' for describing set updates. Thus, the insertion of a fact p(a, b, c) and the removal of a deduction rule $q(x, y) \leftarrow r(x, z) \wedge s(z, y)$ will be denoted by the two facts:

to-be-inserted(p(a, b, c))
to-be-removed(q(x, y) ← r(x, z) ∧ s(z, y))

For the sake of simplicity, and without loss of generality, we assume that modifications are expressed by means of insertions and removals. Thus, changing the address of the employee John from Munich into Paris is expressed as:

to-be-inserted(address(John, Paris))
to-be-removed(address(John, Munich))

We assume a non-sequential, instantaneous semantics for set updates. Given a set update U to a knowledge base KB, the updated knowledge base KB_U is obtained by removing from KB all facts and rules marked in U as 'to-be-removed', and by inserting into KB all expressions 'to-be-inserted'. The following definition formalizes this semantics.

Definition 5:
An update U to a knowledge base KB consists of two (finite) unary relations 'to-be-inserted' and 'to-be-removed' ranging over deduction rules and ground facts in the language $\mathcal{L}$(KB) of KB.

The updated knowledge base KB_U defined by U and KB is defined as

$$KB_U = (KB \setminus \{E \mid \text{to-be-removed}(E) \in U\}) \cup \{E \mid \text{to-be-inserted}(E) \in U\}$$

where a difference A \ B of two knowledge bases A and B contains the facts of A that are not in B, and the deduction rules of A that are not variant of rules in B.

This definition calls for some comments. First, the formalism of Definition 5 only permits one to define set updates, not to specify when to perform them. This second issue requires to extend the declarative specification language we have considered above – for facts, deduction rules, integrity constraints, and update definitions – into a procedural, data manipulation language. Such an extension can be done in various ways. One of them is to keep clearly separated the declarative specification and query language from the procedural manipulation language [Dat85, MKW89]. Other approaches [NK87, NT89, dS88a, dS88b] do not really distinguish the two languages.

Second, as they are defined above, updates may contain both 'to-be-inserted(E)' and 'to-be-removed(E)' for a same fact or rule E. Such updates seem inconsistent. However, they are well-defined because Definition 5 gives priority to insertions over removals: In presence of conflicting updating requirements, the insertion prevails. One could question, whether the opposite priority would not be more convenient for some applications. One could even suggest to consider more sophisticated priorities, depending for example on the predicate defined by the fact or the rule. We do not want to argue in favor of any definition: Which one is more convenient is likely to be application dependent. Modifying Definition 5 appropriately does not affect the issues investigated in this paper.

Finally, Definition 5 interprets set updates as non-sequential, instantaneous changes: No semantics are associated with the intermediate states that might be temporarily reached during the computation of KB_U from KB. In other words, set updates are "atomic transactions": As far as integrity verification is concerned, they are indivisible. If U is a set update of a knowledge base KB, the static integrity constraints express conditions on KB and KB_U but on no intermediate state temporarily reached during the computation of KB_U from KB.

In the rest of this section, we propose to formalize more general updates than set updates in terms of deduction rules. We consider first intensional updates, then sequential updates.

One often needs the capability to refer to a set update by specifying the facts and rules 'to-be-inserted' or 'to-be-removed' through a query on the current knowledge base. For example, let us assume that flights are stored in a ternary relation 'flight' the attributes of which are, successively, the departure airport, the arrival airport, and the flight number. The query

$$\text{flight(Munich, y, z)} \wedge \text{in(y, France)}$$

characterizes the flights from Munich to French airports. Therefore, the removal of these flights is well characterized by the formula:

$$\forall yz\ [\ \text{flight(Munich, y, z)} \wedge \text{in(y, France)} \Rightarrow \text{to-be-removed(flight(Munich, y, z))}\ ]$$

Such a universal formula is, in fact, a deduction rule for the meta-relation 'to-be-removed':

$$\text{to-be-removed(flight(Munich, y, z))} \leftarrow \text{flight(Munich, y, z)} \wedge \text{in(y, France)}$$

To make such rule-based expression of set updates possible, Definition 5 does not need to be modified, provided that the non-sequential, instantaneous semantics of set updates is not abandoned. This means that the non-sequential semantics of Definition 5 can be kept if deduction rules defining the relations 'to-be-inserted' and 'to-be-removed' are not seen as defining triggers or production rules. The non-sequential semantics of Definition 5 requires that bodies of rules for 'to-be-inserted' and 'to-be-removed' are all evaluated on the same knowledge base state, namely on the state prior to any change.

Other rule-based definitions of updates have been considered, e.g., in [AV87, AV88, Abi88], and in the languages LDL [NK87, NT89] and RDL1 [dS88a, dS88b]. In these studies, insertions and removals are immediately triggered, as soon as a rule body is satisfied. Therefore, bodies of updating rules are evaluated on different states. Such a processing leads to complex semantic issues: In particular, updates may be non-deterministic. In contrast, Definition 5 gives a deterministic semantics, for KB_U is always uniquely defined whatever are U and KB.

The notion of range-restricted rules – Definition 4 – must be extended to deduction rules defining updates. Clearly, it would not make sense to restrict rule-based definitions of updates to insertion and removal of ground deduction rules! Definition 6 formalizes range-restricted update rules.

> ***Definition 6:***
> A rule to-be-inserted(E) ← B or to-be-removed(E) ← B is range-restricted if one of the following conditions is satisfied:
>
> - E is an atom and the rule E ← B is range-restricted.
> - E = (H ← C) and the rule H ← (B ∧ C) is range-restricted.

Intuitively, range-restriction for rule-defined updates of deduction rules ensures that those variables not bound by the evaluation of the rule-update body will be bound during the evaluation of the inserted or removed rule.

It is worth noting the close similarity between intensionally defined updates and updates of deduction rules. An intensional update permits one to explicitly affect all the employees with a certain skill to a certain project. Inserting a new rule allows to implicitly define these employees as affected to the considered project. The effect of both updates is the same as far as the semantics is concerned. However, the intensionally defined update results in the insertion of new facts, while these facts remain implicit by the rule insertion.

Updating transactions are often defined as sequences of changes, the subsequent changes being defined in terms of the knowledge base state resulting from the previous changes. If such a sequential transaction is atomic with respect to integrity, i.e., no intermediate state is compelled to satisfy integrity constraints, it cannot be formalized as a sequence of atomic set update transactions. We conclude this section by showing how deduction rules of a new type can be used to express sequential updates in terms of set updates.

In order to express sequential updates, one needs a formalism for distinguishing between various states of a knowledge base. We propose to annotate queries with states as follows. Given two non-sequential updates U_1 and U_2 and a query Q, the expression 'after(U_1, Q)' will denote the query Q against the knowledge base $KB_1 = KB_{U_1}$. Similarly, 'after(U_1, after(U_2, Q))' represents the query Q on the updated knowledge base obtained from KB by U_1 followed by U_2, i.e., on the knowledge base $KB_{U_1\ U_2}$.

> *Definition 7:*
> Let KB be a knowledge base, U an update of KB, and Q a query in the language $\mathcal{L}$(KB) of KB.
>
> The term 'after(U, Q)' denotes the query Q on the knowledge base KB_U.
>
> A sequential update S on KB is a finite ordered list [U_1, ..., U_n] such that U_1 is a non-sequential update of KB, and each U_i (i =2, ..., n) is a non-sequential update of $KB_{U_1 \ldots U_{i-1}}$.

Since Definition 7 makes use of a meta-predicate, namely 'after', it is a rather natural question whether the semantics of sequential updates is first-order or not. One can first remark that updates are as such not, strictly speaking, expressible in classical first-order logic. This logic is untemporal, indeed, and does not provide one with any means for expressing changes of theories. However, one can observe that a query 'after(U, Q)' has indeed a first-order semantics, for it represents the query Q – a first-order formula – on the knowledge base KB_U – a first-order theory. In other words, the meta-predicate 'after' does not need to be interpreted in second-order logic – see [End72] § 4.4, p. 281-289.

In this section, we have proposed a language for defining sequential and non-sequential update intentions by means of special facts and deduction rules. In Section 3.2, we show how to implement a hypothetical reasoning evaluator by means of deduction rules, in order to simulate query evaluation on an updated knowledge base, without actually updating the knowledge base.

2.4 Dynamic Integrity Constraints

The 'after' meta-predicate introduced in Definition 7 can be used for expressing dynamic integrity constraints. Given a – possibly sequential – update U of a knowledge base KB, let us extend the language $\mathcal{L}$(KB) of KB with 'after(U, .)' expressions. A dynamic integrity constraint of KB is simply a closed formula in this extended language. The following definition formalizes this remark.

Definition 9:
Let KB be a knowledge base and U an update of KB. The set of formulas $\mathcal{F}_U(KB)$ is inductively defined as follows:

- F $\in \mathcal{F}_U(KB)$ if F is a formula in $\mathcal{L}(KB)$
- after(U, F) $\in \mathcal{F}_U(KB)$ if F is a formula in $\mathcal{L}(KB)$
- $\neg$ F $\in \mathcal{F}_U(KB)$ if F $\in \mathcal{F}_U(KB)$
- F θ G $\in \mathcal{F}_U(KB)$ if F $\in \mathcal{F}_U(KB)$, if G $\in \mathcal{F}_U(KB)$, and if θ is a logical connective
- Qx F $\in \mathcal{F}_U(KB)$ if F is a formula in $\mathcal{L}(KB)$, x is a free variable in F, and Q denotes $\forall$ or $\exists$

A dynamic integrity constraint is a closed formula F in $\mathcal{F}_U(KB)$ which is not also a formula in the language $\mathcal{L}(KB)$.

Closed formulas in $\mathcal{L}(KB)$ are not dynamic integrity constraints, for they contain no occurrences of 'after' and therefore do not express relationships between two successive knowledge base states.

One could extend Definition 9 by defining constraints related to more than two knowledge base states. We do not give such a definition here because constraints referring to more than two states are not needed for expressing specialized forms of static integrity constraints with respect to updates.

3 Integrity Verification: Principles

In this section, we first consider specializing static integrity constraints with respect to (possibly sequential) set updates. Then, we show how deduction rules can serve to specify – and implement – a hypothetical reasoning that simulates an updated knowledge base. This reasoning propagates the updates through the deduction rules of the knowledge base. Finally, we give a unified view of the two issues: We show how constraint specialization can be viewed as update propagation.

3.1 Specialization of Static Integrity Constraints

For the sake of clarity, we introduce the specialization techniques stepwise, for static integrity constraints of increasing structural complexity. We consider first ground, quantifier-free static integrity constraints, then constraints with universal and existential quantifiers.

The quantifier-free constraints with simplest syntax are ground literals – i.e., facts or negated facts. Although very restricted, this type of constraint is sometimes useful. A factual static constraint like 'department(financial)' ensures, for example, that a company always has a financial department. Let $\neg$p(a, b) be such a constraint. Clearly, it is violated by all updates containing to-be-inserted(p(a, b)). It is also violated by updates U that do not explicitly require the insertion of p(a, b), but result in the truth of p(a, b), i.e., are such that $KB_U \vdash$ p(a, b). For example, if KB contains the deduction rule p(x, y) $\leftarrow$ q(x) $\wedge$ r(y), U = {to-be-inserted(q(a)), to-be-inserted(r(b))} is such an update. This is a general phenomenon: In order to determine the effect of an update on static integrity constraints, one has to take into account all changes resulting from the update. The following definition formalizes the notion of resulting change.

Definition 10:
Let KB be a knowledge base, U an update of KB, and F a ground fact.

F is an insertion resulting from U if $KB_U \vdash F$ and $KB \nvdash F$. F is a removal resulting from U if $KB_U \vdash \neg F$ and $KB \nvdash \neg F$ – i.e., $KB \vdash F$ and $KB_U \nvdash F$.

In the same way as we defined updates by means of the two relations 'to-be-inserted' and 'to-be-removed', we shall consider two relations 'resulting-insertion' and 'resulting-removal' for denoting the changes resulting from an update. Using the meta-predicate 'after' of Definition 7 and the deduction rule formalism, one can reformulate Definition 10 as follows:

$$\begin{aligned} \text{resulting-insertion}(U, F) &\leftarrow \text{after}(U, F) \wedge \neg F \\ \text{resulting-removal}(U, F) &\leftarrow F \wedge \text{after}(U, \neg F) \end{aligned}$$

where F denotes a ground fact in the language $\mathcal{L}(KB)$ of the knowledge base KB under consideration. For negative ground literals ¬F, we have:

$$\begin{aligned} \text{resulting-insertion}(U, \neg F) &\leftarrow \text{resulting-removal}(U, F) \\ \text{resulting-removal}(U, \neg F) &\leftarrow \text{resulting-insertion}(U, F) \end{aligned}$$

Using this formalism, a static integrity constraint ¬p(a, b) which is satisfied by a knowledge base KB is violated by an update U if and only if resulting-insertion(U, p(a, b)) is true. Similarly, an update U violates a positive constraint q(b) satisfied in KB if and only if resulting-removal(U, q(b)). These conditions can be expressed as ¬resulting-insertion(U, p(a, b)) and ¬resulting-removal(U, q(b)), or by the following conditional statements:

$$\begin{aligned} &\text{resulting-insertion}(U, p(a, b)) \Rightarrow \text{false} \\ &\text{resulting-removal}(U, q(b)) \quad\;\; \Rightarrow \text{false} \end{aligned}$$

These update dependent specialized forms of static integrity constraints are dynamic integrity constraints.

Since a conjunctive integrity constraint like p(a, b) ∧ q(c) can be represented by the two constraints p(a, b), q(c), there are no lessons to be learned from considering such compound expressions. Disjunctions are, however, more interesting. Assume that KB satisfies the disjunctive constraint C = p(a, b) ∨ q(c). C is violated in KB_U if both p(a, b) and q(c) are false in this knowledge base. Therefore, if one of the disjuncts of C is a resulting removal, the other disjunct p(a, b) must be true in KB_U. This is expressed by the following implications:

$$\begin{aligned} &\text{resulting-removal}(U, p(a, b)) \Rightarrow \text{after}(U, q(c)) \\ &\text{resulting-removal}(U, q(c)) \quad\;\; \Rightarrow \text{after}(U, p(a, b)) \end{aligned}$$

Before considering quantified constraints, we formalize the specialization of static integrity constraints. The following definition derives dynamic integrity constraints from static integrity constraints.

Definition 11:
Let C be a ground formula. S(C) is the set of formulas obtained from C as follows:
For each atom A occurring in C, S(C) contains the implication

$$\begin{array}{ll} \text{resulting-removal}(U, A) & \Rightarrow \text{after}(U, C_{A/\text{false}}) \text{ if A has positive polarity in C} \\ \text{resulting-insertion}(U, A) & \Rightarrow \text{after}(U, C_{A/\text{true}}) \text{ if A has negative polarity in C} \end{array}$$

where $C_{A/x}$ is the formula obtained from C by replacing A by x.

The following classical implications permit one to simplify the formulas of a set S(C): F ∨ false ⇒ F, F ∧ false ⇒ false, ¬false ⇒ true, F ∧ true ⇒ F, F ∨ true ⇒ true, ¬true ⇒ false, etc.

Consider for example the static integrity constraint: C = (p(a) ∨ ¬q(b)) ∧ ¬(r(c) ∨ r(d)). After simplification the set S(C) consists of:

$$\begin{array}{ll} \text{resulting-removal}(U, p(a)) & \Rightarrow \text{after}(U, \neg q(b) \wedge (\neg r(c) \vee r(d)) \\ \text{resulting-insertion}(U, q(b)) & \Rightarrow \text{after}(U, p(a) \wedge \neg(r(c) \vee r(d))) \\ \text{resulting-insertion}(U, r(c)) & \Rightarrow \text{after}(U, \text{false}) \\ \text{resulting-insertion}(U, r(d)) & \Rightarrow \text{after}(U, \text{false}) \end{array}$$

A further simplification permits one to rewrite 'after(U, false)' as 'false': 'false' is provable in no knowledge base, indeed.

If C is a clause, i.e., if C is a literal or a disjunction of literals, S(C) is obtained by resolving out the literals of C. If L is such a positive literal and if R is the corresponding resolvent, S(C) contains the implication:

$$\text{resulting-removal}(U, L) \Rightarrow \text{after}(U, R)$$

If L is a negative literal ¬A, S(C) contains the implication:

$$\text{resulting-insertion}(U, A) \Rightarrow \text{after}(U, R)$$

Definition 11 may be seen as relying on an extension of unit resolution to general, nonclausal ground formulas.

The following Proposition establishes the correctness of the Definition 11.

<u>*Proposition 1:*</u>
Let KB be a knowledge base, let U be an update of KB, and let C be a static integrity constraint such that KB ⊢ C.

$$KB_U \vdash C \text{ if and only if } KB \cup \text{after}(U, KB) \vdash F, \text{ for all } F \in S(C)$$

where 'after(U, KB)' denotes the extension $\{\text{after}(U, A) \mid KB_U \vdash A\}$ of the relation 'after'.

[*Proof:* By structural induction on C.]

Consider now the universally quantified static constraint: C = ∀x p(x) ⇒ ¬q(x). If resulting-insertion(U, p(a)) holds, then the truth of C in KB_U requires that ¬q(a) holds in KB_U. Similarly, if resulting-removal(U, ¬q(c)), i.e., resulting-insertion(U, q(c)) holds, then after(U, ¬p(c)) must hold as well. This suggests the following update dependent specialized forms for C:

$$\forall x\ [\ \text{resulting-insertion}(U, p(x)) \Rightarrow \text{after}(U, \neg q(x))\ ]$$
$$\forall x\ [\ \text{resulting-insertion}(U, q(x)) \Rightarrow \text{after}(U, \neg p(x))\ ]$$

Note that they are in principle obtained like ground static constraints are specialized. Definition 12 formalizes this observation.

Definition 12:
Let C be a universally quantified range-restricted formula $\forall x\ R[x] \Rightarrow F[x]$ such that $R[x]$ is a range for x, $F[x]$ is a formula containing x as free variable, and no existential quantifiers occur in C.

S(C) is the set of formulas obtained from C as follows. For each atom A occurring in C, S(C) contains a formula:

$$\forall x\ \text{resulting-removal}(U, A) \wedge \text{after}(U, R[x]_{A/\text{false}}) \Rightarrow \text{after}(U, F[x]_{A/\text{false}})$$
if A occurs in C with positive polarity
$$\forall x\ \text{resulting-insertion}(U, A) \wedge \text{after}(U, R[x]_{A/\text{true}}) \Rightarrow \text{after}(U, F[x]_{A/\text{true}})$$
if A occurs in C with negative polarity

where an expression $G_{A/x}$ is obtained from a formula G by replacing the atom A by x.

Applying this definition to the constraint $C = \forall x\ p(x) \Rightarrow \neg q(x)$ produces the formulas:

$$\forall x\ [\ \text{resulting-insertion}(U, p(x)) \Rightarrow \text{after}(U, \neg q(x))\ ]$$
$$\forall x\ [\ \text{resulting-insertion}(U, q(x)) \wedge \text{after}(U, p(x)) \Rightarrow \text{false}\ ]$$

Although the second formula is not identical with the one mentioned above, it is equivalent to it.

Proposition 2 establishes the soundness of Definition 12.

Proposition 2:
Let KB be a knowledge base, let U be an update of KB, and let $C = \forall x\ R[x] \Rightarrow F[x]$ be a static integrity constraint such that $KB \vdash C$.

$$KB_U \vdash C \text{ if and only if } KB \cup \text{after}(U, KB) \vdash F, \text{ for all } F \in S(C)$$

[*Proof:* By structural induction on R and F.]

Consider finally an existential statement $C = \exists x\ p(x)$. Intuitively, any resulting removal of a p fact may affect it, since we do not know for which values C is true in the knowledge base prior to the update. Therefore, the dynamic constraint associated with C is:

$$\forall x\ \text{resulting-removal}(U, p(x)) \Rightarrow \text{after}(U, \exists x\ p(x))$$

Similarly, if $D = \exists x\ p(x) \wedge \neg q(x)$ is a static integrity constraint satisfied in KB, the following statements are necessary conditions for the truth of D in the updated knowledge base KB_U:

(1) $\forall x\ \text{resulting-removal}(U, p(x)) \Rightarrow \text{after}(U, \exists x\ p(x) \wedge \neg q(x))$
$\forall x\ \text{resulting-insertion}(U, q(x)) \Rightarrow \text{after}(U, \exists x\ p(x) \wedge \neg q(x))$

In this case, however, more restrictive premisses can be considered. Assume for example that p(c) is a resulting removal for an update U. If ¬q(c) does not hold in KB, then the removal of p(c) cannot induce a violation of D in KB_U. Therefore, $KB_U \vdash D$ if and only if:

$$\text{(2)} \quad \forall x \ \text{resulting-removal}(U, p(x)) \land \neg q(x) \Rightarrow \text{after}(U, \exists x \ p(x) \land \neg q(x))$$
$$\forall x \ \text{resulting-insertion}(U, q(x)) \land p(x) \Rightarrow \text{after}(U, \exists x \ p(x) \land \neg q(x))$$

It is not an easy matter to decide whether (1) or (2) is more efficient to evaluate. The overhead introduced by computing more restrictive premisses may very well be significantly bigger than the gain it brings. It is beyond the scope of this article to propose a strategy or a heuristic for making this choice. Moreover it seems preferable to define such a choice method for general queries, not only for integrity verification.

In order to be capable to express possible choices, without committing ourselves to one of them, we rely on a meta-predicate 'option'. Our example becomes:

$$\forall x \ \text{resulting-removal}(U, p(x)) \land \text{option}(\neg q(x)) \Rightarrow \text{after}(U, \exists x \ p(x) \land \neg q(x))$$
$$\forall x \ \text{resulting-insertion}(U, q(x)) \land \text{option}(p(x)) \Rightarrow \text{after}(U, \exists x \ p(x) \land \neg q(x))$$

Definition 13 formalizes this notation:

Definition 13:
In a theory T, if $F \Rightarrow H$ holds if and only if $F \land G \Rightarrow H$ holds, then the expression $F \land \text{option}(G) \Rightarrow H$ denotes either formula.

We shall use the notation of Definition 13 with a theory T consisting of a knowledge base KB which satisfies its static integrity constraints. Definition 12 can be generalized as follows:

Definition 14:
Let C be a range-restricted closed formula. S(C) is the set of formulas obtained from C as follows. For each atom A occurring in C, S(C) contains a formula:

$\forall^* \ \text{resulting-removal}(U, A) \Rightarrow \text{after}(U, C_{A/\text{false}})$
if A has positive polarity in C and does not occur within the scope of an existential quantifier

$\forall^* \ \text{resulting-insertion}(U, A) \Rightarrow \text{after}(U, C_{A/\text{true}})$
if A has negative polarity in C and does not occur within the scope of an existential quantifier

$\forall^* \ \text{resulting-removal}(U, A) \land \text{option}(C_{A/\text{true}}) \Rightarrow \text{after}(U, C)$
if A has positive polarity in C and occurs in the scope of an existential quantifier

$\forall^* \ \text{resulting-insertion}(U, A) \land \text{option}(C_{A/\text{false}}) \Rightarrow \text{after}(U, C)$
if A has negative polarity in C and occurs in the scope of an existential quantifier

where $C_{A/x}$ denotes the formula obtained from C by replacing A by x, and where $\forall^* F$ denotes the universal closure of a formula F.

In some cases, the generated option does not constrain the variables occurring in the resulting insertion or resulting removal. This is the case, for example, for $C = \exists x\ p(x) \land \forall y\ q(y) \Rightarrow s(x, y)$. One of the formulas obtained from C according to Definition 14 is:

$$\forall z \text{ resulting-insertion}(U, q(z)) \land \text{option}(\exists x\ p(x)) \Rightarrow \text{after}(U, \exists x\ p(x) \land \forall y\ q(y) \Rightarrow s(x, y))$$

Since z does not occur in option($\exists$x p(x)), this option expression does not constrain z and is therefore not worth considering.

The following result generalizes Propositions 1 and 2.

Proposition 3:
Let KB be a knowledge base, let U be an update of KB, and let C be a range-restricted static integrity constraint such that $KB \vdash C$.

$$KB_U \vdash C \text{ if and only if } KB \cup \{\text{after}(U, A) \mid KB_U \vdash A\} \vdash F, \text{ for all } F \in S(C)$$

[*Proof:* By structural induction on C.]

Query and integrity constraint languages are usually implicitly assumed to be 'rectified' – i.e., they forbid to use the same symbol for denoting two distinct quantified variables in two different parts of a same formula. Rectification forces to rewrite the following static integrity constraint

$$[\forall x \text{ employee}(x) \Rightarrow \text{speaks}(x, \text{English})] \land [\exists x \text{ employee}(x) \land \text{speaks}(x, \text{German})]$$

as, for example:

$$[\forall x \text{ employee}(x) \Rightarrow \text{speaks}(x, \text{English})] \land [\exists y \text{ employee}(y) \land \text{speaks}(y, \text{German})]$$

We implicitly assume here that static integrity constraints are rectified.

3.2 Computation of Resulting Changes

In this section, we show how one can implement a hypothetical reasoning evaluator by means of deduction rules, in order to simulate query evaluation on an updated knowledge base, without actually performing the update. Such a query evaluator is needed, especially if concurrent accesses to the knowledge base are allowed. Indeed, an update might be rejected – for example if it violates some integrity constraints. In order to avoid, or at least to minimize, periods of time during which a knowledge base may be inconsistent and therefore not accessible to other users, one has to delay performing updates until they are proven acceptable. Classical database systems rely on special procedures for simulating updated states. We show in this section, that this is not needed for knowledge bases.

As already observed in Section 3.1, resulting changes can be defined by the following meta-rules:

$$\begin{array}{lcl}
\text{resulting-insertion}(U, F) & \leftarrow & \text{after}(U, F) \land \neg F \\
\text{resulting-removal}(U, F) & \leftarrow & F \land \text{after}(U, \neg F) \\
\text{resulting-insertion}(U, \neg F) & \leftarrow & \text{resulting-removal}(U, F) \\
\text{resulting-removal}(U, \neg F) & \leftarrow & \text{resulting-insertion}(U, F)
\end{array}$$

where F denotes a ground fact in the language $\mathcal{L}$(KB) of the knowledge base KB under consideration and U the considered update of KB.

These rules may, however, be sometimes rather inefficient. The following proposition suggests an interesting improvement.

Proposition 4:
Let KB be a knowledge base and U an update of KB. Let A and B be two formulas in the language $\mathcal{L}$(KB) of KB.

The conjunction A ∧ B is a resulting insertion if A (B, resp.) is a resulting insertion and B (A, resp.) is true in KB_U.

The conjunction A ∧ B is a resulting removal if A (B, resp.) is a resulting removal and B (A, resp.) is true in KB.

[*Proof:* By definition, A ∧ B is a resulting insertion (removal, resp.) if A ∧ B hold (does not hold, resp.) in the updated database KB_U but does not hold (holds, resp.) in KB. Proposition 4 follows.]

By Proposition 4, one can define resulting insertions by means of the following meta-rules:

resulting-insertion(U, F) ← to-be-inserted(F)
resulting-insertion(U, F) ← after(U, rule(F ← B)) ∧ resulting-insertion(B)
resulting-insertion(U, A ∧ B) ← resulting-insertion(U, A) ∧ after(U, B)
resulting-insertion(U, A ∧ B) ← after(U, A) ∧ resulting-insertion(U, B)

where the meta-predicate 'rule' ranges over the deduction rules. In a similar way, one can refine the definition of resulting removals, as well as resulting insertions and removals of general formulas containing disjunctions, quantifiers, etc.

3.3 Specialization as Partial Computation of Changes

In general, static integrity constraints cannot be represented by means of deduction rules since they may express conditions like disjunctive, existential, and negative statements that are not expressible in the formalism of deduction rules. However, the negation of such conditions can be expressed in terms of deduction rules. In other words, deduction rules can be used for expressing forbidden properties.

Consider for example a negative integrity constraint C = ¬p(a) precluding that p(a) holds. Relying on the propositional variable 'false', on could express it by means of the deduction rule:

$$\text{false} \leftarrow p(a)$$

C is violated as soon as 'false' is derivable. Similarly, a disjunctive integrity constraints q(b) ∨ r(c) could be represented by the following deduction rule:

$$\text{false} \leftarrow \neg q(b) \land \neg r(c)$$

Relying on negation in the same way gives rise to expressing existential constraints. For example, the integrity constraint $\exists x\ p(x)$ could be represented by:

$$\text{false} \leftarrow \neg\ [\exists x\ p(x)]$$

If integrity constraints are represented in such a manner as conditions for 'false', detecting if an update U to a knowledge base KB violates integrity constraints corresponds to determining whether 'false' is an insertion resulting from U. Moreover, the specialization of integrity constraints with respect to updates described in Section 3.1 reduces – under this representation of integrity constraints – to a computation of resulting changes.

4 Literature Overview

4.1 From Early Work to General Methods

In this section we want to illustrate the main developments in the field of integrity verification between 1974 and 1983. An evolution can be noticed from the realization that it might be a good idea to centrally impose the integrity of a database, via languages to do so, over special purpose techniques to efficiently test some specific constraints, to generally applicable methods to automatically simplify and evaluate constraints when updates are carried out.

One of the earliest papers addressing the issue of integrity constraints in databases is [Flo74]. The author notices that in computerized databases, the consistency of the stored data with the world that is represented, can best be preserved through the statement of logical constraints. He discerns between static and dynamic integrity constraints and remarks that integrity can best be maintained by checking the constraints when an update is carried out. Moreover, he indicates the possibility of using a kind of theorem prover to derive simplified constraints, but rejects the feasibility of such an approach in practice. Instead, the database designer is given the responsibility for formulating simple tests that can be efficiently evaluated. This process can also involve restructuring the database in such a way that checking certain constraints becomes easier.

Similar ideas can be found in [HM75]. In this paper, a distinction is made between domain and relation constraints. For each type, a rather abstract language, which is independent of any particular database system, is proposed in which constraints can be expressed. In this language, it is possible to state conditions that must be enforced, when to do so and which actions have to be taken when a violation occurs. Interesting for us is the fact that a large part of the responsibility for the decision when a particular constraint must be tested, rests with the person that formulates it. Moreover, no automatic simplification of constraints is incorporated. They must simply be formulated in a way that enables efficient testing.

In [Wil80], an approach is described to enhance existing databases with integrity checking capabilities. Again, the core of the methodology is a language in which integrity constraints, violation actions, costs and several other aspects can be specified. It incorporates facilities to indicate by which updates a constraint can be violated and how it must be instantiated when it is evaluated. It is the constraint designer who uses these facilities and who has to make sure that the actual behaviour of the system is efficient and in agreement with what was intended.

A different approach is taken in [Sto75]. Here, the formalism of the query language QUEL is used to formulate a way of treating a restricted class of integrity constraints with more automatization.

The database designer formulates constraints as universally quantified QUEL assertions. These can be divided in classes of increasing complexity. For each class, Stonebraker gives algorithms to modify insert- or replace-statements in such a way that the resulting database satisfies the constraints (if they were satisfied before the change was effected). Removals do not need attention because they cannot violate the constraints of the restricted form treated (except in the most complex case, and then no transaction modification is proposed anyway). These transformations can be executed automatically and result in quite efficiently executable tests, at least in the less complex cases. Finally, it must be noted that a replace-statement, although specified separately, is in fact more or less treated as a removal followed by an insertion.

It is interesting to compare the point of view taken in [Sto75] with the one we find in [GM79]. In the latter article, a procedural update language is proposed and integrity constraints are formulated as logic statements. Supposing they hold at the start of the transaction program, it is then proposed to use Hoare's program proving logic to investigate whether they can be demonstrated to hold at the end. In other words, at compile-time, transaction specifications would be checked for compliance with the integrity constraints, but they would not be modified. If a transaction program does not pass the test, it is simply not executed. A drawback of such a method seems to be the fact that each user writing transaction programs must explicitly pay attention to all integrity constraints that might be relevant. If he does not, it will be impossible for the system to prove that his program cannot violate integrity. In [Sto75] on the other hand, transaction programs are automatically transformed in such a way that they do not violate constraints. Of course, it must be noted that the class of transactions and constraints treated in [GM79] is more general than those addressed in [Sto75].

Finally, we can also mention [CD83]. In this paper, a more elaborate constraint checking facility for the database system INGRES is discussed and compared with the technique proposed in [Sto75].

An early paper about integrity constraints in deductive databases is [NY78]. Static and dynamic integrity constraints in the form of clauses are discussed. And the issues involved in working with (static) constraints together with derivation rules in one system are addressed. Two types of derivation rules are considered : Query-rules, which leave the defined facts implicit, and generation-rules, which are used in such a way that all derivable information is explicitly stored. As far as checking integrity is concerned, results are stated that allow to determine which constraints have to be checked when updates are performed. But nothing is said about possible simplifications of the identified constraints.

Some authors have considered restricted classes of constraints. And developed specialized techniques to test them efficiently.

[BBC80], e.g., addresses the problem of maintaining constraints that are closed tuple calculus formulas involving two tuple variables, at most one of which is existentially quantified. Moreover, the main part of the constraint is supposed to be a comparison ($\leq$) between some attributes of the tuples involved. As updates, single tuple insertions or removals are considered. The use of Hoare's program verification logic is proposed to verify that certain updates cannot violate certain constraints. For the remaining update-constraint pairs easily verifiable sufficient conditions for consistency are derived. These involve keeping extra aggregate data (minima and maxima) about the data stored in the database. A method is given to decide which data must be stored, how integrity constraints can be tested and how the stored data can be kept up to date.

The subject of [Rei81] is also rather specialized : How to treat type constraints in (a kind of) deductive databases. The author describes how typed formulas can be transformed into a set of formulas which is more manageable in the database. In this process, the constraints on types are also checked.

Finally, [Fdd81] can be mentioned as an example of a paper where an extremely specialized technique is used to treat one particular kind of constraint. Within an entity-relationship framework, it is described how one can efficiently test constraints demanding that for every entity of a certain type E1, there must exist an entity of type E2 that has a certain relation with it.

Other papers address the problem of finding general methods which can be applied to simplify and evaluate a broad class of constraints, independently from any particular system.

A first step in this direction seems to have been taken in [HK78]. There, it is proposed to analyze the influence that given updates can have on given integrity constraints. The result of this analysis would be conditions that can efficiently be tested at run-time. Different tests would be generated and a choice which alternative it is best to actually evaluate under the run-time circumstances would be made by a query optimizer.

Two articles that present more details are [WSK83] and [FW83]. In [WSK83], an extensive classification of integrity constraints is given. All constraints are (universally quantified) tuple calculus formulas. They are classified according to criteria such as : What is constrained, when they have to be invoked, how strict the conditions are they impose, how many tuples are involved. Algorithms are given to decide on whether constraints need testing given certain updates, to facilitate testing of constraints with functions such as sum, average, maximum. Finally, much attention is given to the subject of testing constraints at the end of complex transactions. Different methods to do this, involving early testing and/or protocolling are described and compared. We find different aspects treated in [FW83]. A general purpose declarative constraint language is proposed which would enable formulating (universally quantified) static integrity constraints in a kind of function-free binary predicate logic. The wish to keep the consistency of a set of formulas decidable is the reason for these strong restrictions. The main part of the paper addresses the question of how one can identify and instantiate constraints that need testing when an update is carried out.

But the two probably most influential papers that addressed these issues around the same time, are [Nic79] and [Bla81]. Nicolas treats integrity constraints formulated as first order logic formulas in prenex conjunctive normal form. He is the first one to point out the importance of having the range restricted property for integrity constraints. He proposes algorithms to decide which constraints must be checked when an insertion or removal is effected. And he shows how an affected constraint can often be instantiated and simplified in order to produce conditions which can, in the updated database, be more efficiently checked than the original constraint. Several alternative algorithms are given that produce either more conditions that are more instantiated or less conditions that are less instantiated. And a generalization of the method to transactions consisting of several removals and/or insertions is discussed. Although this paper presented a major step towards clear, general methods for the efficient treatment of integrity constraints, it does of course have its limitations. The main one is probably the fact that no methods are given to derive simplified tests for constraints that are affected through existentially quantified variables (or universally quantified variables within the scope of existentially quantified ones). Or, more precisely stated, Nicolas remarks that it is in general not possible to derive necessary and sufficient conditions for the truth of the constraint by simply instantiating these variables with values from the update. But he does not consider other techniques which might be useful. [Bla81] presents a quite comprehensive approach in a tuple calculus setting. Basically, the proposed techniques are similar to what can be found in [Nic79]. Some methods to treat existential quantifiers are also given. However, in [HI85], it is shown that they are not always correct in all circumstances. Furthermore, the work is restricted to constraints that do not contain several variables ranging over the same relation and only single tuple removals and insertions are considered. As a complement to this work, we can mention [BB82]. In this paper, a query optimization technique is presented. It is particularly useful for evaluating the kind of expressions that are typically produced as a result of the simplification techniques described in [Bla81].

To conclude this section, we give a few comments on [Laf82]. The subject of this paper is not the simplification of integrity constraints but when to test them. And, in this respect, a somewhat special view is taken. "Integrity dependencies" are introduced. These signalize for constraints which relations can be changed and which not, when they are violated (to recover from the violation). And these dependencies can be used for improving the efficiency of automatic integrity checking. For in some cases the verification of a constraint can be delayed until dependent relations occurring in it, are queried. In this way some of the checking work is done at data access time instead of at update time. Some simulations are discussed and the conclusion is that the proposed technique is particularly suitable for applications with a high update rate.

4.2 Specialization Methods in Deductive Databases

In this section, we present a comparative overview of papers presenting methods to deal with integrity constraints in deductive databases. We have made this selection because most of these papers are closely related to each other and together are a good illustration of a number of issues we wish to address. However, it should have become clear to the reader from the rest of this paper that we do not believe there exists any real difference between handling integrity constraints in relational and in deductive databases. Therefore, through future work, certainly other relevant recent papers will be included in this discussion.

In [AMM85], two methods to treat integrity constraints in (negation-free) deductive databases are proposed. The first is simply to evaluate the formulas: They must give "true" as a result. The second consists in modifying the facts and rules in the database in such a way that no information that violates the constraints can be derived. Notice that this can have remarkable consequences. Take, e.g., the constraint that every department must have a manager. If we then have a department in the database for which no manager is known, this method will not signal this fact, but simply pretend that department is not existing!

Most other papers contain a generalization of the method described in [Nic79]: They describe techniques to simplify integrity constraints with respect to updates and evaluate these simplified formulas to determine whether the updates violate constraints. We believe the main differences between most of these methods can be found in two aspects:

- How are resulting changes computed?
- Is constraint simplification and evaluation treated as part of the computation of resulting changes or is it regarded as a separate process?

Therefore, in our discussion, we will concentrate on these two issues. Like in earlier sections of this paper, we will speak about an update U. The database before the update is effected will be denoted KB. And the one that would result from performing the update, KB_U.

[Dec86] presents a method where only real (ground) changes are propagated. For every possible removal or insertion, a check is carried out whether they are in fact realized in KB_U (what we call phantomness test) and not in KB (what we call idleness test). Resulting insertions and removals are all computed, regardless whether they may affect static integrity constraints or not. For each computed resulting change, the relevant static constraints are suitably instantiated, simplified and evaluated.

[TKW85] and [LT85] contain elements of an integrity checking method that has been fully worked out in [LST87]. The method presented here, differs from the one described in [Dec86] in that it does not consider the factbase (the explicitly defined ground facts in KB) while deriving resulting changes. It propagates updates through rules, trying to instantiate resulting changes with constants or terms from the explicitly given update. No idleness or phantomness tests are performed, therefore less constraining premisses can be generated than possible. Consider for example a static integrity constraint $C = \forall x\ p(x) \Rightarrow q(x)$. In presence of the rule $p(x) \leftarrow q(x, y) \wedge r(y)$ and of the insertion of r(a), the method evaluates C in the updated database KB_U, instead of an expression equivalent to $\forall x$ resulting-insertion(U, p(x)) $\Rightarrow$ after(U, q(x)). Resulting change patterns – p(x) in our example – are taken to indicate which constraints must be (possibly) instantiated and evaluated. The authors propose a subsumption test for derived change patterns in order to avoid infinite loops. And they notice even this might not be completely satisfactory when complex terms with functors occur as arguments of predicates.

In [BD88] and [BDM88], an approach is described which has much in common with the previous one. But, for derived change patterns, affected constraints are not simply evaluated. Instead, there is a check first to see whether any instances of the pattern are indeed changes which are phantom nor idle. And only for each such instance is the simplified constraint evaluated. In the example considered above, a dynamic constraint similar – up to the names of predicates – to $\forall x$ resulting-insertion(U, p(x)) $\Rightarrow$ after(U, q(x)) is tested. A further distinguishing characteristic of this approach is the fact that its presentation relies heavily on the use of meta-programs. A short Prolog program is given for generating dynamic constraints from static ones. The hypothetical reasoning of Section 3.2 is given in these articles. The use of meta-programs adds an extra amount of flexibility, for one can imagine different strategies to execute these programs – see further. Finally, we can notice that compilation issues are given a lot of attention. Although only single tuple updates are explicitly cited in [BDM88], [BD88] shows how the very approach and programs given in [BDM88] also apply to set updates.

Another variant is presented in [AIM88]. Here, change patterns which affect constraints are tested for phantomness when they are insertions and for idleness in case of removals. In this way, they get a number of ground updates which are used to instantiate the constraints. For insertions, the instantiated constraints can be further simplified before being evaluated. In [AIM88], the claim is made that this method is superior to both the one proposed in [Dec86] and the one described in [LST87]. Although it is easy to attain a better performance than one of these two on a number of examples, and there might be even cases where both methods are beaten, it should be clear from this paper that we do not believe such a claim can hold true in general.

[DW89] shows yet another possible variant. When changes are propagated, there is always an immediate phantomness test for insertions. Derived deletions are checked for phantomness when it is discovered that they are involved in causing an addition. No idleness tests are performed, and therefore all computation can be done in the updated database KB_U. An interesting aspect of the method is the rewriting of integrity constraints. The negations of integrity constraints are rewritten as derivation rules (according to the method proposed in [LT84]). And when we have an insertion of the head of such a rule, integrity is violated. (This explains their particular choice concerning when to perform phantomness tests.) In this way, the computation of resulting changes and the simplification and evaluation of integrity constraints is unified in the way addressed in Section 3.3.

One of the first approaches with this latter characteristic is described in [SK87] and [KSS87]. Integrity constraints are written as denials and a modified version of SLDNF-resolution is introduced to enable reasoning forwards from updates. Removals are always immediately checked for phantomness and idleness. A phantomness test for insertions is also included. But exactly how and when it is performed depends on the computation rule which would be used. This introduces some flexibility that is lacking in most other proposals.

[MB88] combines the unified view on integrity checking and change propagation from the previous approach with the change propagation method proposed in [Dec86]. Moreover, the method is formulated in a set-oriented query evaluation framework. And compilation aspects are rather extensively addressed.

Within the same unified view, [MK88] stresses the fact that it is impossible to do well in all cases if you fix a global strategy for propagating changes. The authors illustrate with examples that it is necessary to be able to locally (at each particular step) choose between using a form of reasoning like in [Dec86] or like in [LST87] or somewhat in between. And they propose a formalism which would make this possible.

Most methods we discussed here, involve reasoning forwards from updates. Starting from explicitly given insertions and/or removals, resulting changes or patterns of possible resulting changes are computed. And finally, integrity constraints that might be influenced by these changes are evaluated. Whether or not a unified view on these two aspects is taken does of course not make a difference here. The method proposed in [ABI89] on the other hand, at run-time avoids all reasoning starting from updates. Predicate dependencies are pre-compiled and used at run-time to make a first selection of constraints that might be violated. For only those constraints that contain a predicate (with the right polarity) dependent on an updated predicate can be affected. The thus identified constraints are then basically evaluated as any other query. There is one difference: The computation rule used is such that as soon as possible the given update is taken into account. And when there is a branch without the possibility for using the update clause (the goal does not contain any literal with a predicate depending on the update predicate), it is not further elaborated. In fact, the method as it is presented is rather limited and suffers from some inefficiencies. But it does remind one of the fact that it is not always a good idea to reason forwards from updates. In fact, it is preferable not to make a fixed commitment to either way of reasoning. In [BDM88] and [BD88] this independence is achieved by relying on meta-rules that can be evaluated backward as well as forward. The choice of the one or the other evaluation strategy – or of a combination of both – is left to the query evaluator.

5 Conclusion

In this report, we aimed at giving a comprehensive state-of-the-art in integrity verification. A first part of the report was devoted to describe in a unified formalism features that have been previously proposed in different papers and formalized in different ways. We have shown that the notion of deduction rules permits one to express general updates – in particular set and sequential updates. Deduction rules can also be used for describing the hypothetical reasoning necessary for efficiently processing integrity constraints in concurrent knowledge base systems. Finally, we have shown that the specialization of static integrity constraints can be viewed as a partial computation of the meta-rules defining the hypothetical reasoning. A second part of this report was devoted to comparing most of the proposals for integrity verification that have been published over the last years. This study still needs to be refined and, maybe, extended. It is a first attempt towards a better understanding of the extensive amount of literature which has been published on integrity verification.

Acknowledgements

The research reported in this article has been partially supported by the European Community in the framework of the ESPRIT Basic Research Action *Compulog* No. 3012. Bern Martens is supported by the Belgian National Fund for Scientific Research.

We are indebted to our collegues of *Compulog*, ECRC, and the University of Leuven for helpful discussions.

References

[Abi88] S. Abiteboul. Updates, a New Frontier. In *Proc. 2nd Int. Conf. on Database Theory (ICDT)*, 1988.

[ABI89] P. Asirelli, C. Billi, and P. Inverardi. Selective Refutation of Integrity Constraints in Deductive Databases. In *Proc. 2nd Int. Symp. on Math. Fundamentals of Database Theory (MFDBS)*, June 1989.

[AIM88] P. Asirelli, P. Inverardi, and A. Mustaro. Improving Integrity Constraint Checking in Deductive Databases. In *Proc. Int. Conf. on Database Theory (ICDT)*, Sept. 1988.

[AMM85] P. Asirelli, De Santis M., and M. Martelli. Integrity Constraints in Logic Data Bases. *Journal of Logic Programming*, 2(3), 1985.

[AV87] S. Abiteboul and V. Vianu. A Transaction Language Complete for Update and Specification. In *Proc. 6th ACM Symp. on Principles of Database Systems (PODS)*, 1987.

[AV88] S. Abiteboul and V. Vianu. Procedural and Declarative Database Update Languages. In *Proc. 7th ACM SIGACT-SIGMOD-SIGART Symp. on Principles of Database Systems (PODS)*, Austin, Texas, March 1988.

[BB82] P. A. Bernstein and B. T. Blaustein. Fast Methods for Testing Quantified Relational Calculus Assertions. In *Proc. ACM-SIGMOD Int. Conf. on Management of Data (SIGMOD)*, June 1982.

[BBC80] B. Bernstein, B. T. Blaustein, and E. M. Clarke. Fast Maintenance of Semantic Integrity Assertions Using Redundant Aggregate Data. In *Proc. 6th Int. Conf. on Very Large Data Bases (VLDB)*, 1980.

[BD88] F. Bry and H. Decker. Préserver l'Integrité d'une Base de Données Déductive: une Méthode et son Implémentation. In *Proc. 4èmes Journéees Bases de Données Avancées (BDA)*, May 1988.

[BDM88] F. Bry, H. Decker, and R. Manthey. A Uniform Approach to Constraint Satisfaction and Constraint Satisfiability in Deductive Databases. In *Proc. 1st Int. Conf. Extending Database Technology (EDBT)*, March 1988.

[Bla81] B. T. Blaustein. *Enforcing Database Assertions: Techniques and Applications.* PhD thesis, Harvard Univ., Comp. Sc. Dept., Cambridge, Mass., Aug. 1981.

[Bry89a] F. Bry. Logical Rewritings for Improving the Evaluation of Quantified Queries. In *Proc.* 2nd *Int. Symp. on Mathematical Fundamentals of Data Base Theory (MFDBS)*, Visegrád, Hungary, June 1989. Springer-Verlag LNCS 364.

[Bry89b] F. Bry. Towards an Efficient Evaluation of General Queries: Quantifier and Disjunction Processing Revisited. In *Proc. ACM-SIGMOD Int. Conf. on Management of Data (SIGMOD)*, Portland, Oregon, May-June 1989.

[CD83] C. Cremers and G. Domann. AIM – An Integrity Monitor for the Database System Ingres. In *Proc.* 9th *Int. Conf. on Very Large Data Bases (VLDB)*, 1983.

[Dat85] C. J. Date. *A Guide to DB2.* Addison-Wesley, Reading, Massachusetts, 1985.

[Dec86] H. Decker. Integrity Enforcement on Deductive Databases. In *Proc.* 1st *Int. Conf. Expert Database Systems (EDS)*, April 1986.

[Di 69] R. A. Di Paola. The Recursive unsolvability of the Decision Problem for the Class of Definite Formulas. *Jour. of the ACM*, 16(2), 1969.

[dS88a] C. de Maindreville and E. Simon. A Production Rule Based Approach to Deductive Databases. In *Proc.* 4th *Int. Conf. on Data Engineering*, Los Angles, Calif., Feb. 1988.

[dS88b] C. de Maindreville and E. Simon. Modelling Queries and Updates in a deductive Database. In *Proc.* 14th *Int. Conf. on Very Large Data Bases (VLDB)*, Los Angles, Calif., Aug. 1988.

[DW89] S. K. Das and M. H. Williams. A Path Finding Method for Constraint Checking in Deductive Databases. *Data & Knowledge Engineering*, 4(3), 1989.

[End72] H. B. Enderton. *A Mathematical Introduction to Logic.* Academic Press, New York, 1972.

[Fag80] R. Fagin. Horn Clauses and Data Dependencies. In *Proc.* 12th *Annual ACM Symp. on Theory of Computing*, pages 123–134, 1980.

[Fdd81] A. L. Furtado, C. S. dos Santos, and J. M. V. de Castilho. Dynamic Modelling of a Simple Existence Constraint. *Information Systems*, 6, 1981.

[Flo74] J. J. Florentin. Consistency Auditing of Databases. *The Computer Journal*, 17(1), 1974.

[FV83] R. Fagin and M. Y. Vardi. Armstrong Databases for Functional and Inclusion Dependencies. *Information Processing Letters*, 16:13–19, Jan. 1983.

[FW83] R. A. Frost and S. Whittaker. A Step towards the Automatic Maintenance of the Semantic Integrity of Databases. *The Computer Journal*, 26(2), 1983.

[GM79] G. Gardarin and M. Melkanoff. Proving Consistency of Database Transactions. In *Proc.* 5th *Int. Conf. on Very Large Data Bases (VLDB)*, Sept. 1979.

[GMN84] H. Gallaire, J. Minker, and J.-M. Nicolas. Logic and Databases: A Deductive Approach. *ACM Computing Surveys*, 16(2):153–185, June 1984.

[HI85] A. Hsu and T. Imielinski. Integrity Checking for Multiple Updates. In *Proc. ACM-SIGMOD Int. Conf. on Management of Data (SIGMOD)*, 1985.

[HK78] M. Hammer and S. K. Karin. Efficient Monitoring of Database Assertions. In *Proc. Int. Conf. on Management of Data (SIGMOD)*, 1978.

[HM75] H. H. Hammer and D. J. McLeod. Semantic Integrity in a Relational Data Base System. In *Proc.* 1st *Int. Conf. on Very Large Data Bases (VLDB)*, 1975.

[KSS87] R. Kowalski, F. Sadri, and P. Soper. Integrity Checking in Deductive Databases. In *Proc.* 13th *Int. Conf. on Very Large Data Bases (VLDB)*, Brighton, UK, Sept. 1987.

[Kuh67] J. L. Kuhns. Answering Questions by Computer: A Logical Study. Technical Report RM-5428-PR, Rand Corp., 1967.

[Laf82] G. M. E. Lafue. Semantic Integrity Dependencies and Delayed Integrity Checking. In *Proc.* 8th *Int. Conf. on Very Large Data Bases (VLDB)*, 1982.

[LST87] J. W. Lloyd, E. A. Sonenberg, and R. W. Topor. Integrity Constraint Checking in Stratified Databases. *Jour. of Logic Programming*, 4(4), 1987.

[LT84] J. W. Lloyd and R. W. Topor. Making Prolog more Expressive. *Jour. of Logic Programming*, 1(3), 1984.

[LT85] J. W. Lloyd and R. W. Topor. A Basis for Deductive Database Systems. *Jour. of Logic Programming*, 2(2), 1985.

[LT86] J. W. Lloyd and R. W. Topor. A Basis for Deductive Database Systems II. *Jour. of Logic Programming*, 3(1):55–67, 1986.

[MB88] B. Martens and M. Bruynooghe. Integrity Constraint Checking in Deductive Databases Using a Rule/Goal Graph. In *Proc.* 2nd *Int. Conf. Expert Database Systems (EDS)*, April 1988.

[MK88] G. Moerkotte and S. Karl. Efficient Consistency Control in Deductive Databases. In *Proc.* 2nd *Int. Conf. on Database Theory (ICDT)*, 1988.

[MKW89] R. Manthey, V. Küchenhoff, and M. Wallace. KBL: Design Proposal for a Conceptual Language of EKS. Research Report TR-KB-29, ECRC, 1989.

[Nic79] J.-M. Nicolas. Logic for Improving Integrity Checking in Relational Databases. Technical report, ONERA-CERT, Feb. 1979. Also in *Acta Informatica 18(3), Dec. 1982, 227-253.*

[NK87] S. Naqvi and R. Krishnamurthy. Database Updates in logic Programming. In *Proc.* 7th *ACM SIGACT-SIGMOD-SIGART Symp. on Principles of Database Systems (PODS)*, pages 251–262, Austin, Texas, March 1987.

[NT89] S. Naqvi and S. Tsur. *A Logical Language for Data and Knowledge Bases.* Computer Science Press, New-York, 1989.

[NY78] J.-M. Nicolas and K. Yazdanian. *Logic and Data Bases*, chapter Integrity Checking in Deductive Databases. Plenum Press, New York, 1978.

[Rei81] R. Reiter. *Advances in Data Base Theory*, volume 1, chapter On the Integrity of Typed First-Order Data Bases. Plenum Press, New York, 1981.

[Rei84] R. Reiter. *On Conceptual Modelling*, chapter Towards a Logical Reconstruction of Relational Database Theory. Springer-Verlag, Berlin, New York, 1984.

[She88] J. C. Shepherdson. *Foundations of Deductive Databases and Logic Programming*, chapter Negation in Logic Programming, pages 19–88. Morgan Kaufmann, Los Altos, Calif., 1988.

[SK87] F. Sadri and R. Kowalski. A Theorem-Proving Approach to Database Integrity. In *Proc. Workshop on Foundations of Deductive Databases and Logic Programming*, 1987.

[Smu88] R. M. Smullyan. *First-Order Logic*. Springer-Verlag, Berlin, New-York, 1988.

[Sto75] M. Stonebraker. Implementation of Integrity Constraints and Views by Query Modification. In *Proc. ACM SIGMOD Int. Conf. on Management of Data (SIGMOD)*, May 1975.

[TKW85] R. W. Topor, T. Keddis, and D. W. Wright. Deductive Database Tools. Technical Report 84/7, University of Melbourne, 1985.

[Ull88] J. D. Ullman. *Principles of Database and Knowledge-Base Systems*, volume 1. Computer Science Press, Rockville, Maryland, 1988.

[VGT87] A. Van Gelder and R. W. Topor. Safety and Correct Translation of Relational Calculus Formulas. In *Proc. 6th ACM SIGACT-SIGMOD-SIGART Symp. on Principles of Database Systems (PODS)*, pages 317–327, San Diego, Calif., March 1987.

[Wil80] G. A. Wilson. A Conceptual Model for semantic Integrity Checking. In *Proc. 6th Int. Conf. on Very Large Data Bases (VLDB)*, Oct. 1980.

[WSK83] W. Weber, W. Stucky, and J. Karszt. Integrity Checking in Data Base Systems. *Information Systems*, 8(2), 1983.

On Procedural Semantics of Metalevel Negation

S. Costantini   G.A. Lanzarone

Universita' degli Studi di Milano
Dipartimento di Scienze dell' Informazione
Via Comelico 39, I–20135, Milano (+39–2–55006, fax +39–2–55006223)
e–mail costanti@imiucca.bitnet
lanzarone@hermes.unimi.it

Introduction

Metalevel negation (MN for short), presented in [CL91b], enriches the metalogic programming language Reflective Prolog (RP for short, [CL89], [CL91a], [Co90]) with the possibility of declaratively restricting the extension of predicates. This is obtained by defining negative metarules, rather than by using a negation operator. An extended resolution principle, called NRSLD–Resolution, states the role of these negative rules in proofs. Semantics of metalevel negation (described in [CL91b] and summarized in Section 2) is very similar to semantics of Horn–clause languages for definite programs: it is a minimal model semantics, where the model (called the *Iterated Reflective Model*, IRM for short) is characterized as the (iterated) least fixpoint of a suitable mapping. This leads to proofs of correctness and completeness of NRSLD–Resolution very similar to those of classical SLD–resolution [Ll87].

In [CL91b] we have also shown that "Negation as Failure" (NAF for short) can be explicitly represented in RP by means of MN, thus allowing different forms of negation to cohexist in one and the same program. We proved the soundness of this representation R, showing that, given a Horn–clause program P using NAF, and the corresponding RP program P' = P ∪ R, the Well–Founded Model of P ([VRS88], [Pr89]) and the IRM of P' coincide (modulo a natural correspondence). This result on the one hand enlights the ability of the IRM semantics of expressing the Well–Founded Semantics for NAF on the basis of its metalevel representation, and on the other hand is a point in favour of the Well–Founded Semantics being a "natural" semantics for NAF.

Subject of this paper is NRSLD–Resolution and its properties. We introduce the definition of NRSLD–Resolution, together with an effective method that could serve as a basis for its efficient implementation. We prove correctness and completeness of NRSLD–Resolution w.r.t. the IRM, and the Independence of the Computation Rule. These results rely on a *coherence* condition on negative subgoals, which is weaker than non–floundering. This suggests the idea, given the semantic correspondence mentioned above, of using NRSLD–Resolution as a procedural semantics for NAF. In this paper, we argue that: (i) this is easily feasible; (ii) the independence of the computation rule holds for NAF when simulated by MN and subject to NRSLD–Resolution; (iii) there are advantages with respect to the floundering problem.

In Section 1 we briefly illustrate RP and its semantics for *positive* programs (programs without MN). In Section 2 we introduce MN, and the corresponding semantic extensions. In Section 3 we define NRSLD–Resolution, and in Section 4 we present the main results about its properties. In Section 5 we propose a computational schema, and finally in Section 6 we discuss the application of NRSLD–Resolution to NAF. Because of space limitations, we present only a few examples.

1. Reflective Prolog

The metalogic programming and knowledge representation language Reflective Prolog has three basic features. First, a full naming mechanism, which allows the representation of terms and atomic formulas (by means of a new class of terms, *name terms*), and thus allows the representation of metaknowledge in the form of *metalevel clauses* (that contain, by definition, at least a name term). Second, the possibility of specifying *metaevaluation clauses* (which are the metalevel clauses defining the distinguished predicate solve, and form the so–called *metaevaluation level* of a program), that allow to declaratively extend the meaning of the other predicates (called *base predicates,* which form the *base level* of a program). Third, a form of *reflection* which makes this extension effective, both semantically and procedurally [Co90].

The naming mechanism provides *quoted name constants* to denote constants, variables and function symbols (e.g. "c", "V", "fun"); *brackteted name constants* to denote predicate symbols (e.g. <pred>); *function* and *relation name terms* to denote terms and atoms respectively (e.g. "fun"("c","V") and <pred>("c","V")). In the rest of the paper, the name term denoting a term/atom A will be indicated by $\uparrow$A. *Predicate metavariables* (syntax #<Name>, e.g. #P) can be bound to bracketed name constants only (they are intended to denote predicate names). *General metavariables* (syntax $<Name>, e.g. $X) can be bound to metalevel terms in general.

Because of name terms and reflection, RP provides two forms of unification. *Extended unification* is a straightforward extension of usual unification, to deal with name terms and metavariables: given two terms/atoms A and B, it produces an mgu θ such that $A\theta=B\theta$ (for instance, unifying #P($X,"b") and <p>("f"("Y"),$V) we get mgu {#P/<p>, $X/"f"("Y"), $V/"b"}). *Generalized unification*, or g–unification, is used when applying reflection: g–unifying a metalevel term α and a term/atom B, we get a so–called gmgu θ such that $\alpha\theta = \uparrow(B\theta)$ (for instance, g–unifying #R($F("c")) and t(f(X)) we get gmgu { #R/<t>,$F/"f",X/c }). Both mgu's and gmgu's are substitutions, and can be combined as usual.

Procedurally, we have defined a new version of resolution, called RSLD–Resolution (R stands for 'Reflective'): a goal A can be resolved both via clauses with head B, where A and B unify, and via clauses with head solve(α), where α and A g–unify, and vice versa. solve cannot be applied to itself (meta–metaevaluation is not allowed in RP). In the following, when mentioning A and solve($\uparrow$A) we will implicitly mean A $\neq$ solve(...).

An *Extended Herbrand Interpretation* I_E of (the language L defined by) a Reflective Prolog definite program P is an Herbrand Interpretation of P which satisfies the condition that solve($\uparrow$A) belongs to I_E if A belongs to I_E. I_E is characterized, as usual, as a subset of the Herbrand Base B_{PE} of P. A *Reflective Model* for P is an Extended Herbrand interpretation I_E for L which is a Model for

$$P^* = P \cup \{A \leftarrow \text{solve}(\uparrow A) : A \in B_{PE}\}.$$

The axioms A $\leftarrow$ solve($\uparrow$A) are called *reflection axioms*. An atom A is a *Reflective Logical Consequence* of an RP program P if, for every interpretation I_E, I_E is a Reflective Model for P implies that I_E is a Model for A. P* may be considered as the "extended version" of P, i.e. the program which P –due to the reflection capabilities of RP– stands for. Reflective Models are models in the usual sense and, since the *Model Intersection Property* obviously holds for Reflective Prolog definite programs, there exists a Least Reflective Herbrand Model RM_P of a definite program P. It can be characterized as the least fixpoint of a mapping T_{PE} very similar to the usual T_P [Ll87], except that it produces both A and solve($\uparrow$A) whenever one of them can be derived. The definition is:

$T_{PE}(I_E) =$ { A, S $\in B_{PE}$: S = solve($\uparrow$A) $\wedge$
either
A $\leftarrow A_1 \ldots A_n$
or

$S \leftarrow A_1 \dots A_n$
is a clause in $P \wedge \{A_1 \dots A_n\} \subseteq I_E \}$ ♦

We have proved that $RM_P = lfp(T_{PE}) = T_{PE}\uparrow\omega$. RSLD–Resolution has been proved sound and complete w.r.t. RM_P. Semantics of Reflective Prolog is summarized in [Co90], and described in detail in [CL91a]. All the definitions and proofs also appear in [Co89].

2. Metalevel Negation and its Declarative Semantics

The distinguished predicate solve_not, introduced in [CL91b], allows the restriction of the intended meaning of predicates: a goal $p(a_1,\dots,a_n)$ is not a consequence of a Reflective Prolog program if solve_not(<p>($\uparrow a_1,\dots,\uparrow a_n$)) is: rules for solve_not specify when predicates *do not* hold. This metalevel negation can be used to express negative knowledge, which has priority over positive knowledge: both A and solve_not($\uparrow$A) being derivable does not entail that the program is inconsistent, but that A is false. solve_not, being a metapredicate, makes it possible to express not only object–level negative knowledge, but also negative metaknowledge. For instance it is possible to explicitly define negation as failure, as shown in Section 5.

The declarative semantics for MN as introduced in this Section is defined on a wide class of RP programs, called *safe* programs. Conditions for safeness are given in detail in [CL91b]. Basically, since an atom A *implicitly* depends (negatively) on solve_not($\uparrow$A), solve_not($\uparrow$A) cannot depend on A. Safeness is a dynamic condition, to be checked at run–time. An extension of MN semantics has been developed [CL91c], which considers *every* RP program.

In the following, an atom $A \in B_{PE}$ will be called *non–distinguished* if $A \neq$ solve(...) and $A \neq$ solve_not(...). Clauses with head solve_not(...) will be called *negative clauses*, while all the others will be called *positive clauses*. Similarly, atoms solve_not(...) will be called *negative atoms* (w.r.t. *positive atoms*). As a syntactic restriction of the language, the distinguished predicates solve and solve_not apply to non–distinguished atoms only.

The method for defining the declarative semantics SEM(P) of an RP program P with MN is model–theoretic, in the sense of [PP90]: the semantics is determined by choosing an intended model M of a suitably modified version of P, and a formula F is said to be *implied* by SEM(P) if and only if it is satisfied in the intended model, i.e. $M \models F$.

In particular, this intended model is obtained as the iterated least fixpoint of a mapping $T_{.J}(I)$ where I and J are Extended Herbrand Interpretations. $T_{.J}(I)$ is very similar to T_{PE}, since solve_not clauses are treated in the same way as all the other clauses; the only difference is that an atom A cannot be included in $T_{.J}(I)$ if solve_not($\uparrow$A) belongs to J. The definition is the following:

$T_{.J}(I_E) = \{A \in T_{PE}(I_E) : \text{solve\_not}(\uparrow A) \notin J\}$ ♦

The function $T_{.J}$ is clearly monotonic, and we have proved that, independently of J, $T_{.J}$ is continuous. Let ϕ be the empty interpretation. The sequence of *approximate models* of P is defined as follows.

$M_0 = T_{.\phi}\uparrow\omega$
$M_{i+1} = T_{.M_i}\uparrow\omega$

Results reported below assure that the sequence has a fixpoint.

Theorem 2.1. $\forall\ k \geq 0$, if M_{2k} is not the fixpoint, then $M_{2k+2} \subset M_{2k}$. ♦

Theorem 2.2. $\forall\ k \geq 0$, if M_{2k+2} is not the fixpoint, then $M_{2k+1} \subset M_{2k+3}$. ♦

Theorem 2.3. $\forall$ k≥0, if M_{2k+1} is not the fixpoint then $M_{2k+1} \subset M_{2k+2}$. ♦

Corollary 2.1. The sequence M_k, k>0 has a fixpoint. ♦

Therefore, there exists a limit α such that $M = M_\alpha = M_{\alpha+1}$. We take M as the semantics of the program, and call it the *Iterated Reflective Model* of P (IRM_P for short). In [CL91b], Th. 11, we have proved that M is a minimal reflective model of the program P' obtained from P (indifferently, in the instantiated version or not) by replacing every positive clause

A*:–B1,...,Bn (A*=A or A*=solve(α)) with a clause

A*:–B1,...,Bn,~solve_not(A**) (A**= ↑A or A**= α respectively).

Equivalently, IRM_P is a minimal model of the program P& obtained by adding to P' the reflection axioms A:–solve(↑A) (for all non–distinguished atoms A ∈ B_{PE}).

3. NRSLD–Resolution

In this section, we introduce the definition of NRSLD–Resolution, and define an effective algorithm, which could constitute the basis for an efficient implementation. Let us first recall the definition of RSLD–Derivation for positive RP programs Pp and positive goals Gp (i.e., programs and goals not containing solve_not), which is similar to the definition of SLD–Derivation [Ll87]. An *RSLD–derivation* of Pp ∪ {Gp} consists of a (finite or infinite) sequence G_0=Gp,G_1,... of goals, a sequence C_1,C_2,... of variants of program clauses of Pp and a sequence $\theta_1,\theta_2,...$ of substitutions (mgu's and/or gmgu's), such that each G_{i+1} is derived from G_i and C_{i+1} using θ_{i+1}. Let A_1,A_2,... be the selected atoms at each step of the derivation. An *RSLD–Refutation* of Pp ∪ {Gp} is a finite RSLD–Derivation of Pp ∪ {Gp}, which has the empty clause as the last goal.

For convenience of notation in the definition of NRSLD–Resolution, let's introduce the syntactic notion of *negative counterpart* of an atom A.

Definition 3.1. Let A be an atom, θ a substitution (possibly empty or unspecified), and ε the empty goal. The *negative counterpart* $NA\theta$ of A given θ is defined as follows.

– $NA\theta = \varepsilon$ if A = solve_not(...)

– $NA\theta$ = solve_not(A*) otherwise, where A* = ↑(Aθ) if A ≠ solve(...), A* = $\alpha\theta$ if A = solve(α).

If θ is empty or unspecified $NA\theta$ is indicated simply by NA. If A=A_i is a selected atom in an RSLD–derivation, $NA_i\theta$ is shortened as $N_i\theta$. Aθ is called the *positive counterpart* of $NA\theta$. ♦

In the three definitions that follow, let P be any RP program and G any goal.

Definition 3.2. An *NRSLD–Derivation* of P ∪ {G} is composed of:

(i) an RSLD–Refutation of P ∪ {G} with selected atoms $A_1,...,A_n$ and substitutions $\theta_1,...,\theta_n$.

(ii) $\forall$ i ∈ {1,...,n} such that A_i is positive, a NRSLD–Derivation of P ∪ {←$N_i\theta_1...\theta_n$}. ♦

Definition 3.3. An NRSLD–Derivation of P ∪ {G} is *failed* if one of the following conditions holds:

(i) there exists no RSLD–Refutation of P ∪ {G}.

(ii) there exists an RSLD–Refutation of P ∪ {G} with selected atoms $A_1,...,A_n$ and substitutions $\theta_1,...,\theta_n$ and $\exists$ i ∈ {1,...,n}, such that A_i is positive and P ∪ {←$N_i\theta_1...\theta_n$} has a *successful* NRSLD–Derivation. ♦

Definition 3.4. An NRSLD–Derivation of P ∪ {G} is *successful* if the following conditions hold:

(i) there exists an RSLD–Refutation of $P \cup \{G\}$ with selected atoms $A_1,\dots,A_n$ and substitutions $\theta_1,\dots,\theta_n$.

(ii) $\forall i \in \{1,\dots,n\}$ such that A_i is positive, $P \cup \{\leftarrow N_i\theta_1\dots\theta_n\}$ has no successful NRSLD–Derivation. ♦

In the following, the negative counterparts N_i's of the selected atoms A_i's in Definitions 3.2–3.4 are called *implicit negative subgoals*.

Example. Consider the following RP program.

```
p(X,Y):-q(X,Y).                          (1)
q(a,b).                                  (2)
q(b,b).                                  (3)
solve_not(<p>("b","b")):-s(c).           (4)
s(c).                                    (5)
```

The goal ?–p(a,b) has a successful NRSLD–derivation, since it has an RSLD–Refutation via clauses (1)–(2), while solve_not(<q>("a","b")) and solve_not(<p>("a","b")) both fail. On the contrary, ?–p(b,b) has a failed NRSLD–derivation, since it has an RSLD–Refutation via clauses (1) and (3), but solve_not(<p>("b","b")) succeeds via clauses (4) and (5). ♦

4. Properties of NRSLD–Resolution

In the following, let P be a Reflective Prolog program, IRM_P the Iterated Reflective Herbrand Model of P and G a goal. We will often shorten "successful NRSLD–derivation" as sNd.

Definition 4.1. A *computed answer* for $P \cup \{G\}$ is the substitution obtained by restricting the composition $\theta_1\dots\theta_n$ to the variables of G, where $\theta_1,\dots,\theta_n$ is the sequence of mgu's and/or gmgu's used in a successful NRSLD–derivation of $P \cup \{G\}$. ♦

Definition 4.2. The *success set* of P is the set of all $A \in B_{PE}$ such that $P \cup \{\leftarrow A\}$ has a successful NRSLD–derivation. ♦

Definition 4.3. Let G be a goal $\leftarrow A_1,\dots,A_k$ and θ an answer for $P \cup \{G\}$. We say that θ is a *correct answer* for $P \cup \{G\}$ if $IRM_P \models \forall((A_1 \wedge \dots \wedge A_k)\theta)$. ♦

We remind the reader that IRM_P is a minimal model of P& (see Section 2). That means, for every clause C of P&, $IRM_P \models \forall(C)$.

4.1. Completeness

Definition 4.4. An atom A is *coherent* in P if every sNd of $P \cup \{\leftarrow A\}$ produces the identity substitution as computed answer. ♦

If a goal A is coherent in P and has a sNd, then every ground instance of A has a sNd.

Lemma 4.1. Let A be an atom which is coherent in P. Let θ be a substitution such that $A\theta$ is ground. Then there exists a sNd of $P \cup \{\leftarrow A\}$ iff there exists a sNd of $P \cup \{\leftarrow A\theta\}$.

Proof Suppose that A has variables $x_1,\dots,x_n$ (each of which can be either an object variable or a metavariable). Let $\theta = \{x_1/a_1,\dots,x_n/a_n\}$. Since A is coherent, every sNd of $P \cup \{\leftarrow A\}$ does not bind $x_1,\dots,x_n$. Then, given such a derivation, we obtain a sNd of $P \cup \{\leftarrow A\theta\}$ by replacing x_i with a_i in this derivation. Vice versa, given a sNd of $P \cup \{\leftarrow A\theta\}$, we obtain a sNd of $P \cup \{\leftarrow A\}$ by replacing a_i with x_i. ♦

In the following, we assume every implicit negative subgoal $N_i\theta_1\dots\theta_n$ of point (ii) of Definition 3.2 to be *coherent*.

An *unrestricted* NRSLD–derivation is an NRSLD–derivation, except that the substitutions θ_i are no longer required to be most general unifiers or g–unifiers, but are only required to be unifiers or g–unifiers.

Lemma 4.2. (Extended Mgu Lemma) Suppose that $P \cup \{G\}$ has an unrestricted sNd. Then $P \cup \{G\}$ has a sNd of the same length with mgu's and/or gmgu's $\theta'_1,\dots,\theta'_n$ such that, if $\theta_1,\dots,\theta_n$ are the unifiers and/or g–unifiers from the unrestricted derivation, then there exists a substitution γ such that $\theta_1\dots\theta_n = \theta'_1\dots\theta'_n\gamma$. ♦

Lemma 4.3. (Extended Lifting Lemma) Let G be a goal and θ a substitution. Suppose there exists a sNd of $P \cup \{G\theta\}$. Then there exists a sNd of $P \cup \{G\}$ of the same length such that, if $\theta_1,\dots,\theta_n$ are the mgu's and/or gmgu's from the derivation of $P \cup \{G\theta\}$, and $\theta'_1,\dots,\theta'_n$ are the mgu's and/or gmgu's from the derivation of $P \cup \{G\}$, then there exists a substitution γ such that $\theta\theta_1\dots\theta_n = \theta'_1\dots\theta'_n\gamma$. ♦

The proofs of Lemmas 4.2 and 4.3 are exactly like the corresponding ones in [CL91a], except that "RSLD–refutation" is replaced by "successful NRSLD–Derivation".

Lemma 4.4. $\forall$ k≥0, $\forall A \in M_{2k+1}$, if solve_not($\uparrow$A) has no RSLD–refutation, then $P \cup \{\leftarrow A\}$ has a sNd.

Proof By the definition of $T_{P.J}$, $A \in M_{2k+1}$ implies $A \in T_{P.M_{2k}}\uparrow n$, for some n. We prove by induction on n that $P \cup \{\leftarrow A\}$ has a sNd.

base step n =1; by the definition of $T_{P.J}$, $A \in T_{P.M_{2k}}\uparrow 1$ implies that A is a ground instance of a unit clause C_1 of P of the form $A' \leftarrow$, with mgu or gmgu θ_1, where either $A'\theta_1 = A$ or $A' = \text{solve}(\alpha)$ and $\uparrow A = \alpha\theta_1$, or else $A = \text{solve}(\alpha)$ and $\uparrow(A'\theta_1) = \alpha$. Since solve_not($\uparrow$A) has no RSLD–Refutation (and therefore, by definition, has no sNd), by the definition of NRSLD–resolution, $P \cup \{\leftarrow A\}$ has a sNd with selected atom $A_1 = A$, mgu or gmgu θ_1, input clause C_1.

ind. step Suppose that $\forall B \in T_{P.M_{2k}}\uparrow n$, B has sNd. Consider $A \in T_{P.M_{2k}}\uparrow n+1$. By the definition of $T_{P.J}$, $A \in T_{P.M_{2k}}\uparrow n+1$ implies that there is a clause C in P of the form $A' \leftarrow B_1,\dots,B_q$ ($q \geq 0$), and mgu or gmgu θ, where either $A'\theta = A$, or $A' = \text{solve}(\alpha)$ and $\uparrow A = \alpha\theta$, or else $A = \text{solve}(\alpha)$ and $\uparrow(A'\theta) = \alpha$, and $\{B_1\theta,\dots,B_q\theta\} \subseteq T_{P.M_{2k}}\uparrow n$. By the induction hypothesis, $P \cup \{\leftarrow B_i\theta\}$ has a sNd, for i=1,...,q. Since each $B_i\theta$ is ground, these derivations can be combined into a derivation of $P \cup \{\leftarrow (B_1,\dots,B_q)\theta\}$. Thus $P \cup \{\leftarrow A\}$ has an unrestricted NRSLD–derivation which is successful since solve_not($\uparrow$A) has no RSLD–refutation. We can apply the extended lifting lemma to obtain a successful NRSLD–derivation of $P \cup \{\leftarrow A\}$. ♦

Theorem 4.1. IRM_P is contained in the success set of P.

Proof Consider the sequence of the approximate models as defined in section 2. In [CL91a] we have proved that $\forall$ k≥0, $M_{2k+1} \subseteq IRM_P$. We prove that $\forall$ k≥0, $\forall A \in M_{2k+1}$, $P \cup \{\leftarrow A\}$ has a sNd. Clearly, this implies that $\forall A \in IRM_P$ $P \cup \{\leftarrow A\}$ has a sNd, and hence A is in the success set. The proof is by induction on k.

Suppose first that k=0. By the definition of $T_{P.J}$, $A \in M_1$ implies solve_not($\uparrow$A) $\notin M_0$. By the completeness of RSLD–Resolution w.r.t. M_0 (which corresponds to RM_P) [CL91a], solve_not($\uparrow$A) $\notin M_0$ means that solve_not($\uparrow$A) has no RSLD–refutation, and thus by the definition of NRSLD–Resolution, solve_not($\uparrow$A) has no sNd. By Lemma 4.4, A has a sNd.

Now suppose that the result holds for k, i.e. $\forall D \in M_{2k+1}$, $P \cup \{\leftarrow D\}$ has a sNd. We prove that the result holds for k+1, i.e., $\forall A \in M_{2k+3}$, $P \cup \{\leftarrow A\}$ has a sNd. By the definition of $T_{P,J}$, $A \in M_{2k+3}$ implies solve_not(↑A) $\notin M_{2k+2}$. That is, one of the following cases holds:

(i) solve_not(↑A) $\notin M_0$. In this case, by the completeness of RSLD–Resolution w.r.t. M_0, solve_not(↑A) has no RSLD–Refutation, and therefore solve_not(↑A) has no sNd.

(ii) solve_not(↑A) $\in M_0$. In this case, by the completeness of RSLD–Resolution w.r.t. M_0, solve_not(↑A) has a RSLD–Refutation. Let $A_1,...,A_n$ be the selected atoms of such a refutation, and $\theta_1,...,\theta_n$ the resulting substitutions. By Theorem 3 in [CL91a] (which is the counterpart of Theorem 7.4 in [Ll87]), $\forall$ $i \le n$ $[A_i\theta_1...\theta_n] \subseteq M_0$, where $[A_i\theta_1...\theta_n]$ is the set of all the instances of $A_i\theta_1...\theta_n$. For solve_not(↑A) not to belong to M_{2k+2}, there must exist $j \le n$ such that [solve_not(↑(A_j-$\theta_1...\theta_n$))] $\subseteq M_{2k+2}$. By the induction hypothesis and by Lemma 4.1, every instance of solve_not(↑($A_j\theta_1...\theta_n$)) has sNd and therefore, by the definition of NRSLD–Resolution, solve_not(↑A) has no sNd. ♦

In both cases, by Lemma 4.4 A has a sNd.

Theorem 4.2. Suppose that $IRM_P \not\models \forall (P \cup \{G\})$. Then there exists a sNd of $P \cup \{G\}$.

Proof Let G be the goal $\leftarrow A_1,...,A_k$. Since $IRM_P \not\models \forall (P \cup \{G\})$, $\forall(G)$ is false w.r.t. IRM_P. Thus $\{A_1\theta,...,A_k\theta\} \subseteq IRM_P$ for some substitution θ. By Th. 4.1, there is a successful derivation for $P \cup \{A_i\theta\}$, for i=1,...,k. Since each $A_i\theta$ is ground, we can combine these derivations into a successful derivation for $P \cup \{G\theta\}$. Finally, we apply the extended lifting lemma. ♦

Lemma 4.5. Let A be an atom. Suppose that $IRM_P \models \forall(A)$. Then there exists a successful NRSLD–derivation of $P \cup \{\leftarrow A\}$ with the identity substitution as the computed answer.

Proof Suppose that A has variables $x_1,...,x_n$ (each of which can indifferently be either an object variable or a metavariable). Let $a_1,...,a_n$ be distinct constants appearing neither in P nor in A such that the substitution $\theta = \{x_1/a_1,...,x_n/a_n\}$ satisfies the substitution rules given in [CL91a, Def. 4.1]. Then it is clear that $IRM_P \models A\theta$. Since $A\theta$ is ground, Theorem 4.1 shows that $P \cup \{\leftarrow A\theta\}$ has a sNd. Since a_i's appear neither in P nor in A, by replacing a_i with x_i in this derivation, we obtain a sNd of $P \cup \{\leftarrow A\}$ with the identity substitution as the computed answer. ♦

Now the main completeness result, proved in a way similar to classical SLD–resolution.

Theorem 4.3. (Completeness of NRSLD–resolution) For every correct answer θ for $P \cup \{G\}$, there exists a computed answer σ for $P \cup \{G\}$ and a substitution γ such that $\theta = \sigma\gamma$.

Proof Suppose G is the goal $\leftarrow A_1,...,A_k$. Since θ is correct, $IRM_P \models \forall((A_1 \wedge ... \wedge A_k)\theta)$. By Lemma 4.5, there exists a successful NRSLD–derivation of $P \cup \{\leftarrow A_i\theta\}$ such that the computed answer is the identity, for i=1,...,k. We can combine these derivations into a successful NRSLD–derivation of $P \cup \{\leftarrow G\theta\}$ such that the computed answer is the identity.

Suppose that the sequence of mgu's and/or gmgu's of the derivation of $P \cup \{\leftarrow G\theta\}$ is $\theta_1,...,\theta_n$. Then $G\theta\theta_1...\theta_n = G\theta$. By the extended lifting lemma, there exists a successful NRSLD–derivation of $P \cup \{\leftarrow G\}$ with mgu's and/or gmgu's $\theta'_1,...,\theta'_n$ such that $\theta\theta_1...\theta_n = \theta'_1...\theta'_n\gamma'$, for some substitution γ'. Let σ be $\theta'_1...\theta'_n$ restricted to the variables in G. Then $\theta = \sigma\gamma$, where γ is an appropriate restriction of γ'. ♦

In actual application, completeness of NRSLD–Resolution is, of course, affected by the incompleteness of first–order logic (in this sense Theorem 4.3 is not, strictly speaking, a full completeness result). Consider in fact point (ii) of Definition 3.4 (successful derivation). The existence of a sNd for a goal G requires the non–existence of a sNd for every implicit negative subgoal N_i. This, by point (i) of the Definition, is true if N_i has no RSLD–Refutation. It may be however the case that N_i has no RSLD–refutation, but does not fail finitely; in such a case, G has a successful NRSLD–derivation that cannot be found in finite time.

The coherence condition on completeness requires that the implicit negative subgoals succeed with the empty substitution as computed answer, which is a weaker condition than groundness. By the definition of NRSLD–Resolution, the order in which subgoals are executed does not affect coherence, since implicit negative subgoals are instantiated by the computed answer obtained in part (i) of the derivation. This is why we get the independence of the computation rule, as shown in section 4.3.

4.2. Soundness

Theorem 4.4. (Soundness of NRSLD–Resolution). Every computed answer for $P \cup \{G\}$ is a correct answer for $P \cup \{G\}$.

Proof Let G be the goal $\leftarrow A_1,\ldots,A_k$ and $\theta_1,\ldots,\theta_n$ the sequence of mgu's and/or gmgu's used in a successful NRSLD–derivation of $P \cup \{G\}$. We prove that $\forall((A_1,\ldots,A_k)\theta_1\ldots\theta_n)$ is a correct answer by induction on the length of the derivation.

Notice that, being IRM_P a model of P&, and since P& contains the reflection axioms,
$IRM_P \models \forall(A\theta \leftarrow \mathrm{solve}(\uparrow(A\theta)))$, for every atom A and substitution θ.
This means, $IRM_P \models \forall(\mathrm{solve}(\uparrow(A\theta)))$ implies $IRM_P \models \forall(A\theta)$.

Notice also that, by Definition 3.1, for every atoms A and B and substitution θ, if $B\theta$ is an istance of A or if $A = \mathrm{solve}(\alpha)$) and $\uparrow(B\theta)$ is an instance of α, then $N_{B\theta}$ is an instance of N_A.

Suppose first that n = 1. This means: G is a goal of the form $\leftarrow A_1$, P has a unit clause of the form $A\leftarrow$, where either $A_1\theta_1 = A\theta_1$, or else $A = \mathrm{solve}(\alpha)$ and $\uparrow(A_1\theta_1) = \alpha\theta_1$, or else $A_1 = \mathrm{solve}(\alpha)$ and $\uparrow(A\theta_1) = \alpha\theta_1$, and $N_1\theta_1$ (if non–empty) has no successful NRSLD–derivation. If P has the unit clause $A\leftarrow$, then P& has a clause, say C*, of the form $A\leftarrow \sim N_A$. If $N_1\theta_1$ has no sNd, then by the completeness of NRSLD–Resolution, $N_1\theta_1$ is unsatisfiable in IRM_P. Being $N_1\theta_1$ an instance of N_A, $IRM_P \not\models \forall(N_A)$. Since $IRM_P \models \forall(C^*)$, then $IRM_P \models \forall(A)$ which implies $IRM_P \models \forall(A_1\theta_1)$.

Next, suppose that the result holds for computed answers coming from derivations of length n–1, and that $\theta_1,\ldots,\theta_n$ is the sequence of mgu's and/or gmgu's used in a derivation of length n. Let C be the first input clause and A_m the selected atom of G.

Let C be $A\leftarrow B_1,\ldots,B_q$ $(q \geq 0)$, where either $A_m\theta_1\ldots\theta_n = A\theta_1\ldots\theta_n$ or $A = \mathrm{solve}(\alpha)$ and $\uparrow(A_m\theta_1\ldots\theta_n) = \alpha\theta_1\ldots\theta_n$, or $A_m = \mathrm{solve}(\alpha)$ and $\uparrow(A\theta_1\ldots\theta_n) = \alpha\theta_1\ldots\theta_n$. If the overall derivation is successful, then $N_m\theta_1\ldots\theta_n$ (if non–empty) has no successful NRSLD–derivation. If P has the clause $A\leftarrow B_1,\ldots,B_q$, then P& has a clause, say C*, of the form $A\leftarrow B_1,\ldots,B_q,\sim N_A$.

By the induction hypothesis, $IRM_P \models \forall((A_1,\ldots,A_{m-1},B_1,\ldots,B_q,A_{m+1},\ldots,A_k)\theta_1\ldots\theta_n)$. Thus, if $q > 0$, $IRM_P \models \forall((B_1,\ldots,B_q)\theta_1\ldots\theta_n)$. If $N_m\theta_1\ldots\theta_n$ has no successful NRSLD–derivation, then by the completeness of NRSLD–Resolution, $N_m\theta_1\ldots\theta_n$ is unsatisfiable in IRM_P.

Being $N_m\theta_1\ldots\theta_n$ an instance of N_A, $IRM_P \not\models \forall(N_A)$. Since $IRM_P \models \forall(C^*)$, $IRM_P \models \forall((A)\theta_1\ldots\theta_n)$, which implies $IRM_P \models \forall((A_m)\theta_1\ldots\theta_n)$.
Hence $IRM_P \models \forall((A_1,\ldots,A_k)\theta_1\ldots\theta_n)$. ♦

Corollary 4.1. Suppose that there exists a successful NRSLD–derivation of $P \cup \{G\}$. Then $P \cup \{G\}$ is unsatisfiable in IRM_P. ♦

Corollary 4.2. The success set of P is contained in IRM_P. ♦

Theorem 4.5. The success set of P is equal to IRM_P. ♦

Proof The result follows immediately from Theorem 4.3 and Theorem 4.4. ♦

4.3. Independence of the Computation Rule

Consider the Definition 3.2 of NRSLD–Derivation. It is easy to see that the order in which the N_i's are selected is irrelevant. In fact: every N_i is instantiated by the computed answer $\theta_1 \ldots \theta_n$ produced in point (i); their derivations are independent; as soon as an N_i succeeds for some i, the overall derivation fails. Instead, we need a computation rule R to select atoms during the RSLD–Refutation of $P \cup \{G\}$ in point (i). Then, an NRSLD–derivation of $P \cup \{G\}$ via R is an NRSLD–derivation in which the computation rule R is used to select atoms both in the RSLD–Refutation of $P \cup \{G\}$ in point (i), and in the NRSLD–derivations of every N_i in point (ii). An R–computed answer for $P \cup \{G\}$ is a computed answer which comes from a successful NRSLD–derivation of $P \cup \{G\}$ via R. Then it is clear that in order to show the independence of the computation rule for NRSLD–Resolution, it suffices to show the independence of the computation rule for RSLD–Resolution.

Given the following preliminary definition, the proofs of Lemmas and Theorems in the rest of this Section are, if omitted, the same as for corresponding ones in [Ll87, Chapt. 2, Sect. 9].

Preliminary definition. (to hold until the end of this section)
Given a clause C, let C– be the body of the clause. If the head of the clause is $B \neq \mathsf{solve}(\beta)$ then let $C+ = B$, otherwise let $C+ = \beta$. Let A be an atom, and θ a substitution. Assume $C+ = B$, $A \neq \mathsf{solve}(\alpha)$ and θ mgu of C+ and A. In this case, Aθ and C+θ indicate, as usual, the application of θ to A and C+ respectively. Assume instead that $A \neq \mathsf{solve}(\alpha)$ but $C+ = \beta$, and θ gmgu of C+ and A. In this case, Aθ stands for $\uparrow(A\theta)$, while C+θ indicate, as usual, the application of θ to C+. Assume finally that $A = \mathsf{solve}(\alpha)$ but $C+ \neq \beta$, and θ gmgu of C+ and A. In this case, C+θ stands for $\uparrow(C+\theta)$, while Aθ indicate, as usual, the application of θ to A. ♦

Lemma 4.6. (Extended Switching Lemma). Let P be an RP program ad G a goal. Suppose that $P \cup \{G\}$ has an RSLD–Refutation $G_0{=}G, G_1, \ldots, G_q, G_{q+1}, \ldots, G_n$ where G_n is the empty goal, with input clauses $C_1, \ldots, C_n$ and mgu's or gmgu's $\theta_1, \ldots, \theta_n$. Suppose that
G_{q-1} is $\leftarrow A_1, \ldots, A_{i-1}, A_i, \ldots, A_{j-1}, A_j, \ldots, A_k$
G_q is $\leftarrow (A_1, \ldots, A_{i-1}, C_{q}-, \ldots, A_{j-1}, A_j, \ldots, A_k)\theta_q$
G_{q+1} is $\leftarrow (A_1, \ldots, A_{i-1}, C_{q}-, \ldots, A_{j-1}, C_{q+1}-, \ldots, A_k)\theta_q\, \theta_{q+1}$
Then there exists an RSLD–Refutation of $P \cup \{G\}$ in which A_j is selected in G_{q-1} instead of A_i, and A_i is selected in G_q instead of A_j. Furthermore, if σ and σ' are the computed answers for $P \cup \{G\}$ from the given refutation and the new refutation respectively, then Gσ is a variant of Gσ'. ♦

Theorem 4.6. (Independence of the Computation rule for RSLD–Resolution). Let P be an RP program ad G a goal. Suppose there is an RSLD–Refutation of $P \cup \{G\}$ with computed answer σ. Then, for any computation rule R, there exists an RSLD–Refutation of $P \cup \{G\}$ via R with R–computed answer σ' such that Gσ' is a variant of Gσ. ♦

Lemma 4.7. (Independence of the Computation rule for NRSLD–Resolution). Let P be an RP program ad G a goal. Suppose there exists a sNd of $P \cup \{G\}$ with computed answer σ. Then, for any computation rule R, there exists a sNd of $P \cup \{G\}$ via R with R–computed answer σ' such that $G\sigma'$ is a variant of $G\sigma$.
Proof. Immediate from the considerations at the beginning of this Section. ♦

Let the R–success set of a RP program P be the set of all $A \in B_{PE}$ such that $P \cup \{\leftarrow A\}$ has a successful NRSLD–derivation via R.

Theorem 4.7. Let P be an RP program ad R a computation rule. Then the R–success set of P is equal to IRM_P.
Proof. The Theorem follows immediately from Theorem 4.5 and Lemma 4.7 above. ♦

5. A Computational Schema for NRSLD–Resolution

To show that NRSLD–Resolution can be easily implemented, we propose the following computational scheme. Consider the usual resolution step during the derivation of $P \cup \{G\}$, with selected atom A_i, input clause C (where C is $B{:-}B_1,\dots,B_m$), and mgu/gmgu θ_i. From a goal

$:-A_1,\dots,A_{i-1},A_i,A_{i+1},\dots,A_k$

an input clause C, and a unifier θ_i (in case of RP, mgu or gmgu) we usually get

$:-(A_1,\dots,A_{i-1},B_1,\dots,B_m,A_{i+1},\dots,A_k)\theta_i$

In our new scheme, we get

$:-(A_1,\dots,A_{i-1},B_1,\dots,B_n,A_{i+1},\dots,A_k \,\$\, N_i\,;)\theta_i$

In the scheme, '$' acts as a separator between the goals of point (i) of Def. 3.2, and those of point (ii). After more steps (assuming to select atoms before '$'), we get:

$:-(A_1,\dots,A_r \,\$\, N_{i1}\,;\dots;\,N_{is})\theta_{i1}\dots\theta_{is}$

where ';' means that negative subgoals are in disjunction rather than in conjunction. E.g., if selecting N_{i1}, given an input clause C' (where C' is $D{:-}D_1,\dots,D_h$) applicable to N_{i1} with mgu/gmgu δ_1, we get

$:-(A_1,\dots,A_r \,\$\, (D_1,\dots,D_{j-1},D_j,D_{j+1},\dots,D_h \,\$\,)\delta_1\,;N_{i2};\dots;\,N_{is})\theta_{i1}\dots\theta_{is}$

At this step, no negative subgoal is generated after the new $, because N_{i1} is negative. Assuming now to select the positive atom D_j, with input clause C" (where C" is $E{:-}E_1,\dots,E_f$) and mgu/gmgu δ_2 we get:

$:-(A_1,\dots,A_r \,\$\, (D_1,\dots,D_{j-1},E_1,\dots,E_f,D_{j+1},\dots,D_h \,\$\, N_{j1})\delta_1\delta_2\,;N_{i2};\dots;\,N_{is})\theta_{i1}\dots\theta_{is}$

and so on.

The conditions for the derivation to be successful are the following: (i) no further resolution step is possible; (ii) the conjunction of subgoals before the first '$' has been reduced to the empty goal; (iii) none of the goals in the topmost disjunction after the first '$' succeeded. Otherwise (the empty goal has not been derived after the first '$', or one of the topmost N_i's succeeded) the derivation is failed.

As usual, in order to avoid loop problems the derivation should be *fair*, i.e. [Ll87], unless it is failed every atom B in the derivation (in this scheme, every atom B before or after the '$', at any nesting level) ought to be selected within a finite number of steps.

In the context of a prototype implementation, we are experimenting various strategies, including extensions to the usual depth–first strategy.

6. NRSLD–Resolution as a Procedural Semantics for NAF

Both soundness and completeness of SLDNF–Resolution are affected by the floundering of negative subgoals [Ll87]. As for completeness, the problem can be (partially) eliminated by adopting *safe* computation rules, i.e. rules which select negative subgoals only when (and if) ground. Safe computation rules are in contrast with fair search strategies: for instance, the breadth–first strategy is fair, but not safe.

Negation as failure can easily be expressed by means of metalevel negation without using either 'cut' or 'fail, as follows:

```
(i)  solve(<not>$G).
(ii) solve_not(<not>$G):–solve($G).
```

In [CL91b] we have proved that adding clauses (i) and (ii) to a weakly stratified general program P, the IRM of the resulting program P* is equal (modulo a natural correspondence) to the Well–Founded Model (WFM) of P. In [CL91c] we have extended this result to every general program.

Given this semantic correspondence, results about NRSLD–Resolution w.r.t. the IRM of P* immediately extend to corresponding results w.r.t. the WFM of P. That means, NRSLD–Resolution is a sound and complete resolution procedure for NAF, which is independent of the computation rule and can thus replace (with advantage) SLDNF–Resolution.

In order to use NRSLD–Resolution for NAF, some simplifications are possible, observing that:

- ◊ every subgoal not G has an RSLD–Refutation by clause (i) and by the (implicit) reflection axiom; clause (ii) is applicable to every subgoal not G; therefore, from $A_1,\ldots,A_{i-1},\text{not } G,A_{i+1},\ldots,A_k \ \$ \ N_1;\ldots; N_m$ with not G as the selected atom, we derive $A_1,\ldots,A_{i-1},A_{i+1},\ldots,A_k \ \$ \ N_1;\ldots; N_m; G$
- ◊ clause (ii) is applicable to subgoals of the form not G only; therefore, we do not need generating the negative counterpart of the other subgoals.

Consider for instance the following general program P, and the RP program P* obtained by adding (i)–(ii) to P.

```
p(X):–q(X), not r(X).        1
q(a).                        2
```

The computational steps necessary to prove ?–p(X) in P and P* are summarized below, where the input clause is indicated in brackets, and the mgu/gmgu in curly brackets. We assume to adopt the left–to–right computation rule.

```
P  (SLDNF–Resolution)
?–p(X)                    (1)      {}
?–q(X), not r(X)          (2)      {X/a}
?–not r(a)
    ?–r(a)
    fail
X/a
P*  (simplified NRSLD–Resolution)
?–p(X).                   (1)      {}
?–q(X), not r(X) $        (2)      {X/a}
?–$ r(a)
?–$ fail
X/a
```

The example shows that simplified NRSLD–Resolution requires the same number of steps as SLDNF–Resolution. The advantages are the independence of the computation rule, and less problems of floundering. Consider in fact the same program as above, with the order of subgoals in clause 1 reversed.

```
p(X):–not r(X),q(X).                                  1'
```

With SLDNF–Resolution, the goal ?–q(X) flounders. With NRSLD–Resolution, we have:

```
?–p(X).                                   (1')        {}
?–not r(X), q(X) $                        (i)+(ii)    {}
?–q(X) $ r(X)                             (2)         {X/a}
?–$ r(a)
?–$ fail
X/a
```

Moreover, the condiuon of coherence is weaker than floundering. In fact, let's consider this modified version of the program.

```
p(X):–not r(X),q(X).                                  1
q(X).                                                 2'
```

With SLDNF–Resolution, the goal ?–p(X) flounders, whatever the order of subgoals in (1), and whatever the computation rule. With NRSLD–Resolution, we have:

```
?–p(X).                                   (1)         {}
?–not r(X), q(X) $                        (i)+(ii)    {}
?–q(X) $ r(X)                             (2)         {}
?–$ r(X)
?–$ fail
yes
```

References

[CL89] Costantini S. and Lanzarone G.A., *A Metalogic Programming Language*, in: *Logic Programming*, Proceedings of the Sixth International Conference, MIT Press, 1989.

[CL91a] Costantini S. and Lanzarone G.A., *A Metalogic Programming Approach: Language, Semantics, and Applications*, submitted paper.

[CL91b] Costantini S. and Lanzarone G.A., *Metalevel Negation and Non–Monotonic Reasoning*, to appear on *Methods of Logic in Computer Science*. An abridged version appears in: Proceedings of the Workshop on Logic Programming and Non–Monotonic Reasoning, Austin (TX), Nov. 1–2, 1990.

[CL91c] Costantini S. and Lanzarone G.A., *Semantic Considerations on Metalevel Negation and Negation as Failure*, submitted paper.

[Co89] Costantini S., *Formal Definition of Reflective Prolog*, Internal Report 49/89, Computer Science Dept., Univ. of Milano (1989).

[Co90] Costantini S., *Semantics of a Metalogic Programming Language*, International Journal of Foundations of Computer Science, vol.1, n.3, Sept. 1990.

[Ll87] Lloyd J.W., *Foundations of Logic Programming*, (Second, Extended Edition), Springer–Verlag, Berlin, 1987.

[Pr89] Przymusinski T., *Every Logic Program has a Natural Stratification and an Iterated Fixed Point Model*, in: Proceedings of the Eight Symposium on Principles of Database Systems, ACM SIGACT–SIGMOD, 1989.

[PP90] Przymusinska H. and Przymusinski T., *Semantic Issues in Deductive Databases and Logic Programs*, in: R.B. Banerji (ed.) *Formal Techniques in Artificial Intelligence, a Sourcebook*, Elsevier Science Publisher B.V. (North Holland), 1990, 321–367.

[VRS88] Van Gelder A., Ross K., and Schlipf J.S., *Unfounded Sets and Well–Founded Semantics for General Logic Programs*, in: Proceedings of the Symposium on Principles of Database Systems, ACM SIGACT–SIGMOD, 1988.

PROBABILISTIC LOGIC PROGRAMS AND THEIR SEMANTICS

Eugene Dantsin
Department of Mathematics (1),
Electrical Engineering Institute,
Prof. Popov str. 5, St.Petersburg 197376, USSR
Email: dantsin@tor.spb.su

ABSTRACT: The aim of this paper is to generalize logic programs, for dealing with probabilistic knowledge. Using the possible-worlds approach of probabilistic logic ([Nil]), we define *probabilistic logic programs* so that their clauses may be true or false with some probabilities and goals may succeed or fail with probabilities too. Probabilistic logic programs may contain negation, their semantics agrees with negation as failure (unlike probabilistic logic which is based on the standard logical negation).

1. Introduction

The following problem is usual for logic: we know that statements $S_1,\dots,S_n$ are true and want to know whether another statement S is true or not. First-order logic and logic programming may give different answers to this problem because of different meanings of negation. But what can we say about S if we are not sure in truth of $S_1,\dots,S_n$?

There are different ways to formalize such a problem for first-order logic (fuzzy logic, some modal logics, etc.). One of the most appropriate ways is probabilistic logic ([Nil], [FHM]). Its language allows us to formulate statements such as "the probability that F is true is equal to 0.9" where F is a closed first-order formula. Such statements about probabilities have precise and clear semantics based on the possible-worlds approach. Probabilistic logic defines a natural consequence relation on the statements about probabilities and, thereby, formalizes probabilistic reasoning.

As to logic programming, a generalization of logic programs for uncertain clauses has been proposed in [Sha], namely logic programs with uncertainties. This generalization uses certainties factors for clauses and goals, i.e. degrees of belief that a clause is true or a goal succeeds. Computation of certainty factors for goals is based on the rules of fuzzy logic and can be performed by a simple meta-interpreter ([SS]). However, logic programs with uncertainties have no clear non-procedural semantics and, moreover, even have no good procedural semantics if negation appears (see discussion in [NL]).

Using the probabilistic logic approach, we define probabilistic logic programs. Their declarative semantics can be defined in precise and well-understood terms, it generalizes the semantics of logic programs and agrees with negation as failure. Therefore, probabilistic logic programs seem preferable to logic programs with uncertainties from the semantic viewpoint. As to the procedural viewpoint, computation of certainty factors is, generally, more simple than computation of probabilities but computation for probabilistic logic programs can be efficient enough too.

Though we suppose familiarity with main notions of logic programming, we give some definitions which fix the notation and terminology used below. By a *clause* we mean a first-order formula of the form

$$\forall(L_1\&\ldots\&L_m\supset A)$$

where A is an atom, $L_1,\dots,L_m$ are literals (i.e. atoms or negated atoms), $m\geq 0$, and $\forall(F)$ denotes the universal closure of F. Predicates of arity 0 are assumed too and, therefore, some atoms may be proposition letters. Such a clause is written as $A\leftarrow L_1,\dots,L_m$ and in case m=0 it may be written as A. The atom A is called the *head*, the set $L_1,\dots,L_m$ is called the *body* of this clause. A finite set of clauses is called a *logic program* or, for short, a *program*.

The declarative semantics of programs is defined in terms of *canonical models*. The canonical model of a program P is selected among all the Herbrand models of P as the only model which represents the declarative meaning of this program. Canonical models of programs without negation were defined in [EK] as minimal Herbrand models. As to programs with negation, canonical models were defined for stratified programs ([ABW], [CH], [Gel]) and some other classes (for example, [GL], [GRS], etc.), but there is no general description of the canonical model for an arbitrary program. Therefore, if we want to associate any program with its canonical model, we have to restrict the class of programs under consideration below. We shall consider the stratified programs, though we could take other known classes too. The only necessary condition is that if a program P belongs to such a class then any program P' obtained from P by adding clauses with empty bodies also belongs to the same class. The stratified programs obviously satisfy this condition. Thus, when we shall talk about a program or a set of clauses we shall mean a stratified program.

By a *goal* we mean a conjunction of the form $L_1\&\dots\&L_k$ where each L_i is either an atom or a ground negative literal. We are restricted to this class of goals for simplicity, in order to avoid the known difficulty connected with variables in negative literals. Given a program P, we associate a goal G with the problem: does G have an instance which is true in the canonical model of P (a formula is said to be true in a model if the universal closure of this formula is true in the model). We say that G *succeeds* if there exists such an instance. Otherwise we say that G *fails*. This definition also covers the case when G contains symbols not occurring in P, since any atom with such symbols is considered false in every Herbrand model of P.

2. Syntax and semantics of probabilistic programs

DEFINITION. A *probabilistic logic program* (or *probabilistic program* for short) is a triplet $\langle C,U,\pi \rangle$ whose components are defined as follows:

(i) C is a finite set of clauses, they are called *certain* clauses.

(ii) U is a finite set of clauses whose heads are ground atoms and bodies are empty, these clauses are called *uncertain*. The clauses of U must also satisfy another condition: there are no clause A of U and clause $B \leftarrow L_1,\dots,L_m$ of C such that A is an instance of B. (These conditions provide logical independence of the uncertain clauses, i.e. no uncertain clause A can be derived from other certain and uncertain clauses; the meaning of such independence will become clear below).

(iii) π is a function which assigns numbers from the unit interval to the uncertain clauses. The number $\pi(A)$ which is assigned to an uncertain clause A is called the *probability* of A. (Below we shall define the semantics which allows us to interpret $\pi(A)$ as the probability that A is true).

POSSIBLE WORLDS. A probabilistic program $\langle C,U,\pi \rangle$ is viewed as a description of some domain where all the certain clauses are true and the uncertain clauses may be true as well as false, their truth values are unknown. If U contains n clauses then there are 2^n possible combinations of unknown values of the uncertain clauses. Correspondingly, there are 2^n possible sets consisting of true clauses, i.e. sets $C \cup T$ where T runs through all the subsets of U. We think of these sets as axiom sets, as programs which describe *possible worlds* (or possible states of our domain). More precisely, by a *possible world* we mean the canonical model of a program $C \cup T$ where $T \subseteq U$. Thus, each subset of U determines some possible world (briefly, *world*).

Let W be the world corresponding to a subset T of U. Then all the clauses of T are true in W and all the clauses of U\T are false in W. The first part of this statement is obvious and the second part follows from independence of the uncertain clauses (the second item in the definition of probabilistic programs) and the standard semantics of logic programs. Indeed, if A belongs to U\T then A (treated as a goal for the program $C \cup T$) fails. Therefore, A is false in W. In other words, the world corresponding to any assignment of truth values to the uncertain clauses is consistent with this assignment. Consequently, all 2^n worlds are different.

DISTRIBUTIONS ON WORLDS. By a *distribution* on the worlds of $\langle C,U,\pi \rangle$ we mean any discrete probability function μ on the worlds (so that $\mu(W) \geq 0$ for every world W and the sum of all $\mu(W)$ is 1). We interpret $\mu(W)$ as the probability that the world W is *actual*, i.e. corresponds to actual state of the domain under consideration.

A distribution on the worlds allows us to define the probability that a clause (or an arbitrary first-order formula) is true. Given a distribution, we say that a formula F is *true is with probability* p if p is the sum of the probabilities of those worlds where $\forall(F)$ is true.

The certain clauses of $\langle C,U,\pi \rangle$ are true in all the worlds and, therefore, they are true with probability 1 for every distribution. As to the uncertain clauses, their probabilities depend on which distribution is considered. We say that a distribution μ *satisfies* $\langle C,U,\pi \rangle$ if μ determines such probabilities that each uncertain clause A is true with probability $\pi(A)$. Clearly a distribution μ satisfies $<C,U,\pi>$ if and only if μ satisfies the following system:

(1) $\mu(W)\geq 0$ for every world W;

(2) $\Sigma_W \mu(W)=1$ where W runs through all the worlds;

(3) $\Sigma_W \mu(W)=\pi(A)$ for every uncertain clause A, here W runs through those worlds whose axioms contain A, i.e. those worlds which correspond to all such subsets T of U that $A\in T$.

MAXIMUM-ENTROPY DISTRIBUTION. Is there any distribution which satisfies $<C,U,\pi>$? Yes, there is. For example, the following distribution can be constructed for every π. Let W_T be denote the world corresponding to a subset T of U. We define the distribution μ by

$$\mu(W_T) = \prod_{A\in T}\pi(A) \prod_{A\in U\setminus T}(1-\pi(A)) \qquad (*)$$

Such μ satisfies (1)-(3), this fact follows from independence of the uncertain clauses (the second item in the definition of probabilistic programs).

There are other distributions satisfying (1)-(3) too but the distribution defined by (*) is selected as the canonical distribution on the canonical distribution on the worlds of $<C,U,\pi>$. The main reason is that the distribution (*) corresponds to the case when truth values of the uncertain clauses are independent not only logically but also probabilistically. It means that the uncertain clauses $A_1,\ldots,A_n$ get the values $v_1,\ldots,v_n$ correspondingly with the probability which is equal to the product of the probabilities that A_i get the value v_i. Such an approach is natural if we have no additional information about dependencies between uncertain clauses. The distribution (*) maximizes entropy, i.e. maximizes

$$-\Sigma_W \mu(W)\, log\, \mu(W)$$

and is called the *maximum-entropy distribution* for $<C,U,\pi>$. The approach based on maximizing entropy is often applied to formalizations of probabilistic knowledge and reasoning. In particular, the maximum-entropy distribution was proposed for probabilistic logic in [Nil].

PROBABILITIES FOR GOALS. Distributions on the worlds allows us to define not only the probability that a clause is true but also the probability that a goal succeeds. Let G be a goal. If G has an instance which is true in a world W, we call W *successful* for G. Given a distribution on the worlds of $\langle C,U,\pi\rangle$, we say that G *succeeds with probability* p if p is the sum of the probabilities of the worlds successful for G. Clearly the probability that G succeeds and the probability that G is true are equal in case ground G, though in general they may differ.

Below if we talk about the probability that G succeeds and do not mention which distribution is considered, we mean the probability determined by the maximum-entropy distribution. Just such probabilities interests us when we formulate the main problem for probabilistic programs: given a probabilistic program and a goal, what is the (maximum-entropy) probability that G succeeds?

3. Example

We illustrate syntax and semantics of probabilistic programs by a simple example. Suppose we are interested in paths in some directed graph. This graph has four vertices 1,2,3 and 4, but our information about arcs is uncertain. We know that certainly there is an arc from 2 to 3 and know that possibly there are arcs from 1 to 2, from 1 to 3, from 2 to 4 and from 3 to 4. Also we know that certainly there are no other arcs.

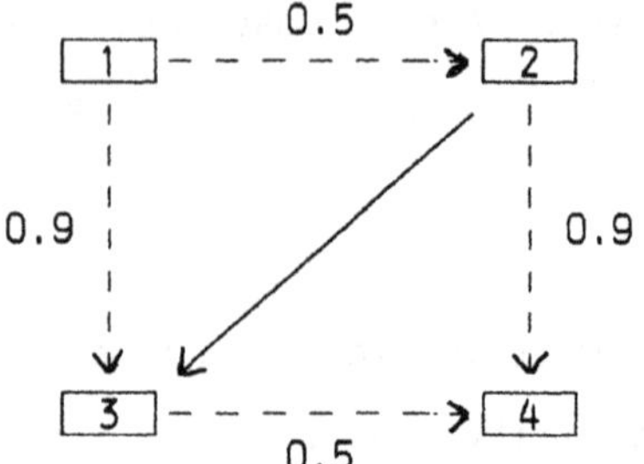

Suppose we also know the probabilities that the possible arcs exist. Namely the probabilities for the arc from 1 to 2 and the arc from 3 to 4 are equal to 0.5 and the probabilities for the arc from 1 to 3 and the arc from 2 to 4 are equal to 0.9. This knowledge can be represented by the probabilistic program $\langle C,U,\pi\rangle$ where

C consists of three clauses:

arc(2,3)

path(X,Y) ← *arc*(X,Y)

path(X,Y) ← *arc*(X,Z),*path*(Z,Y);

U is

{*arc*(1,2),*arc*(1,3),*arc*(2,4),*arc*(3,4)};

π is given by

π(*arc*(1,2)) = π(*arc*(3,4)) = 0.5

π(*arc*(1,3)) = π(*arc*(2,4)) = 0.9.

This probabilistic program describes 2^4 possible graphs and the distribution corresponding to independence of possible arcs. For example, the possible graph with *arc*(2,3) and *arc*(1,2) exists with the probability

$$0.5\cdot(1-0.9)\cdot(1-0.5)\cdot(1-0.9) = 0.0025$$

We are interested in the goal *path*(1.4). To find the probability of this goal, we can sum up the probabilities of those possible graphs in which there is a path from 1 to 4. Straightforward calculation gives the probability of *path*(1,4) to be 0.7. More intelligent ways of finding such probabilities will be discussed below.

4. Computation of probabilities of goals

Given a probabilistic program <C,U,π> and a goal G, how to compute the probability that G succeeds (the probability of G for short)? We propose a way which consists of two steps. The first step deals only with the goal G and the clauses of C and U. Computation at this step is usual enough for logic programming and can be performed by a meta-interpreter. Its result is some propositional formula which can be treated as Boolean combination of independent events. The probabilities of these events are known (they are given by π) and, therefore, the probability of the combination can be computed by means of probability theory methods. Just this computation is executed at the second step.

SUCCESS FORMULA. Now we define a *success formula* which is constructed at the first step. By propositional formula over the set U of uncertain clauses we mean first-order formulas built up from the ground atoms of U and the logical constants *true* and *false* by means of the connectivies $\neg$, &, $\vee$, $\supset$. We view such a formula as a compact representation of the truth values assignments on U, namely those assignments which make the formula true.

Let F be a propositional formula over U and T be a subset of U. We identify T with the assignment of *true* to the clauses of T and *false* to the clauses of U\T. If this assignment make F true then we say that T *satisfies* F and also say that the world W corresponding to T *satisfies* F. The set of all the worlds satisfying F is called the set *represented by* F. Of course, different formulas can represent the same set of worlds.

Let G be a goal. We call a formula F a *success formula for* G if F represents the set of all the worlds which are successful for G. We can illustrate this notion by the example described above. A success formula for the goal *path*(1,4) is

$$(arc(1,2)\&arc(3,4)) \vee (arc(1,2)\&arc(2,4)) \vee (arc(1,3)\&arc(3,4))$$

or any equivalent formula. A success formula for G shows how G depend on the uncertain clauses. In particular, if a success formula is a tautology or unsatisfiable then it means that G does not depend on these clauses. Clearly the probability of G is the sum of the probabilities of the worlds represented by a success formula for G.

FIRST STEP: CONSTRUCTING OF SUCCESS FORMULAS. How to construct a success formula? One of ways is immediately extracted from the definition, namely a straightforward procedure in the manner of "generate and test". This procedure enumerates all 2^n subsets of on U and, for each of them, solves the problem whether the corresponding world is successful for G or not. This problem can be solved by the standard interpreter for logic programs. If the world corresponding to a subset T is successful then the procedure forms and saves the conjunction $A'_1\&...\&A'_n$ where A'_i denotes uncertain clause A_i in case A_i belongs to T and A'_i denotes $\neg A_i$ otherwise. The disjunction of all such conjunctions is obviously a success formula for G.

Of course, running of the logic programs interpreter 2^n times is not efficient in case of large n. It seems more intelligent not to repeat those computations which are independent of uncertain clauses. We give another procedure more satisfactory from this point of view and call it the *interpreter for programs with uncertain clauses*.

This interpreter is described in the form of a logic program including some built-in predicate which is almost the same as the well-known *set_of* predicate. The predicate used here is denoted by *set_of* too and defined as follows. We say that a ground term T is a *solution* for a variable X of a goal G if substitution of T for X transforms G into the goal G′ such that G′ succeeds. The *set_of* predicate computes all the solutions for a given variable of a given goal and returns the disjunction of these solutions (the traditional version of *set_of* returns the list of the solutions). So *set_of*(X,G,D) means that D is a term representing the disjunction of all the solutions for X and G.

INTERPRETER FOR PROGRAMS WITH UNCERTAIN CLAUSES. It is a logic program which consists of clauses defining four predicates:

clause/2

uncertain_clause/2

success/2

success_formula/2

(i) The *clause* predicate is usual for meta-interpreters. This predicate represents the certain clauses of $\langle C,U,\pi\rangle$. Every certain clause $A \leftarrow L_1,\ldots,L_n$ is represented by some clause of our interpreter, namely by

clause$(A, L_1\&\ldots\&L_n)$

in case n>0, or by

clause(A, *true*)

in case n=0.

(ii) Every uncertain clause A of $\langle C,U,\pi\rangle$ is represented by

uncertain_clause(A, *true*)

(iii) An atom *success*(G,F) means that F is a propositional formula which represents one or more worlds successful for the goal G. The meaning of the *success* predicate is more clear from its definition:

success$(G_1\&G_2, F_1\&F_2)$ <- *success*(G_1,F_1), *success*(G_2,F_2)

success(A,F) <- *clause*(A,B), *success*(B,F)

success(A, *true*) <- *clause*(A, *true*)

success(A,A) <- *uncertain_clause*(A,*true*)

success(¬A,¬D) <- *set_of*(F,*success*(A,F),D)

(iv) An atom *success_formula*(G,S) means that S is a success formula for G. This predicate computes all the solutions for F of the goal *success*(G,F) and returns their disjunction:

success_formula(G,S) <- *set_of*(F,*success*(G,F),S)

SECOND STEP: COMPUTING OF PROBABILITIES. If we are given a success formula for G and probabilities of the uncertain clauses then the problem of computing of the probability of G is solved by means of probability theory. Here we say only a few words about it. An obvious approach to solving is based on applying the well-known formula for the disjunction of events:

$$Prob(A \vee B) = Prob(A) + Prob(B) - Prob(A\&B)$$

The result is obtained by applying this formula (more exactly its generalization for n events) to a success formula in disjunctive normal form. Clearly such an algorithm requires exponential time. This problem is NP-hard ([GKP],[FHM]) and, therefore, it seems more reasonable to solve this problem approximately. Just such an algorithm based on Monte-Carlo technique was proposed in [DK]. If we fix some precision of result then the running time of this algorithm depends linearly on the length of a success formula.

References

[ABW] Apt,K.R., Blair,H., Walker,A. Towards a theory of declarative knowledge. In: J.Minker (ed.), Foundations of Deductive Databases and Logic Programming, Morgan Kaufmann Publishers, Los Altos, CA, 1988, 89-148.

[CH] Chandra,A., Harel,D. Horn clause queries and generalizations. The

Journal of Logic Programming 1, 1985, 1-15.

[EK] van Emden,M.H., Kowalski,R.A. The semantics of predicate logic as a programming language. Journal ACM, 23(4), 1976, 733-742.

[FHM] Fagin,R., Halpern,J.Y., Megiddo,N. A logic for reasoning about probabilities. Proc. 3rd IEEE Symp.on Logic in Computer Science, 1988, 277-291. A revised and expanded version of this paper as IBM Research Report RJ 6190, April 1988.

[Gel] van Gelder,A. Negation as failure using tight derivations for general logic programs. In: J.Minker (ed.), Foundations of Deductive Databases and Logic Programming, Morgan Kaufmann Publishers, Los Altos, CA, 1988, 149-176.

[GKP] Georgakopoulos,G., Kavvadias,D., Papadimitriou,C.H., Probabilistic satisfiability, Journal of Complexity 4:1, 1988, 1-11.

[GL] Gelfond,M., Lifschitz,V. The stable model semantics for logic programming. Logic Programming: Proc. 5th International Conference and Symposium, 1988.

[GRS] van Gelder,A., Ross,K., Schlipf,J.S. Unfounded sets and well-founded semantics for general logic programs. Proc 7th on Principles of Database Systems, 1988, 221-230.

[Nil] Nilsson,N.J. Probabilistic logic, Artificial Intelligence 18, 1(1986), 71-87.

[NL] Newsletter of the Association for Logic Programming, v.4/1 (1991), pp.11-12.

[Sha] Shapiro,E., Logic programs with uncertainties: a tool for implementing rule-based systems, Proc. 8th International Joint Conference on Artificial Intelligence, Karlsruhe, Germany, 1983, pp.529-532.

[SS] Sterling,L., Shapiro,E., The art of Prolog, 1986.

Implementation of Prolog as binary definite Programs

Bart Demoen bimbart@kulcs.uucp
K.U.Leuven Department of Computer Science
Celestijnenlaan 200 A B-3030 Leuven Belgium

André Mariën andre@sunbim.uucp
BIM Kwikstraat 4 B-3078 Everberg Belgium

1. Abstract

We describe how Prolog can be implemented with a simplified WAM, by transforming Prolog programs to binary definite programs. This removes all operations from the WAM which are related to environments, as environments are no longer needed for these transformed programs. The obvious catch is that the heap includes the information previously in the environments. We discuss the transformation scheme, including the treatment of disjunctions and cut and the primitives necessary to support the latter. We show how to implement built-in predicates and how to reintroduce optimizations lost in the general transformation. We show best and worst case behavior and a performance analysis for a selected program.

2. Introduction

We assume the reader is familiar with the WAM [War]. We will use the following symbols for the WAM registers:

H :	top of heap or global stack
B :	most recent choicepoint
E :	most recent environment
TR :	top of reset stack (trail)
Ai,Xi :	argument register and temporary variable

A binary definite program is a program of Horn clauses having only two literals, one positive and one negative, or in Prolog terminology: clauses with just one subgoal in the body. In [Tar], it is shown that definite programs have a binary equivalent. The authors introduce the transformation MBIN, which is defined as a Prolog predicate transbin/2 in appendix a. We give some examples of the equivalent binary program for a given 'general' one:

2. 1. append/3

```
append([],L,L) .
append([X|L1],L2,[X|L3]) :-
        append(L1,L2,L3) .
```

```
append([],L,L,Cont) :- call(Cont) .
append([X|L1],L2,[X|L3],Cont) :-
        append(L1,L2,L3,Cont) .
```

2. 2. nrev/2

```
nrev([],[]) .
nrev([X|L],O) :-
        nrev(L,R) ,
        append(R,[X],O) .
```

```
nrev([],[],Cont) :- call(Cont) .
nrev([X|L],O,Cont) :-
        nrev(L,R,
                append(R,[X],O,Cont)) .
```

2. 3. conjunction

```
a(X) :-    b(Y) ,                a(X,Cont) :-  b(Y,
           c(Z) ,                                  c(Z,
           d(T) .                                        d(T,Cont))) .
```

2. 4. fact

```
a(X) .                           a(X,Cont) :-   call(Cont) .
```

Queries must be transformed as well: this is illustrated by an example:

```
?- a(X,Y) , b(Y,Z) .             ?- a(X,Y,b(Y,Z,true)) .
```

The set of binary clauses is interesting because it has the same expressiveness as Prolog and it can be implemented with a reduced WAM: in particular, binary clauses do not need environments, so none of the WAM-instructions ***allocate,deallocate,call*** and the ***Yva****-instructions are needed. It is therefore worthwhile to investigate the runtime behavior of a binary program as compared to its full equivalent. Moreover, the study of binary programs might teach us something about the compilation of full Prolog.

3. Discussion of the transformation

3. 1. The metacall and a specialized version of it.

In the above, the metacall is a bit of an overkill: the argument is never a conjunction or a disjunction; what we need is a binary version of call/1, bcall/1, which can only be called with a single subgoal. bcall/1 can be implemented directly in Prolog: for every callable functor of the program, bcall/1 has one definition, like:

```
bcall(append(X,Y,Z,C)) :- append(X,Y,Z,C) .
```

It is nice that this can be described at the Prolog level, but a specialized built-in predicate performs better: since in ProLog by BIM, there exists such a built-in, our tests use it. On the tested programs, this improved the speed with about 5%.

3. 2. Disjunction

A few examples show how to translate a disjunction:

```
a :- b ; c .                     a(C) :- disj(b(C),c(C)) .

a :- b , (c ; d) , e .           a(C) :- b(disj(c(e(C)),d(e(C)))) .
```

In this way, there must be a binary definition of disj/2:

```
disj(G1,G2) :- bcall(G1) .
disj(G1,G2) :- bcall(G2) .
```

There is a second way to translate disjunction, illustrated by the following example:

```
a :- b , (c ; d) , e .           a(C) :- b(disj(c(true),d(true),e(C))) .
```

We do not like this solution, since disj/3 is not a binary predicate:

```
disj(G1,G2,C) :- bcall(G1) , bcall(C) .
disj(G1,G2,C) :- bcall(G2) , bcall(C) .
```

The most efficient solution, is to introduce a new predicate for each disjunction of the program:

```
a :- b , (c ; d) , e .                a(C) :- b(c_or_d(e(C))) .
                                      c_or_d(C) :- c(C) .
                                      c_or_d(C) :- d(C) .
```

3. 3. An alternative transformation scheme

The idea is illustrated by an example:

full clause:

```
h(X,Y) :- a(X,I) , b(I,J) , c(J,K) , d(K,Y) .
```

binary clause:

```
h(X,Y,C) :- a(X,I,b(I,J,c(J,K,d(K,Y,C)))) .
```

alternative binary clauses:

```
h(X,Y,C) :- a(X,I,env1(I,Y,C)) .
env1(I,Y,C) :- b(I,J,env2(J,Y,C)) .
env2(J,Y,C) :- c(J,K,d(K,Y,C)) . /*last call optimization*/
```

env1/3 and env2/3 are auxiliary predicates: the definition of env1/3 contains the continuation needed when the call to a/3 is finished. In this way, the structure p(I,Y,C) contains the 'permanent' variables and the continuation needed for proceeding with the execution after a/3 has succeeded. This nicely describes in Prolog of how environments are treated in the WAM. But, as Prolog has no destructive assignment on the heap, these environments cannot be updated and must be copied each time, leading to performance loss.

We have not measured the effect of this alternative transformation: we expect it to lead to even higher heap consumption and probably a lower performance for deterministic programs. We will use this idea however to implement built-in predicates.

4. Built-in predicates

A big disadvantage of a binary program, is that quite a few optimizations are lost. To begin with, all the optimizations ProLog by BIM performs on arithmetic are lost because arithmetic cannot be compiled in-line any more. The measurements show this clearly (see section 8). Fortunately, one can regain most of the optimizations in the binary model by making use of the alternative transformation scheme. We illustrate this by an example:

```
a(X) :- b , Y is X + 1 , c(Y) .       a(X,C) :- b(plus1(X,Y,c(Y,C))) .
                                      plus1(X,Y,C) :- X is Y + 1 , bcall(C) .
```

The idea is to define a new predicate for every arithmetic call; this new predicate takes as arguments the variables occurring in the arithmetic call and of course the continuation. The body of the new predicate contains the arithmetic call and the call to the continuation. The measurements show that this technique applied on the test program, makes a lot of difference.

The same technique can be applied to most built-in predicates. The auxiliary predicate is not a binary one, but one can argue that most built-ins can be treated in-line, so that the body of these auxiliary predicates contain only one chunk [Deb]. So, practically speaking, it is a binary clause since no environment is needed. Backtracking built-in predicates - like clause/2 have to be implemented on a lower level.

Meta predicates also present a minor problem: e.g. a goal like

```
bagof(X,G,L)
```

has a binary form that looks like:

```
transbin(G,BG,bagof(X,BG,L,C))
```

transbin/3 is a built-in predicate which transforms a goal G to its binary equivalent BG, and the third argument is of course the continuation. If G is known at transformation time, the transformation of G to BG can be performed then, and the binary form is written as bagof(X,BG,L,C) at once. transbin/3 is the result of applying transbin/2 on itself. Note that transbin/3 is needed only if the programmer does use ordinary goals instead of binary goals.

5. Cut

Cut presents a major problem: it is the inverse problem cut poses during partial evaluation [Ven]. In both cases it is a problem of scope of the cut. In partial evaluation, one needs to restrict the scope, here one needs to expand the scope of a cut.
In ProLog by BIM, cut is implemented the following way (see [Bar] or [Mar] for alternatives): there is a general register, called CUTB, which is dealt with as follows:

call/execute:	CUTB := B
allocate:	E(Y1) := CUTB
retry/trust:	CUTB := B(B)
! in clause with ***allocate***:	B := E(Y1)
! in clause without ***allocate***:	B := CUTB

Within this scheme, we would like to implement cut for binary clauses, and not introduce too many new operations. We propose the following change to the transformation scheme:

A clause like a :- b , ! , c . has the binary form:

```
a(C) :- b(!($cut,c(C))) .
```

A clause like a :- ! , c . has the binary form:

```
a(C) :- !($cut,c(C)) .
```

The abstract machine code generator recognizes the special atom $cut and instead of generating the instruction '*unify_const* $cut' it generates a new instruction that does *H++ := CUTB

Also, instead of generating the instruction '*put_const* $cut,A1' it generates a new instruction which does A1 := CUTB

We also need a definition for !/2:

```
!($cut,C) :- bcall(C) .
```

and the compiler generates instead of '*get_constant* $cut,A1', an instruction that does B := Areg1

We have implemented this modification to the ProLog by BIM system - a matter of an hour or so - and results can be found in section 8.
The implementation of cut in the alternative translation scheme mimics the implementation of cut in a real implementation. An example illustrates this.

```
a :- b, !, c .                a(C) :- b(env($cut,C)) .
                              env(Cut,C) :- !(Cut,c(C)) .
```

This results in the value of CUTB to be saved in the 'environment' env/2 .

6. The control structure (_->_;_)

The implementation of if-then-else serves as an example of how cuts with a restricted scope can be added using the same technology as described above. We give it here without comments:

```
a :- b, ( c -> d ; e ), f .
```

Binary clauses:

```
a(C) :- b(ifthenelse(C)) .
ifthenelse(C) :- c(!($cut,d(f(C)))) .
ifthenelse(C) :- e(f(C)) .
```

or the alternative form:

```
a(C) :- b(env1(C)) .
env1(C) :- c(env2($cut,C)) .
env1(C) :- e(env4(C)) .
env2(Cut,C) :- !(Cut,env3(C)) .
env3(C) :- d(env4(C)) .
env4(C) :- f(C) .
```

7. Rationale for the results

We have to argue that the measurements we will present, are meaningful. The problem is the following: we have compared the execution of 2 languages (one a subset of the other, syntactically and also in terms of abstract machine code) by the same engine - a full implementation of WAM. It could be that the subset can be executed much more efficiently by a specialized version of the general engine. We argue that this is not the case: the main difference between binary and full Prolog is that binary Prolog doesn't need environments. It just means that no ***allocate-deallocate-call-Yva**** instructions are executed and these are the only ones affected by environments: the other instructions do not suffer from environments, neither is it possible to make them more efficient because there are no environments. It means that a full WAM implementation used on binary Prolog, has the performance of a specialized WAM-based implementation.

8. Some measurements

We have taken an N-queens program that works according to principles of domains, forward checking and first failure, but is written in portable Prolog: the program is in appendix b. The original program contains cuts and is called ***cut***. The code is in appendix b. we also tested a version without cuts, called ***nocut***. Its code is in appendix c. Both programs have 2 binary equivalents: they have the extension ***bin***. The opt extension means that arithmetic is optimized as indicated in section 4. The Prolog code was run on a non released version of ProLog by BIM on a SPARCstation 1.

8. 1. Results

The first table shows the result for searching the first solution of the 35-queens problem. The time figures are in seconds. The space requirements can be measured accurately in ProLog by BIM. The space figures given represent the maximal number of entries used while finding the first solution. The local stack contains environments as well as choicepoints.

queens(35)	time	heap	local stack	trail	total space
cut	2.14	29.100	678	1.514	31.292
cut.bin.opt	2.66	58.167	473	1.546	60.186
cut.bin	3.61	67.427	473	1.546	69.446
nocut	2.90	29.100	38.199	2.320	69.619
nocut.bin.opt	4.45	23.2086	30.854	2.637	265.577
nocut.bin	5.48	24.0745	30.854	2.637	274.236

The following table lists the percentages when ***cut*** resp. ***nocut***, is given 100 as reference:

queens(35)	time	heap	local stack	trail	total space
cut	100	100	100	100	100
cut.bin.opt	124	200	70	102	192
cut.bin	169	232	70	102	222
nocut	100	100	100	100	100
nocut.bin.opt	153	798	81	114	381
nocut.bin	189	827	81	114	394

The result is rather dramatic: there is 25% to 50% efficiency loss in the optimal binary program. For the tested program, it seems mandatory that the lost optimizations on arithmetic are regained.

8. 1. 1. The heap

The difference in heap usage between *bin* and *bin.opt*, is entirely due to the deletion of the arithmetic optimizations. It is quite large, but the difference between *bin* and the original version, entirely due to the explicit construction of continuations as heap structures, is even larger. Although heap garbage collection for binary Prolog is easier than for full Prolog, it is doubtful whether it can be made fast enough to outweight the larger heap consumption.

8. 1. 2. The local stack

The figures for local stack for the *bin* programs, are entirely due to choicepoints. For deterministic programs, or programs with cuts, one expect this figure to be lower for a binary program than for the full version. This is indeed the case: *cut.bin* consumes less local stack than *cut.* However, tail recursion optimization is also most effective for deterministic programs, so that one does not expect large differences in local stack consumption between the two models. On the other hand, for a very non-deterministic binary program - or one without cuts like *nocut* - the difference in local stack consumption might be less, because choicepoints of the binary program contain always one more entry.

8. 1. 3. The local stack, heap trade-off

It is clear that there is a local stack, heap trade-off: the binary program does not need environments, at the cost of constructing the continuations on the heap. We will make a best and a worst case estimate for this trade-off.

The best case example has a full Prolog clause which needs an environment which can't be deallocated during execution, and whose binary counterpart needs less extra heap to replace the environment. Here is the example:

the full clause:

```
a(Y1,Y2,Y3,...,Yn) :- b , c(f(Y1,Y2,Y3,...,Yn)) .
```

where b has at least two definitions to prevent the deallocation of the environment.

the binary clause:

```
a(X1,X2,X3,...,Xn,C) :- b(c(f(X1,X2,X3,...,Xn,C))) .
```

The space usage for each clause:

	environment	heap	total
full clause	n+2	n+1	2n+3
binary clause	0	n+4	n+4

So, it seems that in the best case, a binary program would use half the space needed for the full program. This case rarely occurs, since body goals rarely have compound arguments.

The above example hints at an optimization for full WAM: the above full clause has the same meaning as:

a(Y1,Y2,Y3,...,Yn) :- Z = f(Y1,Y2,Y3,...,Yn) , b , c(Z) .

and its environment is only 3 entries long, bringing the total space needed to n+4, exactly what is needed for the binary equivalent. A generalization of this is:

> a permanent variable which will end up on the heap anyway, can as well be put on the heap immediately and never occupy any space in the environment

The worst case is more extreme:

```
a :- g1 , g2 , g3 , ... , gn .
```
versus
```
a(C) :- g1(g2(g3(...gn(C)...))) .
```

The space usage for each clause:

	environment	heap	total
full clause	2	0	2
binary clause	0	2n-2	2n-2

Neither of these extreme cases is likely to occur in realistic Prolog programs.

A typical pattern for Prolog clauses is:

```
a(I,O) :- b(I,O1) , c(O1,O2) , d(O2,O) .
```

with equivalent binary clause:

```
a(I,O,C) :- b(I,O1,c(O1,O2,d(O2,O,C))) .
```

where the total space usage comparison is slightly in favor of the full clause:

	environment	heap	total
full clause	5	0	5
binary clause	0	8	8

8. 1. 4. The trail

It is at first sight surprising that trail usage is higher in ***bin***. There is an explanation, peculiar to ProLog by BIM: there, free permanent variables are always globalized - on the heap - so in ProLog by BIM

```
a :- b , c(X) , d(X) .
b .
b .
c(1) .
```

the binding of X to 1 is not trailed, because X is on the heap at a place more recent than the choicepoint was created. This contrasts with usual WAM, were X is a variable on the local stack, LESS recent than the choicepoint of b/0, so the binding is trailed. The binary program reestablishes this trailing. We expect that there is no difference in trail usage between ***bin*** and the original version, for implementations treating permanent variables the orthodox way.

8. 2. Extreme cases

It is easy to find examples where the binary program performs arbitrarily better or worse than the original program.

8. 2. 1. Binary better than original

Let <Ln> be an abbreviation for a list of n integers
The original program:

```
a :- b , c(<Ln>) .
b .
...
b .
c(_) .
```

The binary equivalent:

```
a(C) :- b(c(<Ln>,C)) .
b(C) :- bcall(C) .
...
b(C) :- bcall(C) .
c(_,C) :- bcall(C) .
```

The complexity of a call to a/0 is O(#(definitions of b) * n) , while the complexity of the equivalent binary call ?- a(true) . is O(#(definitions of b) + n). By choosing n and #(definitions of b) large enough, the binary program becomes arbitrarily faster than the original one.

This is a feature common to all meta-interpreters that construct a complete copy of an activated clause, before starting the execution of the first goal in the body.

8. 2. 2. Original better than binary

A slight modification of the above example gives us the desired result:

```
a :- b(1) , c(<Ln>) .
b(2) .
c(_) .
```

The binary equivalent:

```
a(C) :- b(1,c(<Ln>,C)) .
b(2,C) :- bcall(C) .
c(_,C) :- bcall(C) .
```

The call ?- a . has complexity O(1), while the equivalent call ?- a(true) . has complexity O(n) .

Again, this feature is shared by most meta-interpreters.

8. 2. 3. A lesson from the extreme cases

The lesson is:

> construct structures early in the body, to avoid constructing them several times due to backtracking, and to make memory consumption lower
> construct structures later if failure might make the construction unnecessary

These rules are contradictory: global knowledge - about which predicate calls have zero, one or more solutions - is needed, in order to apply them properly. The lesson can be applied to full Prolog.

9. Conclusion

A study of the runtime behavior of binary definite programs has been made. It indicates that if certain optimizations are carefully reintroduced into the naive binary program, performance can approach closely the full clause performance. This is a bit surprising, because of the much larger heap consumption of binary programs. It is not clear how to make the space requirements lower. The study has revealed some ways to improve traditional WAM compilation. In [Dem], a case study shows that binary programs are very suitable for program transformation. Finally, although binary clauses need only part of the WAM for their execution, they are interesting for the explanation and understanding of full WAM.

10. Acknowledgment

B.D. thanks DPWB for support through project RFO/AI/02.

11. References

[Bar] J. Barklund, 'Efficient interpretation of Prolog Programs' Proc. of the SIGPLAN'87 Symp on Interpreters and Interpretative Techniques, St.Paul, 1987, pp 132-137

[Deb] S. Debray, 'Register Allocation in a Prolog Machine'3SLP 1986

[Dem] B. Demoen, "On the Transformation of a Prolog Program to a more efficient binary Program" April 1991, KUL- CW report 130

[Mar] A. Marien, B. Demoen, 'On the Management of E and B in WAM' Proceedings of NACLP'89 (North American Conference on Logic Programming, Cleveland, Ohio, oct 1989) p. 1030-1047

[Tar] P. Tarau, M. Boyer, 'Elementary Logic Programs', Proceedings of PLILP'90, pp. 159-173, (eds) P. Deransart & J. Maluszynski, Springer-Verlag 1990

[Ven] R. Venken, 'A Prolog meta-interpreter for partial evaluation and its application to source-to-source transformation and query optimisation', Proc. ECAI'84, Pisa 1984

[War] D.H.D. Warren, 'An Abstract Prolog Instruction Set' Technical Report, SRI International, Artificial Intelligence Center, August 1983

Appendix a: MBIN for flat conjunctions

```
bintrans((H :- B),(NH :- NB)) :- ! , bt(H,C,NH) , bintrans(B,C,NB) .
bintrans(H,T) :- bintrans((H :- bcall) , T) .

bintrans((G,B),C,O) :- ! , bt(G,NC,O) , bintrans(B,C,NC) .
bintrans(G,C,O) :- bt(G,C,O) .

bt(G,C,NG) :- G =.. [N|A] , append(A,[C],NA) , NG =.. [N|NA] .
```

Appendix b: the original queens test program with cuts

```
first(N) :-
    genl(N,L) ,
    gen(N,L,Ll,V,N) ,
    queens(Ll) , ! .

genl(0,[]) :- ! .
genl(N,[N|R]) :- M is N - 1 , genl(M,R).

gen(0,_,[],[],_) :- ! .
gen(N,L,[qu(V,N,L,Ln)|R],[V|Rv],Ln) :-
    M is N - 1 , gen(M,L,R,Rv,Ln).

queens([]) .
queens([qu(X,N,Dx,Ldx) | Rest]) :-
    member(X,Dx) ,
    constraintprop(X,N,Rest,[],Newrest) ,
    queens(Newrest) .

constraintprop(X,_,[],L,L) :- ! .
constraintprop(X,Nx,[qu(V,N,Dv,Ldv)|R],In,Out) :-
    Diag1 is X - N + Nx ,
    Diag2 is X + N - Nx ,
    remove(X,Diag1,Diag2,Dv,Ndv,Ldv,Lndv) , Lndv > 0 ,
    insert(qu(V,N,Ndv,Lndv),Lndv,In,Nin) ,
    constraintprop(X,Nx,R,Nin,Out) .

insert(F,Lndv,[],[F]) :- ! .
insert(F,Lndv,[qu(V1,N1,Ndv1,Lndv1)|R],[F,qu(V1,N1,Ndv1,Lndv1)|R])
        :- Lndv =< Lndv1 , ! .
insert(F,Lndv,[F1|R],[F1,F|R]) .
member(X,[X|_]) .
member(X,[_|R]) :- member(X,R) .
```

```
remove(A,B,C,[],[],N,N) :- ! .
remove(A,B,C,[A|R],O,N,M) :- ! ,
    Newn is N - 1 , remove(A,B,C,R,O,Newn,M) .
remove(A,B,C,[B|R],O,N,M) :- ! ,
     Newn is N - 1 , remove(A,B,C,R,O,Newn,M) .
remove(A,B,C,[C|R],O,N,M) :- ! ,
     Newn is N - 1 , remove(A,B,C,R,O,Newn,M) .
remove(A,B,C,[X|R],[X|O],N,M) :- remove(A,B,C,R,O,N,M) .
```

Appendix c: the original queens test program without cuts

Only the clauses that are different are listed:

```
genl(0,[]) .
genl(N,[N|R]) :- N > 0 , M is N - 1 , genl(M,R).

gen(0,_,[],[],_) .
gen(N,L,[qu(V,N,L,Ln)|R],[V|Rv],Ln) :-
          N > 0 , M is N - 1 , gen(M,L,R,Rv,Ln).

constraintprop(X,_,[],L,L) .
constraintprop(X,Nx,[qu(V,N,Dv,Ldv)|R],In,Out) :-
          Diag1 is X - N + Nx ,
          Diag2 is X + N - Nx ,
          remove(X,Diag1,Diag2,Dv,Ndv,Ldv,Lndv) , Lndv > 0 ,
          insert(qu(V,N,Ndv,Lndv),Lndv,In,Nin) ,
          constraintprop(X,Nx,R,Nin,Out) .

insert(F,Lndv,[],[F]) .
insert(F,Lndv,[qu(V1,N1,Ndv1,Lndv1)|R],[F,qu(V1,N1,Ndv1,Lndv1)|R])
          :- Lndv =< Lndv1 .
insert(F,Lndv,[F1|R],[F1,F|R])
          :- F1 = qu(V1,N1,Ndv1,Lndv1) , Lndv > Lndv1.

remove(A,B,C,[],[],N,N) .
remove(A,B,C,[A|R],O,N,M) :-
          Newn is N - 1 , remove(A,B,C,R,O,Newn,M) .
remove(A,B,C,[B|R],O,N,M) :-
          A <> B ,
           Newn is N - 1 , remove(A,B,C,R,O,Newn,M) .
remove(A,B,C,[C|R],O,N,M) :-
          B <> C , A <> C ,
           Newn is N - 1 , remove(A,B,C,R,O,Newn,M) .
remove(A,B,C,[X|R],[X|O],N,M) :-
          A <> X , B <> X , C <> X ,
          remove(A,B,C,R,O,N,M) .
```

PROLOG SEMANTICS FOR MEASURING SPACE CONSUMPTION

A. Ja. Dikovsky

Institute for Applied Mathematics,
Miusskaja sq.,4. Moscow, SU 125047
USSR
Fax: 7(095)9720737

Abstract

A new operational semantics of Prolog is introduced which allows measuring space consumption of Prolog programs. Space complexity classes are introduced in four subsets of Prolog: *kernel Prolog* (no structures, no builtins), *kernel dynamic Prolog* (dynamic clauses control operators allowed), *flat Prolog* (the subset of kernel Prolog with no lists) and *flat dynamic Prolog* (the corresponding subset of kernel dynamic Prolog. Main results show that functional programs can be reduced to deterministic ones, any program in kernel Prolog can be transformed into equivalent program with optimized recursion guaranteeing constant local stack and trail, and any program in dynamic kernel Prolog can be transformed into an equivalent purely iterative program.

1. INTRODUCTION

As it is widely acknowledged SLD-resolution is a successful operational semantics for logic programming [1]. It forms a theoretical basis for Prolog semantics also. As it is well known the standard leftmost-depth-first refutation rule of searching SLD-tree is the rule of interpreting Prolog programs. Though incomplete, it permits highly efficient implementations of Prolog [2,3]. This operational semantics was repeatedly precised or supplemented for solving some metaprogramming problems such as e.g. global optimizations of programs [4,5], automatic extraction of types [6] or modes [7-9], metacompilation [10] an so on. However everyday practice of programming in Prolog also necessitates precision of its standard semantics because with respect to problems related to pragmatic factors it turns out to be inexpressive or even misleading. The most important among such pragmatic factors is consumption of space during program execution. Very often absolutely logically correct and elegant Prolog programs run computer out of space for the reasons not expressible in logical terms. As a result Prolog programming becomes a specific art requiring keen feeling of recursion style compensating for the lack of formal criteria. We illustrate this thesis by very simple but typical examples. First of them gives a definition of screen representation of left-associative conjunctive normal form propositional formulae in the equivalent form without superficial brackets. For example, ((((e;e);e),(e;e)),(e;e)) it represents as (e;e;e),(e;e),(e;e).

Example 1. % wlf(+Lcnf_formula).

```
wlf((C,D)) :-
      wlf(C),
      write(','),
      wdj(D).
wlf(D) :-
      wdj(D).
wdj(D) :-
      write('('),
      wd(D),
      write(')').
wd((D1;D2)) :-
      wd(D1),
      write(';'),
      write(D2).
wd(D) :-
      write(D).
```

This definition seems to be most natural, because it follows the simplest definition of context-free syntax of input formulae. However it will overflow Prolog workspace for sufficiently long input formula, and it is impossible to explain this in logical terms. In the seventies attempts were made to rehabilitate Prolog recursion. For some types of recursive definitions space optimization methods were found [11,12], most well known of them the so called "tail recursion" optimization. For example, instead of pragmatically ineligible definition of wlf/1 above we can use the following tail recursive definition.

Example 2. % t_wlf(+Lcnf_formula).

```
t_wlf((C,D)) :-
      w_deep_c((C,D),Rest),
      write(','),
      !,
      t_wlf(Rest).
```

```
t_wlf(D) :-
      t_wdj(D).
w_deep_c(((C1,C2),D),(Rest,D)) :-
      !,
      w_deep_c((C1,C2),Rest).
w_deep_c((D1,D2),D2) :-
      t_wdj(D1).
t_wdj(D) :-
      write('('),
      t_wd(D),
      write(')').
t_wd((D1;D2)) :-
      w_deep_d((D1;D2),Rest),
      write(';'),
      !,
      t_wd(Rest).
t_wd(D) :-
      write(D).
w_deep_d(((D1;D2);D3),(Rest;D3)) :-
      !,
      w_deep_d((D1;D2),Rest).
w_deep_d((D1;D2),D2) :-
      write(D1).
```

"Tail" property of recursion demands that recursive call should be last in last alternative, and that all subgoals from parent recursive call to child recursive call should be deterministic (i.e. should not have choice points in their proofs). As a matter of fact the definition of t_wlf/1 is logically equivalent to that of wlf/1 but it is not dependent on the length of input formulae. Logical semantics does not explain this. Neither does it explain why (logically superfluous) cuts are needed in the definition of t_wlf/1, namely why would Prolog run out of space again on long formulae without them. Next two examples demonstrate the same effect for predicates not dealing with structures or lists. First of them gives most naive definition of length of an input stream.

Example 3. % strm_len(+Input_stream_handle,-Stream_length).

```
strm_len(H,L) :-
      get0(H,_),
      strm_len(H,L0),
      L is L0+1.
strm_len(_,0).
```

The other one gives an equivalent tail recursive definition.

Example 4. % strm_t_len(+Input_stream_handle,-Stream_length).

```
strm_t_len(H,L) :-
     strm_tl(H,0,L).
strm_tl(H,Acc,L) :-
     get0(H,_),
     A is Acc+1,
     !,
     strm_tl(H,A,L).
strm_tl(_,A,A).
```

And again, strm_len/2 runs Prolog out of space for sufficiently long input streams, whereas t_strm_len/2 always succeeds. It is worth mentioning that such "infinite loop" procedures can always be described equivalently by absolutely iterative backtrack-loops definitions. This is achieved at the cost of using global variables or facts (unit clauses) for parameters passing. Example 5 presents such a definition for input stream length.

Example 5. % strm_i_len(+Input_stream_handle,-Stream_length).

```
strm_i_len(H,L) :-
      ctr_set(0,0), %built-in predicate setting counter 0 to 0
      repeat,
      readchar(H),
      ctr_is(0,L).  %built-in predicate unifying L with counter 0 value
readchar(H) :-
      get0(H,_),
      ctr_inc(0,_), %built-in predicate incrementing counter 0
      !,
      fail.
readchar(_).
```

This definition also succeeds for any input stream, and again this fact cannot be seen from its purely logical interpretation. Though such a style may be regarded as "awful" by those who write mostly metaprolog programs (e.g. interpreters, compilers, partial evaluators and so on) it is widely used in application programming for implementing "infinite" loops, deeply embedded loops and for some other purposes and proves to be rather efficient and helpful.

These examples demonstrate only three styles of writing Prolog programs: purely logical, tail recursive and iterative, and show that standard semantics of Prolog gives no formal criteria of choice between them in concrete situations, though sometimes performance or even fitness of programs depends on this choice.

We propose here simple measures of space consumption as such a formal criteria for objective choice between programming styles. To this end a formal definition of Prolog is introduced (a refined version of that in [13]). It is founded on the notion of a derivation tree, representing a successful branch of SLD-tree. Derivation trees are represented naturally by three stacks: *stack of accessible subgoals (AS), stack of resolvent subgoals (RS)* and *stack of unifiers (US)*. We introduce space metrics on these stacks adequate for Warren abstract machine [2,3] (WAM) implementations of Prolog. These metrics reflect though not directly the sizes of local stack, trail and global stack (heap) of WAM to within a constant factor. For any class C of Prolog programs we introduce complexity classes **C(FA,FR,FU)** of all programs in C with AS,RS and US bounded by functions in classes **FA**, **FR** and **FU** respectively. We consider four classes of programs:

- *kernel Prolog (**KP**)*, i.e. Prolog with lists without structures and with the only built-in feature: the cut operator (!),

- *kernel dynamic Prolog (**KDP**)*, i.e. kernel Prolog enriched by built-in predicates for controlling dynamic clauses (assert/1, retract/1, abolish/1),

- *flat Prolog (**FP**)* and *flat dynamic Prolog (**FDP**)*, i.e. kernel and respectively dynamic kernel Prolog without lists.

For flat and flat dynamic Prolog

FP(*,*,*) = FP(con,con,con)

and respectively

FDP(*,*,*) = FDP(con,con,con)

is true (***** and **con** being the sets of all integer functions and constant integer functions respectively). Above this we find that solution existence problem in **FP** is NP-complete and in **FDP** is PSPACE-complete (it is clear that in **KP** it is unsolvable).

For kernel Prolog we show that **KP** is a conservative expansion of **KP(con,con,*)**, i.e. for any kernel Prolog program an equivalent kernel Prolog program can be constructed with stacks of accessible and resolvent subgoals bounded by constants. This general recursion optimization theorem is somewhat related to the results in [14,15] telling that for any kernel Prolog program there is an equivalent one with all predicates having so called "concluding recursion" definitions only, and that this concluding recursion is always reducible to an iteration scheme. There are two differences. First is that concluding recursion differs from tail recursion in that subgoals intermediate between subsequent recursive calls must be only *functional* (which is weaker than deterministic). Second is that no space consumption estimate is established for the iteration scheme used there. By the way, a technical lemma underlying our recursion optimization theorem shows either that for any functional program an equivalent in a natural sense deterministic program can be constructed.

We define *iterative* programs as those whose all three stacks are bounded by constants. In [13] we have shown that for deterministic kernel dynamic programs recursion can be reduced to iteration in this sense. Here using the above mentioned technical lemma and the result in [13] we show that recursion can be reduced to iteration in **KDP**, i.e. **KDP** is a conservative expansion of **KDP(con,con,con)**.

2. ABSTRACT STACK MACHINE SEMANTICS

In this section we define operational semantics of Prolog programs through the notion of *abstract stack machine (AS-machine)*. States of memory of AS-machine are exactly derivation trees.

Definition 1. A *logical procedure* is a set *lp*/n of definitions of predicates with one distinguished predicate main/n. An *instance* of logical procedure *lp*/n is a program *lp*(t1,...,tn) with the set of definitions *lp*/n and the query

?- main(t1,...,tn).

Let I_+ denote the set of all positive integers and $I_+^* = \{[i_1,\dots,i_n] \mid i_j \text{ in } I_+, n\geq 0\}$ - the set of all finite sequences of integers in I_+. Let v_1v_2 denote the concatenation of sequences v_1,v_2 in I_+^*. We denote by < the complete lexicographic order of sequences in I_+^*.

Definition 2. A finite subset $T \subset I_+^*$ is a *tree* if a) for any v_1v_2 in T v_1 is also in T, and b) for any v and j>1 such that v [j] is in T, v [j-1] is also in T. Sequences in T are called *nodes* of T. The node [] is a *root* of T. For a node v the node v [j] is called jth *son* of v and v - a *parent* of v [j]. A sequence of nodes $v_1,\dots,v_n$, n>0, is a *path (from* v_1 *to* v_n) if for any $1\leq i<n$ v_{i+1} is a son of v_i. For a path $v_1,\dots,v_n$ in T v_1 is called an *ancestor* of v_n and v_n - a *descendant* of v_1. A node is a *leaf* if it has no sons. For a node v_1 in T any node v in T such that v < v1 is a *predecessor* of v_1 (*precedes* v_1). Nodes v such that $v < v_0$ (respectively $v > v_0$) are *to the left* (respectively *to the right*) of v_0.

Definition 3. A pair t=<T,f> where T is a tree and f is a leaf of T is called a *focused tree*; f is called a *focus* of t. Let L be some set, t=<T,f> be a focused tree and l be a function from T to L. Then s=<T,f,l> is called a *labelled focused tree*, l is called a *labelling* and for v in T l(v) is called a *label* of v. A labelled focused tree is a *state* if any node to to the right of its focus is a leaf. Let $pr=lp(\bar{W})$ be a program. s=<T,f,l> is a *state of pr* if for all v < f $l(v) = (a_v,i_v,u_v)$ and for all v ≥ f $l(v) = (a_v,i_v)$, where all a_v are atoms in *pr*, all i_v are nonnegative integers and all u_v are some substitutions of terms for variables in a_v. Labels of nodes of states are called *subgoals*, those to the left of the focus - the *accessible* ones, all others - *resolvent*. The subgoal l([]) is called a *query*; variables in this subgoal (if any) are called *query variables*. We shall not differentiate between nodes of states and occurrences of subgoals corresponding to them in the sequel.

Example 6. The following focused labelled tree is a state of the program wlf((p1,(p2;p3))) resulting from the definition in example 1 by substituting main/1 for wf/1.

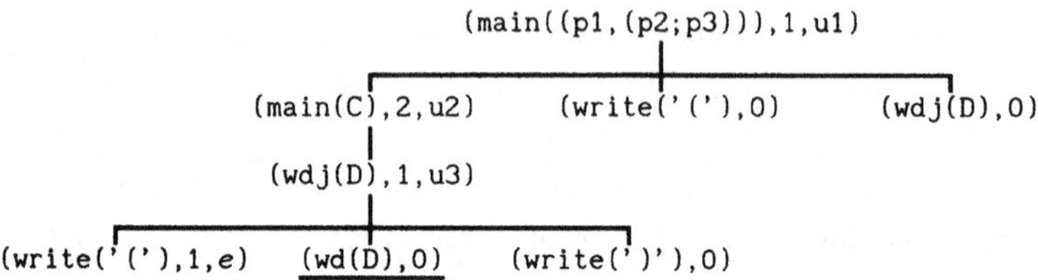

Here u1 is the MGU {C=p1,D=(p2;p3)}, u2 is $\{D^1=C\}$, u3 is $\{D^2=D^1\}$, *e* is the empty substitution. Underlined is the focus subgoal.

A program state can be represented naturally by three stacks.

Definition 4. Let s=<T,f,l> be a state of a program $lp(\bar{W})$ and $(a_1,i_1,u_1),\ldots,(a_k,i_k,u_k),(a_{k+1},i_{k+1}),\ldots,(a_n,i_n)$ be the sequence of all its subgoals in increasing order. Then the sequence $(a_1,i_1),\ldots,(a_k,i_k)$ is called an *accessible subgoals stack* (*AS*), the sequence $u_1,\ldots,u_k$ is called a *unifiers stack* (*US*) and the sequence $(a_{k+1}i_{k+1}),\ldots,(a_n,i_n)$ is called a *resolvent stack* (*RS*). (a_k,i_k) and u_k are the *top elements* of *AS* and *US* respectively. The focus subgoal (a_{k+1},i_{k+1}) is the top element of *RS*. The composition $con(s)=u_1\circ\ldots\circ u_k$ of all substitutions in *US* is called a *context* of the state s.

It is clear that states and stacks defined on them determine each other uniquely, so transformations of states can be described also in terms of transformations of stacks. We define transitions from states to states in *AS*-machines in terms of two operators on states: *unfold* and *backtrack*.

Definition 5. Let *lp/m* be a logical procedure, *pr=lp*(w1,...wm) be some its instance and s be a state of *pr*.

Operator *unfold* applies to *s* if on the top of *RS* (i.e. in focus) there is a subgoal $(p(\bar{V}),d)$ and in the definition of p/n in *lp/m* there is i-th clause, i>d, with the head $p(\bar{U})$ unifyable with $p(\bar{V})\circ con(s)$. Let *u* be the MGU for these two atoms and

$$r :- r_1,\ldots,r_k.$$

k≥0, be an instance of i-th clause not containing variables in *s* . Then the state *unfold(s)* results from *s* when $(p(\bar{V}),d)$ is replaced in *RS* by $(r_1,0),\ldots,(r_k,0)$,

$(p(\bar{V}),i)$ is put on the top of *AS* and *u* is put on the top of *US* (*u eliminates* $p(\bar{V})$ by ith clause for p/n).

Operator *backtrack* applies to *s* when *unfold* cannot be applied to it and if there is an accessible subgoal (r,i) on the top of *AS*, the functor in r is p/n, i-th clause in definition of p/n in *lp/m* is an instance of

$$r :- r_1,\dots,r_k.$$

k≥0, and *RS* contains $(r_1,d_1),\dots,(r_k,d_k),(q_1,d_{k+1}),\dots,(q_l,d_l)$. In this case the state *backtrack(s)* results from *s* when (r,i) is popped from *AS*, topmost MGU *u* is popped from *US* and *RS* is transformed to $(r,i),(q_1,d_{k+1}),\dots,(q_l,d_l)$.

The starting state of AS-machine on the program *pr* is the state s_0 in which *AS* and *US* are empty and *RS* contains (main(w1,...,wm),0) . The *computation* of *pr* is the sequence $comp(pr)=(s_0,s_1,\dots)$ of states in which $s_{i+1}=unfold(s_i)$ or $s_{i+1}=backtrack(s_i)$ for each i.

An accessible subgoal (p(t1,...,tn),d,*u*) in s_i with p/n defined in *lp/m* is called *deterministic* if d equals to the number of clauses in the definition of p/n. otherwise such a subgoal is called a *choice point*.

If $comp(pr)=(s_0,\dots,s_n)$, $s_n = unfold(s_{n-1})$ and *RS* is empty in s_n then *comp(pr)* is *successful*. A *result* of a successful computation of *pr* is the substitution *res(pr)* which is the restriction of $con(s_n)$ to query variables (i.e. variables in w1,...,wm). $comp(pr)= (s_1,\dots,s_n)$ is *unsuccessful* if $s_n = backtrack(s_{n-1})$ and *AS* is empty in s_n.

So as to define AS-machine semantics of a dialect of Prolog with a set of built-in predicates *BIP* we should define for each p/n in *BIP* *unfold(s)* and *backtrack(s)* on those states *s* where a subgoal $(p(\bar{U}),d)$ is on the top of *RS* or a subgoal $(p(\bar{U}),d,u)$ is on the top of *AS* respectively. For example we define completely kernel Prolog if the following definition of cut operator !/0 is added.

If g=(!,0) is on the top of *RS* in *s* and pg is its parent subgoal then in the state *unfold(s)* all accessible subgoals g′ such that pg ≤ g′≤ g become deterministic, g is popped from *RS* and (!,1,*e*) is put on *AS*. For a state *s* with g = (!,1,*e*) on the top of *AS* the state *backtrack(s)* is obtained when g is popped from *AS* and (!,1) is put on *RS*.

3. SPACE COMPLEXITY MEASURES

Space metrics suggested here are designed for settled after a long evolution and most popular today style of implementing Prolog founded on WAM instructions set [2,3], code copying and collecting garbage. Let us introduce the following classification of subgoals in a state of AS-machine.

Definition 6. We call an accessible subgoal g in a state *s* a *hypothesis* if it has a descendant choice point. An accessible subgoal g in *s* is called *founded* if it has no resolvent sons. An accessible subgoal which is not founded is called *unfounded*. A founded subgoal which is not a hypothesis is called *proven*.

Consider the state in the example 6 above. In this state the subgoal (wdj(D),1,u3) is a choice point, hence a hypothesis. Moreover this subgoal is also unfounded because it has the son (write(')'),0) in resolvent stack. On the other hand in the state in the example 7 below all accessible subgoals except (t_wd(D),2,u3) are proven. This example illustrates the effect of execution of the cut operator. It is only for accessible subgoal (!,1,*e*) that the subgoal (t_wd(D),1,u3) becomes deterministic and (because it is founded) proven. As to (t_wdj(D),1,u2), it is not a hypothesis but it is unfounded.

Example 7. The following focused labelled tree is a state of the program t_wlf((p1;p2)) resulting from the definition in example 2 by substituting main/1 for t_wlf/1.

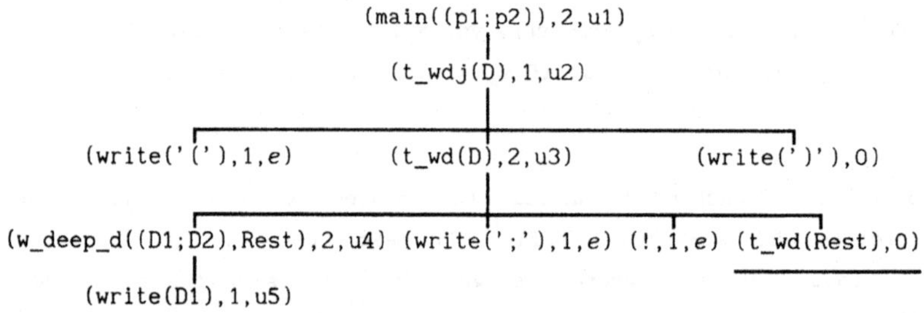

```
                              (main((p1;p2)),2,u1)
                                       |
                               (t_wdj(D),1,u2)
                                       |
             ------------------------------------------------------
             |                         |                          |
     (write('('),1,e)          (t_wd(D),2,u3)             (write(')'),0)
                                       |
             ------------------------------------------------------
             |                         |            |             |
(w_deep_d((D1;D2),Rest),2,u4) (write(';'),1,e) (!,1,e) (t_wd(Rest),0)
             |                                                ‾‾‾‾‾‾‾‾‾‾‾‾‾‾
     (write(D1),1,u5)
```

Here u1 is the MGU {D=(p1;p2)}, u2 is $\{D^1=D\}$, u3 is $\{(D1^2;D2^2)=D^1\}$, u4 is $\{D1^3=D1^2, D2^3=D2^2=Rest^2\}$, u5 is $\{D1^4=D1^3\}$.

Definition 6. Let *pr* be a program, $comp(pr)=(s_1,\dots,s_n)$ be successful and *s* be a state in *comp(pr)*. The *workspace* of *s ws(s)* is defined as:

$$ws(s) = |AS(s)|+|RS(s)|+|US(s)|,$$

where

$$|AS(s)| = \sum_{g \text{ in } s} h(g), \quad h(g) = \begin{cases} 1, & \text{if g is a hypothesis} \\ 0 & \text{for all other subgoals} \end{cases}$$

$$|RS(s)| = \sum_{g \text{ in } s} uf(g), \quad uf(g) = \begin{cases} 1, & \text{if g is unfounded subgoal} \\ 0 & \text{for all other subgoals} \end{cases}$$

$$|US(s)| = \sum_{u \text{ in } US} w(u), \quad w(u) \text{ - the } weight \text{ of unifier } u.$$

Let *u* be the MGU used for elimination of a subgoal p(t1,...,tn) by a clause p(v1,...,vn) :- *body*. Then

$$w(u) = \begin{cases} ts(u), & \text{if p(t1,...,tn) is deterministic} \\ ts(u)+fv(u), & \text{if it is a choice point,} \end{cases}$$

where *ts(u)* is the total size of all structures and lists unified with <u>not query</u> variables and *fv(u)* is the number of free variables in terms t1,...,tn.

We introduce four space complexity characteristics of a program *pr* with successful computation:

$$ws(pr) = \max\{ ws(s) \mid s \text{ in } comp(pr) \},$$

$$as(pr) = \max\{ |AS(s)| \mid s \text{ in } comp(pr) \},$$

$$rs(pr) = \max\{ |RS(s)| \mid s \text{ in } comp(pr) \},$$

$$us(pr) = \max\{ |US(s)| \mid s \text{ in } comp(pr) \}.$$

Definition 7. Let *lp/***m** be a logical procedure. Four partial space complexity functions are connected to it:

$$ws_{lp}(n) = \max\{ ws(pr) \mid pr=lp(t1,...,tm), \sum_{i=1}^{m} |ti| < n \},$$

$$as_{lp}(n) = \max\{ as(pr) \mid pr=lp(t1,...,tm), \sum_{i=1}^{m} |ti| < n \},$$

$$rs_{lp}(n) = \max\{ rs(pr) \mid pr=lp(t1,...,tm), \sum_{i=1}^{m} |ti| < n \},$$

$$us_{lp}(n) = \max\{ us(pr) \mid pr=lp(t1,...,tm), \sum_{i=1}^{m} |ti| < n \}.$$

For a class of logical procedures **C** and classes of integer functions **A**,**R**,**U** we denote by **C**(**A**,**R**,**U**) the class of those *lp* in **C** that there are functions a in **A**, r in **R** and u in **U** such that

$$as_{lp} \le a, \; rs_{lp} \le r, \; us_{lp} \le u .$$

We shall distinguish two classes of functions: ***** - the class of all integer functions and **con** - the class of all constant integer functions and four classes of logical procedures:

KP - the class of all logical procedures in kernel Prolog (i.e. Prolog without structures and only one built-in operator - cut),

KDP - the class of logical procedures in kernel dynamic Prolog (i.e. kernel Prolog enriched by dynamic clauses control predicates assert/1,retract/1,abolish/1),

FP - the class of logical procedures in flat Prolog (i.e. the subset of kernel Prolog without lists),

FDP - the class of logical procedures in flat dynamic Prolog (i.e. the subset of kernel dynamic Prolog without lists).

So, for example, **KP(con,*,*)** is the class of all logical procedures in **KP** with stack of accessible subgoals bounded by constants.

Let us comment shortly these definitions. There is a simple relation between stacks of AS-machine and stacks of Warren abstract machine (i.e. local stack(s), global stack and trail). Stack of accessible subgoals is a model of that region of local stack which contains choice points and frozen activation frames. Resolvent stack corresponds to the part of local stack containing activation frames of active subgoals (i.e. subgoals included into the pointer chain starting from the topmost subgoal on local stack (focus subgoal) and connected by pointers to parent activation frame). Unifiers stack size reflects the size of the part of global stack (heap) not used by actual parameters of logical procedure (component *ts(u)*) and the size of trail. Definitions of complexity measures reflect main optimizations provided by WAM instructions. Namely, *RS* is not increased for founded subgoals, corresponding to last call optimization in WAM; simultaneously it is taken into account that WAM does not create activation frames neither for unit clauses nor for clauses with one call in the body. Above this proven subgoals do not increase the size neither of *AS* nor of *RS*. This reflects tail recursion optimization as well as local stack optimization while executing cut operator. Note that our estimate of *US* size is pessimistic because it reflects growth of global stack but does not reflect its reduction while garbage collection (only while backtracking). So, constant upper bound of *US* is absolute. But if *US* increases unlimited, some superficial for AS-machine semantics factors must be taken into account. For example, for programs not creating structures or lists unlimited growth of *US* reflects inevitable garbage collections and hence a delay. Defined in the introduction predicate t_wlf/1 is a typical example of such procedures. On the other hand procedures generating structures or lists require *US* space proportional to maximal depth of constructed terms. Here is an example of such a procedure.

Example 8. % gen_list(+Depth,-List).

```
gen_list(0,[]).
gen_list(D,[a|T]) :-
      D1 is D-1,
      gen_list(D1,T).
```

AS-machine semantics reflects only those properties of WAM instruction set which are expressible in terms of derivation trees. So it does not reflect, for example, static indexing of clauses heads or activation frames optimization as a result of ordering variables by their occurrences in clause body. Besides this proposed space consumption measures reflect real sizes of WAM stacks only to within a constant factor which may depend on an implementation. Nevertheless this simple model is sufficient for practical Prolog programming. For example, space estimates explain different behavior of logically equivalent procedures in examples 1,2 and 3-5. All the three stacks *AS*, *RS* and *US* grow proportionally to the length of input left associative formulae for the procedure wlf/1. So, it is no wonder that it overflows workspace for sufficiently deep input formula. On the other hand $as_{\text{t\_wlf}}$ and $rs_{\text{t\_wlf}}$ are bounded by constants (cf. example 7). Though $us_{\text{t\_wlf}}$ is not bounded, no terms in *US* are included in a resulting term, so all of them are garbage. This shows that t_wlf will run successfully on any formula fitting Prolog heap. However this also means that real computation will be somewhat delayed by consecutive garbage collections. What shows space consumption analysis of stream length procedures in examples 3-5. $as_{\text{strm\_len}}$ and $rs_{\text{strm\_len}}$ also grow proportionally to the length of the input stream. So, it is not fit for long streams. And again $as_{\text{strm\_t\_len}}$ and $rs_{\text{strm\_t\_len}}$ are bounded by constants. As to $us_{\text{strm\_len}}$ and $us_{\text{strm\_t\_len}}$ our definition of *US* size does not indicate their growth. Nevertheless for some implementations they are growing too. In this case $us_{\text{strm\_i\_len}}$ must be preferred for long streams because its stacks sizes $as_{\text{strm\_i\_len}}$, $rs_{\text{strm\_i\_len}}$ and $us_{\text{strm\_i\_len}}$ are bounded by constants.

4. COMPLEXITY OF FLAT PROLOG

Simple considerations show that all logical procedures in flat Prolog are space bounded.

Theorem 1. **FP**(*,*,*) = **FP**(con,con,con).

Theorem 2. **FDP**(*,*,*) = **FDP**(con,con,con).

We do not list here concrete upper bounds for as_{lp} and us_{lp}. They are sub exponential to the size of *lp*. Using somewhat more precise considerations we estimate algorithmic complexity of logical procedures in flat Prolog.

For a class of logical procedures **C** let *que*(*pr*,**C**) denote the following property: *the computation of the instance pr =* *lp*(t1,...,tn) *of logical procedure lp/*n *in* **C** *is successful.*

Theorem 3. The property *que* is NP-complete in **FP**.

Theorem 4. The property *que* is PSPACE-complete in **FDP**.

For each integer i≥0 we denote by **i** the singleton class containing constant i. Interesting questions arise about existence of various constant space hierarchies in **FP** and **FDP**. E.g. do there exist infinitely many i<j such that

$$\mathbf{FP}(\mathbf{con},\mathbf{i},\mathbf{con}) = \mathbf{FP}(\mathbf{con},\mathbf{j},\mathbf{con})$$

is not true?

5. COMPLEXITY OF KERNEL PROLOG

To classify logic procedures in **KP** and **KDP** we need a notion of strong equivalence.

Let lp/n be a logical procedure and $pr=lp(\bar{W})$ be some its instance. If $comp(pr)$ is successful and $comp(pr)=(s_1,\dots,s_n)$ we set

$$image(pr,1) = \bar{W}\circ con(s_n)$$

and proceed by *backtrack* on s_n. So we obtain new computation $comp(pr,1)$. If it is again successful and $comp(pr,1)=(s_1,\dots,s_n,\dots,s_{n_1})$, we set

$$image(pr,2) = \bar{W}\circ con(s_{n_1})$$

and proceed by *backtrack* on s_{n_1} and so on either infinitely or up to the first i such that $comp(pr,i)$ is infinite or unsuccessful. In first case we set

$$image(pr,j) = \omega$$

for all j≥i. In second case we set

$$image(pr,j) = fail$$

for all j≥i.

Definition 8. A *program* k-*domain* of a logical procedure lp/n (k≥0) is the set $PD_k(lp/n)$ of such its instances $lp(t_1,\dots,t_n)$ that the depth of terms t_i containing variables does not exceed k.

Definition 9. We say that two logical procedures lp_1/n and lp_2/n are *strongly equivalent* ($lp_1/n \equiv lp_2/n$) if for each n-tuple $\bar{W}$ and for all j>0

$$image(lp_1(\bar{W}),j) = image(lp_2(\bar{W}),j). \qquad (1)$$

lp_1/n and lp_2/n are *strongly* k-*equivalent* ($lp_1/n \equiv_k lp_2/n$) if (1) is true for all instances of lp_1/n and lp_2/n in their k-domains.

Definition 10. Let $\mathbf{C}_1$ and $\mathbf{C}_2$ be two classes of logical procedures. $\mathbf{C}_2$ is a *conservative extension* (or *conservative* k-*extension*) of $\mathbf{C}_1$ if $\mathbf{C}_1\subseteq\mathbf{C}_2$ and for each lp_2 in $\mathbf{C}_2$ there is lp_1 in $\mathbf{C}_1$ such that $lp_1\equiv lp_2$ (respectively $lp_1\equiv_k lp_2$).

The next theorem states that in **KP** a single choice point and a bounded number of resolvent subgoals are needed.

Theorem 5. **KP** is a conservative extension of **KP(1,con,*)**.

Let *lp*/1 be a logical procedure and *a* be a term. *lp.a* will denote the logical procedure resulting from *lp*/1 by renaming its main/1 as oldmain/1 and adding the new clause

main(L) :- oldmain([a¦L]).

Theorem 5 relies on the following lemma.

Lemma. There is a logical procedure *solver*/1 in **KP(1,con,*)** such that for each *lp/n* in **KP** and for all $i \geq 1$ and $\bar{W}$

$$image(lp(\bar{W}),i) = image(solver.\{lp\}([[\,]^i ¦ [\bar{W}]]),1).$$

($\{lp\}$ is a coding of logical procedures by ground lists; $[\,]^k$ denotes the k-element list of the form [[],[],...,[]].)

Definition 11. A logical procedure *lp*/n is *functional* if for any its instance $lp(\bar{W})$ and for any i>1 $image(lp(\bar{W}),i)$ equals either to ω or to fail.

The following corollary can be derived from the lemma.

Corollary. For any functional logical procedure lp_1/n in **KP** a deterministic logical procedure lp_2/n exists in **KP(0,con,*)** such that $lp_1/n \equiv_0 lp_2/n$.

Definition 12. Let **C** be a class of logical procedures. We call *lp*/n in **C** *iterative* if it is in **C(con,con,con)**.

Our main optimization result for **KDP** shows that recursion can be eliminated by iteration in this class.

Theorem 6. **KDP** is a conservative extension of **KDP(con,con,con)**.

This theorem is derived from the lemma above and from the same fact proven earlier in [13] for deterministic logical procedures. As to algorithmic complexity of **KP** and **KDP** it is clear that *que*(**KP**) is unsolvable.

6. CONCLUSION

Space optimization results in this paper show that in general both recursion control stacks can be bounded by small integers for logical procedures in **KP** and the whole workspace can be bounded by a constant for any logical procedure in **KDP**. Of course these space optimization theorems must be regarded only as a theoretical background because they cause high delay and lead to processing of a list in global stack as long as the procedure itself. Nevertheless we are convinced that the AS-machine semantics and introduced space consumption measures are adequate for estimating Prolog programs in practice.

7. ACKNOWLEDGEMENTS

We gratefully acknowledge Mats Carlsson from SICS, Paul Tarau from Université de Moncton and Francois Bry from ECRC for stimulating discussion of our Prolog workspace metrics and for kindly criticism of their illustrations.

REFERENCES

1. Lloyd, J.W. *Foundations of logic programming*. Springer Series Symbolic Computation Artificial Intelligence(1987).

2. Warren, D. An abstract Prolog instruction set. - Tech. note 309, SRI Intern.(1983).

3. Tick, E. Warren, D. Towards a pipelined Prolog processor. New Generation Computing,**2**,pp.323-345(1984).

4. Gallagher, J. Codish, M. Shapiro, E. Specialization of Prolog and FCP programs using abstract interpretation. New Generation Computing, **6**,pp.159-186(1988).

5. Janssens, G. Bruynooghe, M. On abstracting the procedural behavior of logic programs. This volume.

6. Kanamory, T. Kawamura, T. Analyzing success patterns of logic programs by abstract hybrid interpretation. ICOT Tech. Report, TR.279(1987).

7. Debray, S.K. Warren, D.S. Automatic mode inference of logic programs". J.Logic Programming,**5**(3),pp.207-229(1988).

8. Mannila, H. Ukkonen, E. Flow analysis of Prolog programs. Proc. 4th Symp.on Logic Programming, pp.205-214, San Francisco(1987).

9. Mellish, C.S. Abstract interpretation of Prolog programs. Proc. 3d Int.Conf. on Logic Programming, pp.181-198,London(1986).

10. Boulanger, D. Deep logic program transformation using abstract interpretation. This volume.

11. Bruynooghe, M. The memory management of Prolog implementations. Logic Programming, N.Y, Academic Press, pp.193-210(1982).

12. Warren, D. An improved Prolog implementation which optimizes tail recursion. Proc. of the Logic Programming Workshop (Tarnlund, S.A. ed.), Debrecen,Hungary(1980).

13. Dikovsky, A.Ja. Space considerations in Prolog. Proc. of the Symp. on Logical Foundations of Comput. Sci. "Logic at Botik'89", (Meyer, A.R. Taitslineds.), pp.101-107(1989).

14. Gavrilenko, Ju. Recursion and iteration on Cuper algebras (Russ.). To publish.

15. Gavrilenko, Ju. Applications of Cuper algebras to the theory of tail recursion in logic programming (Russ.). To publish.

Or-Parallel Prolog with Heuristic Task Distribution

Wai-Keong Foong

Department of Computer Science
University of Melbourne
Parkville, Victoria 3052, Australia.
Tel: +61 3 344-4247 Fax: +61 3 348-1184
Email: foong@cs.mu.oz.au

Abstract

An important determiner of the performance of or-parallel Prolog systems is the order in which the or-nodes in the computation tree are tried. Existing models of or-parallel Prolog select clauses for execution in much the same way as sequential systems do; clauses are tried chronologically (that is, in the order stated in the program) with the added advantage that some can be executed simultaneously on other processors. There is no attempt to examine clauses to assess their suitability in terms of finding answers or select those clauses that have low overhead in task migration for scheduling to other processors. Such systems do not exploit or-parallelism to the fullest and may sometimes have worse performance than sequential systems depending on the structure of the computation tree. Often in an or-parallel environment, it is more desirable to distribute nodes with larger work load to other processors leaving those nodes with lighter work load for local processing. This is so that scheduling costs, including task creation and task migration, are reduced and processors spend more time doing useful work. [8, 5] show that or-parallel Prolog systems are more suited to coarse-grain parallelism. We describe a method of gathering empirical data about clauses in the program. A heuristic function uses the data to deduce the relative "weights" of each clause and the system uses these "weights" to select the clauses for local processing or distributing to other processors.

1 Introduction

Prolog offers two types of parallelism, and-parallelism and or-parallelism. And-parallelism is the simultaneous execution of the goals in the body of a clause, while or-parallelism is the simultaneous execution of alternative clauses that match a goal. In this paper we are concerned only with or-parallelism.

Most modern sequential Prolog systems are implemented according to the WAM architecture [16], where the creation and access of variable bindings are fast constant-time operations (that is, the time for these operations is independent of the amount of work in the processor). To create and bind a variable requires nothing more than a write instruction to a memory word, while accessing its value involves a single read instruction. In the case of binding a variable belonging to an ancestral node (conditional binding), another write operation is needed in the *trail*. This area holds all conditional bindings and these need to be undone when a processor *backtracks* to try an alternative clause. Backtracking in Prolog is somewhat akin to task switching and corresponds to the *try*, *retry* and *trust* operations in the WAM.

In an or-parallel Prolog system, backtracking is replaced by spawning multiple tasks to examine alternative clauses. For an or-parallel Prolog system to be more time-efficient than a sequential Prolog system, the cost of creating multiple tasks must not be so much greater than the cost of the backtrack operation that it outweighs the gains from parallel execution. It is also desirable that any or-parallel implementation of Prolog maintains the efficiency of the operations to bind and access variables.

There are two important aspects of an or-parallel system. One is the selection of tasks for distribution to other processors; the other is the task switching operation itself. Often task switch operations are somewhat affected by the memory architecture of the machine. In shared memory systems like the Aurora model [1], task switching on a processor involves unwinding the private binding array which holds conditional bindings of variables in or-nodes when the processor moves up the tree and restoring bindings as it moves down the tree to the appropriate branch where work is available. In these cases the time taken for task switch operations is dominated by the bind and unbind operations in the binding array. With the Muse model [3], task switching involves copying of stack space (term stack) of the WAM data areas. This approach is more suitable to architectures with non-shared memory models. Evidently the main cost of task switching comes from the copying operations. The larger the heap of the task being distributed, the more time taken to replicate this heap on other processors. Our model is based on the stack copying approach similar to the Muse.

In [5] we measured the performance of two methods of scheduling or-nodes in an or-parallel environment. One involves work distribution giving priority to or-nodes higher up in the computation tree (top-most) while the other gives priority to or-nodes closer to the leaves (bottom-most). The result shows that scheduling top-most nodes first performs better in load balancing, usage of processor resources and overall execution time of the system than bottom-most node scheduling. The model was successful because it was based on the assumption that or-nodes higher up the tree would have acquired a smaller heap and hence have a lower cost of copying than those at lower levels. Another important assumption was that higher nodes have larger workload and hence are more favourable for distribution. Such assumptions may be reasonable for a well-balanced computation trees. However, many Prolog computation trees have highly irregular (unbalanced) shape, and for these trees we can no longer assume that top-most nodes are necessarily heavier. Therefore to ensure optimal scheduling of large work to idle processors we need to determine the "weight" of each clause within the or-node. In oúr system the "weight" of a clause consists of two components; the number of steps it takes to reach it's leave node, the amount of memory cells consumed by the clause. The formal gives an estimate of the actual amount of workload of the clause while the latter estimates the amount of copying overhead necessary to move this clause to another processor. These two components are antithetical to the scheduling decision and the overall weight estimation function must reflect this.

In this paper, we describe a method of static and dynamic analysis of clauses in a or-node for a Prolog program. The static analysis involves collection of clause attributes that pertain to the two components and the dynamic analysis consists of applying a heuristic function to these attributes to give an estimate of the "weight" of the clause. Clauses in an or-node are then ranked according to their weights and the function of the scheduler is to choose the clause whose weight is the highest for distribution. For a full description of the scheduling process and the model, see [4, 5].

In Section 2 we describe the clause attributes relevant for our analysis, the method of gathering these attributes and the heuristic function that applies to the data. Section 3 gives a brief description of the system architecture from [4] incorporating this technique of clause selection. Section 4 gives an analysis of the new model and Section 5 presents the conclusion.

2 Heuristics

In this enhanced scheduling scheme each clause in an or-node is assigned a "weight". This "weight" reflects the favourability of the clause for distribution to other processors. The larger the weight the higher the priority for the clause to move to another processor.

The weight assigned to clauses in or-nodes is computed using an evaluation function which takes several clause attributes as inputs. Clause attributes can be collected during the compilation phase and the execution phase. In the compilation phase, static features of clauses are examined. For example, the number of literals in the clause (subgoals), the type of arguments in the head clause, the groundness of arguments and the number of distinct variables. In the execution phase, dynamic features are collected. This includes examining the arguments of clauses, especially those of type *list* and *structure*. Our analyser, in both phases, consists of two main components. One for analysing the explosiveness of clauses and the other for examining the arguments of the clause and the query (goal) presented. The result generated by them will be fed into the evaluation function to produce a weighted tuple of clause attributes and inserted into a table for easy reference during the execution phase. We discuss the two main components of the analyser below.

2.1 Explosiveness of Clauses

The number of literals is the primary measure of the explosiveness of the clause, and if we can determine the number of steps it takes to resolve all subgoals then an approximation can be made for the execution time of the clause. This gives us a fair indication of what the work load is for this clause. We can deduce the number of steps each subgoal takes with the help of a simple meta-interpreter. An example of a meta-interpreter for pure Prolog is,

```
% solve(Goal, Steps)
solve(true, 0) :- !.
solve((GoalA, GoalB), Steps) :- !,
        solve(GoalA, StepA),
        solve(GoalB, StepB),
```

```
        Steps is StepA + StepB.
solve(Head, Steps) :-
        clause(Head, Body),
        solve(Body, StepB),
        Steps is StepB + 1.
```

The value returned in Steps is the length of the SLD derivation. For a simple program given below without arguments,

```
p :- a, p, b.     b :- b1, b2.          q :- w, z.    z :- z1.
p :- c, q.        b :- b3, b4, b5.      q :- w, q.
                                        q :- x, y.
Unit Clauses,
        a.  b1.  b2.  b3.  b4.  b5.
        c.   x.   y.   w.  z1.
```

our meta-interpreter will return the following results,

```
3?- analyse_cl((p,q,b), RList).

        cl_stats(Steps, Predicate, ClauseNo, Rflag)

RList = [cl_stats(2, b, 1, -), cl_stats(3, b, 2, -),
         cl_stats(6, q, 1, -), cl_stats(9, q, 2, *),
         cl_stats(2, q, 3, -), cl_stats(24, p, 1, *),
         cl_stats(18, p, 2, *)]
```

RList holds the attributes of all or-clauses in the given program. For example, the first element of the list says that the first clause of b takes a total of 2 steps. The "-" indicates that the clause is non-recursive. Similarly, in the last element of RList, it gives the second clause of p a count of 18 and the clause is "*", recursive.

One problem we face with such analysis is the presence of recursive clauses. It is impossible to predict the exact weight of such clauses at compile time when head arguments are not known. Fortunately, for applications like the distribution of tasks in an or-parallel environment, precise information is not crucial. A minor inconsistency in weight estimation does not affect performance drastically. Alternatively, we can use the method proposed in [10] where a weight of 1 is assigned to a tail recursive call In our case, we prefer to delay the computation of the weight until run time when arguments are known and some factor, which is dependent on the size of recursive data structures (see below), can be added to the final load estimate. Future research may involve more complex analysis of recursive arguments using techniques of abstract interpretation to give better load estimates similar to [9] where such analysis was performed for independent and-parallelism.

2.2 Arguments of Clauses

The other attributes may not directly measure the execution time of clauses but they affect the scheduling cost when work is to be distributed to other processors. This scheduling overhead is especially important in our model because of the extensive use of copying during work distribution. The less memory which has to be copied, better the performance of the model. This component estimates the memory space consumed for each clause in the or-node.

The WAM model, on which our implementation is based, is a structure-copying system. Terms are represented by a one-word cell. Each cell consists of a tag which denotes the type of cell and a value or a pointer. In addition, variables are classified as permanent or temporary. Permanent variables take up space in the stack while temporary variables occupy heap space. Stack space is reclaimed when a procedure exits in much the same way as imperative languages, while heap space is only reclaimed by failure or garbage collection. Complex terms and lists are constructed in the heap.

In NU-Prolog, a compound term with N arguments requires (N + 1) cells to represent it, while a list with N arguments requires (2N + 1) cells. Clearly in a system like ours where distribution of work involves copying of heap and trail space, the less memory space the terms occupy, the more efficient and lower the cost of scheduling overhead.

The analysis takes into account of many factors related to arguments such as permanent and temporary variables, the number of cells of compound terms and lists and the groundness of these terms. Of course much of the information can only be collected at run time, however, at compile time, we can do some intermediate analysis to alleviate the task of run time analysis.

The result of the compile time analysis is a tuple stored together with the value for the 'explosiveness' of clauses described in the previous section. We give some examples of how information is collected for different types of arguments in this phase.

In complex terms like structures and lists which may appear at the head of the clause for example,

```
p(a(B, C), x(y(z)), [a,b,c,d]) :- ......
```

for each argument of the clause head, the analyser returns a four-tuple term of the following,

```
attrib(v, 1, struct, 2), attrib(g, 2, struct, 2), attrib(g, 3, list, 4)
```

The fields in the four-tuple term are:

- the mode
- the position
- the type
- number of elements

For nested structures like the second argument of the predicate p, the number of elements field includes the depth of nesting. In the above example, we have one argument + one depth of nesting.

If a *list* type argument is ground, then the number of elements is known at compile time. In cases like this, no run time evaluation is required. In the case of variable terms of type *list*, for example in the clause,

```
p([H|T], _, _) :- .....
```

our analyser will return a term such as,

```
attrib(v, 1, list, _)
```

The fourth argument is left as "unknown" and not applicable at this stage. However at run time when the list is instantiated, this field gives a count of the number of elements in the list much the same as a ground term. It is then appropriate for the evaluation function to give an estimate of the number of cells consumed by the clause. Other data collected includes the number of ground terms. This count includes complex type arguments as well as atoms. Also included in the argument analysis is the number of distinct variables present in the clause which gives a measure of the stack space usage if they are permanent and heap space usage if they are temporary according to the WAM execution model.

2.3 Construction of Table of attributes with Weights

The goal of the static analysis is to gather sufficient information about the clause to enable us to infer the weight of the clause at run time when more information on some of the arguments are known. The two components of the static analyser mentioned above are combined to produce a *table* of clause attributes. Each entry has the following arguments,

```
cl_attr(Ns, Term, Cp, Rflag, A, Ng, Cl, Dv)
....
```

```
where,
           Ns = Total number of steps it takes to resolve all subgoals
         Term = Predicate Name
           Cp = Clause position
        Rflag = Recursive flag, '*' for recursive, '-' non-recursive
            A = Arity of the Predicate
           Ng = Number of ground terms
```

```
Cl = List attributes of complex terms
Dv = Number of distinct variables
```

For example, given the program below :

```
p(a, Y) :- p(Y, a).
p(X, Y) :- q(X), q(Y).
p(X, Y) :- q(X), p(Y, Y).
q(a).
q(b).
member(X, [X|_]).
member(X, [Y|L]) :- member(X, L).
```

the analyser produces the following table :

```
3?- analyse((p(X,Y), member(X, Y)), Table).
 Table = [cl_attr(1, p, 1, *, 2, 1, [], 1),
          cl_attr(4, p, 2, -, 2, 0, [], 2),
          cl_attr(9, p, 3, *, 2, 0, [], 2),
          cl_attr(1, q, 1, -, 1, 1, [], 0),
          cl_attr(1, q, 2, -, 1, 1, [], 0),
          cl_attr(1, member, 1, -, 2, 0, [attr(v, 2, 1, _EHRVG)], 1),
          cl_attr(1, member, 2, *, 2, 0, [attr(v, 2, 1, _EHRWE)], 1)],
```

When the system is presented with the following query,

?- p(a, b), member(a, [b,c,a,d]).

the analyser will collect the attributes of each predicate in the query now relevant for run time analysis. Execution starts with the first call to the p goal. When p encounters choice points, the system will look up the table of clause attributes to find all entries that match the head and arity of p. For each entry and using the run time attributes of goal p, an evaluation function is called to assign the "weight" to each clause after comparing the run time and compile time attributes. Each entry in the *table* is now augmented by an additional "weight" field to reflect their suitability for local or foreign processing. An example of what the *table* is like after this process is as follows,

```
cl_attr(Ns, Term, Cp, Rflag, A, Ng, Cl, Dv) - Weight
Table = [cl_attr(1, p, 1, *, 2, 1, [], 1) - 1,
         cl_attr(4, p, 2, -, 2, 0, [], 2) - 4,
         cl_attr(9, p, 3, *, 2, 0, [], 2) - 7,
         cl_attr(1, member, 1, -, 2, 0, [attr(v, 2, 1, 4)], 1) - 2,
         cl_attr(1, member, 2, *, 2, 0, [attr(v, 2, 1, 4)], 1) - 5]
```

2.4 An Example of an Evaluation Function

We are currently using a very simple (and cheap to compute) evaluation function. However, even with such a simple function, we can achieve considerable overall performance improvement compared to the normal chronological selection of goals for task distribution.

Recall that the final weight assigned to each clause represents the cost effectiveness of distributing the clause. In the context of favouring large grain work load, the more steps it takes for a clause to finish resolving its subgoals, the more desirable it would be for distribution. Contrary to this is the amount of memory cells the clause has consumed. The higher the amount the less favourable it would be for moving to other processors since it involves more copying overhead which is directly proportional to the scheduling costs. However, there are complications involved if the clause we are analysing is recursive. Recursive clauses operate on recursive data structures like lists and structures. Hence the larger these data structures the more the number of steps it takes to finish. We face a contradiction here. On one hand the amount of work load is favourable for distribution if large recursive data structures are encountered, while on the other they consume a large amount of memory space which is adverse for task migration. A more consistent view is to compare the time taken to copy n bytes with the time taken to execute a procedure call since the total number of steps taken by each clause approximates to the total number of procedure calls made by its sub-goals.

Let the variable E represents the number of steps it takes to finish execution for the clause. Let A represents the number of elements of type *list* and/or the (arguments + nested depth) for type *structure*. Let C represents the number of cells consumed by A. For example if A have value 4 for *list* type arguments then C has value 9 (2A + 1). The function subtracts some value expressed by kC and adds A to a recursive clause to reflect the extra load (procedure calls) involve in processing such arguments. Let the variable G represents the *total* of the number of ground atoms and the number of variables in the clause. Again the function subtracts some value expressed in kG to include the amount of data to be copied. In general, the best approximation to the function is given in the following expression,

$W = E + A - kC - kG$

where k measures the amount of the overhead involved in task migration expressed in terms of a procedure call. In the present situation we consider only the time taken to copy n memory cells because this is the most variable portion of overhead involve in task migration. Other overheads include publishing work, finding work, signalling (interrupts) which can be added as some constant value towards the total overhead involves in work migration.

We must stress that a more sophisticated evaluation function would give better estimates of the weight of each clause, but at the expense of greater run time overhead. We envisage that even with this simple scheme, improvements to the task distribution scheme can be realised.

3 The New Model

A detailed description of the execution model and its implementation can be found in [4] and a performance evaluation of the model is in [5]. In this section we give a brief description of the system architecture in general and the scheduling scheme based on this method of heuristic task distribution.

3.1 System Architecture

We assume a multiprocessor architecture where each processor has its own local (private) memory and a shared global memory to store control information. Each processor also implements the standard WAM sequential machine retaining all WAM optimisations and efficiencies. The WAM data areas consist of the stack (environments and choice points), the heap (term stack) and the trail. These are resident in each processor's local memory. In addition, each processor also has a resident copy of the program code.

3.2 Basic Operation

Each processor executes a separate piece of code in the program. When a goal is solved, the result is reported to the user or passed back to the host processor from where the goal was taken. When goal matches multiple clause heads (corresponding to a *Try* instruction in the WAM), it can select any one of the clauses for execution and give the others away to idle processors. The selection criteria is based on the scheme of heuristic weight assignment (described above). This process of clause selection is divided in two stages, *local* and *global*.

The local stage involves interpreting the neck instruction preceding the WAM bytecode of *Try*, *Retry*, *Trust* for a block of alternative clauses. This instruction performs the task of weight evaluation of all clauses with the same clause head. It does this by looking up at the table generated earlier via static analysis and together with the arguments, uses the evaluation function to compute the estimated weight for each clause. Note this process happens only once when a new or-node is encountered and this constitutes the run time phase of the analysis. Presumably if clauses are ordered in increasing order of their weights, then selection of clauses for local processing happens in the normal top-down manner but distribution of clauses are selected bottom-up. Task distribution involves announcing to the system where work can be found. This is done by "publishing" an information node in the global task queue. The information node consists of the processor ID, weight of the clause, position of this node with respect to the computation tree and the address of the choice point. The processor ID together with the address of the choice point uniquely identifies the task suitable for distribution. The weight is used to order the clauses further in the global task queue published by different processors. In the event of a tie in weights, then the position of the node with respect to the computation is used to order the clauses. Currently this favour nodes closer to the root (top-most). The global task queue ensures that the next available task is the most suitable one for distribution system-wide.

3.3 Task Migration

When failure is encountered in a processor, normal WAM-style backtracking is performed. The next alternative clause is selected from the local task queue. If the local task queue is empty, the processor searches the global task queue and selects the first node. If the global task queue is also empty the processor becomes idle. With the available information, the processor interrupts the "host" processor requesting it to release the task. The "host" processor responds to the interrupt and releases the task to the requesting processor. Before releasing the task, it selects the next clause (next heaviest-weight clause) in the same or-node, if one is available, for publishing in the global task queue. If none is available, the pointer to the next alternative clause is set to NULL and stack space will be reclaimed when the processor backtracks to this point. Releasing a task involves copying the heap and trail space relevant to the clause to the destination processor. Once this is done the destination processor restarts from the instruction of the new clause. A processor can select a clause that belongs to itself from the global task queue. This is like sending an interrupt to itself and releasing the clause for its own execution. The overhead for a single processor system will be simply the publishing and deleting of the information node from the global task queue. All other operations of the WAM are maintained.

4 Analysis of the New Model

There are three main principles governing the scheduling policy employed in this model.

First, the matching of idle workers with work should be done very quickly with minimal overhead. By having idle workers looking for work (*receiver-initiated*) instead of busy worker distributing work we minimise the idle time of processors. Also with the global task queue, idle processors can determine accurately where suitable work can be found.

Second, if there is work to be given away then *heavy* nodes (large workload) will be considered before *light* nodes. This minimise the time spent on scheduling and maximises time spent doing useful work. Our evaluation function through weights assignment offers a two stage clause ranking for local processing and global task distribution. This method leads us to a good approximation of best-first search.

Third, the copying of data areas must be kept small. This is related to the above discussion particularly the effectiveness of the evaluation function predicting correct weights for the clauses.

5 Conclusion

To increase overall performance of an or-parallel Prolog system, the algorithm which schedules jobs for distribution to idle processors must favour larger grain-size jobs. We have described a method of gathering information on various clause attributes to help us infer on the work load associated with the clauses. The attributes are input to a heuristic function which assigns a value (weight) to the clause. This weight is an estimate of the amount of work on that branch. Clauses are ranked both locally and globally for effective execution and clause selection. At present a rudimentary implementation of the above heuristic analysis is completed in Prolog. Though the part on evaluation function is simple, it is adequate for this application. A more sophisticated function would risk running into severe run time overhead.

6 Acknowledgements

I would like to thank Lee Naish for many fruitful discussions and John Shepherd for proofreading the draft.

References

[1] Ewing Lusk, David H. D. Warren, Seif Haridi, et al. *The Aurora Or-Parallel Prolog System*, New Generation Computing, 7, 243-271, 1990.

[2] Khayri A. M. Ali and Roland Karlsson, *The Muse Or-Parallel Prolog Model and its Performance*, SICS Research Report, March 1990.

[3] Khayri A. M. Ali and Roland Karlsson, *The Muse Approach to Or-Parallel Prolog*, SICS Research Report, 2 October 1990.

[4] Wai-Keong Foong, *Top-most Scheduling for Efficient Execution of Or-Parallelism in Prolog*, Technical Report 90/22, Department of Computer Science, University of Melbourne, November 1990.

[5] Wai-Keong Foong and John Shepherd, *A Top-Most Scheduling Model for Execution of Or-Parallelism in Prolog and its Performance*, Technical Report 91/7, Department of Computer Science, University of Melbourne, February 1991.

[6] C. B. Suttner and W. Ertel, *Automatic Acquisition of Search Guiding Heuristics*, Proc. of 10th Int'l Conf. on Automated Deduction (CADE). Springer, 1990.

[7] J. Schumann and R. Letz, *PARTHEO: a High Performance Parallel Theorem Prover*, Proc. of 10th Int'l Conf. on Automated Deduction (CADE). Springer, 1990.

[8] M. Furuichi, K. Taki and N. Ichiyoshi, *A Multi-Level Load Balancing Scheme for Or-Parallel Exhaustive Search Programs on the Multi-PSI*, 2nd ACM SIGPLAN Symposium on Principles & Practice of Parallel Programming, 50-59, Seattle, March 1990.

[9] S. K. Debray, N. W. Lin and M. Hermenegildo, *Task Granularity Analysis in Logic Program*, ACM SIGPLAN Conference on Programming Language Design & Implementation, 1990.

[10] E. Tick, *Compile-Time Granularity Analysis for Parallel Logic Programming Language*, 5th FGCS, Tokyo 1988.

[11] Khayri A. M. Ali, *Or-Parallel Execution of Prolog on BC-Machine*, Proc. of the 5th Int. Conf. and Symposium on Logic Programming, 1531-1545, 1988.

[12] R. Butler, T. Disz, E. Lusk, R. Olson, R. Overbeek and R. Stevens, *Scheduling Or-Parallelism: an Argonne perspective*, Proc. of the 5th Int. Conf. on Logic Programming, 1590-1605, MIT, 1988.

[13] D. H. D. Warren, *The SRI Model for Or-Parallel Execution of Prolog - Abstract Design and Implementation Issues*, Proc of the IEEE 1987 Symposium on Logic Programming, San Francisco, 92-102, 1987

[14] D. H. D. Warren, *Or-Parallel Execution Models of Prolog*, TAPSOFT 1987, Pisa, Italy, 243-259, Springer Verlag, 1987.

[15] Wolfgang Bibel, *Automated Theorem Proving*, Friedr. Vieweg & Sohn Verlagsgesellschaft mbH, Braunschweig, 1982.

[16] D. H. D. Warren, An Abstract Prolog Instruction Set, Technical Note 309, SRI International, 1983.

A WAM Compilation Scheme

Hans-Günther Hein*
Computer Science Department
University of Kaiserslautern
P. O. Box 30 49
D-6750 Kaiserslautern
Germany

Manfred Meyer†
German Research Center for
Artificial Intelligence (DFKI)
P. O. Box 20 80
D-6750 Kaiserslautern
Germany

Abstract

In order to ease the compilation of logic programs, source-to-source transformations are applied to obtain still declarative programs. The horizontal compilation consists of grouping clauses together, performing a flattening process to remove nested compound terms, partially evaluating them, and reordering the constraints followed by variable classification. The constraints are mapped onto a single $\doteq$-primitive.

For the resulting, very concise representation we present the vertical compilation rules for generating WAM code. They can be further optimized taking type inference information into account.

1 Introduction

The WAM [26] is a well elaborated machine model for executing logic programs. When compiling PROLOG, the unification algorithm is basically *specialized* according to the arguments occurring in the head of the program clauses. During head unification, the variables are either stored in the environment or held in registers to be moved to their appropriate argument positions in the body calls. This compilation technique is normally working on a clause based view. Global properties of variables together with e.g. a determinism analysis can help to specialize the unification algorithm, so that code can be generated which is even more efficient than C code [20, 21].

The trend towards native code generation is obvious: Some instructions for moving a variable to an argument position—a frequently used instruction in the clause body—can

*Please, send correspondence to the first author; e-mail: hein@informatik.uni-kl.de

†The work of the second author is carried out as part of the ARC-TEC project, supported by BMFT under grant ITW 8902 C4.

(usually) be performed by *one* assembly language statement. The interpretation overhead, caused by a byte code emulator for these instructions is tremendous. Unfortunately, a *simple* macro expansion of WAM instructions tends to be rather memory consuming. By incorporating program analysis techniques ([27, 2, 19]) both memory and execution time resources can be minimized by generating native code. Furthermore, a lot of machine independent optimizations can be applied concerning the unification of compound terms ([22, 16]), shallow backtracking ([22, 4]), improved indexing techniques ([6, 24, 10]) and general optimizations [8] .

2 Horizontal compilation

In this section we will introduce horizontal (*source-to-source*) transformations to be performed before vertical compilation (*source-to-instruction*). The horizontal transformation results in a program which is still declarative but much better suited for vertical compilation.

2.1 Definite Equivalences

First, we will apply Clark's completion procedure [5]. A severe disadvantage of standard PROLOG compilation techniques is the clause based view of the compiler. This procedure groups clauses together by a merely syntactical transformation, replacing implications by equivalences. Thus, the whole procedure is coded into one definite equivalence. This resembles the WAM level, where there is only one entry point into any procedure, thus easing global compilation techniques [10].

Def 1 (Clark's completion procedure) *Let the clause*

$$\mathsf{p}(\mathsf{t}_1,\ldots,\mathsf{t}_n) \leftarrow \mathsf{B}_1,\ldots,\mathsf{B}_n.$$

be part of a program P. In the first step, we replace the constraints in the head by new variables $X_1,\ldots,X_n$, writing

$$\mathsf{p}(\mathsf{X}_1,\ldots,\mathsf{X}_n) \leftarrow X_1 \doteq t_1,\ldots,X_n \doteq t_n, \mathsf{B}_1,\ldots,\mathsf{B}_n.$$

We obtain a set of clauses with the same head:

$$\mathsf{p}(\mathsf{X}_1,\ldots,\mathsf{X}_n) \leftarrow \mathsf{E}_1$$
$$\vdots$$
$$\mathsf{p}(\mathsf{X}_1,\ldots,\mathsf{X}_n) \leftarrow \mathsf{E}_k$$

with $E_i \equiv [X_1 \doteq t_1,\ldots]$ for$1 \leq i \leq n$. Now, we group the clauses together and obtain:

$$\mathsf{p}(\mathsf{X}_1,\ldots,\mathsf{X}_n) \leftrightarrow \mathsf{E}_1 \vee \ldots \vee \mathsf{E}_n.$$

Example: (Clark's Completion for rev/2)

rev(nil,nil).
rev([L1|L2],L3) ← rev(L2,L), app(L,[L1],L3).
is transformed to:
rev(X1,X2) ← $X1 \doteq nil$, $X2 \doteq nil$.
rev(X1,X2) ← $X1 \doteq [L1|L2]$, $X2 \doteq L3$, rev(L2,L), app(L,[L1],L3).
results in:
rev(X1,X2) ↔ ($X1 \doteq nil$, $X2 \doteq nil$) ∨
($X1 \doteq [L1|L2]$, $X2 \doteq L3$, rev(L2,L), app(L,[L1],L3))

Flattening compound terms, which is normally done in the code generator, is performed as a horizontal precompilation step. We can write a list as a structure `cons(L1,L2)`, where *L1* is the head of the list and *L2* is its tail. This `cons` functor is handled by a special tag in the WAM. Thus, the pointer to a `cons` functor is not needed.

Def 2 (WAM flattening) *Let $p(t_1, \ldots, t_n)$ be an atom in the definite equivalence. If t_i $(1 \leq i \leq n)$ is not a variable, replace $p(t_1, \ldots, t_n)$ by*

$$Var \doteq t_i, p(t_1, \ldots, t_{i-1}, Var, t_{i+1}, \ldots, t_n),$$

where Var is a new variable not yet occurring in the definite equivalence.

Let $s_1 \doteq s_2$ be a $\doteq$-constraint. If s_2 is a variable and s_1 is not a variable, replace $s_1 \doteq s_2$ by $s_2 \doteq s_1$. Let s_1 be a variable and s_2 be a structure $f(t_1, \ldots, t_n)$ of arity n: If t_i $(1 \leq i \leq n)$ is neither a variable nor a constant, replace $s_1 \doteq f(t_1, \ldots, t_{i-1}, t_i, t_{i+1}, \ldots, t_n)$ by

$$Var \doteq t_i, s_1 \doteq f(t_1, \ldots, t_{i-1}, Var, t_{i+1}, \ldots, t_n)$$

where Var is a new variable not occurring in the definite equivalence.

If neither s_1 nor s_2 are variables, they are either constants or structures. If $s_1 = const_1$ and $s_2 = const_2$ with $const_1 \neq const_2$, replace the constraint by a `fail`, *if $const_1 = const_2$, delete the equation. If $s_1 = f(t_1, \ldots, t_n)$ and $s_2 = f(u_1, \ldots, u_n)$ are structures with same arity and functors, replace the $\doteq$-constraint by n $\doteq$-constraints of the form*

$$t_1 \doteq u_1, t_2 \doteq u_2, \ldots, t_n \doteq u_n;$$

otherwise replace the $\doteq$-constraint by a `fail`.

WAM flattening continues until no more rules can be applied.

Note that the flattening order is subject to change in the constraint reordering process. The resulting definite equivalences have a restricted syntactical form:

- The head is an atom whose arguments are variables.
- The arguments of the literals in the body are variables.
- The $\doteq$-constraints have one of the forms:

- $Var_1 \doteq Var_2$
- $Var \doteq constant$
- $Var \doteq f(t_1,\ldots,t_n)$, where t_i are either constants or variables ($1 \leq i \leq n$).

- The or-operator $\vee$ separates the or-branches.
- A special atom `fail` indicates compile time failure.

Example: (rev/2 after WAM flattening)

rev(X1,X2) $\leftrightarrow$
  $(X1 \doteq nil,\ X2 \doteq nil)\ \vee$
  $(X1 \doteq cons(L1, L2),\ X2 \doteq L3,$ rev(L2,L), $X3 \doteq cons(L1, nil),$ app(L,X3,L3))

Herein, the arguments occurring in the procedure head are temporary variables. The finite tree constraints will only be handled by the code generation for $\doteq$ for both head and tail.

2.2 Variable classification

One very important classification issue is the recognition of the variable's class [7], which is either *permanent* or *temporary*. Permanent variables (Y-variables) have to be stored in the environment while the temporary ones (X-variables) can be held in a register.

Before a literal is called, there is a certain assignment of X-registers. After a register-destroying literal is called, nothing can be said about the assignment. After a register-preserving literal is executed, the assignment of the X-registers has not changed or only changed deterministically (known at compile time). Examples of register preserving literals are unification over finite trees ($t1 \doteq_{T(\Sigma,Var)} t2$), boolean constraints ($b1 \doteq_{Bool(\oplus,\wedge)} b2$) [3], finite domain unification ($d1 \doteq_{FD} d2$) [23], cut, and input/output operations.

Def 3 (chunk, pseudo-chunk) *A chunk is a sequence of literals $p_1,\ldots,p_n,q$, where $p_1,\ldots,p_n$ are register-preserving literals and q is a register-destroying literal. A pseudo-chunk is a sequence of literals $p_1,\ldots,p_n$, where $p_1,\ldots,p_n$ are register-preserving literals, p_n is the last literal in the clause. Chunks are the basic entities for the code generator.*

2.3 Variable Types

In order to keep the definition of temporary and permanent variables simple, we assume that the disjunctions are not nested arbitrarily in the definite equivalences, but only occur at the outermost level.

Def 4 (X-variable, Y-variable) *A (temporary) X-variable is a variable which occurs only in* **one** *chunk, otherwise it is a (permanent) Y-variable.*

The register associated with an X-variable is referred to by $Var^{\rightarrow \text{X-reg}}$, which ranges over $\{1, \ldots, \#reg\}$, where #reg is the maximum number of temporary registers in the abstract machine. A Y-variable has a slot in the environment, whose number is denoted by $Var^{\rightarrow \text{Y-reg}}$.

Example: (rev/2 after chunk separation and X/Y-variable distinction)

rev(X1,X2) $\leftrightarrow$
 $([\ll X1 \doteq nil,\ X2 \doteq nil \gg])\ \vee$
 $([\ll X1 \doteq cons(\underline{L1}, L2),\ X2 \doteq \underline{L3},$ rev(L2,$\underline{\text{L}}$)$\gg,\ \ll X3 \doteq cons(\underline{L1}, nil),$app($\underline{\text{L}}$,X3,$\underline{\text{L3}}$)$\gg$])

The $\ll,\gg$-brackets indicate the beginning resp. the end of a chunk. X1, X2, L2, X3 are X-variables. If one of the $\doteq$-primitives fails in the first or-branch, backtracking occurs, resetting the variables which need to be unbound and restoring the register-variables X1, X2 from the choicepoint. "rev/2" and "app/3" are register-destroying calls which are terminating a chunk. L1, L3, L occur in both chunks of the last OR-branch, so they are Y-variables (underlined in the example listed above).

Def 5 (Type) *A variable Var has an attribute* $Var^{\rightarrow \text{type}}$*, denoting whether the variable is an X-variable (*`temp`*) or a Y-variable (*`perm`*).*

For both variable types one must know whether they occur for the first time or not. A variable occurring for the first time has to be created and initialized.

Def 6 (Occurrence) *A variable Var has an attribute* $Var^{\rightarrow \text{occurrence}}$*, indicating whether the variable occurred for the first time (*`first`*) or not (*`nonfirst`*).*

Last call optimization allows to remove the stack frame before the last call has been done. This reduces the space needed by tail-recursive procedures, but induces other problems: One has to ensure that no references to unbound variables in the actual environment are established. These unbound Y-variables have to be globalized. Another problem is that references from the heap into the stack are forbidden. Since a procedure being called may have an X-register bound to a Y-variable located in the caller's environment and the callee may use this X-register in a compound term, a forbidden reference from the heap to the stack may result.

Def 7 (Safeness) *A variable Var has the attribute* $Var^{\rightarrow \text{safeness}}$*, indicating whether the variable is potentially unsafe or not. A Y-variable is potentially unsafe if it occurs for the last time and has not been statically bound to a non-variable term, and it has not been bound to a variable occurring in the head. An X-variable is unsafe, if it is used in a compound term, potentially referring to an environment which may be removed.*

In the code generation process we will use the attributes *type*, *occurrence* and *safeness*, which will be noted in this particular order as subscripts to the variables. The notation $Z_{perm,nonfirst,safe}$ indicates that the variable Z is a (permanent) safe Y-variable not occurring for the first time. Sometimes the safeness-attribute is omitted to indicate that the safeness needs not to be known.

Type inference techniques also allow to state the type and whether a variable needs to be dereferenced or trailed [19]. In the following section these will be ignored.

Def 8 (drf, trl, type) *A variable Var has the feature $Var^{\rightarrow drf}$ being true iff this variable is already dereferenced. A variable Var has the feature $Var^{\rightarrow trl}$ being true iff the variable binding must not be trailed. A variable Var has the feature $Var^{\rightarrow type}$ indicating the inferred type (ranging over a type lattice).*

Example: (rev/2 after classification)

$$
\begin{array}{l}
\mathrm{rev}(\mathtt{X1}^{\rightarrow \text{X-reg=1}}_{\text{temp,first,safe}}, \mathtt{X2}^{\rightarrow \text{X-reg=2}}_{\text{temp,first,safe}}) \leftrightarrow \\
\quad \ll \mathtt{X1}^{\rightarrow \text{X-reg=1}}_{\text{temp,nonfirst,safe}} \doteq \mathit{nil}, \mathtt{X2}^{\rightarrow \text{X-reg=2}}_{\text{temp,nonfirst,safe}} \doteq \mathit{nil} \gg \\
\vee \\
\quad \ll \mathtt{X1}^{\rightarrow \text{X-reg=1}}_{\text{temp,nonfirst,safe}} \doteq \mathit{cons}(\mathtt{L1}^{\rightarrow \text{Y-reg=3}}_{\text{perm,first,safe}}, \mathtt{L2}^{\rightarrow \text{X-reg=1}}_{\text{temp,first,safe}}), \\
\quad\quad \mathtt{X2}^{\rightarrow \text{X-reg=2}}_{\text{temp,nonfirst,safe}} \doteq \mathtt{L3}^{\rightarrow \text{Y-reg=1}}_{\text{perm,first,safe}} \\
\quad\quad \mathrm{rev}(\mathtt{L2}^{\rightarrow \text{X-reg=1}}_{\text{temp,nonfirst,safe}}, \mathtt{L}^{\rightarrow \text{Y-reg=2}}_{\text{perm,first,safe}}) \gg, \\
\quad \ll \mathtt{X3}^{\rightarrow \text{X-reg=2}}_{\text{temp,first,safe}} \doteq \mathit{cons}(\mathtt{L1}^{\rightarrow \text{Y-reg=3}}_{\text{perm,nonfirst,safe}}, \mathit{nil}), \\
\quad\quad \mathrm{app}(\mathtt{L}^{\rightarrow \text{Y-reg=2}}_{\text{perm,nonfirst,unsafe}}, \mathtt{X3}^{\rightarrow \text{X-reg=2}}_{\text{temp,first,safe}}, \mathtt{L3}^{\rightarrow \text{Y-reg=1}}_{\text{perm,nonfirst,safe}}) \gg .
\end{array}
$$

Classification of variables is not an easy job [7, 14, 15], since tricky variable assignments to registers have to be made, which also determine the quality of the resulting code.

3 Vertical Compilation

The $\doteq$ constraints are not context-sensitive, since we do not have to distinguish between head and body of a clause. This simplifies the implementation of the code generator. We can generate standard WAM code from the $\doteq$-primitives as shown in tables 1 to 5 (The register *freereg* is a register not used in the current chunk). The rules we present determine how to replace the $\doteq$-constraints by WAM instructions. The instructions `put_value_temp` Reg_i, Reg_i and `get_variable_temp` Reg_i, Reg_i are instructions with no effect; hence, they can be omitted.

Example: (WAM code generation for rev/2)

$$\frac{Var1_{\text{temp,first}} \doteq Var2_{\text{temp,first}}}{\texttt{put\_variable\_temp}\ Var1^{\to\text{X-reg}},\ Var2^{\to\text{X-reg}}} \quad (1)$$

$$\frac{Var1_{\text{temp,first}} \doteq Var2_{\text{temp,nonfirst}}}{\texttt{get\_variable\_temp}\ Var1^{\to\text{X-reg}},\ Var2^{\to\text{X-reg}}} \quad (2)$$

$$\frac{Var1_{\text{temp,nonfirst}} \doteq Var2_{\text{temp,nonfirst}}}{\texttt{get\_value\_temp}\ Var1^{\to\text{X-reg}},\ Var2^{\to\text{X-reg}}} \quad (3)$$

$$\frac{Var1_{\text{perm,first}} \doteq Var2_{\text{temp,first}}}{\texttt{put\_variable\_perm}\ Var1^{\to\text{Y-reg}},\ Var2^{\to\text{X-reg}}} \quad (4)$$

$$\frac{Var1_{\text{perm,nonfirst,safe}} \doteq Var2_{\text{temp,first}}}{\texttt{put\_value\_perm}\ Var1^{\to\text{Y-reg}},\ Var2^{\to\text{X-reg}}} \quad (5)$$

$$\frac{Var1_{\text{perm,nonfirst,unsafe}} \doteq Var2_{\text{temp,first}}}{\texttt{put\_unsafe\_value\_perm}\ Var1^{\to\text{Y-reg}},\ Var2^{\to\text{X-reg}}} \quad (6)$$

$$\frac{Var1_{\text{perm,first}} \doteq Var2_{\text{temp,nonfirst}}}{\texttt{get\_variable\_perm}\ Var1^{\to\text{Y-reg}},\ Var2^{\to\text{X-reg}}} \quad (7)$$

$$\frac{Var1_{\text{perm,nonfirst,safe}} \doteq Var2_{\text{temp,nonfirst}}}{\texttt{get\_value\_perm}\ Var1^{\to\text{Y-reg}},\ Var2^{\to\text{X-reg}}} \quad (8)$$

$$\frac{Var1_{\text{temp,nonfirst}} \doteq Var2_{\text{perm,first}}}{\texttt{get\_variable\_perm}\ Var2^{\to\text{Y-reg}},\ Var1^{\to\text{X-reg}}} \quad (9)$$

$$\frac{Var1_{\text{perm,nonfirst,unsafe}} \doteq Var2_{\text{temp,nonfirst}}}{\begin{array}{l}\texttt{put\_unsafe\_value-perm}\ Var1^{\to\text{Y-reg}}, freereg\\ \texttt{get\_value\_temp}\ Var2^{\to\text{X-reg}}, freereg\end{array}} \quad (10)$$

$$\frac{Var1_{\text{perm,first}} \doteq Var2_{\text{perm,first}}}{\begin{array}{l}\texttt{put\_variable\_perm}\ Var1^{\to\text{Y-reg}}, freereg\\ \texttt{get\_variable\_perm}\ Var2^{\to\text{Y-reg}}, freereg\end{array}} \quad (11)$$

$$\frac{Var1_{\text{perm,first}} \doteq Var2_{\text{perm,nonfirst,safe}}}{\begin{array}{l}\texttt{put\_value\_perm}\ Var2^{\to\text{Y-reg}}, freereg\\ \texttt{get\_variable\_perm}\ Var1^{\to\text{Y-reg}}, freereg\end{array}} \quad (12)$$

$$\frac{Var1_{\text{perm,first}} \doteq Var2_{\text{perm,nonfirst,unsafe}}}{\begin{array}{l}\texttt{put\_unsafe\_value\_perm}\ Var2^{\to\text{Y-reg}}, freereg\\ \texttt{get\_variable\_perm}\ Var1^{\to\text{Y-reg}}, freereg\end{array}} \quad (13)$$

$$\frac{Var1_{\text{perm,nonfirst,safe}} \doteq Var2_{\text{perm,nonfirst,safe}}}{\begin{array}{l}\texttt{put\_value\_perm}\ Var1^{\to\text{Y-reg}}, freereg\\ \texttt{get\_value\_perm}\ Var2^{\to\text{Y-reg}}, freereg\end{array}} \quad (14)$$

$$\frac{Var1_{\text{perm,nonfirst,safe}} \doteq Var2_{\text{perm,nonfirst,unsafe}}}{\begin{array}{l}\texttt{put\_unsafe\_value\_perm}\ Var2^{\to\text{Y-reg}}, freereg\\ \texttt{get\_value\_perm}\ Var1^{\to\text{Y-reg}}, freereg\end{array}} \quad (15)$$

$$\frac{Var1_{\text{perm,nonfirst,unsafe}} \doteq Var2_{\text{perm,nonfirst,unsafe}}}{\begin{array}{l}\texttt{put\_unsafe\_value\_perm}\ Var2^{\to\text{Y-reg}}, freereg\\ \texttt{put\_unsafe\_value\_perm}\ Var1^{\to\text{Y-reg}}, freereg\\ \texttt{get\_value\_perm}\ Var2^{\to\text{Y-reg}}, freereg\end{array}} \quad (16)$$

Table 1: Compilation of $\doteq$-constraints for variables

$$\frac{\mathit{Var1}_{\text{temp,first}} \doteq \mathit{const}}{\texttt{put\_constant}\ \mathit{const},\ \mathit{Var1}^{\to\text{X-reg}}} \quad (17)$$

$$\frac{\mathit{Var1}_{\text{temp,first}} \doteq \texttt{nil}}{\texttt{put\_nil}\ \mathit{Var1}^{\to\text{X-reg}}} \quad (18)$$

$$\frac{\mathit{Var1}_{\text{temp,nonfirst}} \doteq \mathit{constant}}{\texttt{get\_constant}\ \mathit{Var1}^{\to\text{X-reg}}} \quad (19)$$

$$\frac{\mathit{Var1}_{\text{temp,nonfirst}} \doteq \texttt{nil}}{\texttt{get\_nil}\ \mathit{Var1}^{\to\text{X-reg}}} \quad (20)$$

$$\frac{\mathit{Var1}_{\text{perm,first}} \doteq \mathit{constant}}{\begin{array}{c}\texttt{put\_constant}\ \mathit{constant},\mathit{freereg}\\ \texttt{get\_variable\_perm}\ \mathit{Var1}^{\to\text{Y-reg}},\mathit{freereg}\end{array}} \quad (21)$$

$$\frac{\mathit{Var1}_{\text{perm,first}} \doteq \texttt{nil}}{\begin{array}{c}\texttt{put\_nil}\ \mathit{freereg}\\ \texttt{get\_variable\_perm}\ \mathit{Var1}^{\to\text{Y-reg}},\mathit{freereg}\end{array}} \quad (22)$$

$$\frac{\mathit{Var1}_{\text{perm,nonfirst,safe}} \doteq \mathit{constant}}{\begin{array}{c}\texttt{put\_value\_perm}\ \mathit{Var1}^{\to\text{Y-reg}},\mathit{freereg}\\ \texttt{get\_constant}\ \mathit{constant},\ \mathit{Var1}^{\to\text{Y-reg}}\end{array}} \quad (23)$$

$$\frac{\mathit{Var1}_{\text{perm,nonfirst,safe}} \doteq \texttt{nil}}{\begin{array}{c}\texttt{put\_value\_perm}\ \mathit{Var1}^{\to\text{Y-reg}},\mathit{freereg}\\ \texttt{get\_nil}\ \mathit{Var1}^{\to\text{Y-reg}}\end{array}} \quad (24)$$

$$\frac{\mathit{Var1}_{\text{perm,nonfirst,unsafe}} \doteq \mathit{constant}}{\begin{array}{c}\texttt{put\_unsafe\_value\_perm}\ \mathit{Var1}^{\to\text{Y-reg}},\mathit{freereg}\\ \texttt{get\_constant}\ \mathit{constant},\mathit{freereg}\end{array}} \quad (25)$$

$$\frac{\mathit{Var1}_{\text{perm,nonfirst,unsafe}} \doteq \mathit{nil}}{\begin{array}{c}\texttt{put\_unsafe\_value\_perm}\ \mathit{Var1}^{\to\text{Y-reg}},\mathit{freereg}\\ \texttt{get\_nil}\ \mathit{freereg}\end{array}} \quad (26)$$

Table 2: Compilation of $\doteq$-constraints for constants

Remember that we will not produce code for the head of `rev/2`. We will use a label as an entry point for the corresponding definite equivalence. In this paper, we only give the compilation rules for the instructions which are related to unification. The control instructions [26] are used here without further notice.

`rev/2:`

$\Rightarrow$ `try_me_else rev/21,2`

$\texttt{X1}^{\to\text{X-reg}=1}_{\text{temp,nonfirst,safe}} \doteq nil,$

$\Rightarrow^{20}$ `get_nil 1`

$\texttt{X2}^{\to\text{X-reg}=2}_{\text{temp,nonfirst,safe}} \doteq nil$

$\Rightarrow^{20}$ `get_nil 1`

$\Rightarrow$ `proceed`

$\vee$

`rev/21:`

$\Rightarrow$ `trust_me_else_fail`

$\Rightarrow$ `allocate 3`

$\texttt{X1}^{\to\text{X-reg}=1}_{\text{temp,nonfirst,safe}} \doteq cons(\text{L1}^{\to\text{Y-reg}=3}_{\text{perm,first,safe}}, \text{L2}^{\to\text{X-reg}=1}_{\text{temp,first,safe}}),$

$\Rightarrow^{33}$ `get_list 1`

$\Rightarrow^{42}$ `unify_variable_perm 3`

$\Rightarrow^{42}$ `unify_variable_temp 1`

$\texttt{X2}^{\to\text{X-reg}=2}_{\text{temp,nonfirst,safe}} \doteq \text{L3}^{\to\text{Y-reg}=1}_{\text{perm,first,safe}}$

$\Rightarrow^{9}$ get_variable_perm 1,2

$\text{rev}(\text{L2}^{\to\text{X-reg}=1}_{\text{temp,nonfirst,safe}}, \text{L}^{\to\text{Y-reg}=2}_{\text{perm,first,safe}})$

$$\frac{\mathit{Var1}_{\text{temp,first}} \doteq \mathit{structure}(\mathit{position}_1, \ldots, \mathit{position}_n)}{\begin{array}{c}\texttt{put\_structure}\ \mathit{structure},\ \mathit{Var1}^{\rightarrow\text{X-reg}} \\ \text{build-instruction for } \mathit{position}_1 \\ \ldots \\ \text{build-instruction for } \mathit{position}_n\end{array}} \quad (27)$$

$$\frac{\mathit{Var1}_{\text{temp,nonfirst}} \doteq \mathit{structure}(\mathit{position}_1, \ldots, \mathit{position}_n)}{\begin{array}{c}\texttt{get\_structure}\ \mathit{structure},\ \mathit{Var1}^{\rightarrow\text{X-reg}} \\ \text{unify-instruction for } \mathit{position}_1 \\ \ldots \\ \text{unify-instruction for } \mathit{position}_n\end{array}} \quad (28)$$

$$\frac{\mathit{Var1}_{\text{perm,first}} \doteq \mathit{structure}(\mathit{position}_1, \ldots, \mathit{position}_n)}{\begin{array}{c}\texttt{put\_structure}\ \mathit{structure}, \mathit{freereg} \\ \text{build-instruction for } \mathit{position}_1 \\ \ldots \\ \text{build-instruction for } \mathit{position}_n \\ \texttt{get\_variable\_perm}\ \mathit{Var1}^{\rightarrow\text{Y-reg}}, \mathit{freereg}\end{array}} \quad (29)$$

$$\frac{\mathit{Var1}_{\text{perm,nonfirst,safe}} \doteq \mathit{structure}(\mathit{position}_1, \ldots, \mathit{position}_n)}{\begin{array}{c}\texttt{put\_value\_perm}\ \mathit{Var1}^{\rightarrow\text{Y-reg}}, \mathit{freereg} \\ \texttt{get\_structure}\ \mathit{structure}, \mathit{freereg} \\ \text{unify-instruction for } \mathit{position}_1 \\ \ldots \\ \text{unify-instruction for } \mathit{position}_n\end{array}} \quad (30)$$

$$\frac{\mathit{Var1}_{\text{perm,nonfirst,unsafe}} \doteq \mathit{structure}(\mathit{position}_1, \ldots, \mathit{position}_n)}{\begin{array}{c}\texttt{put\_unsafe\_value\_perm}\ \mathit{Var1}^{\rightarrow\text{Y-reg}}, \mathit{freereg} \\ \texttt{get\_structure}\ \mathit{structure}, \mathit{freereg} \\ \text{unify-instruction for } \mathit{position}_1 \\ \ldots \\ \text{unify-instruction for } \mathit{position}_n\end{array}} \quad (31)$$

$$\frac{\mathit{Var1}_{\text{temp,first}} \doteq \texttt{cons}(\mathit{car}, \mathit{cdr})}{\begin{array}{c}\texttt{put\_list}\ \mathit{Var1}^{\rightarrow\text{X-reg}} \\ \text{build-instruction for } \mathit{car} \\ \text{build-instruction for } \mathit{cdr}\end{array}} \quad (32)$$

$$\frac{\mathit{Var1}_{\text{temp,nonfirst}} \doteq \texttt{cons}(\mathit{car}, \mathit{cdr})}{\begin{array}{c}\texttt{get\_list}\ \mathit{Var1}^{\rightarrow\text{X-reg}} \\ \text{unify-instruction for } \mathit{car} \\ \text{unify-instruction for } \mathit{cdr}\end{array}} \quad (33)$$

$$\frac{\mathit{Var1}_{\text{perm,first}} \doteq \texttt{cons}(\mathit{car}, \mathit{cdr})}{\begin{array}{c}\texttt{put\_list}\ \mathit{freereg} \\ \text{build-instruction for } \mathit{car} \\ \text{build-instruction for } \mathit{cdr} \\ \texttt{get-variable-perm}\ \mathit{Var1}^{\rightarrow\text{Y-reg}}, \mathit{freereg}\end{array}} \quad (34)$$

$$\frac{\mathit{Var1}_{\text{perm,nonfirst,safe}} \doteq \texttt{cons}(\mathit{car}, \mathit{cdr})}{\begin{array}{c}\texttt{put\_value\_perm}\ \mathit{Var1}^{\rightarrow\text{Y-reg}}, \mathit{freereg} \\ \texttt{get\_list}\ \mathit{freereg} \\ \text{unify-instruction for } \mathit{car} \\ \text{unify-instruction for } \mathit{cdr}\end{array}} \quad (35)$$

$$\frac{\mathit{Var1}_{\text{perm,nonfirst,unsafe}} \doteq \texttt{cons}(\mathit{car}, \mathit{cdr})}{\begin{array}{c}\texttt{put\_unsafe\_value\_perm}\ \mathit{Var1}^{\rightarrow\text{Y-reg}}, \mathit{freereg} \\ \texttt{get\_list}\ \mathit{freereg} \\ \text{unify-instruction for } \mathit{car} \\ \text{unify-instruction for } \mathit{cdr}\end{array}} \quad (36)$$

Table 3: Compilation of Compound Terms

Unify instructions for arguments in structures

$$\frac{\mathit{structure}(\ldots,*_{i-1},\mathit{constant},*_{i+1},\ldots)}{\texttt{unify\_constant}\ \mathit{constant}} \quad (37)$$

$$\frac{\mathit{structure}(\ldots,*_{i-1},\texttt{nil},*_{i+1},\ldots)}{\texttt{unify\_nil}} \quad (38)$$

$$\frac{\mathit{structure}(\ldots,*_{i-1},\mathit{Var}_{\text{temp,first}},*_{i+1},\ldots)}{\texttt{unify\_variable\_temp}\ \mathit{Var}^{\to\text{X-reg}}} \quad (39)$$

$$\frac{\mathit{structure}(\ldots,*_{i-1},\mathit{Var}_{\text{temp,nonfirst,safe}},*_{i+1},\ldots)}{\texttt{unify\_value\_temp}\ \mathit{Var}^{\to\text{X-reg}}} \quad (40)$$

$$\frac{\mathit{structure}(\ldots,*_{i-1},\mathit{Var}_{\text{temp,nonfirst,unsafe}},*_{i+1},\ldots)}{\texttt{unify\_local\_value\_temp}\ \mathit{Var}^{\to\text{X-reg}}} \quad (41)$$

$$\frac{\mathit{structure}(\ldots,*_{i-1},\mathit{Var}_{\text{perm,first}},*_{i+1},\ldots)}{\texttt{unify\_variable\_perm}\ \mathit{Var}^{\to\text{Y-reg}}} \quad (42)$$

$$\frac{\mathit{structure}(\ldots,*_{i-1},\mathit{Var}_{\text{perm,nonfirst,safe}},*_{i+1},\ldots)}{\texttt{unify\_value\_perm}\ \mathit{Var}^{\to\text{Y-reg}}} \quad (43)$$

$$\frac{\mathit{structure}(\ldots,*_{i-1},\mathit{Var}_{\text{perm,nonfirst,unsafe}},*_{i+1},\ldots)}{\texttt{unify\_local\_value\_perm}\ \mathit{Var}^{\to\text{Y-reg}}} \quad (44)$$

Unify instructions for arguments in structures

$$\frac{\mathit{structure}(\ldots,*_{i-1},\mathit{constant},*_{i+1},\ldots)}{\texttt{build\_constant}\ \mathit{constant}} \quad (45)$$

$$\frac{\mathit{structure}(\ldots,*_{i-1},\texttt{nil},*_{i+1},\ldots)}{\texttt{build\_nil}} \quad (46)$$

$$\frac{\mathit{structure}(\ldots,*_{i-1},\mathit{Var}_{\text{temp,first}},*_{i+1},\ldots)}{\texttt{build\_variable\_temp}\ \mathit{Var}^{\to\text{X-reg}}} \quad (47)$$

$$\frac{\mathit{structure}(\ldots,*_{i-1},\mathit{Var}_{\text{temp,nonfirst,safe}},*_{i+1},\ldots)}{\texttt{build\_value\_temp}\ \mathit{Var}^{\to\text{X-reg}}} \quad (48)$$

$$\frac{\mathit{structure}(\ldots,*_{i-1},\mathit{Var}_{\text{temp,nonfirst,unsafe}},*_{i+1},\ldots)}{\texttt{build\_local\_value\_temp}\ \mathit{Var}^{\to\text{X-reg}}} \quad (49)$$

$$\frac{\mathit{structure}(\ldots,*_{i-1},\mathit{Var}_{\text{perm,first}},*_{i+1},\ldots)}{\texttt{build\_variable\_perm}\ \mathit{Var}^{\to\text{Y-reg}}} \quad (50)$$

$$\frac{\mathit{structure}(\ldots,*_{i-1},\mathit{Var}_{\text{perm,nonfirst,safe}},*_{i+1},\ldots)}{\texttt{build\_value\_perm}\ \mathit{Var}^{\to\text{Y-reg}}} \quad (51)$$

$$\frac{\mathit{structure}(\ldots,*_{i-1},\mathit{Var}_{\text{perm,nonfirst,unsafe}},*_{i+1},\ldots)}{\texttt{build\_local\_value\_perm}\ \mathit{Var}^{\to\text{Y-reg}}} \quad (52)$$

Build instructions for arguments in structures

Table 4: Compilation of arguments in structures

$$\frac{\mathit{literal}(\ldots,*_{i-1},\mathit{Var}_{\text{temp,first}},*_{i+1},\ldots)}{\texttt{put\_variable\_temp}\ \mathit{Var}^{\to\text{X-reg}}, i} \quad (53)$$

$$\frac{\mathit{literal}(\ldots,*_{i-1},\mathit{Var}_{\text{temp,nonfirst}},*_{i+1},\ldots)}{\texttt{put\_value\_temp}\ \mathit{Var}^{\to\text{X-reg}}, i} \quad (54)$$

$$\frac{\mathit{literal}(\ldots,*_{i-1},\mathit{Var}_{\text{perm,first}},*_{i+1},\ldots)}{\texttt{put\_variable\_perm}\ \mathit{Var}^{\to\text{Y-reg}}, i} \quad (55)$$

$$\frac{\mathit{literal}(\ldots,*_{i-1},\mathit{Var}_{\text{perm,nonfirst,safe}},*_{i+1},\ldots)}{\texttt{put\_value\_perm}\ \mathit{Var}^{\to\text{Y-reg}}, i} \quad (56)$$

$$\frac{\mathit{literal}(\ldots,*_{i-1},\mathit{Var}_{\text{perm,nonfirst,unsafe}},*_{i+1},\ldots)}{\texttt{put\_unsafe\_value\_perm}\ \mathit{Var}^{\to\text{Y-reg}}, i} \quad (57)$$

Table 5: Simplified Code Generation for User-Literals

$\Rightarrow^{54}$ `put_value_temp 1,1` — omitted

$\Rightarrow^{55}$ `put_variable_perm 2,2`

$\Rightarrow$ `call 3,rev/2`

$X3^{\rightarrow X\text{-reg}=2}_{temp,first,safe} \doteq cons(L1^{\rightarrow Y\text{-reg}=3}_{perm,nonfirst,safe}, nil)$,

$\Rightarrow^{32}$ `put_list 2`

$\Rightarrow^{51}$ `build_value_perm 3`

$\Rightarrow^{46}$ `build_nil`

$app(L^{\rightarrow Y\text{-reg}=2}_{perm,nonfirst,unsafe}, X3^{\rightarrow X\text{-reg}=2}_{temp,nonfirst,safe}, L3^{\rightarrow Y\text{-reg}=1}_{perm,nonfirst,safe})$

$\Rightarrow^{57}$ `put_unsafe_value_perm 2,1`

$\Rightarrow^{54}$ `put_value_temp 2,2` — omitted

$\Rightarrow^{56}$ `put_value_perm 1,3`

$\Rightarrow$ `deallocate`

`execute app/3`

3.1 Realizing Structure Sharing

Compound terms which are local to an or-branch need not to be reconstructed. Thus, compound terms can be shared introducing an additional Y-variable. However, structure sharing must be handled with care if an $\doteq$-primitive is generated by the transformation rules, where two variables are both `nonfirst` and hence a general unification may be invoked.

Example: (Structure sharing)

Consider the following clause:

```
d(a(X,Y),W):-
    d(X,a(X1,X2)),
    d(Y,a(Y1,Y2)),
    no(a(a(X1,X2),a(Y1,Y2)),W).
```

In a conventional compiler, this is compiled into:

`d(U1,U2):-`
  $U1 \doteq a(X,Y)$, $U2 \doteq W$, $V1 \doteq a(X1,X2)$, `d(X,V1)`,
  $V2 \doteq a(Y1,Y2)$, `d(Y,V2)`,
  $V3 \doteq a(X1,X2)$, $V4 \doteq a(Y1,Y2)$, $V5 \doteq a(V3,V4)$, `no(V5,W)`.

Applying our scheme using structure sharing, this is compiled into:

`d(U1,U2)` $\leftrightarrow$ ... $\vee$
($U1 \doteq a(X,Y)$, $U2 \doteq W$, $V1 \doteq a(X1,X2)$, `d(X,V1)`,
  $V2 \doteq a(Y1,Y2)$, `d(Y,V2)`,
  $V3 \doteq a(V1,V2)$, `no(V3,W))` $\vee$ ...

The resulting code is more efficient concerning both memory and time resources.

3.2 Incorporating mode and type information

It is easy to use the additional information in the $\doteq$-primitives to refine WAM instructions. However, this results in an explosion of the WAM instructions as well as of the transformation rules. Moreover, an extension to unification leads to a redesign of the complete WAM instructions semantics. WAM instructions can be separated into control (`try`, `retry`, `trust`, `allocate`, `call`, etc.) and unification instructions (`get`, `put`, `unify`). But even the unification instructions have a control aspect, since they deal with the early removal of environments to avoid dangling references. We developed another implementation point of view based on the following primitives: A primitive making potentially unsafe Y-variables safe, another primitive globalizing dereferenced variables, if necessary and the control instructions. The code for compiling the $\doteq$-primitives can be derived by specializing the basic unify routine.

3.3 Constraint Reordering

The idea of constraint reordering, according to some computational metric, intends to get a fail as soon as possible. In our scheme argument reordering can be applied at the horizontal, declarative level. For example, a constraint is *cheap*, if one of the arguments of an $\doteq$-primitive is a constant and the variable is already dereferenced and ground, while an *expensive* $\doteq$-constraint invokes the general unification routine.

4 Future work

We are using the compilation scheme in a LISP-based implementation generating "LISP native code", integrating finite domain unification. Several constraint schemes have been developed extending the basic unification routine either doing the extensions by a proper interface in a C based environment [13], which induces overhead by the separation between the solver and the WAM, or using Meta-terms [17], defining an "unification event handler" in PROLOG. Constraint extensions by defining horizontal transformations into basic constraint and extending the code generator by compilation rules for these basic constraints may serve as an alternative on a very deep implementation layer.

5 Conclusion

We have presented a constraint based view on WAM-compilation. The program as a collection of procedures is taken as a definite equivalence and thereafter split into $\doteq$-constraints and control code (disjunctions, user's literals). Then, partial evaluation, type inference and variable classification are applied resulting in a still declarative representation. Subsequently, we presented the vertical compilation scheme in a very condensed rule-based manner where we do not have to distinguish between head and tail compilation.

Acknowledgements

This paper would never have been written without the work carried out in the Knowledge Compilation Group within the ARC-TEC project at DFKI. The idea of horizontal compilation has been initially discussed and implemented in the RFM group of Harold Boley. Special thanks to Thomas Krause who implemented a variable classification scheme which influenced our work to a great deal.

References

[1] H. Aït-Kaci and R. Nasr. LOGIN: A Logic Programming Language with Built-in Inheritence. *Journal of Logic Programming*, 3:185–215, 1986.

[2] M. Bruynooghe and G. Jennsens. An Instance of Abstract Interpretation Integrating Type and Mode Inferencing. In R. A. Kowalski and K. A. Bowen, editors, *Proceedings of the Fifth International Conference and Symposium on Logic Programming*, pages 669–683, 1988.

[3] W. Büttner and H. Simonis. Embedding Boolean Expressions into Logic Programming. *Journal of Symbolic Computation*, 4:191–205, 1987.

[4] M. Carlsson. On the Efficiency of Optimising Shallow Backtracking in Compiled Prolog. In G. Levi and M. Martelli, editors, *Proc. of the Sixth ICLP*, pages 3–16. MIT Press, 1989.

[5] K. Clark. Negation as Failure. In H. Gallaire and J. Minker, editors, *Logic and Databases*, pages 293–322. Plenum Press, New York, 1978.

[6] R. M. Colomb. Enhancing Unification in Prolog through Clause Indexing. *Journal of Logic Programming*, 10:23–44, 1991.

[7] S. K. Debray. Register Allocation in a Prolog Machine. In *Symposium on Logic Programming*, pages 267–275. IEEE, 9 1986.

[8] S. K. Debray. A Simple Code Improvement Scheme for Prolog. In G. Levi and M. Martelli, editors, *Proc. of the Sixth ICLP*, pages 17–32. MIT Press, 1989.

[9] S. K. Debray and D. S. Warren. Detection and optimization of functional computations in prolog. In E. Shapiro, editor, *Third International Conference on Logic Programming*, number 225 in Lecture Notes in Computer Science, pages 490–504. Springer, 1986.

[10] T. Hickey and S. Mudambi. Global Compilation of Prolog. *Journal of Logic Programming*, 7:193–230, 1989.

[11] M. Höhfeld and G. Smolka. Definite Relations over Constraint Languages. LILOG-Report 53, IBM Deutschland GmbH, WT LILOG, Dept. 3504, P.O. Box 800880, D-7000 Stuttgart 80, Germany, October 1988. To appear in Journal of Logic Programming.

[12] J. Jaffar and J.-L. Lassez. Constraint Logic Programming. In *Proceedings of Principles of Programming Languages*, pages 111–119, 1987.

[13] J. Jaffar, S. Michaylov, P. Stuckey, and R. Yap. The CLP($\mathcal{R}$) Language and System. Technical Report CMU–CS–90–181, School of Computer Science, Carnegie Mellon University, Pittsburgh, PA, October 1990.

[14] G. Janssens, B. Demoen, and A. Marten. Improving the Register allocation in WAM by Reordering Unification. In R. A. Kowalski and K. A. Bowen, editors, *Proceedings of the Fifth International Conference and Symposium on Logic Programming*, pages 1388–1402, 1988.

[15] L. Matyska, A. Jergová, and D. Toman. Register Allocation in WAM. In K. Furukawa, editor, *Proc. of the Eighth ICLP*, pages 142–156, 1991.

[16] M. Meier. Compilation of Compound Terms in Prolog. In S. Debray and M. Hermenegildo, editors, *North American Conference on Logic Programming*, pages 63–79. MIT Press, October 1990.

[17] U. Neumerkel. Extensible Unification by Metastructures. In M. Bruynooghe, editor, *Proceedings of the Second Workshop on Meta-programming in Logic*, pages 352–364, 1990.

[18] D. Sahlin. *An Automatic Partial Evaluator for Full Prolog.* PhD thesis, The Royal Institute of Technology, Stockholm, March 1991.

[19] A. Taylor. Removal of Dereferencing and Trailing in Prolog Compilation. In G. Levi and M. Martelli, editors, *Proc. of the Sixth ICLP*, pages 48–62. MIT Press, 1989.

[20] A. Taylor. Quicksort Speed in Prolog. USENET electronic article in comp.lang.prolog, July 1990.

[21] A. Taylor. *High Performance Prolog Implementation.* PhD thesis, University of Sydney, June 1991.

[22] A. K. Turk. Compiler Optimizations for the WAM. In E. Shapiro, editor, *Third International Conference on Logic Programming*, number 225 in Lecture Notes in Computer Science, pages 657–662. Springer, 1986.

[23] P. van Hentenryck. *Constraint Satisfaction in Logic Programming.* MIT Press, 1989.

[24] P. van Roy. A Prolog Compiler for the PLM. Technical Report UCB/CSD 84/203, University of California, Computer Science Division, University of California, Berkeley, California 94720, November 1984.

[25] P. van Roy. An Intermediate Language to Support Prolog's Unification. In E. L. Lusk and R. A. Overbeek, editors, *Proceedings of the North American Conference on Logic Programming*, pages 1148–1164, 1989.

[26] D. H. D. Warren. An Abstract Prolog Instruction Set. Technical Note 309, Artificial Intelligence Center, SRI International, 333 Ravenswood Ave, Menlo Park, CA 94025, October 1983.

[27] R. Warren, M. Hermenegildo, and S. K. Debray. On the Practicality of Global Flow Analysis of Logic Programs. In R. A. Kowalski and K. A. Bowen, editors, *Proc. of the Fifth International Conference and Symposium on Logic Programming*, pages 684–699, 1988.

Safe Positive Induction in the Programming Logic TK

Martin C. Henson,
Department of Computer Science, University of Essex, Colchester, Essex, ENGLAND.
Department of Computer Science, University of Otago, Dunedin, NEW ZEALAND.
hensm@uk.ac.sx

§1 Abstract

We describe an alternative schema of induction for the programming logic TK based on safe positive induction. This replaces the original schema based on the well founded part of a relation. We show how the new schema can be included into the realizability definition and how the soundness of realizability can be extended to allow for the derivation of recursive programs from proofs of specifications which use the new schema. We further show how systems of mutual induction can be handled naturally with the new schema. In particular we show how useful systems of mutually recursive combinators can be derived which realize the principles of mutual induction.

§2 Introduction

The apparatus which a programming logic provides for inductive types is, perhaps, its most important component. This is because we rely on inductive types for the definition of many recursive types ubiquitous in programming languages: natural numbers, lists, trees, and so on. Moreover, it is from proofs which involve inductive types that we obtain recursive programs.

The programming logic TK has been designed, and is being constantly redesigned, in order to provide a natural vehicle for the derivation of programs from their specifications. In this paper we investigate some of the consequences of replacing the notion of inductive type outlined in [HeT 88] (which owes much to that of [Fef 79]) with a principle of restricted monotone induction. Some work has been done with similar schemata [HaN 87] for example and we will discuss some of the connections with the work reported here in §6.

In §3 we shall introduce the schema of safe positive induction we propose and show that it subsumes the principle used in [HeT 88] [Hen 89a]. In §4 we extend the soundness proof for TK-realizability in order to accommodate the new schema. This provides us with a means for obtaining recursive programs from proofs by safe positive induction. In §5 we show how the principle provides, naturally, for systems of mutually inductive types. That such a principle is powerful enough, with product types, to express mutual induction is well known [Mos 74] but the novelty here is to see how this interacts with realizability and, most importantly, how naturally such realizers can be expressed. These realizers dictate the form of programs which are obtained from proofs and we continue to require that these programs arise naturally in a high level programming notation: the appearance of devious codings in these realizers, while mathematically adequate, would constitute a disaster from the programming perspective. This is to say that in the design of TK we are interested in certain *intensional* aspects of the theory as well as *extensional* adequacy. The issue of intensional adequacy and accordance with practice for formal theories has been discussed in [Fef 79] and these issues are certainly as important for formal theories of program development as they are in the foundations of mathematics. In §6 we discuss future work and provisional conclusions and then in §7 and §8 we provide acknowledgements and references to the literature. Finally, we include an appendix §9 which includes a summary of the logic TK and one or two basic results and definitions.

§3 Safe Positive Induction

We begin by reminding ourselves of the three rules of TK which allow for the articulation of inductive types:

$$\frac{z \in \mathrm{A}}{z \in \mathrm{I(A,}\varphi)}\ (\mathrm{I(A,}\varphi)\text{-intro(i)}) \qquad \frac{(\forall y)(\varphi(z,y) \Rightarrow y \in \mathrm{I(A,}\varphi))}{z \in \mathrm{I(A,}\varphi)}\ (\mathrm{I(A,}\varphi)\text{–intro(ii)})$$

$$\frac{\begin{array}{c}(\forall x)(x \in \mathrm{A} \Rightarrow \psi(x)) \\ (\forall x)(x \in \mathrm{Clo(I(A,}\varphi)) \Rightarrow (\forall y)(\varphi(x,y) \Rightarrow \psi(y)) \Rightarrow \psi(x))\end{array}}{(\forall x)(x \in \mathrm{I(A,}\varphi) \Rightarrow \psi(x))}\ (\mathrm{I(A,}\varphi)\text{-elim})$$

The type I(A,φ) is the well founded part of the relation φ with minimal elements in the type A. This principle of induction was explored in many example derivations in [Hen 89] and the reader of that paper would be forgiven for having found some of the definitions less than perspicuous! The clumsiness has repercussions: the realizing combinators specialising the general realizer for the induction principle I(A,φ)-elim are also often awkward and this leads inexorably to rather unattractive programs. This we believe is largely avoided by moving to a more general notion of induction: safe positive induction. The rules replacing the three above are as follows:

$$\frac{z \in \mathrm{A}}{z \in \Xi(\mathrm{A},\lambda X.\mathrm{B})}\ (\Xi\text{-intro(i)}) \qquad \frac{z \in \mathrm{B}(\Xi(\mathrm{A},\lambda X.\mathrm{B}))}{z \in \Xi(\mathrm{A},\lambda X.\mathrm{B})}\ (\Xi\text{–intro(ii)})$$

$$\frac{\mathrm{A} \subseteq \mathrm{T} \qquad \mathrm{B(T)} \subseteq \mathrm{T}}{\Xi(\mathrm{A},\lambda X.\mathrm{B}) \subseteq \mathrm{T}}\ (\Xi\text{-elim})$$

This specialises to the following useful induction rule:

$$\frac{(\forall x\in A)\psi(x) \quad (\forall x\in B(\{w\in \Xi(A,\lambda X.B) \mid \psi(w)\})) \Rightarrow \psi(x))}{(\forall x\in \Xi(A,\lambda X.B))\psi(x)}$$

when T = $\{w\in\Xi(A,\lambda X.B) \mid \psi(w)\}$. (When context allows we often suppress the typing in the comprehension types $\{w\in\Xi(A,\lambda X.B) \mid \psi(w)\}$ writing them simply $\{w \mid \psi(w)\}$ for ease of presentation). This form of Ξ-elim is so useful that we will devote our attention to it exclusively in the sequel.

The type Ξ(A,λ*X*.B) is, informally, to be thought of having been built up from a base type A by repeated application of the *type operation* λ*X*.B. In this paper we investigate this rule in the case where both A and B are *safe types* and where $\{w\in\Xi(A,\lambda X.B) \mid \psi(w)\}$ *is realizably safe*. [Hen 90]. Secondly, the operation λ*X*.B must be monotone in *X*. However, full monotone induction is impredicative (since the statement that λ*X*.B is monotone in *X* involves a quantification over all types including the putative type Ξ(A,λ*X*.B) itself) and we have not been persuaded that theories of program development need to be so strong. A standard, predicative, solution is to restrict λ*X*.B syntactically in such a way that it is inevitably monotone in X. So we shall insist that *X* occurs in B in *positive* positions and this is sufficient to ensure that λ*X*.B is monotone in *X* [Mos 74]. Our principle is close to that for primitive recursion over the least fixpoint of a functor. In fact what we have is deliberately weaker: TK is being developed for practical program development and also to study formal program derivation. We are trying to keep the theory as weak as is possible to support our requirements. We have also deliberately separated a base type A from the functor B. It would be quite simple to have used a principle such as Ξ(λ*X*.A ∪ B) with *X nfi* A. In order to extract programs from proofs which separate recursive and non-recursive clauses we have found it preferable to distinguish the base and inductive cases in the definition of Ξ(A,λ*X*.B). We shall see in proposition 4.5 that A and B(Ξ(A,λ*X*.B)) must be recursively separable. This is very reasonable and indeed is built into the definition of many functional languages (e.g. Miranda™[1] - see the example below).

3.1 Proposition

$$\Xi(A,\lambda X.B) \equiv A \cup B(\Xi(A,\lambda X.B))$$

proof (⊆) by Ξ-elim (⊇) by Ξ–intro •

The introduction rules Ξ-intro(i) and Ξ–intro(ii) are quite intuitive but Ξ-elim rule may require a little motivation. We illustrate it with a simple example: the S-expressions of LISP.

In Miranda we might write:

```
sexp * ::= Leaf * | Node (sexp *) (sexp *)
```

1. ™ Miranda is a trademark of Research Software Limited.

In TK this would be translated as follows: $Sexp = \Xi(\{\text{Leaf}\} \times A, \lambda X.\{\text{Node}\} \times X \times X)$ (with A a free type parameter). Let $B(X) = \{\text{Node}\} \times X \times X$. The inductive assumption, $x \in B(\{w \in \Xi(A,\lambda X.B) \mid \psi(w)\})$, in this case amounts to:

$$x \in (\{\text{Node}\} \times \{w \in \Xi(A,\lambda X.B) \mid \psi(w)\} \times \{w \in \Xi(A,\lambda X.B) \mid \psi(w)\}) \text{ or:}$$

$(x)_0 = \text{Node} \wedge ((x)_1)_0 \in \{w \in \Xi(A,\lambda X.B) \mid \psi(w)\} \wedge ((x)_1)_1 \in \{w \in \Xi(A,\lambda X.B) \mid \psi(w)\}$ that is:

$$(x)_0 = \text{Node} \wedge \psi(((x)_1)_1) \wedge \psi(((x)_1)_1) \wedge x \in (\{\text{Node}\} \times \Xi(A,\lambda X.B) \times \Xi(A,\lambda X.B)).$$

So Ξ-elim specialises to:

$$\frac{(\forall x \in A)\psi(<\text{Leaf}, x>) \quad (\forall x \in Sexp)(\forall y \in Sexp)(\psi(x)) \wedge \psi(y) \Rightarrow \psi(<\text{Node}, x, y>)}{(\forall x \in Sexp)(\psi(x))}$$

which is the expected principle of S-expression induction.

The inductive type schema $I(A,\varphi)$ [HeT 88] can be modelled with $\Xi(A,\lambda X.B)$ in the following way.

3.2 Proposition

Let TK^- be the programming logic TK without rules for inductive types. Let T_I be the rules for the inductive type $I(A,\varphi)$ and T_Ξ be those for $\Xi(A,\lambda X.B)$:

$$TK^- + T_\Xi \vdash T_I$$

proof Take $I(A,\varphi) =_{def} \Xi(A,\lambda X.\{z \mid (\forall y)(\varphi(z,y) \Rightarrow y \in X\})$ and it is then a routine matter to derive the three $I(A,\varphi)$ rules. •

§4 Extracting Programs from Safe Positive Inductive Proofs

Programs are recovered from proofs in TK by means of a realizability interpretation. The essential idea is to associate a TK formula $e\ \rho\ \varphi$ with each TK formula φ. $e\ \rho\ \varphi$ is to be read, roughly, as "program e meets specification φ". The full definition is given by induction over the structure of TK formulae. Further motivation, explanation and examples can be found in [Hen 89a] and [Hen 89b].

4.1 Definition (TK-realizability)

(ia)	$e\ \rho\ \varphi$	is	φ	($Safe(\varphi)$)
(ib)	$e\ \rho\ (x \in T)$	is	$<e, x> \in T^*$	
(iia)	$e\ \rho\ (\varphi \wedge \psi)$	is	$\varphi \wedge (e\ \rho\ \psi)$	($Safe(\varphi)$)
(iib)	$e\ \rho\ (\varphi \wedge \psi)$	is	$(e\ \rho\ \varphi) \wedge \psi$	($Safe(\psi)$)

(iic)	$e\,\rho\,(\varphi \wedge \psi)$	is	$(e)_0\,\rho\,\varphi \wedge (e)_1\,\rho\,\psi$	
(iiia)	$e\,\rho\,(\varphi \vee \psi)$	is	$e = \mathbf{t} \wedge \varphi \vee e = \mathbf{f} \wedge \psi$	$(Safe(\varphi) \wedge Safe(\psi))$
(iiib)	$e\,\rho\,(\varphi \vee \psi)$	is	$(e)_0 = \mathbf{t} \wedge \varphi \wedge (e)_1\,\rho\,\varphi \vee (e)_0 = \mathbf{f} \wedge \psi \wedge (e)_1\,\rho\,\psi$	
(iv)	$e\,\rho\,(\varphi \Rightarrow \psi)$	is	$(\forall x)(x\,\rho\,\varphi \wedge \varphi \Rightarrow (e\,x)\downarrow \wedge (e\,x)\ \rho\,\psi$	
(va)	$e\,\rho\,(\forall x)\varphi$	is	$(\forall x)((e\,x)\downarrow \wedge (e\,x)\ \rho\,\varphi)$	
(vb)	$e\,\rho\,(\forall x)(\varphi \Rightarrow \psi)$	is	$(\forall x)(\varphi \Rightarrow (e\,x)\downarrow \wedge (e\,x)\,\rho\,\psi)$	$(Safe(\varphi))$
(vib)	$e\,\rho\,(\exists x)\varphi$	is	$\varphi(e)$	$(Safe(\varphi))$
(via)	$e\,\rho\,(\exists x)\varphi$	is	$(e)_0\,\rho\,\varphi(x \leftarrow (e)_1) \wedge \varphi(x \leftarrow (e)_1)$	
(vib)	$e\,\rho\,(\exists x)\varphi$	is	$\varphi(e)$	$(Safe(\varphi))$
(vii)	$e\,\rho\,(\forall X)\varphi$	is	$(\forall X)(e\,\rho\,\varphi)$	
(viii)	$e\,\rho\,(\exists X)\varphi$	is	$(\exists Y)((e)_1 = Y \wedge (e)_0\,\rho\,\varphi(X \leftarrow (e)_1) \wedge \varphi(X \leftarrow (e)_1)$ •	

Where there is an overlap in the above the most specific clause is taken. In particular clause (ib) only applies when T is not safe. Since $\Xi(A,\lambda X.B)$ is always a safe type we shall not require clause (ib) and we shall avoid altogether any discussion of the star-types T*. A full analysis of star-types and the arguments which justify the safety of $\Xi(A,\lambda X.B)$ appear in [Hen 90].

It is instructive to see how $e\ \rho\ \varphi$ expresses the intuitive meaning "e meets specification φ". Consider a standard pre/post condition specification $(\forall x)(\varphi(x) \Rightarrow (\exists y)(\psi(x, y)))$ where the pre and post conditions are decidable (and therefore safe: see 9.5) and suppose that $t\,\rho\ (\forall x)(\varphi(x) \Rightarrow (\exists y)(\psi(x,y)))$ we see that $(\forall x)(\varphi(x) \Rightarrow (t\,x)\downarrow \wedge (t\,x)\ \rho\ (\exists y)(\psi(x,y)))$ which is just $(\forall x)(\varphi(x) \Rightarrow (t\,x)\downarrow \wedge\ \psi(x, (t\ \ x)))$, that is, t is a function which, when given an input satisfying the precondition, produces an output which, together with that input, satisfies the postcondition.

Programs are obtained from proofs of such specifications by means of the following, crucial, proposition.

4.2 Proposition (Soundness of Realizability)

If $\mathrm{TK}^- \vdash \varphi$ *then there is a* TK *term* t *for which* $\mathrm{TK} \vdash t\,\rho\,\varphi \wedge t\downarrow$

proof By induction on the length of the derivation $\vdash \varphi$. Essentially we need to provide a combinator for every rule which transforms realizers of the premises into realizers of the conclusion. •

We have stated Proposition 4.2 for TK^- although it is true for $\mathrm{TK}^- + \mathrm{T}_\mathrm{I}$. In this section we show that it can be extended to the theory $\mathrm{TK}^- + \mathrm{T}_\Xi$. We begin by stating precisely the class of safe positive types from which our type operations may be selected.

4.3 Definition (Safe positive types and formulae)

We define the *SP_X-types* and *SP_X-formulae* simultaneously:

(i)				$SP_X(X)$
(ii)	*If*	$SP_X(\varphi)$	*then*	$SP_X(\{z \mid \varphi\})$
(iii)	*If*	$SP_X(A)$ *and* $SP_X(B)$	*then*	$SP_X(\Xi(A,\lambda Y.B))$
(iv)	*If*	$SP_X(T)$	*then*	$SP_X(t \in T)$
(v)	*If*	$SP_X(\varphi)$ *and* $SP_X(\psi)$	*then*	$SP_X(\varphi \wedge \psi)$
(vi)	*If*	X *nfi* φ *and* $SP_X(\psi)$	*then*	$SP_X(\varphi \Rightarrow \psi)$
(vii)	*If*	$SP_X(\varphi)$	*then*	$SP_X((\forall x)\varphi)$
(viii)	*If*	$SP_X(\varphi)$	*then*	$SP_X((\forall Y)\varphi)$
(ix)	*If*	X *nfi* φ *and* $Safe(\varphi)$	*then*	$SP_X(\varphi)$ •

Now, of course, we need to know that such types are monotonic. The proof of this allows us to associate with every safe positive operation a combinator which witnesses the fact that it is indeed monotonic. These combinators are crucially important in extending the soundness of realizability to the rule $\Xi(A,\lambda Y.B)$-elim. Our condition on type operations is, then, that they be not just monotonic, but witnessed monotonic.

4.4 Definition

A *type operator,* B, *is witnessed-monotonic in* X *(*$WM_X(B)$*) iff it is monotonic in* X *and we can associate with it a TK term (ambiguously written)* B *so that* $(B\,f)\ \rho\ B(A) \subseteq B(C)$ *whenever* $f\ \rho\ A \subseteq C$. •

Whilst it is necessary to know that every safe positive type operation is witnessed monotonic it is not immediately possible to prove it. Showing that every safe positive operation is monotonic is an elementary induction over the structure of SP_X-types and we wish to obtain the witnessing information by means of the soundness of realizability. However, one of the clauses in the definition of SP_X-types (clause (iii)) involves the inductive type itself and so we appear to require the soundness of realizability for the theory $TK^- + T_\Xi$ in order to proceed; but we cannot extend the soundness of realizability to $\Xi(A,\lambda Y.B)$-elim without knowing that the type operation $\lambda Y.B$ is witnessed monotonic.

The reasons for this difficulty are not hard to find: the use of the inductive schema can be *fully iterated.* This is a very useful feature and it is worth looking at a simple example. Consider the type of trees which is constructed from a base type A by means of an operation which allows a list of subtrees at each node. This would be formalised as follows:

$$Tree = \Xi(\{\text{Leaf}\} \times A, \lambda X.\{\text{Node}\} \times List(X)) \text{ where:}$$
$$List(X) = \Xi(\{\text{Nil}\}, \lambda Y.\{\text{Cons}\} \times X \times Y)$$

the recursion combinator for *Tree*-elim will clearly require that for *List*-elim and both are instances of $\Xi(A,\lambda X.B)$-elim.

The solution rests on the observation that such iteration of inductive types is well-founded (by the syntax of the language) and thus we can build up the realizers by induction on the structure of the expressions. All we require in order to show that every safe positive operator is witnessed monotonic is that $\Xi(A,\lambda X.B)$-elim is realized *on the assumption that* B *is witnessed monotonic*. We begin with that result.

4.5 Proposition

If $RSafe(\psi)$ and $WM_X(B)$ then the following rule is sound for realizability:

$$\frac{(\forall x\in A)\psi(x)\quad(\forall x\in B(\Xi(A,\lambda X.B)))(x\in B(\{w\in\Xi(A,\lambda X.B)\mid\psi(w)\})\Rightarrow\psi(x))}{(\forall x\in\Xi(A,\lambda X.B))\psi(x)}$$

proof This is just a case in the proof of 4.2 extended to the system $TK^{-}+T_{\Xi}$ where type operations are known to be witnessed monotonic so we may assume *ex hypothesi* that we have a term f so that $(f\,x)\ \rho\ \psi(x)$ whenever $x\in A$ and a term g so that $(g\,x\,q)\ \rho\ \psi(x)$ whenever $x\in B(\Xi(A,\lambda X.B))$ and $q\ \rho\ x\in B(\{w\mid\psi(x)\})$. We need an operator, ***irec***, so that: $(\textbf{\textit{irec}}\,f\,g)\ \rho\ (\forall x\in\Xi(A,\lambda X.B))\psi(x)$. This amounts to $(\forall x\in\Xi(A,\lambda X.B))(\textbf{\textit{irec}}\,f\,g\,x\ \rho\ \psi(x))$. We now postulate properties of such an operation ***irec***, show these are satisfactory by $\Xi(A,\lambda X.B)$-induction and then discuss the construction of a suitable TK term with these properties.

$$\begin{array}{ll} \textbf{\textit{irec}}\,f\,g\,x=f\,x & (x\in A)\\ \textbf{\textit{irec}}\,f\,g\,x=g\,x\,(B\,(\textbf{\textit{irec}}\,f\,g)\,x) & (x\in B(\Xi(A,\lambda X.B))) \end{array}$$

Base case $(x\in A)$: immediate.

Induction case $(x\in B(\Xi(A,\lambda X.B)))$: we need to show that $g\,x\,(B\,(\textbf{\textit{irec}}\,f\,g)\,x)\ \rho\ \psi(x)$ and this follows immediately if we can show that $(B\,(\textbf{\textit{irec}}\,f\,g)\,x)\ \rho\ x\in B(\{w\mid\psi(w)\})$. Now *ex hypothesi* (inner) we may assume that:

$x\in B(\{w\in\Xi(A,\lambda X.B)\mid(\textbf{\textit{irec}}\,f\,g\,w)\ \rho\ \psi(w)\})$.

So we have to show that:

$(\forall x\in B(\{w\in\Xi(A,\lambda X.B)\mid(\textbf{\textit{irec}}\,f\,g\,w)\ \rho\ \psi(w)\}))((B\,(\textbf{\textit{irec}}\,f\,g)\,x)\ \rho\ x\in B(\{w\mid\psi(w)\}))$ and since $RSafe(\psi)$ this can be rewritten by realizability as:

$(B\,(\textbf{\textit{irec}}\,f\,g))\ \rho\ (\forall x\in B(\{w\in\Xi(A,\lambda X.B)\mid(\textbf{\textit{irec}}\,f\,g\,w)\ \rho\ \psi(w)\}))(x\in B(\{w\mid\psi(w)\}))$ or as $(B\,(\textbf{\textit{irec}}\,f\,g))\ \rho\ B(\{w\in\Xi(A,\lambda X.B)\mid(\textbf{\textit{irec}}\,f\,g\,w)\ \rho\ \psi(w)\})\subseteq B(\{w\mid\psi(w)\})$. Since we have $WM_X(B)$ this will follow if we can show that:

$(\textbf{\textit{irec}}\,f\,g)\ \rho\ \{w\in\Xi(A,\lambda X.B)\mid(\textbf{\textit{irec}}\,f\,g\,w)\ \rho\ \psi(w)\}\subseteq\{w\mid\psi(w)\}$ but this reduces to a logical truth. It remains to construct a TK term with these properties. This is trivial since the term language of TK contains the untyped lambda calculus. •

4.6 Proposition

If SP_X(B) then WM_X(B) for every X free in B.

proof Suppose that f ρ A ⊆ C, we proceed by SP_X-induction.
Base case B = X: This is immediate and from the proof we obtain $(B\,f) = f$.
Induction case 1 B = {z | φ(z,X)}: There are six cases. We illustrate with one.
Ad $\varphi(z,X) = \varphi_1(z,X) \wedge \varphi_2(z,X)$: We need to construct a TK term B so that: $(B\,f)\ \rho\ \{z \mid \varphi(z,A)\} \subseteq \{z \mid \varphi(z,C)\}$. *Ex hypothesi* we may assume that $\varphi_1(z,X)$ and $\varphi_2(z,X)$ are monotonic and so we have two terms B_1 and B_2 so that $(B_1 f)\ \rho\ \{z \mid \varphi_1(z,A)\} \subseteq \{z \mid \varphi_1(z,C)\}$ and $(B_2 f)\ \rho\ \{z \mid \varphi_2(z,A)\} \subseteq \{z \mid \varphi_2(z,C)\}$. We now simply prove that $\{z \mid \varphi(z,A)\} \subseteq \{z \mid \varphi(z,C)\}$ from the assumptions $\{z \mid \varphi_1(z,A)\} \subseteq \{z \mid \varphi_1(z,C)\}$ and $\{z \mid \varphi_2(z,A)\} \subseteq \{z \mid \varphi_2(z,C)\}$. By the soundness of realizability we obtain the necessary witness for this proof which turns out to be: $(B\,f\,x) = \langle (B_1 f\,x),(B_2 f\,x)\rangle$.
Induction case 2 $B = \Xi(D,\lambda Y.E)$: We may assume *ex hypothesi* that we have a D so that $(D\,f)\ \rho\ D(A) \subseteq D(C)$. Also we have $SP_X(E)$ and $SP_Y(E)$ and *ex hypothesi* $WM_X(E)$ and $WM_Y(E)$. Let us write Ξ_A and Ξ_C for $\Xi(D(X{\leftarrow}A),\lambda Y.E(X{\leftarrow}A))$ and $\Xi(D(X{\leftarrow}C),\lambda Y.E(X{\leftarrow}C))$. We have to construct a term B so that $(B\,f)\ \rho\ \Xi_A \subseteq \Xi_C$. We now show that $\Xi_A \subseteq \Xi_C$ and then the term B arises by the soundness of realizability. We proceed by means of the elimination rule of 4.5. Note that, *RSafe*(Ξ_C) and $WM_Y(E)$ so the conditions for the soundness of the elimination rule are met.
Base case $x \in D(A)$: D is monotonic in X and so $x \in D(C)$ whence, by Ξ-intro(i), $x \in \Xi_C$ as required.
Induction case $x \in E(Y{\leftarrow}\Xi_A, X{\leftarrow} A)$: We may assume *ex hypothesi* (inner) that $x \in E(Y{\leftarrow}\Xi_A \cap \Xi_C, X{\leftarrow} A)$. But E is monotonic in Y and so $x \in E(Y{\leftarrow}\Xi_C, X{\leftarrow} A)$. E is also monotonic in X and so $x \in E(Y{\leftarrow}\Xi_C, X{\leftarrow}C)$. Finally, by Ξ-intro(ii), $x \in \Xi_C$ as required. The realizer of the this proof is obviously an instance of ***irec*** as defined in 4.5. In fact a careful calculation from the proof provides:
$B\,f = \textbf{\textit{irec}}\ (\lambda x.\langle t,(D\,f\,x)\rangle)\ (\lambda xq.\langle f,(E(Y{\leftarrow}\Xi_C)\,f\,x\ (E(X{\leftarrow} A)\ (\lambda uv.vu)\ q))\rangle)$ •

4.7 Corollary

The following rule is sound for realizability when RSafe(ψ):

$$\frac{(\forall x \in A)\psi(x) \quad (\forall x \in B(\Xi(A,\lambda X.B)))(x \in B(\{w \in \Xi(A,\lambda X.B) \mid \psi(w)\}) \Rightarrow \psi(x))}{(\forall x \in \Xi(A,\lambda X.B))\psi(x)}$$

proof This follows immediately as a consequence of 4.5 and 4.6. •

We will finish the section by returning briefly to our example type *Tree* and illustrating the specialisation of ***irec*** in this case.

The recursion combinator for *Tree* is:

trec $f\ g$ <Leaf, x> $= f\,x$
trec $f\ g$ <Node, x> $= g\ x\ ($***lmap*** $($***trec*** $f\ g)\ x)$

and it is clear from this how the recursion operator for *Tree* requires the combinator ***lmap*** which is itself the specialisation (as determined by the proof of 4.5) of the recursion operator ***lrec*** for the type *List*.

§5 Mutually Inductive Systems of Induction

In this section we set up a general framework for solving families of mutually inductive monotone type equations in the theory $TK^{-} + T_{\Xi}$. That mutual induction is expressible using a principle of monotone induction is well known [Mos 74] but what we have to demonstrate is that this can be integrated smoothly with realizability. What we require is that each family of mutually inductive types gives rise to a mutually recursive system of realizing combinators in a natural fashion. For ease of presentation we shall attack binary systems of mutual induction; the generalisation to arbitrary systems is then quite obvious and unenlightening.

5.1 Definition

Let φ_0, φ_1 be TK formulae; $\mu_0 = (\varphi_{00}, \dots \ \varphi_{0n} / \psi_0)$, $\mu_1 = (\varphi_{10}, \dots \ \varphi_{1n} / \psi_1)$ be TK (possibly derived) rules; t_0, t_1 be TK terms. Let $\bar{x}$ be the sequence of all the free variables occurring in indicated the expressions below.

(i)	$(\varphi_0 \times \varphi_1)(\bar{x})$	=	$\varphi_0(\overline{x \leftarrow (x)_0}) \wedge \varphi_1(\overline{x \leftarrow (x)_1})$
(ii)	$(\mu_0 \times \mu_1)(\bar{x})$	=	$((\varphi_{00} \times \varphi_{10})(\bar{x}), \dots (\varphi_{0n} \times \varphi_{1n})(\bar{x}) / (\psi_0 \times \psi_1)(\bar{x}))$
(iii)	$(t_0 \times t_1)(\bar{x})$	=	$\langle t_0(\overline{x \leftarrow (x)_0}), t_1(\overline{x \leftarrow (x)_1})\rangle$
(iv)	$T_0 \times T_1$	=	$\{\langle x, y\rangle \mid x \in T_0 \wedge y \in T_1\}$ •

The expression $T_0 \times T_1$ is now ambiguous since T_0 and T_1 may be types or terms. Unless we indicate otherwise we shall always take T_0 and T_1 in such expressions to be types.

5.2 Fact

$$t_0(\bar{x}) \ \rho \ \varphi_0(\bar{x}) \ \textit{and} \ t_1(\bar{x}) \ \rho \ \varphi_1(\bar{x}) \quad \textit{iff} \quad (t_0 \times t_1)(\bar{x}) \ \rho \ (\varphi_0 \times \varphi_1)(\bar{x})$$

proof. Suppose that $\bar{x}$ are the free variables of t_0 and t_1. Then from the assumptions we get $t_0(\overline{x \leftarrow (x)_0}) \ \rho \ \varphi_0(\overline{x \leftarrow (x)_0}) \wedge t_1(\overline{x \leftarrow (x)_1}) \ \rho \ \varphi_1(\overline{x \leftarrow (x)_1})$ which is equivalent to $\langle t_0(\overline{x \leftarrow (x)_0}), t_1(\overline{x \leftarrow (x)_1})\rangle \ \rho \ \varphi_0(\overline{x \leftarrow (x)_0}) \wedge \varphi_1(\overline{x \leftarrow (x)_1})$ which is just $(t_0 \times t_1)(\bar{x}) \ \rho \ (\varphi_0 \times \varphi_1)(\bar{x})$ •

The outline of the rest of the section is as follows. First we consider a pair of mutually inductive type equations for types T_0 and T_1. We then define a type T so that $T \equiv T_0 \times T_1$. Next we show that there are two elimination rules μ_0 and μ_1 so that $\mu_0 \times \mu_1$ follows from T-elim (similarly for the introduction rules). Finally we define two functions h_0 and h_1 so that $h_0 \times h_1$ is a realizer of T-Elim. Note that $h_0$and h_1 realize the rules μ_0 and μ_1 as a consequence of fact 5.2.

5.3 Proposition

$TK^{-} + T_{\Xi}$ supports the solution of *binary systems of mutual induction.*

The general form of a binary system is, for some specified types A_0 and A_1 and type operations $B_0(X,Y)$ and $B_1(X,Y)$ (safe and monotone in both X and Y):

(i) $\quad T_0 \equiv A_0 \cup B_0(T_0,T_1)$

(ii) $\quad T_1 \equiv A_1 \cup B_1(T_0,T_1)$

proof

We construct a type $\Xi(A,\lambda X.B)$ where $A = A_0 \times A_1$ and $B(X) = B_0((X)_0,(X)_1) \times B_1((X)_0,(X)_1)$. and we set $T_0 = (\Xi(A,\lambda X.B))_0$ and $T_1 = (\Xi(A,\lambda X.B))_1$. It is then a routine matter to check that T_0 and T_1 so defined satisfy (i) and (ii). We will write T for $\Xi(A,\lambda X.B)$ •

5.4 Proposition

The following system of elimination rules for types T_0 and T_1 satisfying the equations in the preamble to Proposition 5.3 are sound.

$$\frac{(\forall x\in A_0)\psi_0(x) \quad (\forall x\in B_0(T_0,T_1))(x\in B_0(\{w|\ \psi_0(w)\},\{w|\ \psi_1(w)\}) \Rightarrow \psi_0(x))}{(\forall x\in T_0)\psi_0(x)} \text{(i)}$$

$$\frac{(\forall x\in A_1)\psi_1(x) \quad (\forall x\in B_1(T_0,T_1))(x\in B_1(\{w|\ \psi_0(w)\},\{w|\ \psi_1(w)\}) \Rightarrow \psi_1(x))}{(\forall x\in T_1)\psi_1(x)} \text{(ii)}$$

proof

Consider the rule T-elim where T is defined as in the proof of Proposition 5.3.

$$\frac{\begin{array}{c}(\forall x\in A)(\psi_0(x)\times\psi_1(x))\\ (\forall x\in B(\Xi(A,\lambda X.B)))(x\in B(\{w \mid \psi_0(z)\times\psi_1(z)\ \}) \Rightarrow (\psi_0(x)\times\psi_1(x)))\end{array}}{(\forall x\in \Xi(A,\lambda X.B))(\psi_0(x)\times\psi_1(x))} \text{(iii)}$$

Assume the four premises of (i) and (ii) above. From the minor premises $(\forall x\in A_0)\psi_0(x)$and $(\forall x\in A_1)\psi_1(x)$ we obtain $(\forall x\in A_0\times A_1)(\psi_0((x)_0) \wedge \psi_1((x)_1))$ or $(\forall x\in A)(\psi_0(x)\times\psi_1(x))$ which is the minor premise of (iii). Furthermore, from:
$(\forall x\in B_0(T_0,T_1)(x\in B_0(\{w|\ \psi_0(w)\},\{w|\ \psi_1(w)\}) \Rightarrow \psi_0(x))$ and from
$(\forall x\in B_1(T_0,T_1)(x\in B_1(\{w|\ \psi_0(w)\},\{w|\ \psi_1(w)\}) \Rightarrow \psi_1(x))$ we obtain the major premise of (iii): $(\forall x\in B((\Xi(A,\lambda X.B)))(x\in B(\{w\in\Xi(A,\lambda X.B) \mid \psi_0(z)\times\psi_1(z)\}) \Rightarrow (\psi_0(x)\times\psi_1(x)))$ since their conjunction amounts to:
$(\forall x\in B_0(T_0,T_1)\times B_1(T_0,T_1))(x\in B_0(\{w|\ \psi_0(w)\},\{w|\ \psi_1(w)\}\times B_1(\{w|\ \psi_0(w)\},\{w|\ \psi_1(w)\})$
$\Rightarrow (\psi_0(x)\times\psi_1(x))$ and because $B_0(T_0,T_1)\times B_1(T_0,T_1) \equiv B(\Xi(A,\lambda X.B))$ and

$B_0(\{w|\ \psi_0(w)\},\{w|\ \psi_1(w)\}XB_1(\{w|\ \psi_0(w)\},\{w|\ \psi_1(w)\} \equiv B(\{w \mid \psi_0(z)X\psi_1(z)\}$. Now applying (iii) we obtain $(\forall x\in \Xi(A,\lambda X.B))(\psi_0(x)X\psi_1(x))$ which decomposes into the required conjuncts $(\forall x\in T_0)\psi_0(x)$ and $(\forall x\in T_1)\psi_1(x)$ immediately. •

5.5 Proposition

The following system of introduction rules for the types T_0 and T_1 defined in the preamble to Proposition 5.3 are sound:

$$\frac{x\in A_0}{x\in T_0}\ \text{(i)} \qquad\qquad \frac{x\in A_1}{x\in T_0}\ \text{(iii)}$$

$$\frac{x\in B_0(T_0,T_1)}{x\in T_0}\ \text{(ii)} \qquad\qquad \frac{x\in B_1(T_0,T_1)}{x\in T_1}\ \text{(iv)}$$

proof

We note that $x_0\in A_0 \wedge x_1\in A_1$ allows us to conclude that $<x_0,x_1> \in \Xi(A,\lambda X.B)$ hence $x_0\in T_0$ and $x_1\in T_1$ as required. Similarly $x_0\in B_0(T_0,T_1) \wedge x_1\in B_1(T_0,T_1)$ implies that $<x_0,x_1> \in B(\Xi(A,\lambda X.B)$ hence $x_0\in T_0$ and $x_1\in T_1$ as required.

5.6 Proposition

Let $f_0\ \rho\ (\forall x\in A_0)\psi_0(x)$, $f_1\ \rho\ (\forall x\in A_1)\psi_1(x)$, $g_0\ \rho\ (\forall x\in B_0(T_0,T_1))(x\in B_0(\{w|\ \psi_0(w)\},\{w|\ \psi_1(w)\}) \Rightarrow \psi_0(x))$ and $g_1\ \rho\ (\forall x\in B_1(T_0,T_1))(x\in B_1(\{w|\ \psi_0(w)\},\{w|\ \psi_1(w)\}) \Rightarrow \psi_1(x))$.

The following system of mutually recursive combinators, ***irec₀*** and ***irec₁*** realize the rules (i) and (ii) defined in the preamble to Proposition 5.4.

$$\textbf{\textit{irec}}_0\ f_0\ g_0\ f_1\ g_1\ a_0 = f_0\ a_0$$
$$\textbf{\textit{irec}}_0\ f_0\ g_0\ f_1\ g_1\ x_0 = g_0\ x_0\ (B_0(\textbf{\textit{irec}}_0\ f_0\ g_0\ f_1\ g_1,\ \textbf{\textit{irec}}_1\ f_0\ g_0\ f_1\ g_1)\ x_0)$$
$$\textbf{\textit{irec}}_1\ f_0\ g_0\ f_1\ g_1\ a_1 = f_1\ a_1$$
$$\textbf{\textit{irec}}_1\ f_0\ g_0\ f_1\ g_1\ x_1 = g_1\ x_1\ (B_1(\textbf{\textit{irec}}_0\ f_0\ g_0\ f_1\ g_1,\ \textbf{\textit{irec}}_1\ f_0\ g_0\ f_1\ g_1)\ x_1)$$

proof

Now we know that $(\forall x\in A_0)\psi_0(x) \wedge (\forall x\in A_1)\psi_1(x)$ or $(\forall x\in A)(\psi_0X\psi_1(x))$ so we obtain $(f_0Xf_1)\ \rho\ (\forall x\in A)(\psi_0X\psi_1(x))$ by the assumptions above, the definition of realizability and corollary 5.2. Similarly, we reasoned earlier that:

$(\forall x\in B_0(T_0,T_1))(x\in B_0(\{w|\ \psi_0(w)\},\{w|\ \psi_1(w)\}) \Rightarrow \psi_0(x)) \wedge (\forall x\in B_1(T_0,T_1))(x\in B_1(\{w|\ \psi_0(w)\},\{w|\ \psi_1(w)\}) \Rightarrow \psi_1(x)) \Rightarrow (\forall x\in B(\Xi(A,\lambda X.B)))(x\in B(\{w \mid \psi_0(z)X\psi_1(z)\}) \Rightarrow (\psi_0(x)X\psi_1(x)))$. Combining this with the realizability assumptions above we obtain:

$\lambda<u_0,u_1>.((g_0u_0)X(g_1u_1))\rho(\forall x\in B(\Xi(A,\lambda X.B)))(x\in B(\{w \mid \psi_0(z)X\psi_1(z)\}) \Rightarrow (\psi_0(x)X\psi_1(x)))$ because if $<u_0,u_1>\in B(\Xi(A,\lambda X.B))$ then $u_0\in B_0(T_0,T_1)$ and $u_1\in B_1(T_0,T_1)$ and so $(g_0\ u_0) \in B_0(C_0,C_1)$ and $(g_1\ u_1) \in B_1(C_0,C_1)$ where C_i is the type of realizers of $\psi_i(z)$ $(i\in\{0,1\})$. Thus

$(g_0\ u_0)\times(g_1\ u_1) \in B_0(C_0,C_1)\times B_1(C_0,C_1) \to (C_0\times C_1)$ and $B_0(C_0,C_1)\times B_1(C_0,C_1) \to (C_0\times C_1) = B(C_0\times C_1) \to (C_0\times C_1)$ as is required for the major premise of $\Xi(A,\lambda X.B)$-elim. From this and the soundness of realizability for the $\Xi(A,\lambda X.B)$-elim rule we find that:
$\textbf{\textit{irec}}\ f_0\times f_1\ \lambda\langle u_0,u_1\rangle.((g_0u_0)\times(g_1u_1))\ \rho\ \ (\forall x\in\Xi(A,\lambda X.B))(\psi_0(x)\times\psi_1(x))$.
We now show, by $\Xi(A,\lambda X.B)$-induction, that:
$(\textbf{\textit{irec}}_0\ f_0\ g_0\ f_1\ g_1)\times(\textbf{\textit{irec}}_1\ f_0\ g_0\ f_1\ g_1) = \textbf{\textit{irec}}\ (f_0\times f_1)\ (\lambda\langle u_0,u_1\rangle.(g_0u_0)\times(g_1u_1))$.
base step
$(\textbf{\textit{irec}}_0\ f_0\ g_0\ f_1\ g_1)\times(\textbf{\textit{irec}}_1\ f_0\ g_0\ f_1\ g_1)\langle a_0,a_1\rangle = \langle(f_0\ a_0),(f_1\ a_1)\rangle = (f_0\times f_1)\ \langle a_0,a_1\rangle =$
$\textbf{\textit{irec}}\ (f_0\times f_1)(\ \lambda\langle u_0,u_1\rangle.(g_0u_0)\times(g_1u_1))\ \langle a_0,a_1\rangle$.
Induction step
We shall write δ_0 for $B_0(\textbf{\textit{irec}}_0\ f_0\ g_0\ f_1\ g_1,\ \textbf{\textit{irec}}_1\ f_0\ g_0\ f_1\ g_1)$ and δ_1 for $B_1(\textbf{\textit{irec}}_0\ f_0\ g_0\ f_1\ g_1,\ \textbf{\textit{irec}}_1\ f_0\ g_0\ f_1\ g_1)$ for convenience.
$((\textbf{\textit{irec}}_0\ f_0\ g_0\ f_1\ g_1)\times(\textbf{\textit{irec}}_1\ f_0\ g_0\ f_1\ g_1))\ \langle x_0,x_1\rangle = \langle(g_0\ x_0\ (\delta_0\ x_0)),\ (g_1\ x_1\ (\delta_1\ x_1))\rangle =$
$((g_0\ x_0)\times(g_1\ x_1))\ \langle(\delta_0\ x_0),(\delta_1\ x_1)\rangle = ((g_0\ x_0)\times(g_1\ x_1))((\delta_0\times\delta_1)\ \langle x_0,\ x_1\rangle) =$ (*ex hypothesi*)
$(g_0\ x_0)\times(g_1\ x_1)(B(\textbf{\textit{irec}}\ (f_0\times f_1)(\lambda\langle u_0,\ u_1\rangle.(g_0\ u_0)\times(g_1\ u_1)))\ \langle x_0,\ x_1\rangle) =$
$(\lambda\langle u_0,\ u_1\rangle.(g_0\ u_0)\times(g_1\ u_1))\ \langle x_0,\ x_1\rangle\ (B(\textbf{\textit{irec}}\ (f_0\times f_1)(\lambda\langle u_0,\ u_1\rangle.(g_0\ u_0)\times(g_1\ u_1)))) =$
$\textbf{\textit{irec}}\ (f_0\times f_1)\ (\lambda\langle u_0,\ u_1\rangle.(g_0\ u_0)\times(g_1\ u_1)))\ \langle x_0,\ x_1\rangle$ as required, and this completes the induction and the proof. •

§6 Conclusions and Future Work

Principles of positive induction have been investigated within the context of Martin-Löf's theory of types [Mar 84] for example in [Dyb 87] a useful isomorphism equation is shown to be satisfiable for strictly positive set operations but this does not work in the intensional version of the theory. In [Dyb 90] a schema which determines a collection of theories is utilised to determine an *inversion principle* by which equality and elimination rules can be obtained from formation and introduction rule. This builds on earlier work of Martin-Löf [Mar 71]. Mendler [Men 87] considers very general forms of induction based on certain fixpoint operators but these require the *subtype* [Pet 84] an addition to type theory which has caused more than a little argument in recent years. Most recently [Sal 90] shows that the subtype does not in general even satisfy an obvious principle of separation and this may turn out to be the final word on that topic. More recent work which appears to be free from the problems of the subtype has been undertaken [CoP 89]. The PX system [HaN 87] is based on Feferman's theory T_0 and includes a general principle of induction, conditional inductive generation (CIG) which is a generalisation of that provided by Feferman. Perhaps because PX is based on the language LISP the system seems very baroque and the CIG induction principle is no exception. Our own approach in TK to positive induction seems very much more perspicuous.

The most recent version of the TK program derivation environment TKE [SaG 90] has been constructed with sufficient flexibility to allow for a variety of experiments with the basic rules of the theory to be undertaken. The principle of safe positive induction has been included in TKE successfully and programs whose specifications utilise it (examples from [Hen 90a]) have been undertaken. It is now becoming increasingly clear that we need to develop *higher level* programming logics which are based on higher level programming languages. In particular a translation from the type system of such a programming language (for example the algebraic types of Miranda [Tur 85] or the corresponding types in Haskell [WaH 90]) to specialisations of $\Xi(A,\lambda X.B)$ is an urgent theoretical and practical issue.

§7 Acknowledgements

Thanks as usual are owed to the members of the Essex Constructive Set Theory and Functional Programming Group.

Part of this work was carried out while I was a visiting lecturer in the Department of Computer Science at the University of Otago in New Zealand. I should like to take this opportunity to extend my gratitude to my hosts for making my visit so worthwhile.

This work was supported by an SERC research grant: GR/F/02809.

§8 References

[CoP 89] Coquand, T., Paulin, C., **Inductively defined types,** *Proc. of the workshop on programming logic,* 1989.

[Dyb 87] Dybjer, P., **Inductively defined sets in Martin-Löf's type theory,** *Proc. of the workshop on General Logic, Edinburgh,* 1987.

[Dyb 90] Dybjer, P., **An inversion principle for Martin-Löf's type theory,** Manuscript, 1990.

[Fef 79] Feferman, S., **Constructive theories of functions and classes,** *Logic Coll. '78,* pp 159-224, North Holland, 1979.

[HaN 87] Hayashi S. & Nakano, H., **The PX system – A computational logic,** Publications of the Research Institute for Mathematical Sciences, Kyoto University, Tokyo, 1987.

[Hen 89a] Henson, M. C., **Program development in the constructive set theory TK,** *Formal Aspects of Computing,* **1,** pp 173-192, 1989.

[Hen 89b] Henson, M. C., **Realizability models for program construction,** *Proc. Conf. on Mathematics of Program Construction,* Gröningen, LNCS **375**, pp 256-272, Springer, 1989.

[Hen 90] Henson, M. C., **Information Loss in the Programming Logic TK,** *Proc. IFIP TC2 Working Conf. on Programming Concepts and Methods,* pp 509-545, Elsevier, 1990

[HeT 88] Henson, M. C. & Turner, R., **A constructive set theory for program development,** *Proc. 8th Conf. on FST & TCS,* Bangalore, LNCS **338**, pp 329-347, Springer, 1988.

[Mar 71] Martin-Löf, P., **Haupsatz for the intuitionistic theory of iterated inductive definitions,** *Proc. 2nd Scandinavian Logic Symp.*, pp 179-216, 1971.

[Mar 84] Martin-Löf, P., **Intuitionistic type theory,** *Bibliopolis*, 1984.

[Mos 74] Moschovakis, Y. N., **Elementary induction on abstract structures.** North Holland, 1974.

[Pet 84] Petersson, K., **The subset type former and the type of small types in Martin-Löf's theory of types,** Tech. rep. 3, University of Göteborg, Programming methodology group, 1984.

[SaG 90] Sanderson, M. T. & Ghosh-Roy, R., **TKE: the TK proof development environment,** *private communication,* University of Essex, 1990.

[Sal 89] Salvesen, A., **On information discharging and retrieval in Martin-Löf's type theory,** Ph.D. thesis, University of Oslo, 1989.

[Tur 85] Turner, D. A, **Miranda - A non-strict functional language with polymorphic types,** *in: Proc. IFIP Int. Conf. on functional programming languages and computer architecture,* Nancy, LNCS **201**, Springer Verlag, pp 445-472, 1985.

[WaH 90] Wadler, P. , Hudak, P. (*eds.*) *et al*, **Report on the programming language Haskell,** University of Glasgow, Technical Report, 1990.

§9 Appendix: The theory TK

In this appendix we provide a brief overview of the programming logic TK. In 9.1 we describe the language of the theory, and in 9.2 the proof rules.

9.1 The language of the theory TK

The theory consists of a language of terms, sets and formulae. The term language consists of the untyped lambda notation extended with constants and types as objects. We assume rudimentary infrastructure of terms coded in well known ways, for example for pairing and selecting from pairs $(<\_,\_>; (\_)_0; (\_)_1)$.

$$t \in TERM; \quad t \mid x \mid c \mid \lambda x.t \mid (t\, t) \mid \mathrm{T}$$

Sets are called types or kinds. Although both classify objects of the theory the kinds also classify those objects which are constructed from types.

$$\mathrm{T} \in TYPE; \quad X \mid \{x \mid \varphi'\} \mid (\mathrm{I}(\mathrm{A},\varphi) \text{ or } \Xi(\mathrm{A},\lambda X.\mathrm{B}))$$

$$\mathrm{K} \in KIND; \quad \mathrm{T} \mid \{x \mid \varphi\}$$

Sets can be formed from properties by comprehension and by induction. There are two alternative forms of inductive types: Either I(A,φ) the well founded part of the relation φ with minimal elements in A or, alternatively, the safe positive schema Ξ(A,λX.B) which is investigated in this paper. Atomic formulae consist of membership, equality, definedness and absurdity assertions:

$$\alpha \in ATOM; \quad t\in \mathrm{T} \mid t\in \mathrm{K} \mid t{=}t \mid t{\downarrow} \mid \bot$$

Those involving just equality, definedness and absurdity we call the *simple atoms*. We can then close off the assertions in the normal way, although we include quantifiers over types as well as objects.

$$\varphi \in WFF; \quad \alpha \mid \varphi \wedge \varphi \mid \varphi \vee \varphi \mid \varphi \Rightarrow \varphi \mid (\forall x)\varphi \mid (\forall X)\varphi \mid (\exists x)\varphi \mid (\exists X)\varphi$$

The set *WFF'* over which φ' ranges is that subset of *WFF* which includes no element involving a bound type variable (by quantification). This completes the presentation of the basic language of the theory.

9.2 The rule system of TK

The logical apparatus of TK falls into two parts. First we have the ordinary rules of intuitionistic predicate logic of partial terms.

$$\frac{\varphi \quad \psi}{\varphi \wedge \psi}\ \wedge\text{–intro} \qquad \frac{\varphi \wedge \psi}{\varphi}\ \wedge\text{–elim(i)} \qquad \frac{\varphi \wedge \psi}{\psi}\ \wedge\text{–elim(ii)}$$

$$\frac{\psi}{\varphi \vee \psi}\ \vee\text{–intro(i)} \qquad \frac{\psi}{\varphi \vee \psi}\ \vee\text{–intro(ii)} \qquad \frac{\varphi \vee \psi \quad \begin{matrix}[\varphi]\\ \vdots \\ \eta\end{matrix} \quad \begin{matrix}[\psi]\\ \vdots \\ \eta\end{matrix}}{\eta}\ \vee\text{–elim(ii)}$$

$$\frac{\begin{matrix}[\varphi]\\ \vdots \\ \psi\end{matrix}}{\varphi \Rightarrow \psi}\ \Rightarrow\text{-intro} \qquad \frac{\varphi \quad \varphi \Rightarrow \psi}{\psi}\ \Rightarrow\text{–elim}$$

We take $\neg\varphi =_{def} \varphi \Rightarrow \bot$. $\neg$-intro and $\neg$-elim are then special cases of $\Rightarrow$-intro and $\Rightarrow$-elim respectively.

$$\frac{\varphi(x)}{(\forall x)\varphi}\ \forall\text{-intro} \qquad \frac{(\forall x)\varphi \quad t{\downarrow}}{\varphi(x{\leftarrow}t)}\ \forall\text{-elim}$$

For $\forall$-intro we require that x is not free in any assumption on which φ depends.

$$\frac{\varphi(x{\leftarrow}t) \quad t{\downarrow}}{(\exists x)\varphi}\ \exists\text{-intro} \qquad \frac{(\exists x)\varphi \quad \begin{matrix}[\varphi(x{\leftarrow}y)]\\ \vdots \\ \eta\end{matrix}}{\eta}\ \exists\text{-elim}$$

For $\exists$-elim we require that y is not free in η, $(\exists x)\varphi$ or any assumption (except $\varphi[x{\leftarrow}y]$) on which η depends.

$$\frac{\varphi(X)}{(\forall X)\varphi}\ \forall X\text{-intro} \qquad \frac{(\forall X)\varphi}{\varphi(X{\leftarrow}\mathrm{T})}\ \forall X\text{-elim}$$

For $\forall X$-intro we require that X is not free in any assumption on which φ depends.

$$\frac{\varphi(X\leftarrow \mathrm{T})}{(\exists x)\varphi}\ \exists X\text{-intro} \qquad\qquad \frac{(\exists X)\varphi \quad \begin{matrix}[\varphi(X\leftarrow Y)]\\ \vdots \\ \eta\end{matrix}}{\eta}\ \exists X\text{-elim}$$

For $\exists X$-elim we require that Y is not free in η, $(\exists X)\varphi$ or any assumption (except $\varphi[X\leftarrow Y]$) on which η depends.

$$\frac{\perp}{\varphi} \quad \text{intuitionistic-absurdity}$$

Next we have the logic of terms. For convenience we introduce the meta-notation $t \approx t' =_{\mathrm{def}} t{\downarrow} \vee t'{\downarrow} \Rightarrow t = t'$.

$$x{\downarrow} \quad \text{var-denotes} \qquad\qquad c{\downarrow} \quad \text{const-denotes}$$

$$\mathrm{T}{\downarrow} \quad \text{type-denotes} \qquad\qquad t \approx t \quad \text{eq-reflexive}$$

$$\frac{t \approx t'}{t' \approx t} \quad \text{eq-symmetric} \qquad\qquad \frac{t = t'}{t{\downarrow}} \quad \text{eq-denotes}$$

$$\frac{(t\,t')}{t{\downarrow}} \quad \text{ap-denotes(i)} \qquad\qquad \frac{(t\,t')}{t'{\downarrow}} \quad \text{ap-denotes(ii)}$$

$$\frac{t \in K}{t{\downarrow}} \quad \text{mem-denotes} \qquad\qquad \frac{t \approx t' \quad \varphi(t)}{\varphi(t')} \quad \text{substitution}$$

$$\lambda x.t \approx \lambda y.t(x\leftarrow y) \quad \text{alpha } (y\ nfi\ t) \qquad\qquad \frac{t'{\downarrow}}{(\lambda x.t)t' \approx t(x\leftarrow t')} \quad \text{beta}$$

$$\frac{t \approx t'}{\lambda x.t \approx \lambda x.t'} \quad (\xi) \qquad\qquad \lambda x.t{\downarrow} \quad \text{abs-denotes}$$

Transitivity of equality and the remaining congruence rules of lambda equality are consequences of substitution.

Finally, the rules for sets:

$$\frac{z \in \{x \mid \varphi(x)\}}{\varphi(x\leftarrow z)} \ (\{\}\text{–elim}) \qquad\qquad \frac{\varphi(x\leftarrow z)}{z \in \{x \mid \varphi(x)\}} \ (\{\}\text{–intro})$$

$$\frac{z \in \mathrm{A}}{z \in \mathrm{I}(\mathrm{A},\varphi)} \ (\mathrm{I}(\mathrm{A},\varphi)\text{-intro(i)}) \qquad\qquad \frac{(\forall y)(\varphi(z, y) \Rightarrow y \in \mathrm{I}(\mathrm{A},\varphi))}{z \in \mathrm{I}(\mathrm{A},\varphi)} \ (\mathrm{I}(\mathrm{A},\varphi)\text{–intro(ii)})$$

$$\frac{\begin{array}{c}(\forall x)(x \in A \Rightarrow \psi(x)) \\ (\forall x)(x \in \mathrm{Clo}(I(A,\varphi)) \Rightarrow (\forall y)(\varphi(x,y) \Rightarrow \psi(y)) \Rightarrow \psi(x))\end{array}}{(\forall x)(x \in I(A,\varphi) \Rightarrow \psi(x))} \quad (I(A,\varphi)\text{-elim})$$

where $\mathrm{Clo}(I(A,\varphi)) =_{\mathrm{dcf}} \{z \mid (\forall y)(\varphi(z,y) \Rightarrow y \in I(A,\varphi))\}$. The two premises of the I(A,φ)-elim rule have been written vertically. When T is a specialisation of I(A,φ) we will refer to the rules above as T-intro(i), T-intro(ii) and T-elim.

The safe positive induction schema is governed by the following three rules:

$$\frac{z \in A}{z \in \Xi(A, \lambda X.B)} \quad (\Xi\text{-intro(i)}) \qquad \frac{z \in B(\Xi(A, \lambda X.B))}{z \in \Xi(A, \lambda X.B)} \quad (\Xi\text{-intro(ii)})$$

$$\frac{(\forall x \in A)\psi(x) \quad (\forall x \in B((\Xi(A,\lambda X.B)))(x \in B(\{w \in \Xi(A,\lambda X.B) \mid \psi(w)\}) \Rightarrow \psi(x))}{(\forall x \in \Xi(A,\lambda X.B))\psi(x)} \quad (\Xi\text{-elim})$$

9.3 Definition (Safe formulae and types)

(i)	*If*	*α is a simple atom*	*then*	*Safe*(α)
(ii)	*If*	*Safe*(φ) *and Safe*(ψ)	*then*	*Safe*(φ ∧ ψ)
(iii)	*If*	*Safe*(ψ)	*then*	*Safe*(φ ⇒ ψ)
(iv)	*If*	*Safe*(φ)	*then*	*Safe*(∀x)φ
(v)	*If*	*Safe*(T)	*then*	*Safe*(z∈T)
(vi)	*If*	*Safe*(φ)	*then*	*Safe*({x | φ})
(vii)	*If*	*Safe*(φ) *and Safe*(A)	*then*	*Safe*(I(A,φ))
(viii)	*If*	*Safe*(A) *and Safe*(B)	*then*	*Safe*(Ξ(A, λX.B)) •

9.4 Definition (Rsafe fomulae and types)

(i)	*RSafe*(φ)	*iff*	*Safe*(*e* ρ φ)
(ii)	*RSafe*(T)	*iff*	*RSafe*(x∈T) •

9.5 Proposition

If φ is decidable then φ is equivalent to a safe formula.

Proof. If φ is decidable then it is possible to rewrite it immediately as a safe formula by double negation (using 9.3(iii) above). •

WAM SPECIFICATION FOR PARALLEL EXECUTION ON SIMD COMPUTER

S.Ivanets, N. Ilinsky, M. Krylov
Department of Cybernetics,
Moscow Engineering-Phisics Institute,
Kashirskoe sh., 31, Moscow, USSR.

Abstract. We have been working out a compiler from *Prolog* into special code, targeted in a two-computer media. The media consists of conventional *host*-computer and parallel *SIMD*-computer like *Connection Machine*. Our purpose is to construct an effective programming system preserving the conventional *Prolog* semantics. We managed to design a system with total parallelism of logic program execution.

Keywords. Prolog, parallel unification, hypercube, SIMD-computer, Warren Abstract Machine, compiler.

1. Introduction

Now we investigate the problem of building a highly effective compiler for *Prolog*. There is a special two-computer media at our dispose, were we intend to execute logic programs. This media is composed of host-computer with *SISD* architecture and parallel *SIMD*-computer with hypercube topology.

Our compiler produce modified *WAM* instructions, different efficiently from that on *SISD*-computer. All instructions of modified *WAM* operate on parallel objects in memory of *SIMD*-computer. This allows us to design a system with total parallelism of logic program execution and efficiently less run-time.

Besides that the usage of *WAM* made it possible for us to keep unchanged the semantics of conventional *Prolog*. All accumulated for *SISD* methods of *Logic Programming* can be applied here as well.

2. The Connection Machine

As a parallel computer in our project we use a *Connection Machine*[1] [1, 7]. It is a modern well known computer system which popularity is growing constantly.

---

1 *Connection Machine* is a registered trademark of Thinking Machines Corporation, designer W. Daniel Hillis, 1985.

The *Connection Machine* is a fine-grained, highly parallel computer whose first version consists of 65536 processing elements. Each of these elements has 4096 bits of memory and a 1-bit-wide *ALU*. The system is controlled by microcontroller, which executes macroinstructions originating in a host-computer. A machine cycle (3 microcontroller states) takes about 750 ns, yielding a raw system throughput exceeding two billion 32-bit operation per second. When operating on data smaller then 32 bits, proportionally higher processing rates are obtained.

All processing elements execute the same program with each operating on the contents of its own memory. The *ALU* has a context flag that allows processors to be selectively disabled. Communication is provided by a hypercube network and same other internal networks. The hypercube supports full packet-switched communication between arbitrary processors.

3. Our interpretation of SIMD-computer

Because of the peculiarity of conventional Prolog we use a restricted, linear scheme of processor element connections. For ordered processor elements we ignore all there other hypercube connections.

Target *SIMD*-computer memory is shown on the picture.

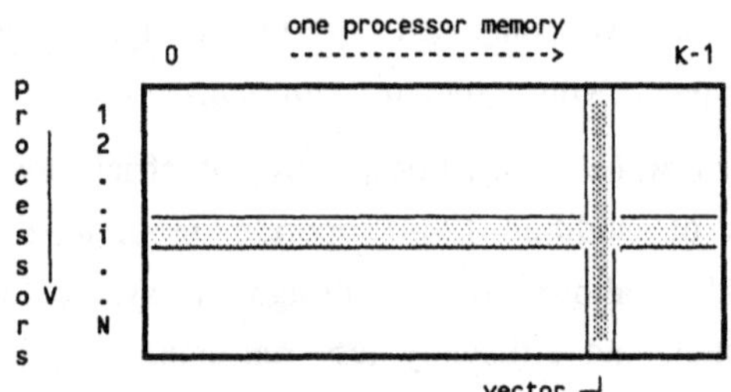

picture 1. Memory of SIMD-computer

N is a number of processor elements, in the case it is equal to 65536. *K* is a memory of one processor element, now 4096 bits. All data and control structures are placed vertically across all the processor elements. So we could treat them simultaneously.

4. Modified Warren Abstract Machine

A modified version of *Warren Abstract Machine (WAM)* [2] is a basis for our project. On the picture below there are shown *WAM's* main data areas and registers with remarks on data that can be processed in parallel.

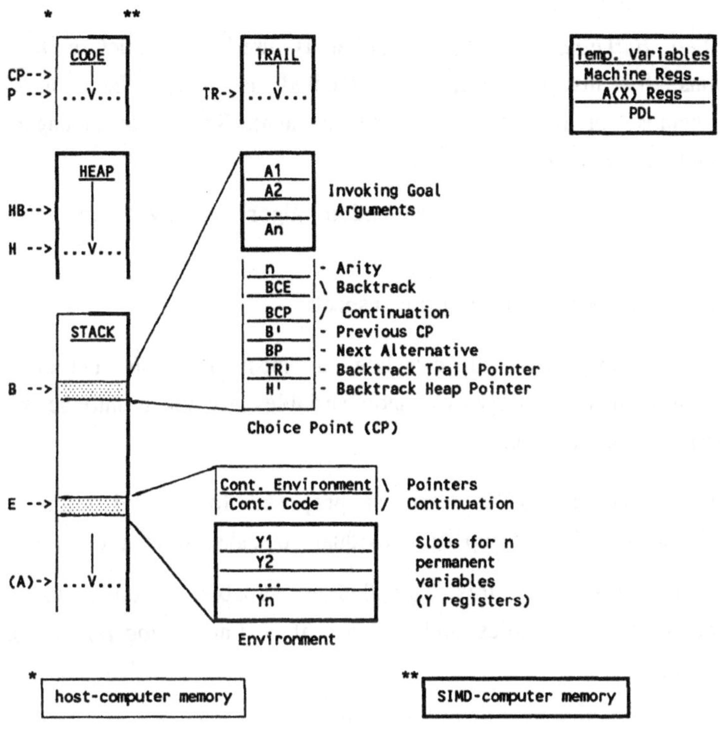

picture 2. Data and control areas of WAM

We have not changed the most of *WAM's* data and control structures. But they were divided into two groups. Almost all control structures do not assume parallel treatment. So they are stored in a *host*-computer memory. Other objects, for example predicate arguments, could be treated in parallel. That's why they are stored in *SIMD*-computer memory in a so-called "**vector**". Vector consists of memory cells with the same memory address through all the processor elements (see picture 1).

As the result of compilation all the predicates have only one parameter. The parameter is a reference to vector, that consists of predicate arguments, stored in *SIMD*-computer memory. At a run-time the unification of the call argument vector with the head argument vector of an applicable clause could be done in parallel (so-called *parallel unification*).

A modified *Barklund's* [3, 4] algorithm is used for structure treatment. All structured terms, for instance lists, are presented as matrix. Each matrix row is placed in a memory of individual processor element. Such data arrangement makes possible parallel access and data treatment.

In many cases this parallel unification algorithm gives an *exponential* of productivity.

Parallelism is used in two following aspects:

1. There is a possibility to calculate a most common unifier for some subexpressions simultaneously. Common variables are taken into account here. It is a distributed unification.

2. A parallel processing of subterms is proposed for nested terms. Unification process is devided into group of individual subtasks. It is a deep unification.

Under certain conditions this algorithm is capable to unify lists and trees (compound terms) with N nodes and A arcs at a time $O(\log N)$, using $O(N+A)$ processor elements.

5. Main memory areas

To solve the problem we proposed a memory management scheme shown on the picture below.

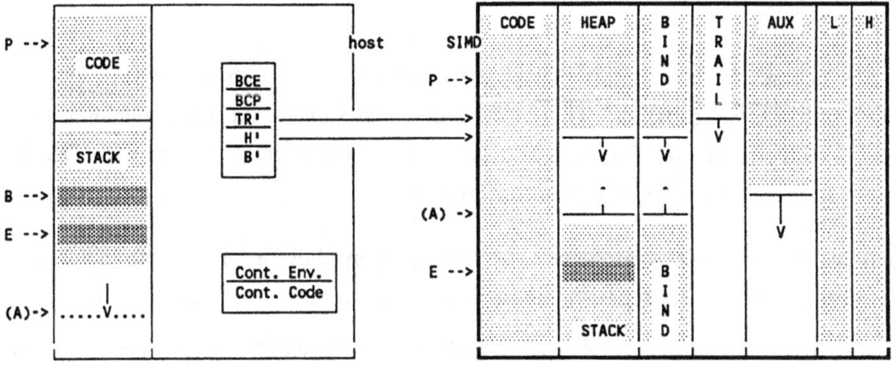

picture 3. Data and control objects
in a two-computer media

All major areas of *SIMD* memory are arranged in stacks. The exclusions are *Code* area intended for vector part of initial program code and a working registers for unification procedure *L* and *H*.

Heap and *Stack* data are of the same structure. This allows simultaneous processing with the same unification algorithm. More over this one-vector arrangement makes possible *variable trimming* and *Tail Recursion Optimization*.

Bind is intended to store variable bindings obtained in the course of unification.

Trail store several fragments of *Bind*, received in calculations and is intended to restore *Bind*.

Auxiliary area *Aux* is intended for special data structures like strings. Thanks to various reasons strings are very suitable for parallel processing, especially for unification.

We propose a method for quick strings unification, based on same particular features of hypercube topology of *Connection Machine* [8]. One of this features consists in that hypercube of *n* dimension could be present as 2^k hypercubes of *n-k* dimension. The nodes of arbitrary pare of cubes are on the same distance one from the other. If string elements are placed in such hypercubes, all string could be processed at a time. Number of steps here is equal to the maximum distance between the cubes. It is less then dimension of cube. So for the arbitrary amount of strings with arbitrary length we got the fixed number of unification steps.

6. Instruction set

In the coarse of compilation same sequence of special instructions is put in accordance with *Prolog* program. This instructions are formed on host-computer and should be executed on *SIMD*-computer. They have much in common with WAM instruction but there are same essential differences:

- Each instruction is in accordance with predicate symbol or all its arguments;
- All program terms are represented in a common, general manner.

As the result the group of *WAM* instructions, for example (get_variable, get_value, get_structure ...), is reduced to a single instruction, for example get. A list of basic instructions is shown in the table.

get Head, Number	Load pointers to the head arguments in a working register H
put Call, Number	Load pointers to the goal arguments in a working register L
allocate Env, Number, Base	Create clause environment
deallocate Env, Number	Delete clause environment
call Name	Transfer control to the call procedure
unify	unification (parallel Barklund's algorithm)

Table 1. Modified WAM instruction set

So we have reduced *WAM*, makes it more easy and general. That should increase the *WAM* execution effectiveness on SIMD [11, 12].

7. Conclusions

We present a new version of *WAM* targeted to parallel *SIMD*-computer execution. It has some essential features like distribution of data structures between *host*-computer and *SIMD*-computer, fully unchanged conventional *Prolog* semantics, highly effective structure unification algorithm, the possibility to work out quick built-in predicates for strings processing.

Linear growth of logic program execution time with dependence of data amount is typical for conventional model. We expect the *logarithmic* growth of this index for our model [8, 9, 10, 11, 12].

8. Related and future work

An example of logic programming language on *ICL DAP* was proposed by Kacsuk and Bale [9]. *Prolog* had been essentially modified by sets and matrix. Programmer himself is responsible for parallelism in his program. This approach considerably restricts the style of logic programming and the set of solving problem.

Another implementation technique could be found in works of Tarnlund's group from Sweden [3, 5]. It is based on vectorising of programs with automatic parallelism at compilation and without language distortion. This approach cover much wider set of problems and could be used as a basis for *SIMD*-computer compiler.

We are going to proceed with further improvements of characteristics of Prolog compiler for *SIMD*-computer. This work is a part of the "Parallel Computer" project at LOGOS research group.

Acknowledgments

The authors want to thank their colleags at Moscow Engineering-Phisics Institute, for their continuous support. We thank J. Barklund and H. Millroth from Uppsala University and S. Debray from University of Arizona for the materials kindly passed to us. We have very useful communications with Mats Carlson from Uppsala University.

References

[1] The Connection Machine. W.Daniel Hillis, The MIT Press, Cambridge, Massachusetts, London, England 1986

[2] An Abstract Prolog Instruction Set. David H.D. Warren. Technical Note 309, SRI International, AI Center, Compute Science and Technology Division,1983.

[3] Parallel Unification. J. Barklund, Uppsala Theses in Computing Science 9, UPMAIL, Computing Science Dept, UPPSALA 1990

[4] Data Parallel Algorithm. W.Daniel Hillis and Guy L. Steele, Communications of the acm December 1986, Vol 29, Numb 12

[5] Reforming Compilation of Logic Programs. H. Millroth, Uppsala Theses in Computing Science 10, UPMAIL, Computing Science Dept, UPPSALA 1990

[6] McBryan O., The Connection Machine: PDE Solution on 65536 Processors, Parallel Computing 9, 1988

[7] Hockney R.W., Parallel Computers: Architectures and Performance, Int. Conf. "Parallel Computings", 1986

[8] Hillis W., D., The Connection Machine A Fine Grained Multi-Processor, Second AIAA/NASA/USAF Symp. on Automation, Robotics and Advanced Computing for the National Space Program, 1987

[9] P.Kacsuk and A. Bale, DAP Prolog: A Set-oriented Approach to Prolog., The Computer Journal, October 1987, Vol 30, Numb 5

[10] McBryan O., New Architectures: Performance Highlights and New Algorithms, Parallel Computing 7, 1988

[11] Nilsson M., Tanaka H., Fleng Prolog - The Language Which Turns Supercomputers into Parallel Prolog Machines, Proc., Japanese Logic Programming Conf., ICOT, Tokyo, June 1986 ed. Wada

[12] J. Barklund, N. Hagner, M. Wafin, KL1 in Condition Graphs on a Connection Machine., Proc of the 5th Int Conf on Fifth Generation Computer Systems 1988, ed by ICOT

[13] Eugene Albert, Kathleen Knobe, Joan D. Lukas, Guy L. Steele Jr, Compiling Fortran 8x Array Features for the Connection Machine Computer System. , ACM 1988

On abstracting the procedural behaviour of logic programs.

G. Janssens[*], *M. Bruynooghe*[**]

Department of Computer Science,
K.U.Leuven
Celestijnenlaan 200A
B-3001 Heverlee
Belgium

ABSTRACT

Abstract interpretation is a widely applied method for doing static analysis of logic programs. A diversity of formalisms and applications have appeared in the literature. This paper describes at a rather informal level our formalism based on AND-OR-graphs and compares it with the approach based on denotational semantics.

* Supported by the project RFO-AI-02 : "Logic as a basis for artificial intelligence : control and efficiency of deductive inference - parallelism "

** Supported by the Belgian National Fund for Scientific Research.

Introduction

In general abstract interpretation aims at computing at compile time certain features of the concrete execution of a program. Applying abstract interpretation to extract features of the concrete execution of programs involves two levels of abstraction. The first one consists of the development of a framework, a parametrized construction for the analysis of programs in a given language, together with theorems that ensure the soundness and termination of the analysis provided that the domains and operations - the parameters - supplied to complete the construction obey certain safety requirements. During this process the actual application does not have to be taken into account - it determines the second level of abstraction. Frameworks can be formalized for particular languages or language classes. Our framework, for example, is developed for logic programs [2]. Such a framework offers a backbone for the development of a new application, which is concerned with the second level of abstraction. We have been using abstract interpretation of logic programs for deriving information which can be used during code generation, especially for specializing versions of procedures [3, 4, 15]. The desired code optimizations determine the required expressive power of the latter abstraction : it must offer a formalism to describe sets of concrete substitutions. Such a description is called an *abstract substitution* and the set of abstract substitutions an *abstract domain*.

This paper informally discusses both levels of abstractions. We explain what the rationale is behind abstract interpretation of logic programs and in particular behind our approach, such that the reader gets an idea of what abstract interpretation is about and what it can offer him. For a more formal and precise specification of our framework, one is referred to [2-4, 15].

In section 1 we briefly sketch the general idea of abstract interpretation. Section 2 points out what kind of information about the execution behaviour is useful for code generation and thus should be derivable by abstract interpretation. The next two sections deal with the two levels of abstraction. The issues raised in section 2 allow us to formulate some basic requirements and characteristics of a framework, as is done in section 3. We illustrate these arguments by referring to our framework and to frameworks based on the denotational semantics. In section 4, we discuss some important issues of the design of an abstract domain a.o. a property of abstract domains which makes that the abstract interpretation can be implemented efficiently. The conclusion is found in section 5.

1. Basic principles of abstract interpretation

Abstract interpretation as considered here is intended to compute a description of the execution behaviour. It is important to be aware of the need to use approximations and of the implied loss of precision. Consider the deterministic finite automaton in figure 1.1. The concrete behaviour of such an automaton can be characterized by its language, namely the set L_1 of strings accepted or recognized by the automaton :

$$L_1 = \{0, 10, 110, ..., 111...1110, \cdots\}$$

Note that this set is infinite and thus impractical as a characterization. The accepted strings can be partitioned according to the characters appearing in the string : the strings made out of 0's, denoted by the class <0>, the strings made out of 1's, denoted by the class <1>, and the strings made out of 0's and 1's, denoted by <0, 1>. Suppose that we only want to distinguish between these classes.[1] Then sets built from these classes can be used as a description space or abstract domain and the set $AL_1 = \{<0>, <0, 1>\}$ is the most precise description - called the *abstraction* - of L_1. The *abstraction function* α_1 maps the set of the accepted strings into the minimal set of classes needed to describe it : $\alpha_1(L_1) = AL_1$. The abstraction provides us with a finite and compact description of the language, but in general implies some loss of information. The *concretization function* γ_1 maps the set AL_1 to the set of languages $\{L_1, L_2, \cdots\}$ described by it. AL_1 does not only characterizes L_1, but infinitely many other languages such as e.g.

$L_2 = \{0, 00,..., 00...0, 010, 01010,..., 0101...010\}$.

This is illustrated in figure 1.2.

The intended use of the derived information strongly affects the final degree of abstraction that is still acceptable. Suppose we only want to know which characters appear in the accepted strings. Now the abstract domain consists of sets built from 0 and 1 and $AL_2 = \{0, 1\}$ is the best abstraction of L_1 and also of L_2, although even more precision is lost : AL_2 also describes

$L_3 = \{0, 00, 000,..., 000...000,...,1, 11, 111,..., 111...111,...\}$, while AL_1 did not.

The requirement that the abstraction function yields the "best" description closely ties it with the concretization function. In [7] rather natural conditions have been imposed on their relationship. Before giving them, we introduce some terminology :

C, the concrete domain[2]
A, the abstract domain[3]

$\alpha : 2^C \rightarrow A$, the abstraction function
where 2^S denotes the powerset of the set S
$\gamma : A \rightarrow 2^C$, the concretization function

The functions α and γ must have the following properties :

1. α and γ are monotonic.
2. α and γ are adjoint, i.e.
   a. $\forall c \in C : c \subseteq \gamma(\alpha(c))$
   b. $\forall a \in A : a = \alpha(\gamma(a))$

These properties impose a structural similarity between the concrete and the abstract domain. However, this is not essential. The framework in [2] introduces only a

1. It is well known that one can develop a finite description, which is an exact characterization for the language of such an automaton, namely the automaton itself.
2. in this example, the language of the automaton (its set of accepted strings)
3. in this example, the set of classes, or the set of characters

concretization function that provides the elements of the abstract domain with a meaning (a semantics).

The application of the abstract interpretation technique to computer programs results in a tool for global (or interprocedural) analysis. The term *program point* is used in the literature to name a position in the program source at which we wish to infer properties of the program state when control reaches that point. An example of such properties could be the values of the program variables. During the execution of a program, such program points are visited in a certain order. Thus a sequence of program points is determined. Program points can be reached more than once, but the property of interest does not necessarily has the same value at each visit. So, one has to determine the property of the set of values. The properties of interest are modeled as points in the abstract domain.
Abstract interpretation mimics the execution of the program by replacing concrete operations by their abstract counterparts. Suppose i is the concrete property at some program point and after the execution of a sequence P of instructions, property j holds. Let i_a be $\alpha(i)$, P_a the abstract counterpart of the sequence P and assume that P_a computes j_a from i_a. The abstract interpretation P_a is correct if $j \in \gamma(j_a)$ holds.

Cousot and Cousot [7] have formalized the abstract interpretation for procedural languages. Their theory is based on denotational semantics. They define program points as points in the flowcharts and they approximate the value assignments that exist each time the execution reaches that program point. The abstract interpretation paradigm has for a long time been recognized as a powerful global analysis tool for procedural languages and has recently been adapted for the analysis of logic programs.

2. The motivation for deriving the information

The information gathered by abstract interpretation is intended to allow code optimizations, in particular safe specializations of PROLOG predicates [21]. Consider the well-known append/3 predicate. Depending on the actual query,

— it **tests** whether a specific list is the concatenation of two given lists :
?- append([1,2], [3], [1,2,3]).

— it **constructs** a list that is the concatenation of two given lists :
?- append([1,2], [3], L).

— it **splits** a given list into two :
?- append(L1, L2, [1,2,3]).

The code for append/3 must be general enough to deal with these classes of queries and many more; it must have the power of general unification. Nevertheless, a typical characteristic of a WAM-based PROLOG compiler [29] is that it seldom uses the general unification algorithm for unifying a call and a head, but performs a case analysis. Special instructions are generated which are adapted to the arguments appearing in calls and clause heads. Further improvements of the code are possible if more information about the arguments of a query or a call is available. It is well known that mode declarations allow for a substantial speed-up [28]. In the same vein, information about the possible values for the actual arguments is very useful [15,21]. Methods for gathering this additional information at compile time are based on abstract interpretation : they compute safe descriptions of **all possible call patterns of the predicates.** Indeed, the same predicate can be called in different ways and thus have distinct call patterns. Therefore, the *context* of a call is relevant. Each call pattern can be linked to a specific version of the predicate definition if the difference is substantial enough for the compiler. Consider the following example.

P(...) :- ..., { θ_1 } append(X, Y, Z), ... {construct a list}
Q(...) :- ..., { θ_2 } append(X, Y, Z), ... {splits a list}

append(...) .
append(...) :- ..., { θ_3 } append(U, V, W).

Here θ_1, θ_2 and θ_3 are the concrete substitutions describing the values of the variables just before the call of append/3 in the body of P, just before the call of append/3 in the body of Q and just before the recursive call of append/3 in its recursive clause. Suppose that the properties of θ_1 and θ_2 are substantially different. One approach could be to have a single version of append/3 that can deal with both call patterns. This would lead towards the general case. The other approach is to have two different versions of append/3 : one corresponding to the θ_1 call pattern and the other to the θ_2 call pattern. The former one will be used if the initial call to append/3 has the same properties as θ_1 and the latter if its properties are the same as θ_2, as the properties of θ_3 depend on the initial call to append/3. The latter approach leads to the idea of *multiple specialization* [31].
Summarizing, we can say we need all possible call patterns for the predicates and also information about their context.

3. The first level of abstraction : the framework

3.1 Some requirements

The aim of abstract interpretation is to obtain valid descriptions of the procedural behaviour of logic programs. The basic idea is to start from a formal description of the procedural semantics and derive an abstraction from it, which is finite, compact and informative.

There are some important differences with the approach based on flow-charts which has been used for procedural languages. Firstly, the dataflow is bi-directional as unification is more powerful than the parameter binding mechanisms of procedural languages. Secondly, we must be able to deal with *nondeterminism* : the call of a predicate can be solved by more than one of its defining clauses. The backtracking mechanism assures that all clauses are tried. Moreover, once we use descriptions of sets of concrete substitutions, it is even more plausible that more than one clause is applicable. Thirdly, most of the data structures in logic programming languages are *recursive*. Due to the logic programming paradigm, simple operations on them, which in the procedural languages would be implemented by loops (iterations), now are defined recursively.

Each (recursive) call to a predicate uses freshly renamed versions of the defining clauses. During the execution of a recursively defined predicate the set of variables increases with each call. In other words, each call has a different context. So, it is difficult to detect similar or identical calls. This is a crucial operation in the course of the abstract interpretation, which allows to construct a finite description (more details are found in section 3.2). Whereas, the code in the loops of procedural languages has always access to the same set of variables, namely those in its scope.

Let us try to identify the aspects of the procedural behaviour that have to be described by the abstract interpretation method. The execution of a program P for a given query Q θ_q^c results in a (possibly empty) set of solutions $\{ \theta_{q_1}^s, \ldots, \theta_{q_n}^s \}$, which are enumerated during backtracking :

Q	θ_q^c	$\{ \theta_{q_1}^s, \ldots, \theta_{q_n}^s \}$

where θ_q^c is the concrete *call-substitution* of Q and $\theta_{q_i}^s$ a concrete *success-substitution* of Q.

For code generation we want this kind of information about all called predicates and even about all possible ways a predicate is called in. This is called the *extended* execution [23].

Q	θ_q^c	$\{ \theta_{q_1}^s, \ldots, \theta_{q_n}^s \}$
⋮		
P	$\theta_P^{c_1}$	$\{ \theta_{P_1}^{s_1}, \ldots, \theta_{P_{m_1}}^{s_1} \}$
P	$\theta_P^{c_2}$	$\{ \theta_{P_1}^{s_2}, \ldots, \theta_{P_{m_2}}^{s_2} \}$
⋮		

It is important to notice that it is sufficient to have the values of the variables appearing in the call, so e.g. the domain of $\theta_P^{c_1}$ and of $\theta_{P_i}^{s_1}$ can be limited to var(P).[4]

The code will be specialized for a *class of similar* queries Q $\{\theta^{c_1}, \ldots, \theta^{c_k}\}$, rather than for just one query Q θ^{c_i}. The design of an abstract domain is aimed at finding an adequate description β of a set of concrete substitutions : $\{\theta^{c_1}, \ldots, \theta^{c_k}\}$ is described by the abstract call-substitution β^c with $\gamma(\beta^c) \supseteq \{\theta^{c_1}, \ldots, \theta^{c_k}\}$. It is not always possible to find an abstract substitution that describes exactly a specific set. It is the responsibility of the designer to balance the expressive power (granularity) of the abstract domain and its complexity. For each β^c the corresponding abstract success-substitution β^s will be computed such that $\gamma(\beta^s) \supseteq \{\theta_1^{s_1}, \ldots, \theta_{m_1}^{s_1}, \ldots, \theta_1^{s_k}, \ldots, \theta_{m_k}^{s_k}\}$. We lose information about which success-substitution corresponds to which call-substitution, but this is

4. For any syntactic object o, *var(o)* denotes the set of variables occurring in o.

reasonable as similar queries will have similar results. The outcome of an *abstract* execution can be characterized by a table of the following form :

Q	β_q^c	β_q^s
⋮		
P	$\beta_p^{c_1}$	$\beta_p^{s_1}$
P	$\beta_p^{c_2}$	$\beta_p^{s_2}$
⋮		

For specializing versions this linear table is not appropriate in the sense that it does not say which *versions* of the predicates appearing in the body of a clause *correspond* to a specific call of a predicate. The information about the *calling hierarchy* is indispensable (it can be obtained by an extra pass through the program).

3.2 Our framework

We will sketch how our method of abstract interpretation satisfies the above requirements in a natural way. We start from the procedural semantics defined in terms of the SLD-derivation [19]. The concrete execution of a program for a given query is completely described by the corresponding SLD-tree.

Traditionally, a step in the SLD-derivation is expressed as follows (assuming Prolog's left to right computation rule) :

> Given the goal $\leftarrow (A_1, A_2, \ldots, A_n)\theta$
> and the clause $B \leftarrow B_1, \ldots, B_m$.
> Let σ be the mgu of $A_1\theta$ and B.
> Then the next goal in the derivation is $\leftarrow(B_1, \ldots, B_m, A_2, \ldots, A_n)\theta\sigma$.

Applying the same procedure for a number of additional derivation steps leads to the goal $\leftarrow (A_2, \ldots, A_n)\theta'$. During the derivation the subsequent goals and the corresponding substitutions involve an increasing set of variables, which is undesirable in the case of abstract interpretation.

Therefore, we introduce the idea of *subderivation* :

> Given the goal $\leftarrow (A_1, A_2, \ldots, A_n)\theta$
> and the clause $B \leftarrow B_1, \ldots, B_m$.
> Let σ be the mgu of $A_1\theta$ and B, restricted to var(B).
> Then a subderivation can be started for the goal $\leftarrow (B_1, \ldots, B_m)\sigma$.
> Suppose its answer substitution is $\tau = \sigma\rho$.
> Let μ be the mgu of $A_1\theta$ and $B\tau$, restricted to var($A_1\theta$).
> Then the next goal is $\leftarrow (A_2, \ldots, A_n)\theta\mu$.

In [15] we show that $\leftarrow (A_2, \ldots, A_n)\theta'$ and $\leftarrow (A_2, \ldots, A_n)\theta\mu$ are identical. Notice that a single call now requires two unification based operations, one corresponding to the **call,** as in standard SLD-resolution, and one corresponding to **return** which updates the callers environment.

For us this approach has the advantage that

— activations of the same clause that are actually renamings, can be easily detected as they appear as subderivations that are identical upto renaming. (This is crucial to guarantee termination of the analysis).

— an alternative representation for the SLD-tree emerges by treating the subderivations separately : the *AND-OR-graph*. The alternative clauses of a predicate are assembled in an *OR-node*. For each clause, an *AND-node* contains the information concerning its subderivation. The derivation sketched above gives rise to the partial AND-OR-graph of figure 3.1.

The abstract execution can be described in the same way :

Given the goal $\leftarrow (A_1, A_2, \ldots, A_n)\beta^1$
and the clause $B \leftarrow B_1, \ldots, B_m$.
Then the subderivation $\leftarrow (B_1, \ldots, B_m)\beta^2$ is computed, where β^2 is a safe approximation of the concrete substitutions that can occur at run time : If $\theta \in \gamma(\beta^1)$ and if $\sigma = mgu(A_1\theta, B)|_{var(B)}$ then it must be that $\sigma \in \gamma(\beta^2)$.

Suppose β^3 is the answer substitution of a subderivation.[5]
Then the next goal is $\leftarrow (A_2, \ldots, A_n)\beta^4$, where β^4 is again a safe approximation of the possible concrete substitutions :
If $\theta \in \gamma(\beta^1)$, if $\sigma = mgu(A_1\theta, B)|_{var(B)}$,
if $\sigma\rho \in \gamma(\beta^3)$ and $\mu = mgu(A_1\theta, B\sigma\rho)|_{var(A_1\theta)}$,
then it must be that $\theta\mu \in \gamma(\beta^4)$.

In figure 3.2, the abstract substitutions are added to the *abstract* AND-OR-graph. β^1 (respectively β^2, β^4) is the abstract call-substitution of A_1 (respectively B_1, A_2) and β^3 (respectively β^4) is the abstract success-substitution of B_m (respectively(A_1).

The framework deals with the construction of the AND-OR-graph. The computation of β^2 from β^1 is called the *procedure-entry* and abstracts the call-operation. The computation of β^4 from β^1 and the β^3's is called the *procedure-exit* and abstracts the return-operation. The application dependent aspects are dealt with by the *primitive* operations such as the procedure-entry and procedure-exit. To isolate parameter passing from unification, it is possible to introduce a third operation *abstract unification* which abstracts the concrete unification of two terms. More details about the framework are given in [2].
The use of abstract substitutions as descriptions of concrete substitutions makes it perfectly possible that the same predicate appears twice (or more) on the same branch with the same abstract call-substitution, especially in the case of recursively defined predicates. This results in an infinite repetition : in figure 3.3a the unspecified subtrees

5. Each clause in the definition of the predicate has its own subderivation which yields its own answer substitution.

are identical upto renaming. However a finite representation as an AND-OR-graph (fig 3.3b) results when the number of different (upto renaming) subtrees is finite. The idea is similar to that of rational trees introduced by Colmerauer [6] and illustrated in fig 3.4 : if a tree contains a finite number of distinct subtrees, then it is rational and it has a finite representation.

The development of a finite representation is based on the characteristics of the abstract domain and uses the comparison of abstract substitutions. This is denoted by the $\leq$-operation. $\beta_1 \leq \beta_2$ holds if β_2 describes a larger or equal set of concrete substitutions than β_1.

The absence of ascending chains[6] is sufficient to guarantee termination. If, in addition, the abstract domain is finite, then the AND-OR-tree can only have a finite number of different subtrees and can be finitely represented without giving up any precision. With an infinite domain of finite width (there can be infinite descending chains - descending chains being the dual property of ascending chains), one has to exclude subtrees labeled with a predicate P and call-substitution β_2 of trees labeled with the same predicate P and call-substitution β_1, where $\beta_2 < \beta_1$. In this case, the subtree rooted at $\beta_2 P$ has to be safely approximated by the tree rooted at $\beta_1 P$. Under this condition, the ascending chain property guarantees that every branch of the AND-OR-tree will have only a finite number of different subtrees. Under the weaker condition that only the ascending chain property holds, one has to restrict the number of different call-substitutions on a branch by an arbitrary number. Suppose this number is two for the above example and that $\beta_1 \neq \beta_2 \neq \beta_3$. Thus the subtree rooted at β_3 P may not appear in the AND-OR-graph. If $\beta_3 < \beta_2$, then the subtree rooted at β_2 P is a safe approximation and we are in the case of figure 3.5a.[7] Otherwise the computation of the second call of P must be redone with a call-substitution that covers β_2 and β_3 (see figure 3.5b). Therefore, we compute the upper bound of β_2 and β_3, upp(β_2, β_3), with the property upp(β_2, β_3) $\geq \beta_2$ and upp(β_2, β_3) $\geq \beta_3$. The treatment of the third call of P consists of comparing its call-substitution with upp(β_2, β_3) and both cases are possible. As there are no ascending chains, eventually the third call substitution will be $\leq$ than the second.

Imposing such a depthlimit is reasonable. Generating specialized versions is particularly worthwhile if they are used more than once e.g. by recursive calls. If we first have some non-recursive versions, which reflect the initialization of a data structure, followed by a recursive version, which deals with the final description of the general data structure, then it is better , from code generation point of view, to suppress the first versions and to deal only with the general recursive version. This can be enforced by the abstract interpretation method by imposing a depthlimit of one. So, one often uses depthlimit one, even in the case of finite domains. Another reason for this is to avoid excessive computation times.

---

6. An ascending chain for $\leq$ is an infinite sequence of elements $\{a_1, a_2, \ldots, a_n, \ldots\}$ such that $a_1 < a_2 < \cdots < a_n < \cdots$.
7. Alternatively, one could decide to compare β_3 with β_1.

The above discussion already mentioned some of the algebraic properties the abstract domain (the set of abstract substitutions) must have. Some formalization is needed in order to define the algebraic structure. The set of variables in the clause/query that an abstract substitution β adorns, is called the *domain D* of that abstract substitution. A_D is the set of all abstract substitutions with the same domain. We use the convention that elements of A_D are denoted by β or δ, with or without subscripts or superscripts. The abstract interpretation framework requires A_D to have some algebraic structure. In fact, one can slightly relax the conditions imposed by [2]. It suffices to have :

— a reflexive transitive relation $\leq$ (a preorder)
  with the property that $\forall \beta, \delta \in A_D$: if $\beta \leq \delta$ then $\gamma(\beta) \subseteq \gamma(\delta)$.
  So, there is also an equivalence relation $\equiv$
  with the property that $\forall \beta, \delta \in A_D$: if $\beta \equiv \delta$ then $\gamma(\beta) = \gamma(\delta)$.
  ($\leq$ modulo equivalence classes is a partial order)
— an upper bound operator *upp*[8] with the property that
  $\forall \beta_1, \beta_2 \in A_D : \exists\, upp(\beta_1,\beta_2) \in A_D : \beta_1 \leq upp(\beta_1,\beta_2)$ and $\beta_2 \leq upp(\beta_1,\beta_2)$.
— a maximal element β_{max} such that $\forall \beta \in A_D : \beta \leq \beta_{max}$.
— a minimal element $\perp$ such that $\gamma(\perp) = \phi$ and $\forall \beta \in A_D : \perp \leq \beta$.
— There exist no ascending chains for $\leq$ in abstract domains A_D annotating recursive predicates.

The last property guarantees termination of the abstract interpretation procedure. If the failure of a call is detected during abstract interpretation, the minimal element $\perp$ denotes its abstract success-substitution. The minimal element can also be used during abstract interpretation as a first (and naive) value of the success-substitution of a recursive call.

At this point, we can give an overview of characteristics of our framework, which try to capture its functionality and which can be used to compare them with other approaches :

— The framework uses an AND-OR-graph to *represent* the *results* of the analysis. At the same time, it addresses the *computation* of the result by defining a procedure for constructing the graph. The graph is a better interface towards a compiler than the table as sketched in section 3.1, especially when there is an interest in multiple specialization of procedures.

— *Correctness* can be proven by showing that the final graph rooted at βQ represents all goal statements appearing in SLD-derivations starting from queries $Q\theta$ with $\theta \in \gamma(\beta)$ [2]. The proof depends on the assumption that the abstract operations procedure-entry, procedure-exit, and unification are correct approximations of their concrete counterparts.

---

8. Commutativity and associativity of the upper bound are desirable. To obtain as much precision as possible, upp should return a value as small as possible, e.g. the least upper bound in case the abstract domain is a lattice.

— *Termination* can be proven under the assumption that the abstract domain associated with recursive predicates do not have ascending chains [2].

— The algorithm avoids that work is spend on irrelevant call-success pairs. Moreover, the graph structure is a good basis to discuss further optimizations :

- identical subgraphs should only appear (or be computed) once :
  Suppose a predicate has as call-substitutions β^{c_i} and β^{c_j}. If $\beta^{c_i} = \beta^{c_j}$, then only β^{s_i} or only β^{s_j} will have to be computed by constructing a subgraph corresponding to its subderivation.

- If in the above case, $\beta^{c_i} \leq \beta^{c_j}$, then β^{s_j} can be initialized with β^{s_i} (could adversely affect precision in case of non-monotonicity of one of the abstract operations).

- When a success-substitution is updated, we know exactly which parts of the AND-OR-graph have to be computed again.

3.3 The denotational approach

We do not consider here so called bottom up frameworks [5, 13, 22] which aim at abstracting the T_P operator, but only those frameworks aiming at computing the call-success pairs sketched in section 3.1. Within the denotational approach, one typically starts with redefining the standard semantics by formulating a function whose least fixed point is the result of the extended execution sketched in section 3.1. This function also has the call-return pair of operations as sketched in section 3.2. In approaches as [17, 23], the total function is computed i.e. all possible call-success pairs. The abstraction then consists of defining a second function, the non-standard one, which safely approximates the standard one and uses an abstract domain which is chosen in such a way that the fixed point can be computed in finite time. Efficiency is only implicitly addressed. Other authors [12, 30] address efficiency somewhat more explicit by using the minimal function graph approach originally developed in the context of functional programming [16]. They try to compute only the part of the function that is relevant for the class of queries being studied, i.e. the part represented by an AND-OR-graph. The idea is to compute the fixed point of a partial minimal function starting from a query, (Q, θ^c_q, $\perp$) with $\perp$ as approximation of its success-substitution. Each step re-computes the minimal partial function based on its currently known call-success patterns and using the clauses of the program. Consider the following program :

Q(...) :- R(...).
Q(...) :- S(...).
Q(...).

We start from a query or from a class of queries depending on the semantics we are interested in, e.g.

Q	β^c_q	$\perp$

This query determines the call patterns of the predicates R and S. The third clause for Q gives rise to a first approximation of β^s_q. So, at this point the minimal partial function is defined by the following table :

Q	β_q^c	$\beta_q^{s_0}$
R	β_r^c	$\perp$
S	β_s^c	$\perp$

The following iterations will continue with updating this definition. During this process, the first and second clause for Q contribute to the resulting β_q^s. Finally, given the appropriate structure of the abstract domain, it is guaranteed that a fixed point is reached.

There is a problem with detecting redundant entries in the table. This is illustrated by the following example.

Q(...) :- R(...), S(...).
Q(...).
R(...).
R(...) :- ..., R(...), ...

The call-substitution of Q determines β_r^c and (R, β_r^c, $\perp$) is added to the table. The first approximation of β_r^s is then determined by its nonrecursive clause and gives rise to a call-substitution for S which is only temporal, because the approximation of β_r^s is not yet the final value. The temporal call-substitution for S remains in the table and causes unnecessary extra work during the computation of the fixed point. The detection of this kind of redundant entries is not trivial. Whereas, our approach first deals with all the clauses of R such that a number of this temporal versions are simply avoided. However, temporal versions are still possible (e.g. Q(...) :- ..., Q(...), S(...).), but are easily removed. Consider for example calls that appear after a recursive call and more than one approximation is computed for the recursive calls. In the AND-OR-graph, the temporary versions are discarded as a new approximation gives rise to a new call-substitution and the previous one can be removed.

The characteristics of the denotational approach are :

— the standard semantics is defined as the fixed point of a function; the abstract semantics is defined as a function which safely approximates the standard semantics one. Defining and proving correctness of the standard semantics function is a somewhat nontrivial task, especially when one is not familiar with denotational semantics.

— correctness depends on the correctness of the standard semantics function and consists of showing that the abstract function yields a safe approximation. This proof depends on the assumption that the call-return pair of concrete operations is correctly approximated by the corresponding abstract operations as in the case of our AND-OR-graph framework.

— termination relies on showing that the fixed point computation terminates. This requires slightly stronger requirements than in the AND-OR-graph framework. Apart from the ascending chain property of the abstract domain, the operations need to be monotone and for every pair of elements in the abstract domain, a least upper bound must exist, as combining results from different clauses requires an upper bound operator which also has to be monotone. The minimal function graph construction may also require the so called downward closure of the abstract domain [12].

— efficiency issues are only considered to a very limited extent. The minimal function graph (the table with call-success pairs) needs extra processing to be suited as input for a compiler considering multiple specialization.

4. The abstract domain

The second level of abstraction is concerned with the design of an abstract domain for each application, such as modes [2,11,20,24], types [18], combined modes and types [1,3,15], compile time garbage collection [25,26] and independence analysis for AND-parallelism [8,14,27]. Now, the creativity of the designer plays a crucial role, as there exist no guidelines which automatically yield "the" definition of the abstract domain. However, there are some important points that must be considered.

1. During the exploration phase, we consider alternative formalisms that model in some way (parts of) the *properties of interest* for the application at hand.

   a. The application itself usually allows a whole **range** of possible descriptions : from a minimal approach to a meticulous description. It must be decided how this descriptions can be expressed as abstract substitutions.
   In a concrete substitution θ the variables in its domain, dom(θ), are bound to Prolog terms. In the same vein, an abstract substitution $\beta = \alpha(\theta)$ can associate with such a variable a description of the properties of interest. This approach is for instance used in the case of modes describing the instantiation state of the values of program variables. Let $\{m_1, \ldots, m_i, m_j, ,m_k, ...\}$ be the set of possible instantiation states and $\{X, Y, Z\}$ the domain of β. $\beta = \{X \leftarrow m_i, Y \leftarrow m_j, Z \leftarrow m_k\}$ is a possible abstract substitution and in $\theta \in \gamma(\beta)$ the values bound to X (Y, Z) must be terms described by m_i (m_j, m_k).
   Alternatively, an abstract substitution can enumerate classes of related program variables. For instance, all variables belonging to the same class have a common free variable in their values [14,27].

   b. Besides the properties of interest, information can be added in order to keep trace of *dependencies between values* of program variables. Again a number of valid approaches exist.
   An interesting issue is to see how precisely we want to cover the **effect of unification** of program variables, e.g.
   $$\beta^c = \{'X \leftarrow m_i, Y \leftarrow m_j\} \quad \mathbf{X = Y} \quad \beta^s = \{'X \leftarrow m_{ij}, Y \leftarrow m_{ij}\}$$
   After unification, X and Y have the same mode. In $\gamma(\beta^s)$ they will have values described by mode m_{ij}, but information is lacking about the fact that they actually have the same value. If this information is made explicit - added to the abstract substitution as in [15] -, higher precision can be obtained.

Another crucial question is whether aliasing must be taken into account. **Aliasing** refers to the situation where the same object appears in the value of 2 or more variables - the object is shared - and changing the instantiation of one of the variables might affect the value of another variable. This problem is also known as the **sharing of free variables** between values of program variables. For example, in $\theta = \{X \leftarrow f(g(U)), Y \leftarrow U, Z \leftarrow Z\}$ X and Y share the free variable U. If the value of X changes (its subterm U becomes more instantiated), the value of Y changes too.

The question is whether an abstract operation that changes the description of a variable in an abstract substitution, necessarily affects the descriptions of other variables. Remember that the descriptions must be safe approximations of the values that can occur at run time and those values can thus change. If for all $\theta \in \gamma(\beta)$ holds that $\theta\sigma \in \gamma(\beta)$ for any concrete substitution σ, then all possible instantiations are safely approximated by β. The abstract domain is called *closed under substitution* [9, 10]. An example of such a domain is the one that associates with each program variable a mode from the set $\{g(round), a(ny)\}$. Consider again $\theta = \{X \leftarrow f(g(U)), Y \leftarrow U, Z \leftarrow Z\}$. $\beta = \alpha(\theta) = \{X \leftarrow a, Y \leftarrow a, Z \leftarrow a\}$. Suppose abstract interpretation derives that X becomes ground, then $\beta' = \{X \leftarrow g, Y \leftarrow a, Z \leftarrow a\}$ is a valid abstract substitution : the new value of Y, which is also ground, is safely described by the mode a and the value of Z does not change, so its mode obviously remains valid.

If the abstract domain is not closed under substitution, the sharing between program variables must be taken into account. This is the case with the abstract domain based on the modes $\{g, a, f(ree)\}$. Now, $\beta = \alpha(\theta) = \{X \leftarrow a, Y \leftarrow f, Z \leftarrow f\}$. Suppose abstract interpretation derives that X becomes ground, then Y has no longer mode f. If no additional information about sharing is available, we must assume the worst case i.e. sharing is possible between all program variables. Thus, β' should be $\{X \leftarrow g, Y \leftarrow a, Z \leftarrow a\}$, although the value of Z does not change. In order to have more **selective** updates, additional information is needed. We could trace which program variables are allowed to share, as in [15]. In our example X and Y possibly share and thus β' becomes $\{X \leftarrow g, Y \leftarrow a, Z \leftarrow f\}$.

Note that for code generation the mode f is important information, and therefore the latter solution is preferred.

This explorative study leaves us with a number of alternatives.

2. The frameworks impose an algebraic structure on the abstract domain. Thus, we must see how the required operations can be defined for (a number of) the considered abstract domains and show that they satisfy certain properties. The more restrictive these properties are, the less alternative formalisms can be retained. If the framework imposes for instance a lattice structure, the existence of a least upper bound is necessary. This might be a property that drastically reduces the granularity of your abstract domain, and some alternatives might be excluded although they have the right level of expressive power. Therefore, it is better to

leave as much freedom as possible for the designer. Another design option is the existence of canonical forms for abstract substitutions : if $\gamma(\beta_1) = \gamma(\beta_2)$, then β_1 must syntactically identical to β_2 if the designer chooses for canonical forms. In this case, the $\equiv$-operation is very simple and results of complex operations can be stored and reused afterwards during the abstract interpretation process. But the computation of canonical forms is sometimes not trivial, especially in the presence of redundant information.

3. We must also check how difficult it is to define the primitive operations and what level of precision can be obtained.

4. Up till now we did not discuss the effect of the abstract domain A on the efficiency of the abstract interpretation process. Debray [10] identifies the following factors determining the complexity of the abstract interpretation :

— the time required, in the worst case, to decide whether two arbitrary elements of A are equal, $T_\equiv$.

— the number of distinct call patterns : at each argument position of a predicate can occur |A|[9] different abstract substitutions.

— the cost of processing a call pattern, which depends on $T_\equiv$ and $|A|^a$ with a the maximal arity of the predicates in the program.

— the cost of the primitive operations.

All these factors clearly depend on the complexity of the abstract substitutions : e.g. checking whether two modes are identical, takes constant time, whereas for checking whether two regular expressions denote the same language, only exist exponential time algorithms. The presence of dependency information increases $T_\equiv$ and the costs of all operations involving abstract substitutions. Moreover, this information increases |A| substantially. Debray [10] identifies the abstract domains closed under substitution as the class of abstract domains for which abstract interpretation can be implemented efficiently. Given that in practice the number of call-success patterns for any predicate are usually bound by a (small) constant, the complexity of the abstract interpretation process reduces to $O(N \times T_\equiv)$, with N the size of the program. Once the abstract domain is no longer substitution closed, the aliasing information becomes necessary for precision reasons and it makes all the operations on the abstract domain, the primitive operations and thus the abstract interpretation process potentially much more complex. Note that from code generation point of view the latter domains are preferable.

5. Finally, a difficult decision must be made : the precision and efficiency must be balanced taking into account the needs of the application at hand.

---

9. The size of the abstract domain A, |A|, is equal to the number of abstract substitutions in A.

5. Conclusion

Abstract interpretation computes a description of the procedural behaviour of logic programs, which can be used by a compiler to generate optimized code. The approximation of the procedural behaviour consists of two levels. The first one is general in the sense that it does not depend on the actual application. The frameworks for abstract interpretation deal with this level. Our framework is based on the AND-OR-graph. It is presented as an abstraction of a (set of) SLD-tree(s), which are a formal description of the concrete procedural semantics. We have argued that it is good interface to a compiler, especially if multiple specialization is wanted. Moreover, the AND-OR-graph contains information that makes it possible to optimize the computation during abstract interpretation. The last two characteristics do not hold any more for the approaches based on the denotational semantics. There are some other minor differences concerning correctness proofs and termination.
The second level of abstraction must allow us to express the information needed for the application at hand. Therefore, we must design the formalism, the abstract domain. This needs a lot of engineering : in order to obtain more precise results, the abstract domain usually becomes more complex; this implies that the abstract interpretation process becomes less efficient. The designer must take up the challenge to find "the" abstract domain, which balances precision and efficiency for his application.

References

[1] A. Bansal and L. Sterling, ''An abstract interpretation scheme for logic programs based on type expression'' pp. 422-429 in *Proc. Int. Conf. on Fifth Generation Computer Systems*, Tokyo (1988).

[2] M. Bruynooghe, ''A practical framework for the abstract interpretation of logic programs'' *to appear in J.Logic Programming* (1990).

[3] M. Bruynooghe and G. Janssens, ''An instance of abstract interpretation integrating type and mode inferencing'' pp. 669-683 in *Proc. 5th Int. Conf. and Symp. on Logic Programming*, Seattle (1988).

[4] M. Bruynooghe, G. Janssens, A. Callebaut, and B. Demoen, ''Abstract interpretation : towards the global optimization of Prolog programs'' pp. 192-204 in *Proc. 4th Symp. on Logic Programming*, San Francisco (1987).

[5] M. Codish, D. Dams, and E. Yardeni, *Bottom-up abstract interpretation of sequential and concurrent logic programs*, draft (June 1990).

[6] A. Colmerauer, ''Prolog and infinite trees'' pp. 231-251 in *Logic programming*, ed. K.L. Clark and S.-A. Tärnlund, Academic Press (1982).

[7] P. Cousot and R. Cousot, ''Abstract interpretation : A unified lattice model for static analysis of programs by construction of approximation of fixpoints'' pp. 238-252 in *Proc. 4th ACM POPL Symp.* (1977).

[8] S.K. Debray, "Static analysis of parallel logic programs" pp. 711-732 in *Proc. 5th Int. Conf. and Symp. on Logic Programming*, Seattle (1988).

[9] S.K. Debray, "Efficient dataflow of logic programs" pp. 260-273 in *Proc. 15th Ann. ACM Symp. Principles of Programming Languages*, San Diego (1988).

[10] S.K. Debray, *Efficient dataflow of logic programs*, Draft, Dept. of Computer Science, The university of Arizona, Tucson (1990).

[11] S.K. Debray and D.S. Warren, "Automatic mode inference of logic programs" *J.Logic Programming*, Vol.5 (3) , pp. 207-229 (1988).

[12] J. Gallagher and M. Bruynooghe, "The derivation of an algorithm for program specialisation" pp. 732-746 in *Proc. 7th Int. Conf. on Logic Programming*, Jerusalem (1990).

[13] N. Heintze and J. Jaffar, "A finite presentation theorem for approximating logic programs " in *Proc. 17th Ann. ACM Symp. Principles of Programming Languages*, San Francisco (1990).

[14] D. Jacobs and A. Langen, "Accurate and efficient approximation of variable aliasing in logic programs" pp. 154-165 in *North American Conference on Logic Programming* (1989).

[15] G. Janssens and M. Bruynooghe, "Deriving descriptions of possible values of program variables by means of abstract interpretation" *to appear in J.Logic Programming*.

[16] N. Jones and A. Mycroft, "Data flow analysis of applicative programs using minimal function graphs : abridged version" pp. 296-306 in *Conference record of the 13th Ann. ACM Symposium on Principles of Programming Languages*, New York (1986).

[17] N. Jones and H. Sondergaard, "A semantics-based framework for the abstract interpretation of Prolog" pp. 123-142 in *Abstract interpretation of declarative languages*, ed. C. Hankin, Ellis Horwood (1987).

[18] T. Kanamori and T. Kawamura, *Analyzing success patterns of logic programs by abstract hybrid interpretation*, ICOT Technical Report, TR. 279 (1987).

[19] J.W. Lloyd, *Foundations of Logic Programming*, Springer Series Symbolic Computation - Artificial Intelligence (1987).

[20] H. Mannila and E. Ukkonen, "Flow analysis of Prolog programs" pp. 205-214 in *Proc. 4th Symp. on Logic Programming*, San Francisco (1987).

[21] A. Mariën, G. Janssens, A. Mulkers, and M. Bruynooghe, "The impact of abstract interpretation on code generation : an experiment in efficiency" pp. 33-47 in *Proc. 6th Int. Conf. on Logic Programming*, Lisbon (1989).

[22] K. Marriot and H. Sondergaard, "Bottom-up abstract interpretation of logic programs" pp. 733-748 in *Proc. 5th Int. Conf. and Symp. on Logic Programming*, Seattle (1988).

[23] K. Marriot, H. Sondergaard, and N. Jones, *Denotational abstract interpretation of logic programs*, draft (1990).

[24] C.S. Mellish, "Abstract interpretation of Prolog programs" pp. 463-474 in *Proc. 3th Int. Conf. on Logic Programming*, London (1986), revised pp. 181-198 in *Abstract Interpretation of Declarative Languages*, eds. S. Abramsky and C. Hankin, Ellis Horwood (1987).

[25] A. Mulkers, W. Winsborough, and M. Bruynooghe, "Analysis of shared data structures for compile-time garbage collection in logic programs" pp. 747-762 in *Proc. 7th Int. Conf. on Logic Programming*, Jerusalem (1990).

[26] A. Mulkers, W. Winsborough, and M. Bruynooghe, *Analysis of shared data structures for compile-time garbage collection in logic programs (Extended version)*, Report CW117, Department of Computer Science, Katholieke Universiteit Leuven (1990).

[27] K. Muthukumar and M. Hermenegildo, "Determination of variable dependence information at compile time through abstract interpretation" pp. 166-185 in *North American Conference on Logic Programming* (1989).

[28] D.H.D. Warren, *Implementing Prolog - compiling predicate logic programs*, DAI research reports 39 40, Dept. of Artificial Intelligence, University of Edinburg (1977).

[29] D.H.D. Warren, *An abstract Prolog instruction set*, Technical Report, SRI international, Artificial intelligence center (1983).

[30] W. Winsborough, *Multiple specialization using minimal-function graph semantics*, Department of Computer Science, The Pennsylvania State University, to appear in J.Logic Programming.

[31] W. Winsborough, "Path-dependent reachability analysis for multiple specialization" pp. 133-153 in *North American Conference on Logic Programming*, Cleveland (1989).

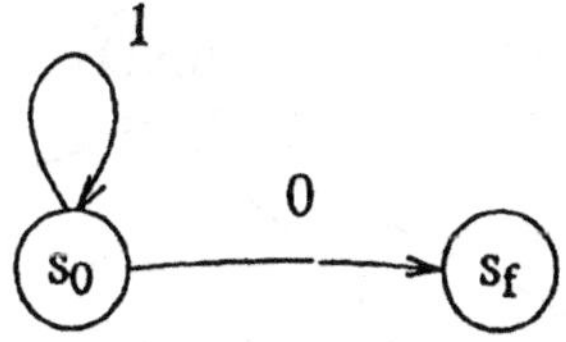

Figure 1.1. A finite deterministic automaton.

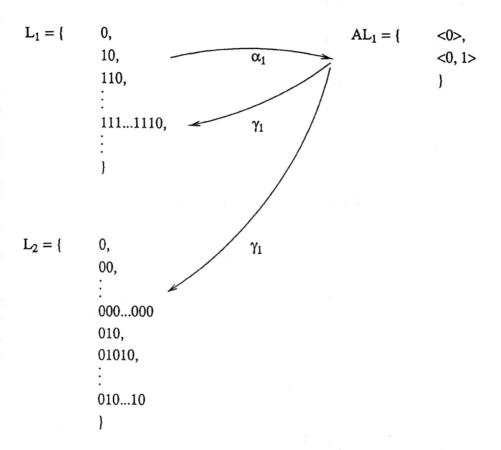

L_1 = { 0, 10, 110, ⋮ 111...1110, ⋮ }

α_1

γ_1

AL_1 = { <0>, <0, 1> }

γ_1

L_2 = { 0, 00, ⋮ 000...000 010, 01010, ⋮ 010...10 }

Figure 1.2. Languages and a possible abstraction.

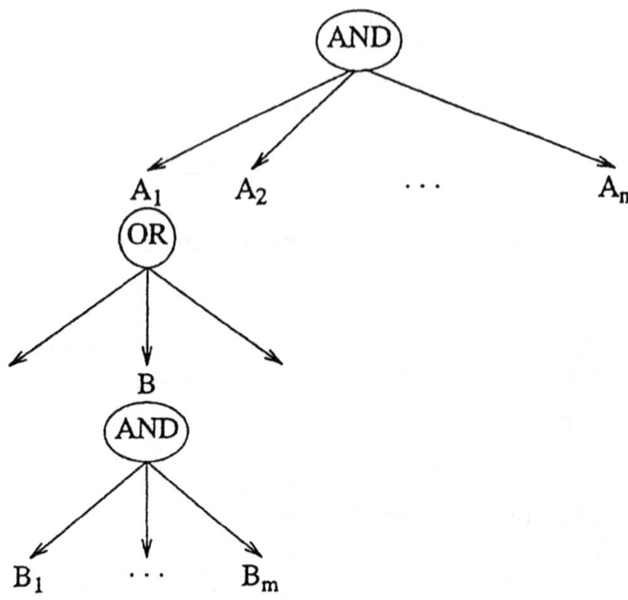

Figure 3.1. Part of an AND-OR-tree.

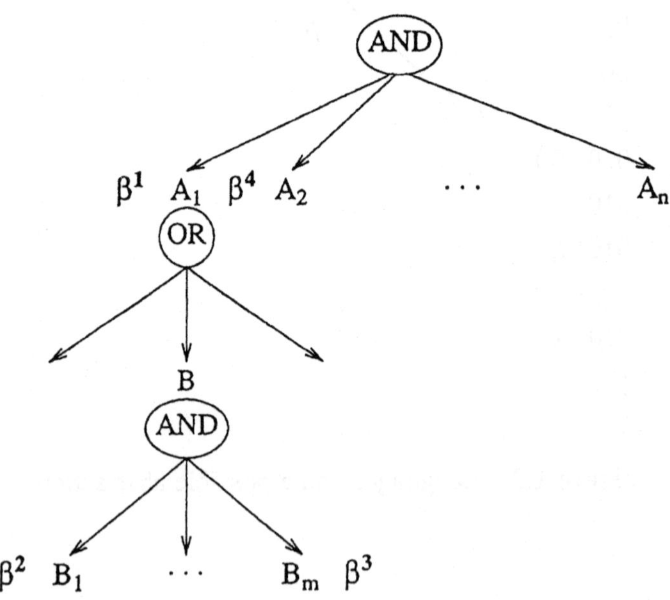

Figure 3.2. Part of an abstract AND-OR-tree.

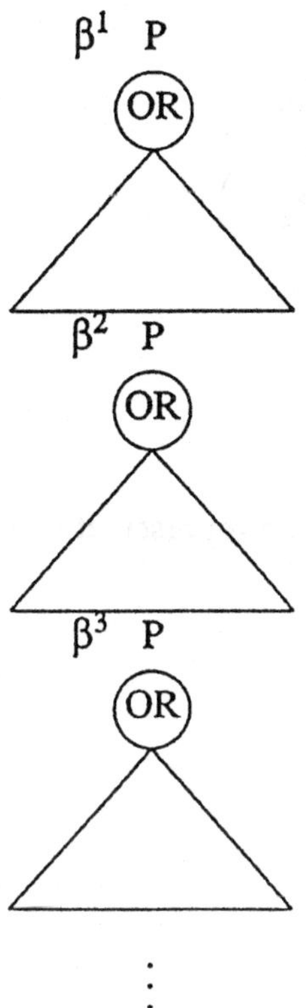

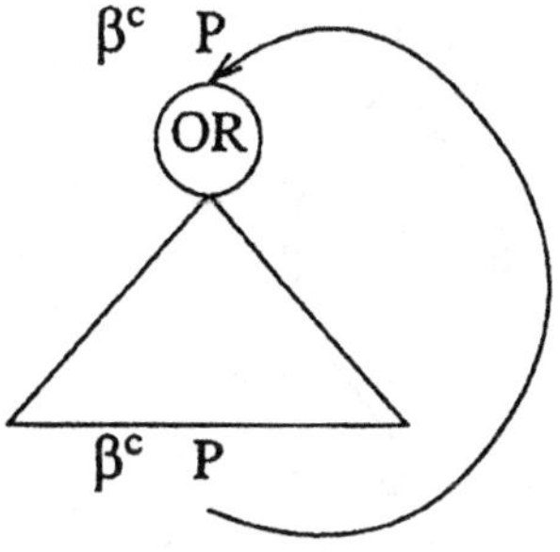

3.3.b A corresponding abstract AND-OR-graph with $\beta^c \geq \beta^1$, $\beta^c \geq \beta^2$, $\beta^c \geq \beta^3$

3.3.a Infinite branch in an abstract AND-OR-tree

Figure 3.3. AND-OR-graph as representation of an infinite AND-OR-tree.

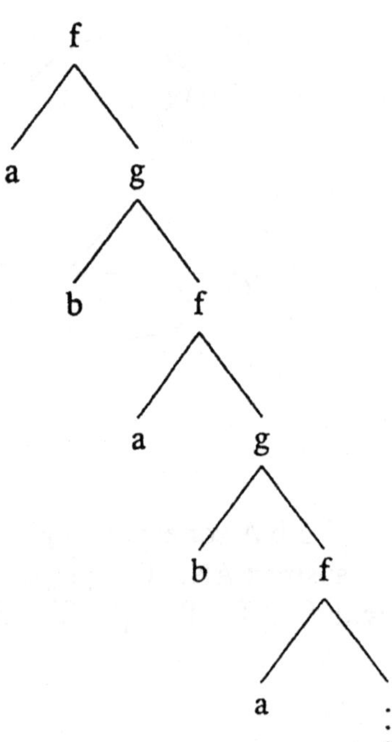

An infinite tree with a finite number of distinct subtrees

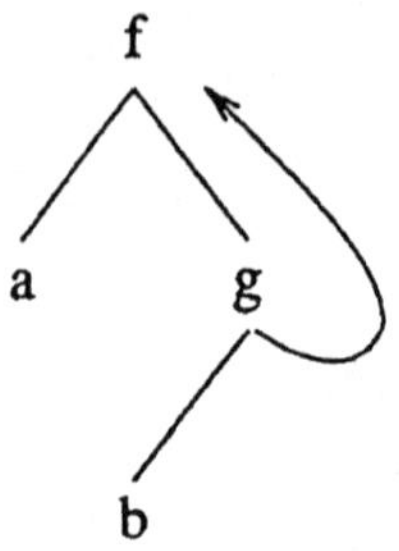

Corresponding finite representation

Figure 3.4. A rational tree.

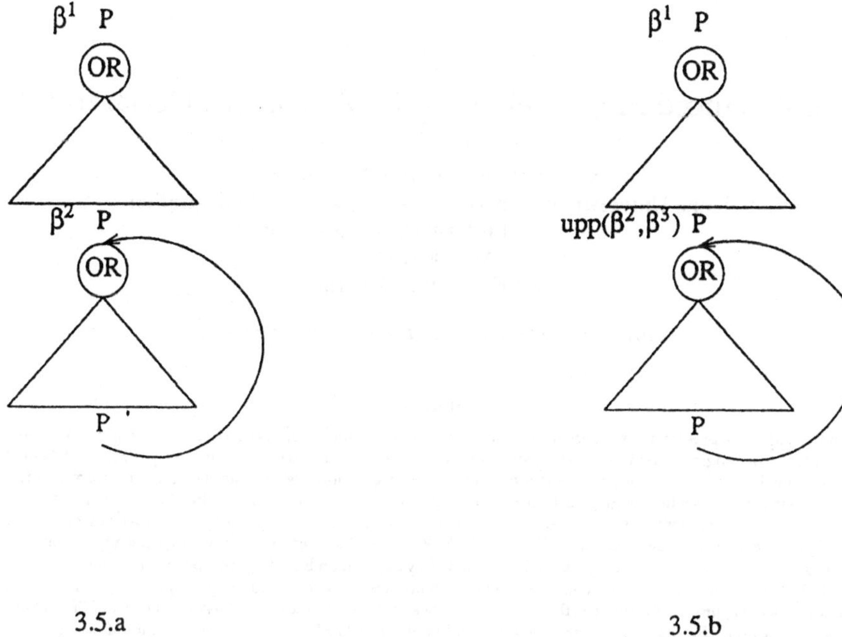

Figure 3.5. Some abstract AND-OR-graphs for the infinite AND-OR-tree of figure 3.3.a.

Treating Enhanced Entity Relationship Models in a Declarative Style

Norbert Kehrer, Gustaf Neumann
Vienna University of Economics and Business Administration
Institute of Information Processing
Augasse 2–6
A–1090 Vienna, Austria

kehrer@wu-wien.ac.at, neumann@wu-wien.ac.at

Abstract

In this paper we present an approach to represent schema information, application data and integrity constraints in form of a logic program. An information system is specified by an enhanced entity relationship (EER) model which is transformed by means of a one-to-one mapping into a set of ground facts. The application data corresponding to the schema is represented by ground facts called observations. In order to check whether the application data conforms to the given schema, a set of general integrity rules is defined which expresses the dependencies (functional, inclusion and exclusion dependencies) implied by the EER model. In order to check whether the application EER model is a valid EER model a meta EER model is defined. Any application EER diagram appears as an instance of the meta EER diagram and can be represented using observations. This way the same set of integrity constraints can be used to check the conformance between the application data and the application EER diagram, the meta EER diagram and the application EER diagram. Since the representation of the meta EER diagram is an instance of the meta EER diagram, the validity of the meta EER diagram can be checked as well. The resulting logic program is composed of the application data, the application schema, the meta schema, a general set of constraints plus optionally additional application specific constraints and deduction rules.

1 Introduction

The ER approach [Che76] was introduced 15 years ago as a diagrammatical technique for unifying different database models. Several extensions to Chen's basic formalism were proposed to capture concepts like generalization [SS77] or categories [EWH85]. We will follow the enhanced entity relationship (EER) flavor as presented in [EN89]. In this paper we will not be concerned with certain EER constructs such as composite, derived or multi-valued attributes and predicate defined categories, or sub-/superclasses.

As the ER approach was seen primarily as a design tool mappings were developed for the implementation of ER models (eg. mapping to the relational model by [TYF86]). As a result the coupling between the ER models and its implementation was rather loose. In this paper we will follow a different approach where the ER model is viewed as an executable specification of an information system. An EER model is mapped by a simple one-to-one transformation from its graphical representation to a set of ground facts. This simple transformation can be performed either by hand or by a program we have developed which takes as its source the output of a public available graphical editor. In this paper we present a set of general integrity rules to check the application data against the schema. By using a meta EER diagram the schema of the EER model can also be checked.

In addition to the general EER specific integrity constraints, additional application specific constraints or further deduction rules may be added .

2 A Meta EER Diagram

The meta EER model in Figure 1 gives a short overview of the EER methodology and introduces the basic terminology. The diagram can be read as follows: The central EER concepts are *entity type* and *relationship type*. Both of these concepts are generalized to so called *modelled types*. Since a modelled type is either an entity type or a relationship type a disjoint generalization was used. A modelled type might be described by *attributes*. An attribute is identified by a *name* (identifying attribute) and characterized by its *type* (simple, identifying or multivalued). Composite attributes are constructed using the composite relationship type). Since the names of attributes are only unique per modelled type, attributes are modelled as weak entities with the modelled type as owner.

Entity types and relationship types can be connected via *roles*, which are identified through a *name*, and which have a *cardinality* and a *participation* value. The role names are unique per relationship type, each occurrence of a relationship type participates in the *participates* relation (total participation). Each *weak entity type* (a subtype of entity type) is identified by an *identifying relationship type* (subtype of relationship type) and vice versa. The enhanced ER

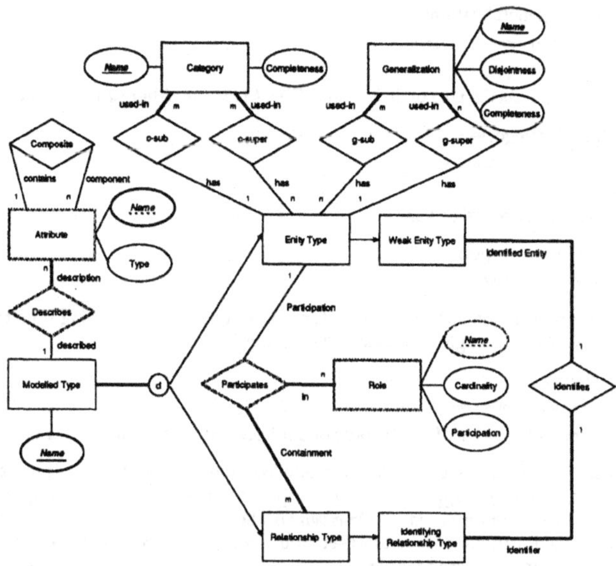

Figure 1: A Meta EER Diagram

constructs *category* and *generalization* are used to define hierarchies of entity types. A generalization (identified by *name*, characterized by the attributes *disjointness* and *completeness*) has one entity type, a supertype and might have several (one or many) entity types as subtypes. A *category* on the contrary has one subtype and might have several supertypes.

3 The Mapping of the Schema and the Data of EER Diagrams into Logic

The information contained in an EER diagram can be separated into two components.

1. An extensional part containing the names of the concepts used in the EER model, a specifuc classification of these concepts (attribute, entity type, relationship type), the links between these basic concepts and the definition of certain properties of the concepts, and

2. an intensional part containing integrity constraints and deduction rules. In this paper we are concerned primarily with integrity constraints expressible within an EER model.

The extensional part of an application consists of the extensional part of the schema as specified by the EER diagram plus application data. The intensional part of the application is composed of the intensional part of the schema plus optionally additional application specific constraints over the data that cannot be expressed in an EER model (see section 5). The integrity rules will be used to check the conformance between the schema and the data. In order to check the wellformedness of a schema the meta EER diagram can be used.

The extensional part of an EER diagram can be obtained by performing a simple one-to-one mapping from the EER diagram to a set of facts. In our representation we represent an EER diagram in terms of the links between the basic EER concepts. These links are either roles, attributes, generalizations, or categories. In addition, a predicate is needed to identify weak entities. We are using in the following a Prolog like syntax for the logic formulas where constants start with lower case letters and where variables are capitalized. For better readability dashes are used to separate words within names.

1. One-to-one mapping of roles:

   ```
   role(Role-name, Rel-name, Ent-name, Cardinality, Participation)
   ```

   where *Cardinality* is either *one* or *many* and *Participation* is either *partial* or *total*.

2. One-to-one mapping of attributes:

   ```
   attribute(Att-name, Mt-name, Type)
   composite(Mt-name,  Att-name,  Att-name-component)
   ```

   where *Type* is one of *simple, identifying* or *multivalued*. *Mt-name* stands for the name of a modelled type of the EER methodology, i.e. an entity type or a relationship type.

3. One-to-one mapping of generalizations:

```
generalization(Gen-name, Ent-name, Disjointness, Completeness)
g-sub(Gen-name, Ent-name)
```

   where *Disjointness* is either *overlapping* or *disjoint* and *Completeness* is either *partial* or *total*.

4. One-to-one mapping of categories:

```
category(Cat-name, Ent-name, Completeness)
c-super(Cat-name, Ent-name)
```

   where *Completeness* is either *partial* or *total*.

5. Identification of weak entity types:

```
identifies(Rel-name, Weak-ent-name)
```

The facts in items 1 to 5 keep the schema information in form of the one-to-one mapping from the diagram. All data of the application will be kept in the single predicate *"observation/4"*:

6. Instances for the schema:

```
observation(Att-or-Role-name, Mt-name, Tupid, Value)
```

   where *Att-or-Role-name* is the name of an attribute or a role, *Mt-name* is the name of a modelled type.

Tupid is an tuple identifier that uniquely determines a modelled type in the database. It is also used to group the various *Observations* to a certain tuple (aggregation). The tuple identifier is a concept comparable to the surrogate in [Cod79]. Note that a representation based on *Observations* allows us to cope with null values (no observation available) or with multivalued attributes (several observations with identical first three arguments and different fourth arguments) in a simple and uniform manner.

A typical application consists of schema information (one-to-one mapping) and corresponding data in form of observations:

(1) $schema_{application\ EER} + observations_{application\ EER}^{describing\ application\ data} = application\ program$

By using the integrity rules introduced in the next section it is possible to check the data against the schema. In order to check the wellformedness of the application EER diagram the meta EER diagram can be used where the application EER diagram is given in form of observations for the meta EER diagram:

(2) $schema_{meta\ EER} + observations_{meta\ EER}^{describing\ application\ EER} = schema_{application\ EER}$

(3) $schema_{meta\ EER} + observations_{meta\ EER}^{describing\ application\ EER} + observations_{application\ EER}^{describing\ application\ data} = application\ program$

In order to check the wellformedness of the meta EER diagram the meta EER diagram itself can be expressed in terms of observations:

(4) $schema_{meta\ EER} + observations_{meta\ EER}^{describing\ meta\ EER} = schema_{meta\ EER}$

(5) $observations_{meta\ EER}^{describing\ meta\ EER} + observations_{meta\ EER}^{describing\ application\ EER} + observations_{application\ EER}^{describing\ application\ data}$
$= application\ program$

Item (5) shows that in principle the whole application could be specified only in terms of observations plus a single set of EER specific integrity rules. However, when the system is maintained *"manually"*, it appears to be very hard to distinguish the various abstraction layers and to comprehend the observations. To reduce this disadvantage a simple set of rules can be given which deduces the representation of the one-to-one mapping from a set of observations. The integrity constraints in the next section assume the schema to be in the representation of the one-to-one mapping plus a single set of observations (like the items (1), (2) and (4)).

4 General Integrity Constraints of the EER Model

4.1 Functional Dependencies

Marking attributes as identifying and the specification of cardinalities of 1 in relationship types in an EER model are possible ways to express functional dependencies on the modelled data.

A functional dependency (FD) is a constraint on a relation R which states that the values of a tuple on one set of attributes X uniquely determine the values on another set Y of attributes. It is written as $X \Rightarrow Y$ and is formally defined by the following implication [GV89]:

$$t_1(X) = t_2(X) \rightarrow t_1(Y) = t_2(Y)$$

t_1 and t_2 are two different tuples of R. If the values on the set of attributes X are the same in t_1 and t_2 then the values on the attribute set Y have to be the same, too. In other words, a FD $X \Rightarrow Y$ is violated if there exist two tuples which have the same values in X and different values in Y. This can be expressed by the following rule:

```
not-determine(value(Att-LHS,Mt-LHS), value(Att-RHS,Mt-RHS)) :-
      two(value(Att-RHS,Mt-RHS), T1,T2),
      not two(value(Att-LHS,Mt-LHS), T1,T2).
```

This rule defines the violation of a functional dependency of the type *value(attribute-LHS)* $\Rightarrow$ *value(attribute-RHS)* (where the *attributes* are atomic attributes). Since in our representation both the values and the tuple identifiers are accessible in the same way, we could express dependencies of the form *value(attribute-LHS)* $\Rightarrow$ *tupid(attribute-RHS)* or *tupid(attribute-LHS)* $\Rightarrow$ *value(attribute-RHS)* or *tupid(attribute-LHS)* $\Rightarrow$ *tupid(attribute-RHS)* with the same ease. We could generalize the rule as follows:

```
not-determine(LHS, RHS) :-
      two(RHS, T1,T2),
      not two(LHS, T1,T2).
```

In this clause *LHS* and *RHS* stand for *value(Att,Mt)* or *tupid(Mt)* (*tupid(Mt)* is a shorthand for *tupid(_Mt)*). The *two* predicate is defined below. It uses the predicate *obs*, which is identical to *observeration* except for weak entities (see later).

```
two(value(Role,Type), Tuple0,Tuple1, V0,V1) :- obs(Role,Type,Tuple0,V0), obs(Role,Type,Tuple1,V1), not V0 == V1.
two(tupid(Role,Type), Tuple0,Tuple1, V0,V1) :- obs(Role,Type,Tuple0,V0), obs(Role,Type,Tuple1,V1), not Tuple0 == Tuple1.
```

In cases where the left hand side of a functional dependency is not atomic we use a clause to define the left hand side attributes and proceed as follows:

```
not-determine(FullLhs, RHS) :-
      extract_goal(FullLhs,Forall,LHS),
      two(RHS, T1,T2),
      not ( (Forall, two(LHS, T1,T2)) ).
```

This way the formulation for the violation *"for any two different RHS, all elements of the LHS must be equal"* is reformulated as *"...no element of the LHS is allowed to be different"*.

Identifying attribute determines tuple identifier

For each identifying attribute *Att* of a modelled type of the EER schema there exists a functional dependency between *Att* and the tuple identifier of the form:

value(identifying-att,modelled-type) $\Rightarrow$ *tupid(modelled-type)*

This corresponds to the definition of an identifying attribute as an attribute whose values can be used to identify an entity uniquely, because in our approach an entity is represented by its tuple identifier.

```
violated(value(Att,Mt) => tupid(Mt)) :-
    attribute(Att, Mt, identifying),
    not-determine(value(Att,Mt), tupid(Mt)).
```

It has to be noted that the identifying attribute of weak entity types does not determine the weak entity [Che76], but together with the owner entities it does. This could be expressed informally as

value(identifying-att + tupids-of-owners,weak-ent) $\Rightarrow$ *tupid(weak-ent)*

where + is used as a constructor. The concatenation of value and tuple identifiers of owners is achieved in the implementation within the *obs* predicate used by *two* and *not-determine*.

Tuple identifier determines singlevalued attributes

A (singlevalued) attribute is a function which maps from an entity set or a relationship set into a value set. For our representation this means that the value of each singlevalued attribute of the modelled type *Mt* is determined by the tuple identifier of *Mt*:

tupid(modelled-type) $\Rightarrow$ *value(any-attribute,modelled-type)*

We need not check the constraint that the identifying attribute determines the values of the other attributes, because it follows from the two previous constraints:

value(identifying-att,modelled-type) $\Rightarrow$ *tupid(modelled-type)* $\wedge$
tupid(modelled-type) $\Rightarrow$ *value(any-attribute,modelled-type)* $\rightarrow$
value(identifying-att,modelled-type) $\Rightarrow$ *value(any-attribute,modelled-type)*

Entities participating in a relationship type with cardinality *One*

Each role *R* of a relationship type *Rel* in which an entity type participates with cardinality *One* is determined by all other roles of *Rel* together [TYF86]:

$$value(other\ roles,rel\text{-}type) \Rightarrow value(one\text{-}role,rel\text{-}type)$$

This constraint is independent of the degree of the relationship type.

```
violated(value(other-roles,Rel) => value(Role1,Rel)) :-
    role(Role1,Rel,_,1,_),
    not-determine(other_role(Role1,Rel,Role2) ^ value(Role2,Rel), value(Role1,Rel)).
```

This is an example of a FD where the left hand side is not atomic, but consists of all the roles of a relationship type. The goal in front of ^ computes each role of the LHS of the functional dependency.

All roles determine tuple identifier

All roles of a relationship type *r* together determine the tuple identifier of *r*:

$$value(all\ roles,rel\text{-}type) \Rightarrow tupid(rel\text{-}type)$$

This constraint can be expressed as:

```
violated(value(all-roles,Rel) => tupid(Rel)) :-
    relation(Rel),
    not-determine(role(Role,Rel,_,_,_) ^ value(Role,Rel), tupid(Rel)).
```

As above, this is an example of a FD with a not atomic left hand side.

4.2 Inclusion Dependencies

The use of relationship types, generalizations, and specializations in EER models indicates that entity or relationship sets are subsets of some other entity or relationship set. The property of being a subset of another set is covered by inclusion dependencies.

Inclusion dependencies (ID) specify that each member of some set *A* must also be a member of a set *B*. An inclusion dependency $A \subseteq B$ is violated iff there is an occurrence (value or tuple identifier) of *A* which is not an occurrence of *B*. The following rule expresses this content:

```
not-included(LHS,RHS) :-
    extract_goal(LHS,Goal,L),
    one(L,V),
    not (Goal,one(RHS,V)).
```

Like the rule for FD violations this rule may be used to check inclusion dependencies between tuple identifiers and attribute or role values in any combination. The predicate *one* is defined similar to the predicate *two* and returns either a tuple identifier or a value depending on its first argument.

Participating entities included in entity type

The values of a role of a relationship type must be tuple identifiers of the entity type participating in that role:

$$value(role,rel\text{-}type) \subseteq tupid(entity\text{-}type)$$

This ID is checked by the following rule:

```
violated(value(Role,Rel) << tupid(Ent)) :-
    role(Role,Rel,Ent,_,_),
    not-included(value(Role,Rel),tupid(Ent)).
```

Totally participating entity types

For entity types which participate totally in a relationship type the previous ID has to hold in the other direction, too. Each tuple identifier of an entity type *e* must be a value of a role in which *e* participates totally:

$$tupid(entity) \subseteq value(role,rel\text{-}type)$$

Generalizations

A generalization may be total or partial. A total generalization specifies the constraint that every entity in the superclass must be a member of some subclass in the specialization [EN89]. For our representation this means that in a total generalization with supertype *super* there must be at least one subclass *sub* for each tuple identifier T of *super*, where T is included in *sub*:

$$tupid(supertype\text{-}in\text{-}total\text{-}gen) \subseteq tupid(at\text{-}least\text{-}one\text{-}subtype)$$

This constraint is valid only for total generalizations. There is also an inclusion dependency $tupid(subtype) \subseteq tupid(supertype)$ for both partial and total generalizations. In our representation we guarantee through the use of deduction rules that the tuple identifiers of supertypes are also tuple identifiers of the subtypes. Therefore this ID needs not to be checked.

Categories

Similar to generalizations we have to assure that for total categories tuple identifiers of the superclasses are also tuple identifiers of the subclass in a category and that the attributes are inherited. This mechanism may not be applied to partial categories, because not every entity of a supertype has to be member of the subclass. Instead the members of the subclass in partial categories have to be stated explicitly. Therefore we will have to check if the tuple identifiers and attribute values of a subclass in a partial category occur in one of the superclasses specified for the category, which is expressed by the inclusion dependencies:

$$tupid(subclass) \subseteq tupid(some\text{-}superclass)$$
$$value(att,subclass) \subseteq value(att,some\text{-}superclass)$$

The mechanism of attribute inheritance will be described in more detail in a later section.

All roles in a relationship must be specified

In each relationship instance the associated entities have to be specified. This constraint is violated if there are two different roles *role1* and *role2* in a relationship type *rel* and the set of tuple identifiers of *rel* which have a value for *role1* is a proper subset of the tuple identifiers of *rel* having a value for *role2*. A is a proper subset of B if $A \subseteq B$ and not $B \subseteq A$. So the constraint for the violation using the $\subseteq$ operator may be written as:

$$tupid(role1,rel\text{-}type) \subseteq tupid(role2,rel\text{-}type) \wedge \neg\, tupid(role2,rel\text{-}type) \subseteq tupid(role1,rel\text{-}type)$$

4.3 Exclusion Dependencies

An exclusion dependency (ED) is the constraint indicating that no member of a set A is a member of a set B (empty intersection). An ED $A \rightleftharpoons B$ is violated iff a tuple identifier of A is also a tuple identifier of B. We define this constraint violation only for tuple identifiers and not for attribute values or roles because this is not expressible in the EER methodology. This constraint can be formulated simply as:

```
not-excluded(LHS,RHS) :-
    one(LHS,V),
    one(RHS,V).
```

Disjoint subclasses

A disjointness constraint on a generalization specifies that the subclasses in the generalization must be disjoint [EN89]. This constraint can be expressed by mutual exclusion dependencies between all the subclasses. Let *sub1* and *sub2* be two different subclasses of a disjoint generalization. If a tuple identifier of *sub1* is also a tuple identifier of *sub2* the ED of disjoint subclasses is violated:

$$tupid(disjoint\text{-}subclass\text{-}1) \rightleftharpoons tupid(disjoint\text{-}subclass\text{-}2)$$

Unique Tuple Identifiers

Two different modelled types *mt1* and *mt2* may not contain the same tuple identifier unless *mt1* is a subtype of *mt2* or *mt2* is a subtype of *mt*. The goal *subtype(X,Y)* checks if one entity type is a subtype (via a subclass or category) of another entity type.

```
violated(tupid(Mt1) =-= tupid(Mt2)) :-
      not-excluded(tupid(Mt1),tupid(Mt2)),
      Mt1 /== Mt2,
      not subtype(Mt1,Mt2),
      not subtype(Mt2,Mt1).
```

4.4 Schema Conformity

The conformity of the database with the schema – i.e. the attributes, roles, and modelled types appearing in *Observation* facts must be specified in the schema – cannot be checked by one of the above dependencies, because the rules only refer to data integrity whereas the schema conformity may be viewed as an inclusion dependency between data and schema representation. Therefore we use a separate consistency rule for this dependency.

An observation containing an attribute or role *ra* of a modelled type *mt* does not conform to the schema if *ra* or *mt* have not been specified in the schema correspondingly.

```
violated(not-in-schema(Ra,Mt)) :-
    observation(Ra, Mt, T, V),    not attribute-or-role(Ra, Mt).
```

4.5 Type Hierarchy

The concepts of the generalization and category allow the construction of a hierarchy of entity types. In our representation we use a mechanism for the inheritance of attributes in that hierarchy and for the inclusion of tuple identifiers of one entity type in other entity types. It is built upon the following rules:

- In a generalization the tuple identifiers and attributes which were stated in an *Observation* for a subtype become tuple identifiers and attributes of the supertype. The entity – represented by the tuple identifier – belongs to both types. The name of the supertype is an alias for the name of the subtype. Therefore we call this process "aliasing".
- In a total category the attributes specified for a superclass are inherited by the subclass, and the tuple identifiers of the superclass become tuple identifiers of the subclass (aliasing).

Nonetheless, additional *Observations* may be specified for the superclass in a generalization and for a subclass of a category. Therefore the corresponding inclusion dependencies, which we described earlier, have to be checked.

The inheritance of attributes and the aliasing of tuple identifiers are performed by deduction rules for a predicate called *Observation-In-Hierarchy*. This predicate has the same arguments as *Observation* and covers all *Observations* plus the ones that result from the inheritance mechanism. Actually the integrity constraint definitions are based upon this predicate except in the cases where *Observation* is used explicitly.

5 Application specific Integrity Constraints

For certain applications, it will be necessary to specify integrity constraints which are not expressible within an EER diagram. Let us look at the meta EER diagram again to identify such application specific constraints. In the meta EER diagram there are essentially two types of constraints missing: domain restrictions and exclusion of certain (recursive) definitions. An example of a missing domain restriction would be to specify that the *participation* of a *role* is either *total* or *partial* (or that the *role* in which a weak entity type participates in an *identifying Relationship* must be *total*, the names of roles and attributes of a relation must be disjoint). An example of an EER construction that should be forbidden would be if an entity-type is *owner* of itself (or if a *subtype* is its own *supertype*, two supertypes in a generalization do not have a common root). Another constraint would be that each relationship type must have at least two roles attached. We call such integrity constraints *"application specific"* (as opposed to the general constraints valid for all EER models), since they are only needed for testing the schema (i.e. the application meta-EER).

6 Using the Integrity Constraints and Stratification Problems

When the consistency rules of the last section are used the observation facts are checked with respect to the schema information. Within this test the rules are assumed to be correct, some of the facts might be invalid or missing (constraint satisfaction problem). A straightforward implementation of the integrity constraints (eg. in Prolog) will lead to the naïve and exhaustive approach where all constraints are applied on all observations each time the set of observations is modified. This is no practical solution for applications with realistic sizes and further investigations are necessary to reduce the search space.

In addition the integrity constraints might be used to enforce integrity and to trigger from an update of the facts additional, so called *"induced"* updates (see for example [BDM87]).

Another interesting problem is to reason about inconsistencies and to try to derive consistent subsets of the set of observations even if some observations are invalid. A convenient formulation of the predicate *"consistent-observation"* is typically a non-stratified problem. To illustrate this problem we will use a small example using a single entity type *a* with two attributes *i* (for an identifying attribute) and *o* (for another non-identifying attribute) and two constraints:

```
observation(a,i,t1,1).
observation(a,o,t1,x1).
```

```
observation(a,i,t2,2).
observation(a,o,t2,x2).
observation(a,i,t3,1).
observation(a,o,t3,x3).

[1]  fdv(E,i,T,V)  :- obs(E,i,T,V), obs(E,i,T2,V), not (T==T2).
[2]  nmv(E,A,T,V)  :- obs(E,A,T,V), other-att(A,A1), not obs(E,A,T,V1).

[3]  consistent-observation(E,A,T,V) :-
         observation(E,A,T,V), not fdv(E,A,T,V), not nmv(E,A,T,V).

[4a] obs(T,A,V)  :- observation(E,A,T,V).
[4b] obs(T,A,V)  :- consistent-observation(E,A,T,V).
```

The consistency rule *fdv* stands for functional dependency violated, *nmv* means no-missing-violated (all attributes must be specified). For simplicity *other-att* is defined here as:

```
[5] other-att(i,o).
[6] other-att(o,i).
```

The clauses [1,2,3,4a,5,6] represent the stratified approach where the integrity constraints are based directly on the observations. It can be easily seen that *fdv* holds for *t1* and *t3* and that *nmv* never holds. If the integrity constraints are based upon *consistent-observation* instead of *observation* by using [4b] instead of [4a], the program is non-stratified, since there is now a *"recursion going through a negation"*. The intended semantics is that now the observations of the *o*-attributes are also removed from the set of *consistent-observations*. The set of consistent observations that is true regardless of the order in which the consistency rules are applied consists only the observations containing *t2*.

For this example the same results could be obtained using a layered stratified program, where the first consistency rule *fdv* would be based on the given observations resulting in a set *obs1*. The set *obs1* could be used as a basis for the second consistency rule *nmv* leading to the final consistent set. Different results will be obtained when the order of the consistency rules is changed. The second disadvantage of the layering approach using stratified programs is that in the general case, where *N* consistency rules are given, *N!* different layerings are possible, which will result in different models. The intended *"save subset"* of consistent observations should contain only the intersection of these models.

References

[ABW87] C. Apt, H. Blair, A. Walker: *"Towards a Theory of Declarative Knowledge"*, in Minker (ed.): *"Foundations of Deductive Databases and Logic Programming"*, Morgan Kaufmann, Los Altos 1987.

[BDM87] F. Bry, H. Decker, R. Manthey: *"A Uniform Approach to Constraint Satisfaction and Constraint Satisfiability in Deductive Databases"*, Proceedings of EDBT 88, May 1988, Venice.

[Che76] P. Chen: *"The Entity Relationship Model – Toward a Unified View of Data"*, Transaction of Database Systems, 1:1, March 1976.

[Che91] W.C. Cheng: *"Tgif 2.6 - A Xlib based drawing facility under X11"*, available via anonymous ftp from export.lcs.mit.edu, May 1991.

[Cod79] E. Codd: *"Extending the Database Relational Model to Capture More Meaning"*, Transactions of Database Systems, 4:4, December 1979.

[DZ88] P.W. Dart, J. Zobel: *"Conceptual Schemas Applied to Deductive Database Systems"*, Information Systems, Vol. 13, pp. 273–287, 1988.

[EN89] R. Elmasri, S.B. Navathe: *"Fundamentals of Database Systems"*, Benjamin/Cummings, Redwood City, Calif. 1989.

[EWH85] R. Elmasri, J. Weeldreyer, A. Hevner: *"The Category Concept: An Extension to the Entity-Relationship Model"*, International Journal on Data and Knowledge Engineering, 1:1, May 1985.

[GV89] G. Gardarin, P. Valduriez, *"Relational Databases and Knowledge Bases"*, Addison-Wesley, Reading 1989.

[MS89] V. M. Markowitz, A. Shoshani: *"On the Correctness of Representing Extended Entity-Relationship Structures in the Relational Model"*, in J. Clifford, B. Lindsay, D. Maier: *"Proceedings of the 1989 ACM SIGMOD International Conference on the Management of Data"*, ACM, New York 1989, pp. 430–439.

[TYF86] T.J. Teorey, D. Yang, J.P. Fry: *"A logical Design Methodology for Relational Databases Using the Extended Entity-Relationship Model"*, ACM Computing Surveys 18, 2, June 1986.

[Teo90] T.J. Teorey: *"Database Modeling and Design: The Entity-Relationship Approach"*, Morgan Kaufmann, San Mateo, Calif. 1990.

[SS77] J. Smith, D. Smith: *"Database Abstractions: Aggregation and Generalization"*, Transactions of Database Systems, 2:2, June 1977.

Processing of ground regular terms in Prolog

Evgeny L. Kitaev
Institute for Informatics Problems
USSR Academy of Science
30/6 Vavilov str., Moscow 117900, USSR

Abstract: This paper deals with the problem of efficient implementation of the interface between Prolog and conventional programming languages. We show how types may be used in Prolog to support various run-time representations of terms. In the presence of certain restrictions it is possible to represent ground terms by means of data structures used in conventional programming languages. Applied to the interface, it allows to minimize data conversions.

1. Introduction

Prolog is a logic programming language encouraging fast development and debugging of programs. Simple and powerful features of Prolog make it a convenient language for writing prototype systems. At the same time, there remains a problem of efficient implementation of Prolog capable to satisfy the requirements of application programming.

In recent years there has been a considerable interest in application of the framework for abstract interpretation of logic programs [1] for improvement of Prolog implementation. Static inference of mode information seems to be one of the main achievements of this approach. Some advanced algorithms have been proposed recently [2],[3].

Informally the mode of a predicate in a logic program is an assertion about which of its arguments are input arguments and which are output arguments in any call to that predicate arising from that program. In general there are situations when a predicate argument is used both as an input and an output argument at the same time, which is a distinctive feature of Prolog.

However it turns out that a great part of predicates in programs encountered in practice are used in a strictly moded way. Then it makes sense to consider the class of well-moded programs [4] comprising programs that have any argument of each predicate moded either as an input or an output argument. Well-moded programs have close similarity in the paradigm of functional programming, which is used for integration of logic and functional programming languages [5].

More expressive forms of mode information include assertions about types of predicate arguments. The general issue of types in logic programming languages has been addressed by many authors and a variety of type systems has been proposed. Types was found helpful for semantic improvements of logic programming languages, program validation and useful optimizations of the implementation [2],[6],[9].

However it seems that not all benefits of the type concept for logic programming have been appreciated yet. In this paper we propose to use types for flexible organization of internal representation of terms.

Though Prolog is said to be an untyped language, the run-time representation of terms contains typing information in tag fields of data cells. This is in contrast with typed languages, where types are primarily a compile time notion and not a part of run-time data structures.

Then the embodiment of types in Prolog allows to modify the representation of terms by excluding typing information from their run-time representation, if such information can be accessed at compile time. This allows us to represent terms by means of data structures similar to ones found in conventional programming languages.

Development of integrated systems with Prolog, such as Prolog&DBMS coupling, shows an increased demand for efficient interface between Prolog and conventional programming languages. Concerning this problem, a principal incompatibility in internal data representations seems to be the major disadvantage, which requires complex conversions of data at the interface level. Then it may be helpful to expand Prolog with the ability to process terms which are represented internally by means of data structures used in conventional programming languages. To distinguish these terms from terms represented in a "standard" way, we call them regular terms.

In this paper we restrict our attention only with regular representation of ground (variable-free) terms. For this reason we define a subclass of Prolog programs called GRT-programs. The informal meaning of the GRT-class is that it is an adaptation of well-modedness in presence of types. However limited this class is, it covers a wide range of programs (program parts) used in practice.

In this class the restriction for ground terms is always guaranteed to be satisfied. This paper addresses all aspects of the GRT-programs implementation, while major attention is paid to the possibility of "non-standard" run-time representation of ground terms.

2. Example

To illustrate the shortcomings of Prolog external interface let us consider the following example. Suppose we have a relational database containing the relation EMPLOYEE which lists a manufacture employees together with their positions and salaries.

EMPLOYEE(NAME:CHAR[15],POSITION:CHAR[20],SALARY:INTEGER).

All salaries are divided to finite number of groups by the *Top* and *Step* parameters which are the highest salary and the division step. Given a salary group we assume that a query for a number of employees with salary in that group may be expressed in the language of the underlying DBMS.

The following program creates an integer list of the EMPLOYEE distribution over the required salary division and passes it to the procedure which draws a line bar diagram of that distribution. In addition the program prints a share of manager employees in every salary group. An employee is considered to be a manager if the name of his position contains the substring "Manager" (for example "Division Manager", "Department Manager").

```
:- salary_distribution(10000,1000).

salary_distribution(Top,Step):-
     create_distribution(Top,Step,D_List),
     show_line_bar(Top,Step,D_List).

create_distribution(0,_,[]).
create_distribution(From,Step,[Emp_N|D_List]):-
     To is From - Step,
     total_emps(From,To,Emp_N),
     manager_emps(From,To,Emp_N,Manager_Share),
     report_managers(From,To,Manager_Share),
        create_distribution(To,Step,D_List).


manager_emps(From,To,Emp_N,Manager_Share):-
     emp_positions(From,To,Pos_List),
     manager_numb(Pos_List,0,Manager_Numb,From,To),
     Manager_Share is Manager_Numb/Emp_N.

manager_numb([],N,N,_,_):-!.
manager_numb([Pos|Pos_List],K,Res,From,To):-
    substring(Pos,"Manager",_),!,
        emp_managers(From,To,Pos,N),
        K1 is K + N,
         manager_numb(Pos_List,K1,Res,From,To).
manager_numb([_|Pos_List],K,Res,From,To):-
    manager_numb(Pos_List,K,Res,From,To).

report_managers(From,To,Share):-
    write("Salary from "),write(From), write(" to "),
    write(To), write(" Manager_Share="), write(Share),nl.
```

The predicates total_emps/3, emp_position/3 and emp_managers/4 are mapped to the queries for the database. Note that a query corresponding to total_emps/3 computes a number of employees with salary in the range of *From,To* (returns integer), a query for emp_position/3 gives a projection on the attribute POSITION for the salary-ranged group of employees (returns an array of fixed-length strings char[20]) and a query for emp_managers/4 restricts that of total_emps/3 in that it accounts for a numbers of employees at a particular position only.

We assume that these queries and the procedure show_line_bar/3 which draws a line bar diagram are implemented externally.

```
total_emps(From,To,Emp_N):-
   set_of(Name,Pos^Salary^(employee(Name,Pos,Salary),
                Salary < From, Salary >= To), Name_List),
   list_length(Name_List,Emp_N).
/* TOTAL_EMPS(Integer,Integer,Integer) */
```

```
emp_positions(From,To,Pos_List):-
   set_of(Position,Name^Salary^(employee(Name,Pos,Salary),
                  Salary < From, Salary >= To), Pos_List).
/* EMP_POSITIONS(Integer,Integer,Array_of Char[20]) */

emp_managers(From,To,Pos,Emp_N):-
   set_of(Name,Salary^(employee(Name,Pos,Salary),
                Salary < From, Salary >= To), Name_List),
   list_length(Name_List,Emp_N).
/* EMP_MANAGERS(Integer,Integer,Integer,Integer) */

/* SHOW_LINE_BAR(Integer,Integer,Array_of Integer) */
```

When this program is executed, it is not possible to avoid heavy conversions of data at the interface level. While the conversion of integers is quite simple (just append or eliminate a tag field), others are very expensive.

. A conversion between an array of fixed-length strings char[20] and a Prolog list of strings after the query for emp_position/3 is evaluated. (All strings must be loaded to the Symbol Table of Prolog and a Prolog list must be created).

. A counterpart conversion of Prolog string to the fixed-length string before the query for emp_managers/4 can be evaluated.

. A Prolog integer list must be converted to an array of integers before show_line_bar/3 is executed.

Needless to say that these conversions become a reason of inefficiencies in integrated application which interfaces with Prolog. It appears to be a serious drawback if conversions of complex data structures are involved.

3. Regular Types

In this section we define a notion of regular types. A type system is introduced with definitions of a type expression language and a mapping from type expressions to subsets of the universe of all values. A universe is restricted here to be the set of all ground terms **T**.

We assume a predefined set of type names, while ω and ε will be used to denote its elements. Given an expression for type ω, there exists a mapping $\omega \to T_\omega$, where $T_\omega \subseteq$ T. The subset T_ω of T is called a domain of type ω. A ground term t has type ω iff $t \in T_\omega$. To denote a fact that a term t has type ω we use an annotation expression $t::\omega$. A suffix $::\omega$ is called a type annotator of term t. (To be correct, we exclude the operator ::/2 from a set of functional symbols used to construct elements of T).

Suppose pre-existence of a finite set of basic types $\varepsilon_1,\dots,\varepsilon_k$. Basic types are used to describe constants which have a regular representation. A set of basic types should be wide enough to cover all the necessary simple data types used in conventional typed programming languages. For example, there must be a type of integer *int* and floating-point *float* numbers, fixed-length *char(n)* and variable-length *char(*)* strings.

A set of ground terms T is expanded to include constants of basic types and compound terms constructed from these constants. For example, a constant 123::int is a regular integer and "Division Manager"::char(20) is a regular fixed-length string (blank-padded).

We also introduce a predefined type *universal* used to collect all ground terms retaining their "standard" representation. Terms of *universal* type are called universal terms. For example, constants 123 and "Division Manager" are universal terms.

It should be mentioned that the domains of basic types and the domain of *universal* are mutually disjoint:

$\forall i,j\ i\neq j:\ T_{\varepsilon_i} \cap T_{\varepsilon_j} = \emptyset$ and $\forall i:\ T_{\varepsilon_i} \cap T_{universal} = \emptyset$

While the basic types and the type *universal* are considered to be predefined, all other types must be declared by the following statements:

<type_name>::<type_body> [::<implementation>]

A type declaration consists of a type name, a type body and an optional *implementation* part (it will be described later). The type body is an expression of the following form:

$$f_1(\omega^1_1,\ldots,\omega^1_{k_1}),\ldots,f_n(\omega^n_1,\ldots,\omega^n_{k_n})$$

where $n\geq 1$, $f_1,\ldots,f_n$ are pairwise different function symbols, $1\leq i\leq n$: $k_i>0$, and ω^i_j are type names. A subexpression $f_i(\omega^i_1,\ldots,\omega^i_{k_i})$ is called an element of the type body.

The declared types are called compound regular types. These types are used to describe sets of compound ground terms which have a regular representation.

Given a declaration of compound regular type $\omega{::}f_1(\omega^1_1,\ldots,\omega^1_{k_1}),\ldots,f_n(\omega^n_1,\ldots,\omega^n_{k_n})$, a ground term t has type ω if t is a compound term $f(t_1,\ldots,t_p)$ and there is an element $f_i(\omega^i_1,\ldots,\omega^i_{k_i})$ in the type body of ω such that $f/p=f_i/k_i$ and $\forall j=1,\ldots,p$: $a_j \in T_{\omega^i_j}$. Additionally a domain T_ω of every compound type ω contains the constant nil::ω.

Let us give some examples. The compound type intList::.(int,intList) collects integer lists. The term list1::int.2::int.3::int.nil::intList is of that type. The type char20List::.(char(20),char20List) collects lists of fixed-length strings.

The type btree :: tree(btree,char(*),btree) describes binary trees of names. The following term belongs to it.

tree(nil::btree,a::char(*),tree(nil::btree,b::char(*),nil::btree))

AND-OR tree is described by the following types:

```
and_rooted_tree::and(and_node_descriptor,list_of_or_trees).
or_rooted_tree::or(or_node_descriptor,list_of_and_trees).
list_of_or_trees::.(or_rooted_tree,list_of_or_trees).
list_of_and_trees::.(and_rooted_tree,list_of_and_trees).
and_node_descriptor:: .......
or_node_descriptor:: .......
```

Given a finite set Ω of type declarations for $\omega_1,\ldots,\omega_q$, the domains of these compound regular types are pairwise disjoint if any element in the body of any type has no occurences in the bodies of other types of Ω.

$$\forall i,\forall j\ \ \omega_i{::}..,f^i_j(\omega^j_1,..,\omega^j_{k_j}),..\ \ \forall i':i\neq i',\forall j'\ \ \omega_{i'}{::}..,f^{i'}_{j'}(\omega^{j'}_1,..,\omega^{j'}_{k_{j'}}),..$$
$$f^i_j(\omega^j_1,..,\omega^j_{k_j}) \neq f^{i'}_{j'}(\omega^{j'}_1,..,\omega^{j'}_{k_{j'}}).$$

If a finite set of type declarations obeys to this property, it is said to be a set of disjoint type declarations.

A ground term $t \in T$ is a regular term (relative to Ω), if it belongs either to basic or compound type. If a ground term does not belongs to any of basic or compound types (declared in Ω) then it is considered to be of the type *universal*.

Given a set of disjoint type declarations, the universe of all ground terms is divided to a finite set of pairwise disjoint domains: $T_{\varepsilon_1} \cup \ldots \cup T_{\varepsilon_k} \cup T_{\omega_1} \cup \ldots \cup T_{\omega_q} \cup T_{universal} = T$ In addition we need to restrict the construction of universal compound terms from the regular ones. It is assumed that all allowed combinations are specified in the declaration for the special type *semi_regular*.

semi_regular::$f_1(\omega^1_1,\ldots,\omega^1_{k_1}),\ldots,f_n(\omega^n_1,\ldots,\omega^n_{k_n})$.

A domain of universal terms $T_{universal}$ is divided by the declaration of the *semi_regular* to the disjoint subsets of well-formed universal terms $T^{wf}_{universal}$ and bad-formed ones $T^{bf}_{universal}$, while $T_{universal} = T^{wf}_{universal} \cup T^{bf}_{universal}$.

A universal term t belongs to the $T^{wf}_{universal}$ if:

1. t is a constant in $T_{universal}$.

2. t is a universal compound term $f(t_1,\ldots,t_p)$, a functional symbol f/p does not construct any element in the body of *semi_regular*, and every $t_1,\ldots,t_p$ belongs to $T^{wf}_{universal}$.

3. t is a universal compound term $f(t_1,\ldots,t_p)$, there is an element $f(\omega_1,\ldots,\omega_p)$ in the type body of *semi_regular* and $\forall i=1,\ldots,p$ $t_i \in T_{\omega_i}$.

4. Mode information and GRT-programs

The process of program execution in Prolog can be understood in terms of the SLD-resolution [7]. Consider a particular state in an SLD-derivation, i.e. $\leftarrow A_1,A_2,\ldots,A_k$. At this point, a renamed clause $A \leftarrow B_1,\ldots,B_q$. is selected and a substitution $\theta = mgu(A,A_1)$ is computed resulting to the next state of the derivation is $\leftarrow B_1,\ldots,B_q,A_2,\ldots,A_k\theta$.

If a subgoal A_1 is an atom $p(a_1,\ldots,a_k)$, the current step of an SLD-derivation is referred to as a call to the predicate p/k. Applying the execution procedure for a number of additional derivation steps eventually leads to a state $\leftarrow A_2,\ldots,A_n\theta^*$, where θ^* is the accumulated substitution. This state is referred to as a successful exit from the call.

A call pattern $\rho_{p/k}$ (subscript may be omitted) of a predicate p/k is an expression $p(\pi_1,\ldots,\pi_k)$, where every π_i is either an integer $(1 \le \pi_i \le k)$ or a type name. We denote $z(\rho)=\{\pi_i \mid \pi_i$ is an integer$\}$ a set of integers occurring in ρ and $types(\rho)=\{\pi_i \mid \pi_i$ is a type name$\}$ a set of type names in ρ. A call pattern must satisfy the following restrictions:

1. For any $z \in [1,q]$, where $q=max(z(\rho))$ is a maximal integer occurring in ρ, $\exists j$: $\pi_j = z$, $\pi_j \in z(\rho)$.

2. For any i: $\pi_i \in z(\rho)$ if $\exists j$: $j<i$, $\pi_i < \pi_j$ then $\exists m$: $m<i$, $\pi_m = \pi_i$.

A call $A=p(a_1,..,a_k)$ belongs to a call pattern $\rho=p(\pi_1,..,\pi_k)$ if:

1. $\forall i$: $\pi_i \in types(\rho)$, a_i is a ground term and

a). If π_i= universal, then $a_i \in T^{wf}_{universal}$.

b). If $\pi_i \neq$ universal, then $a_i \in T_{\pi_i}$.

2. $\forall i$: $\pi_i \in z(\rho)$, a_i is a free variable and $\forall i,\forall j$: $\pi_i \in z(\rho)$, $\pi_j \in z(\rho)$ it appears $\pi_i=\pi_j$ iff a_i and a_j are equal.

A substitution pattern τ_ρ of a call pattern ρ is a finite list of type names $\tau_\rho=[\omega_1,..,\omega_q]$, where q=max(z($\rho$)). Suppose a call A belongs to a call pattern ρ. Let vars(A)=$[v_1,..,v_q]$ be a list of different variables of A taken in the order of their first occurrence in A. Suppose a call A is exited returning a substitution $\theta=\{v_1/t_1,..,v_q/t_q\}$ over variables of vars(A). We say that the exit satisfies a substitution pattern τ_ρ if $\forall i$, $t_i \in T_{\omega_i}$ when $\omega_i \neq$ universal and $t_i \in T^{wf}_{universal}$ when ω_i= universal.

Consider a program P consisting of predicate definitions $p_1,\ldots,p_n$. Assume Ω is a set of disjoint type declarations which defines types $\omega_1,\ldots,\omega_s$ (basic types and a *universal* are included). We denote $\Delta_{p_i,\Omega}$ a set of call patterns of a predicate p_i constructed from types $\omega_1,\ldots,\omega_s$. Obviously, it is a finite set. We denote $\nabla_{\rho,\Omega}$ a set of substitution patterns of a call pattern ρ constructed from types $\omega_1,\ldots,\omega_s$, which is a finite set too.

Definition: A program P is called to be a GRT-program relative to a set of disjoint type declarations Ω, if there exists a call pattern $\rho \in \Delta_{p_i,\Omega}$ and a substitution pattern $\tau_\rho \in \nabla_{\rho,\Omega}$ for any predicate definition p_i of P, such that any call of p_i arising from the execution of P belongs to ρ and any successful exit from that call satisfies τ_ρ.

5. Analysis and Partial Evaluation of GRT-programs

Given a Prolog program and a set of type declarations, we must be able to prove that it is a GRT-program relative to the given set of types. This is achieved with global analysis of a program.

Static inference of mode information for Prolog appears to be a well-known application of abstract interpretation in logic programming [1]. Some advanced algorithms of mode inference have been proposed recently [2],[3]. This method is easily adapted in our case.

In this section we describe a procedure of analysis aimed to prove that a program belongs to the GRT-class. In addition, program analysis is combined with Partial Evaluation which is a well-known method of logic program optimization [8]. We use a specific variant of Partial Evaluation which transforms a source program to the special representation called further a residual program. It allows to reveal some important properties of GRT-programs discussed in the next section.

An abstract variable is an expression α_ω, where ω is a type name and α is a name of abstract variable (we use letters α,β,γ,.. to denote abstract variables). We say that an abstract variable α has type ω. Abstract variable is used to denote a term belonging to its type.

An abstract term is a term constructed from constants, functional symbols, abstract and logic variables. An abstract term is ground if it does not contain logic variables. We define unification of abstract terms by extending the "standard" unification with rules of unification of abstract variables. A successful unification of abstract terms gives two results. The first is a substitution over the logic variables of terms involved in unification, and the second is a residual unification which is a sequence of matching operations over the abstract variables.

1. If an abstract variable α_ω is unified with a logic variable v, then v is assigned to α_ω.

2. If an abstract variable α_ω is unified with a constant c, then the residual unification is appended with expression α_ω=c which is a matching between a term represented by α_ω and c.

3. If an abstract variable α_ω is unified with a compound term $f(t_1,\ldots,t_k)$, then the residual unification is appended with a decomposition $<\beta^1_{\omega_1},\ldots,\beta^k_{\omega_k}>:\alpha=f/k$, where $\beta^1_{\omega_1},\ldots,\beta^k_{\omega_k}$ are new abstract variables. We define types $\omega_1,\ldots,\omega_k$ in the following way:

a).If ω is a *universal* and a functional symbol f/k does not construct an element in the body of type *semi_regular*, then every ω_i is *universal*.

b).If ω is *universal* and there is an element $f(\varepsilon_1,\ldots,\varepsilon_k)$ in the body of *semi_regular*, then $\forall i=1,\ldots,k : \omega_i = \varepsilon_i$.

c).If ω is a compound regular type which has a body element $f(\varepsilon_1,\ldots,\varepsilon_k)$, then $\forall i=1,\ldots,k : \omega_i = \varepsilon_i$.

d).In other cases unification fails.

Following a successful decomposition, every $\beta^i_{\omega_i}$ should be unified with t_i.

4. If two abstract variables α_ω and β_ε are unified, then the residual unification is appended with a matching operation $\alpha_\omega=\beta_\varepsilon$.

We denote the construction of compound term in the residual form of a program by a composition expression $<\alpha_\omega>:\alpha_\omega \Leftarrow f(t_1,\ldots,t_k)$, where $f(t_1,\ldots,t_k)$ is an abstract ground term which has no occurrences of α_ω and α_ω is a new abstract variable. A type ω is a type of $f(t_1,\ldots,t_k)$.

Applied to a program, the analysis procedure must infer call patterns for predicate definitions of this program and for each call pattern it must find the corresponding substitution pattern. It may happens that the analysis will find several different call patterns for a single predicate definition in the program. Although this is a violation of GRT-program definition, we can easily overcome this by transforming the program to include a separate version of the original predicate definition for each computed call pattern.

For example, suppose a program contains a predicate definition member/2 and the analysis infers two call patterns ρ_1=member(universal,universal) and ρ_2=member(1,intList). Then two versions of member/2 must be created which are D^{ρ_1}:member_universal_universal/2 and D^{ρ_2}:member_1_intList/2. Note that applying this transformation, we must also rename the subgoals for member/2 appropriately.

The above transformation is called a definition specialization. If D is an original definition for member/2, then D^{ρ_1} and D^{ρ_2} are specializations of D.

Suppose ρ is a call pattern for a predicate definition $D_{p/k}$. An abstract call A_ρ is an expression built from ρ by replacing each occurrence of $\pi_i \in types(\rho)$ to a new abstract variable α_{π_i} and by replacing each $\pi_i \in z(\rho)$ to logic variables v_i, such that $\forall \pi_i \in z(\rho), \forall \pi_j \in z(\rho)$: $\pi_i = \pi_j$ iff v_i and v_j are the same. Then $vars(A_\rho)=[v_1,..,v_q]$, where $q=max(z(\rho))$.

The analysis is combined with Partial Evaluation which creates a residual form of the specialization $D^{\rho}_{p/k}$ of $D_{p/k}$. A residual form of a definition is the result of Partial Evaluation of each clause in the definition. Given a clause $h:-b_1,\dots,b_n$. of $D_{p/k}$ and a call pattern ρ for p/k, evaluation proceeds as follows:

1. An abstract call A_ρ and a clause head h are unified. If it fails, this clause is excluded from specialization $D^{\rho}_{p/k}$. A successful unification returns a substitution θ_0 over logic variables of $vars(A_\rho)$ and variables used in the clause. In addition, it returns a residual unification.

2. Subgoals $b_1,\dots,b_n$ of a clause body are evaluated stepwise starting with the first one. Let $\theta_0\theta_1\dots\theta_{i-1}$ be a composite substitution over logic variables computed immediately before the i-th subgoal is evaluated. Suppose $b_i\theta_0\theta_1\dots\theta_{i-1}$ gives a call $d(a_1,\dots,a_m)$.

a).To satisfy the GRT-program property it should be $\forall j=1,..,m$: a_j is either an unbound logic variable or a ground abstract term. If a_j is a compound ground abstract term $f(t_1,\dots,t_k)$, then the residual form of clause is appended with a composition $<\alpha_\omega>:\alpha_\omega \Leftarrow f(t_1,\dots,t_k)$, where α_ω is a new abstract variable. A call pattern ρ_i for d/m is created in an obvious way.

b).Application of the analysis and Partial Evaluation to the predicate definition d/m and the call pattern ρ_i creates a specialization $D^{\rho_i}_{d/m}$ and computes a substitution pattern $\tau_{\rho_i}=[\varepsilon_1,..,\varepsilon_s]$, where $s=max(z(\rho_i))$.

c).Let $vars(d(a_1,..,a_m))=[u_1,\dots,u_s]$. A substitution $\theta_i=\{u_1/\alpha^1_{\varepsilon_1},\dots,u_s/\alpha^s_{\varepsilon_s}\}$ is defined to bind all logic variables to new abstract variables of corresponding types.

d).A residual form of a clause is appended with the expression $<\alpha^1_{\varepsilon_1},\dots,\alpha^s_{\varepsilon_s}>:d^*(a^*_1,\dots,a^*_m)$, where d^* is a specialization of d corresponding to a call pattern ρ_i, and a^*_i are defined in the following way. An argument a^*_i equals to $\alpha^j_{\varepsilon_j}$ if a_i is a logic variable u_j. If a_i is a ground compound abstract term, then a^*_i equals to α_ω which is an abstract variable assigned in the composition of a_i. Otherwise a^*_i is equivalent to a_i.

3. Suppose a clause body has been evaluated resulting to a composite substitution $\theta_0\theta_1\dots\theta_n$. To satisfy the GRT-program property, the projection of this substitution over the logic variables of $vars(A_\rho)$ should appear as $\{v_1/t_1,\dots,v_q/t_q\}$ such that for every $i=1,..,q$ a variable v_i is assigned to a ground abstract term t_i. An expression $exit([t_1,..,t_q])$ becomes the last element of the residual form of a clause. Types of terms t_i constitutes a substitution pattern $\tau_\rho=[\omega_1,..,\omega_q]$. To satisfy the GRT-program definition, evaluation of other clauses in the definition should give the same substitution pattern.

Let us consider the following example. The predicate definition insert/3 is used to append a node into the binary tree of names.

```
insert(X,nil::btree,tree(nil::btree,X,nil::btree)).
insert(X,tree(Left,Y,Right),tree(Left1,Y,Right)):-
   X<Y, insert(X,Left,Left1).
insert(X,tree(Left,Y,Right),tree(Left,Y,Right1)):-
   X>Y, insert(X,Right,Right1).
```

Analysis and Partial Evaluation of this definition relative to a call pattern ρ=insert(string,btree,1) creates a specialization insert_string_btree_1/3 which has the following residual form.

A_ρ=insert(α_{string},β_{btree},v_1) τ_ρ=[btree]

Clause 1: β=nil; <γ_{btree}>:γ ⇐ tree(nil,α,nil); exit([γ]);

Clause 2: <γ_{btree},δ_{string},φ_{btree}>:β=tree/3;

α<δ; <ψ_{btree}>: insert_string_btree_1(α,γ,ψ);

<χ_{btree}>: χ ⇐ tree(ψ,δ,φ); exit([χ]);

Clause 3: <γ_{btree},δ_{string},φ_{btree}>:β=tree/3;

α>δ; <ψ_{btree}>: insert_string_btree_1(α,φ,ψ);

<χ_{btree}>: χ ⇐ tree(γ,δ,ψ); exit([χ]);

Recall from the previous section that there exists only a finite number of different call patterns for a predicate definition, provided that a set of type declarations is finite. Then a finiteness of the analysis procedure defined above is achieved, if the information computed on the first evaluation of a call pattern is reused on subsequent calls. Details can be found in [3].

It is important to note that when the analysis procedure attempts to make an assumption about argument types in call patterns and substitution patterns, it is directed by type annotations of constants appearing in the program. For example, the annotated constant *nil::btree* in the definition of *insert/3* accounts for the substitution pattern is determined to include the *btree* type.

To make type annotation of constants more convenient, we can use unification to propagate type information between different annotation expressions. For example, the first clause of *insert/3* can be rewritten to be:

```
insert(X,nil::Type,tree(nil::Type,X,nil::Type)).
```

In that case the definition of *insert/3* may be used to append a node to any binary tree (not just to a binary tree of names *btree*).

Further improvements are of interest. One of the most important is that an automatic assignment of type annotations is possible provided that desirable call patterns and substitution patterns for some predicate definitions in the program are known in advance. The simplest way to satisfy the desirable patterns will be to try all the available annotations (obviously it appears to be very inefficient). The elaboration of this possibility is beyond the scope of this paper.

6. Prolog-machine for GRT-programs

The effect of Partial Evaluation for a GRT-program is that a residual program is an essential improvement of the original one. The most advantageous is that unification is evaluated completely by the transformation. A residual unification consists only of simple matching, decomposition and composition operations over ground terms.

Notice that logic variables are no longer present. They are replaced by abstract variables in the residual program. The semantics of abstract variables is principally different from that of logic variables and is very close to the use of variables in conventional programming languages (except for destructive assignment). Abstract variables are always assigned to ground terms of corresponding type. A state of an abstract variable always follows from the context; a variable α appearing in the expression prefix $<\ldots,\alpha,\ldots>$:*expression* becomes assigned when *expression* is evaluated.

Now we discuss an impact of the above mentioned issues on the design of Prolog machine for GRT-programs. The terminology of WAM [10] will be used.

When a WAM-like architecture is used to implement a GRT-program, a considerable amount of superficial work is consumed by the logic variables management. They are a variable initialization, the operations of trailing for variable instantiation and of restoring the instantiation state on backtracking, and the Trail stack maintenance itself.

If the residual representation is used to implement a GRT-program, only the Local and the Global stacks are really necessary. Abstract variables are always allocated in the Local stack. The Global stack is used to store compound ground terms.

There is no restriction on the location of the stacks in memory. Moreover, we are not limited to allocate the stacks in contiguous storage areas. This gives a real advantage if there is no support for virtual memory. A dynamic management of physical memory can be implemented in a very economical way, which appears to be important in the integrated systems.

Considering data representation in various Prolog implementations, the general principle is that an internal representation of terms always include some typing information stored in tags, which makes it a self-contained. It means that, unlike the conventional programming languages where types are used only at compile time, Prolog terms always have a complete type information as a part of their run-time representation.

In our case, only terms of the *universal* type should be represented in that way. The following format may be used.

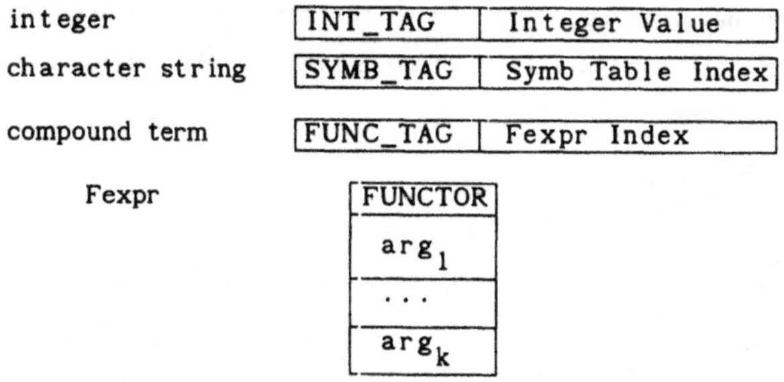

integer	INT_TAG	Integer Value
character string	SYMB_TAG	Symb Table Index
compound term	FUNC_TAG	Fexpr Index

Fexpr

FUNCTOR
arg_1
...
arg_k

In contrast, we have no need to include type information into the regular terms representation, because it is always known at compile time (note that all abstract variables are typed). This allows to represent regular terms in a format which is closely compatible to the data structures used in conventional programming languages. To emphasize the compatibility issue, an affordable representation format for some regular types is defined below by means of variable declarations written in the C programming language (here α stands for a variable name).

Regular Type / **C Type**

```
int          int α;
float        double α;
char(20)     char α[20];
char(*)      char *α;
intList      struct intList{
                 int head;
                 struct intList *tail;
             } *α;
btree        struct btree{
                 struct btree *left_node;
                 char *name;
                 struct btree *right_node;
             } *α;
```

Different representations are possible. To specify the desired format, an optional *implementation* part of regular type declaration may be used.

A compiler is responsible for generation of a sufficient set of procedures to manipulate terms of existing types. At least matching, composition and decomposition routines are required for each type appearing in the program. It should be mentioned that matching of *universal* terms must take into account that *semi-regular* function may appear as a subterm, which requires switching to the matching routines defined for types given in the *semi_regular* type declaration.

In some restricted cases, regular terms may be represented with arrays. If there is a type declaration type:: f(elem_type,type), and a residual program contains no composition operations over the terms of this type, we may represent such terms with what is called read-only arrays (roArray). In the presence of compositions a more involved representation should be used, which is called read-write arrays (rwArray).

Considering an example in the beginning of this paper, two types may be declared:

dList:: .(int,dList)::woArray.

posList:: .(char(20),posList)::roArray.

An abstract variable which has type represented with roArray should be assigned to a structure containing an array size and a pointer to the first element of that array.

Regular Type	**C Type**
posList	`struct posList{ int length; char arrPointer[20][]; } α;`

Decomposition $<\beta_{char(20)},\gamma_{posList}>:\alpha_{posList} = ./2$ binds β with the first element of the array and assigns γ with the posList structure, where γ.length=α.length-1 , and γ.arrPointer=α.arrPointer+20. The decomposition is considered successful if α.length≠0. Test $\alpha_{posList}$=hil::posList succeeds if α.length=0.

Read-only arrays may be used to describe arrays created with external procedures, such as emp_positions/3 from the example. It should be understood that in general it is preferable to use tree structures to represent ters in Prolog. In the presence of compositions we can use arrays only if the corresponding terms are always constructed in a "linear" way.

$\alpha^1 \Leftarrow .(\beta^1,nil::dList),\ \alpha^2 \Leftarrow .(\beta^2,\alpha^1),..,\alpha^k \Leftarrow .(\beta^k,\alpha^{k-1})$

For example, terms of the dList type are constructed starting with the composition over the terminal term nil::dList, following with strictly sequental creation of next levels of that term. It is not allowed that after a k-th level composition $\alpha^k \Leftarrow .(\beta^k,\alpha^{k-1})$ there appears a composition of the same or earlier levels $\gamma \Leftarrow (\delta,\alpha^m)$, where m<k.

To ensure correct organization of read-write arrays we use the following representation.

Regular Type	**C Type**
dList	`struct dList{ int free; int arrPointer[]; } α;`

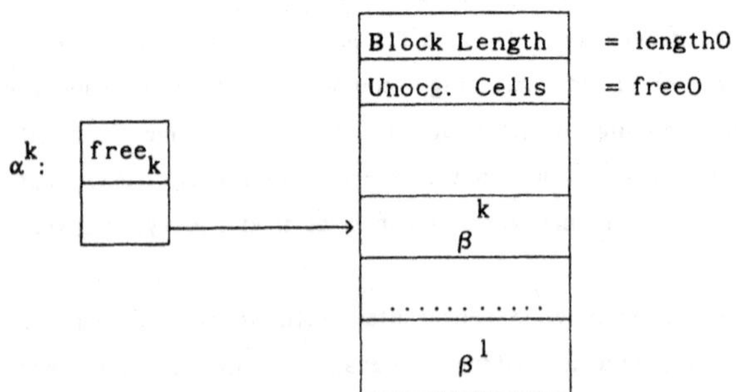

Field α_k.free contains a number of unoccupied positions in the array when a k-th level composition is executed. Two data fields are reserved in the array, which are maximum array size length0 and a current number of unoccupied positions free0. When composition $\gamma \Leftarrow .(\beta,\alpha)$ is executed, it tests that α.free=free0 and free0≠0. If this test fails, an error occurs. Otherwise the following assignments are performed γ.free=α.free-1, free0=free0-1 and β becomes a new element of the array. Field γ.arrPointer is assigned with a pointer to that element.

A memory block for array is reserved in stack when composition $\alpha^1 \Leftarrow .(\beta^1$,nil::dList) is performed. Its length may be defined statically by the declaration type::ω::rwArray(n). It it possible to define array length dynamically in the type annotator of the constant nil::ω::rwArray(n).

Then we may rewrite an example in the following way, which allows to simplify data conversions at the interface level.

```
salary_distribution(Top,Step):-
      ArrSize is Top div Step + 1,
      create_distribution(Top,Step,D_List,ArrSize),
      show_line_bar(Top,Step,D_List).

create_distribution(0,_,[]::_::Size,Size).
create_distribution(From,Step,[Emp_N|D_List],Size):-
      To is From - Step,
      total_emps(From,To,Emp_N),
      manager_emps(From,To,Emp_N,Manager_Share),
      report_managers(From,To,Manager_Share),
         create_distribution(To,Step,D_List,Size).
```

An analysis of GRT-programs is relatively simple, which may be used in the compiler to perform some deep optimizations. The following optimization seems to be of interest.

In the absence of additional information, a composition operation must always use the Global stack. In contrast, the Local stack may be used, if it is known in advance that the created term does not appear as an output argument (or as a subterm of an output argument). This clears the Global stack from garbage. The relevant information can be easily gathered with the global analysis of a residual program.

7. Conclusion

In this paper we have shown how types may be used in Prolog to combine various run-time representations of terms. In particular we was interested with the possibility to represent terms by means of data structures used in conventional programming languages. To restrict the consideration only with ground terms, we defined a class of GRT-programs, which is an adaptation of the well-modedness in the presence of types. All aspects of the GRT-programs implementation have been considered.

The presented approach is supposed to be useful for efficient interface between Prolog and conventional programming languages, because it allows to minimize data conversions at the interface level.

The author would like to thank Dmitri Boulanger and Alexander Zuravlov for helpful discussions on this material.

8. References

1. Bruynooghe M. A practical framework for the abstract interpretation of logic programs// Proc. of Fifth Int. Conf. and Symp. on Logic Programming, Univ. of Washington, Seatle, Aug. 15-19. 1988. Tutorial N 2. p.23.

2. Janssens G. and Bruynooghe M. An instance of abstract interpretation integrating type and mode inferencing. Proc. of Fifth Int. Conf. and Symp. on Logic Programming, Univ. of Washington, Seattle, Aug. 15-19. 1988. pp. 289-298.

3. Debray S.K., Warren D.S. Automatic mode inference for logic programs// Logic Programming. 1988. N 5. p.207-229.

4. Plumer L. Terminating Proofs for Logic Programs, Lecture Notes in AI, 446, Springer-Verlag, Berlin, 1990.

5. Reddy U.S. Transformation of logic programs into functional programs, In Proc. of the 1984 Int. Symp. on Logic Programming, Atlantic City, N.J., Feb. 1984, IEEE, New York, 1984, pp. 187-196.

6. Kanamori T. and Horiuchi K. Type inference in Prolog and Its Application, Proc 9th IJCAI, pp. 704-707, W.Kaufmann, 1985.

7. Lloyd J.W. Foundations of Logic Programming Springer-verlag, Berlin, 1984. p.284.

8. Venken R., Demoen B. A Partial Evaluation System for Prolog: some Practical Considerations// New Generation Computing. 1988. N 6. p.279-290.

9. Mycroft A. and O'Keefe R.A. A Polymorphic Type System for Prolog, Artificial Intelligence, Vol. 23, pp. 259-307, 1984.

10. Warren D.H.D. An Abstract Prolog Instruction Set, Technical Note 309, SRI International, 1983.

Compiling Flang

Andrei Mantsivoda, Vyacheslav Petukhin
Department of Mathematics
Irkutsk University
Irkutsk 664003
USSR
email: cdp!glas!bem

1 Introduction

In this paper we consider the main aspects of implementation of functional-logic language Flang [1]. We use *Warren Abstract Machine* (WAM) [2] as the basic technique for this purpose, because WAM is most convenient for implementation of Flang's main tools. Unfortunately, the 'pure' WAM does not have all means necessary for implementation of the language, since it does not support such important features of Flang as, say, functionality. In order to adapt WAM to our purpose we shall change WAM itself and introduce some transformations of Flang source programs to make them 'edible' for WAM. This modification of WAM is called the Flang Abstract Machine (FAM). The main additional features of FAM are
(i) support of valued terms;
(ii) optimizations for parameters management of functional style;
(iii) implementation of special technique for algebraic computations;
(iv) support of constraint satisfaction technique [3];
(v) additional instructions to support special tools (e.g. conditional expressions);

The compiler fulfils the following steps:
(i) transformation of Flang source program into standard form;
(ii) global analysis and type checking;
(iii) translation of transformed Flang program into intermediate code of FAM;
(iv) local optimization of intermediate code and translation to native code of a target computer.

This compiler from Flang program to the native code of the IBM PC has been implemented in C. The performance of executable code is very high. For example, for some benchmarks the performance of Flang programs is close to that of corresponding programs written in TurboPascal. On some benchmarks the Flang Compiler generates code which 8-10 times faster than the code generated by the Arity/Prolog compiler (*N-Queens* problem, some sorting algorithms).

2 Description of Flang

Flang [1] is a functional-logic language intended for symbolic data processing, solving combinatorial problems, for algebraic computations, as well as for training students in methods of artificial intelligence. Flang makes it easy to write programs in functional, logic and mixed styles. The reason is that these two kinds of programming are unified in Flang on some natural basis.

Flang can be considered as an extension of Prolog, but really it came from the functional style of programming. It is impossible to separate logic and functional parts of Flang: there is one idea for the two styles.

Flang contains two main types of functions:
(i) non-determinate functions;
(ii) algebraic functions;

There is the following distinction between them. When Flang machine fails to reduce a non-determinate goal it uses the backtracking procedure, but when the goal is algebraic, the system uses a kind of self-quotation (leaves the goal unreduced).

Non-determinate functions are the generalization of 'usual' functions in the following way:
(i) the evalution of functions with unground arguments is permitted;
(ii) depth-first strategy of computation of functions is used: if the system can not reduce a goal, it uses the backtracking procedure looking for alternative ways of execution;

This generalization of functions includes usual Prolog relations (they are represented by functions with only one possible value — *true*). On the other hand, we can treat non-determinate functions as usual functions and, thus, to write purely functional programs. Let us consider some examples of non-determinate definitions in Flang. First, purely functional definition - a definition of function factorial:
$factorial(0) <=> 1;$
$factorial(X) <= X > 0 \;\; X * factorial(X-1);$

Functional definition with logic variables (*quicksort*):
$partition(X, [], [], []) <=> true;$
$partition(X, [Y|Z], [Y|W1], W2) <= X >= Y -> partition(X, Z, W1, W2);$
$partition(X, [Y|Z], W1, [Y|W2]) <= partition(X, Z, W1, W2);$
$qsort([], X) <=> X;$
$qsort([X|Y], Z) <= partition(X, Y, W1, W2) \;\; qsort(W1, [X|qsort(W2, Z)]);$

Logic definitions:
$parent(paul, john) <= true;$
$parent(john, george) <= true;$
$grandparent(X, Y) <= parent(X, Z) \;\; parent(Z, Y);$

Functional-logic definition:
$ancient(X, Y) <= parent(X, Y) \;\; [X, Y];$
$ancient(X, Y) <= parent(X, Z) \;\; [X|ancient(Z, Y)];$

Some remarks on the programs: The sign '<=' plays in Flang the role of Prolog's ':-'; '− >' is the sign for cut in Flang; '<=>' is a shorthand for '<= − >' (': −!' in Prolog). The function factorial has purely functional definition. The definition of *qsort* is functional too (for example, it does not need backtracking), but variables $W1$ and $W2$ in rules are logic variables. The system uses ***unification*** to treat them. Relations parent and grandparent are defined just like those in Prolog. The function ancient (which returns a list of relatives between X and Y) is non-determinate (the system can use the backtracking procedure while executing it).

Algebraic functions are introduced to facilitate writing programs in the field of algebraic computations. The distinction between usual (non-determinate) and algebraic functions can be demonstrated in the following simple example. Let us define a function f:
$f(0) <= 1;$

Then,

***f* is non-determinate**		***f* is algebraic**	
Goal:	$f(0)$	**Goal:**	$f(0)$
Answer:	1	**Answer:**	1
Goal:	$f(1)$	**Goal:**	$f(1)$
Answer:	false	**Answer:**	$f(1)$

An example of algebraic program ('+' and '*' are declared as algebraic):

Goal:	$(a+1)*b;$
Answer:	$(a+1)*b$
Rule:	$(X+Y)*Z <= X*Z+Y*Z;$
Goal:	$(a+1)*b;$
Answer:	$a*b+1*b;$
Rule:	$1*X <= X;$
Goal:	$(a+1)*b;$
Answer:	$a*b+b$

and so on.

Using the idea of algebraic function we can define in Flang constructors with constraints, for instance, the constructor of ordered list. Given the atom ' :' declared as algebraic function and infix operator, the sorting program consists of only one logically pure and simple rule:

$X:Y:Z <= X > Y -> Y:X:Z;$

Now,

Goal: $3:2:1:6:5:4:9:8:7:0:nil;$
Answer: $0:1:2:3:4:5:6:7:8:9:nil;$

Algebraic functions can be compiled into extremely fast code. For example, the sorting program above is translated into native code having about the same speed as the equivalent program written in TurboPascal (the Pascal program contains 61 lines).

Recently, tools for solving combinatorial problems were incorporated into Flang. We used some modification of the methods offered in [3]. Special instructions and new tools for memory management which support constraint satisfaction technique were introduced in FAM.

3 Outline of compilation process

In this section we describe the main steps of compiling Flang programs.

The memory management of FAM is similar to the memory management in WAM. The main memory segments are (local) *Stack*, *Heap*, *Trail* and *registers.* Some special segments of memory intended to support *constraint satisfaction technique* are also introduced.

Terms in Flang are valued. Registers are used in FAM to return values of functions. Thus, after the computation of a function call, the system puts the result of computation in *Ri*.

The main types of instructions of FAM are analogous to those of WAM. There are *GET-*, *PUT-*, *TRY*-instructions, *etc.* Any of these groups were extended and refined to support many optimizations. Besides, special instructions were introduced, *e.g.* for compilation of constraint satisfaction programs.

The first step of compilation process is

Transformation of Flang source program into a standard form.
We demonstrate this step, using the definition of the function *factorial*
$fact(0) <=> 1;$
$fact(x) <= x > 0 \;\; x * fact(x-1);$

The main hereditary defect of FAM is that it can not manage nested calls of functions. So, before translating into FAM code, the compiler has to transform the source program to get rid of terms of the form $f(\ldots g(\ldots)\ldots)$, where the call of a function g is the argument of the call of f. In the case of factorial we have to transform the term $X * fact(X-1)$. This procedure of deliverance from nested calls is known as *flattening* [4].

To demonstrate the main idea of flattening we apply the built-in relation *unify (=)*. Using it, the second rule of the definition will be transformed into

$$fact(x) <= x > 0, \_v1 = x - 1, \_v2 = fact(\_v1), \_v3 = x * \_v2, \_v3;$$

where $\_v1, \_v2$ and $\_v3$ are some new variables. The result of computation is saved in the variable $\_v3$. In general case, given a term of the form $f(\ldots g(\ldots)\ldots)$, the compiler transforms it into something like

$\_v = g(\ldots)\ldots f(\ldots \_v \ldots)$

where $\_v$ is a new variable. Variables like $\_v$ have special features which allow to fulfil some important optimizations.

Global Analysis and Type Checking
is the next main step of the compilation process. Global analysis contains the following most important parts:

(a) Analysis of arguments and variables.
This analysis gives the information which is very important for optimizations. The analyser assigns types to all arguments and variables of a program. For any type of a variable and an argument the compiler later uses the special method of compilation. We can choose many different algorithms for the type computations with different levels of complexity (some of them were described in [5, 6]). But in any case, we should distinguish arguments of the following types:
(i) free - argument is a free variable;
(ii) unground - the argument is a term with uninstantiated variables;
(iii) ground - argument does not contain free variables;
(iv) unknown - different occurences have different types;

Analogous information on variables is derived too. This information can lead to very good improvements of the compiled code. For example, when all variables of a program are either free or ground, then this program can be executed without the use of *TRAIL*! This is the situation when any variable is used for only one of the two purposes - to pass parameters or to return results.

(b) Analysis of determinate and non-determinate rules.
This analysis can reduce the number of choice points drastically. In the case of purely functional programs this number sometimes can be reduced to zero. The system distinguishes three types of backtracking:
(i) branching - *no need to restore values and no choice point;*
(ii) near backtracking - *the system restores variables but not registers;*
(iii) far (usual) backtracking.

This part also includes the analysis of *cuts*.

In the following program
$fact(x) <= x > 0 -> x * fact(x-1);$
$fact(0) <= 1;$

applying the first rule to some goal, Flang-system does not hurry to set choice point. If the relation $x > 0$ is *true* then the system has to go through the cut and there is no need for the choice point. Otherwise, ($x > 0$ is *false*) the system uses branching and tries to apply the second rule. So, the first rule in this program is determinate.

This example is very simple and requires only local analysis of the rule. But there are many more sophisticated examples of this kind, requiring global analysis to prove that a particular rule is determinate.

The information received from global analysis permits to make many significant optimizations. There are some examples of optimizations:

If a rule has the form
$f(\ldots) <= \ldots g(X,\ldots), h(X,\ldots),\ldots;$
and X has only two occurences in the rule, then X in many cases is a temporary variable. So, Flang-system can save its contents not in the stack but in a register. Another possible optimization is the registers optimization. The idea of the registers optimization can be demonstrated by the following example:

$f(X,Y) <= Z = fact(X)\ \ h(X,Y,Z);$
where *fact* is the factorial function (see above). Easy to see that it is not necessary to save variables X, Y and Z in *Stack*, since the execution of *fact* requires very limited number of registers, so we can save these variables in the rest of registers. The execution of *fact* does not spoil them. Thus, all these variables are temporary.

Another possible optimization is the optimization of the last use of a structure. Sometimes the system can prove that in some part of a program a structure is used for the last time. In this case, the system may use destructive assignment to change the structure itself instead of copying. In the *quicksort* program (see section 4) the list $[3,2,1,6,5,4,9,8,7]$ is used only once. So, the system can use the list itself to build the result.

The spectrum of these and many other optimizations often lead to very effective compiled code. For example, in the case of purely functional program, the global analysis permits to do 'purely functional compilation'- the improvement of the general scheme with the following features:
(i) replacement of unification by matching and allocation;
(ii) avoiding choice points;
(iii) avoiding saving the address of the father activation record in Stack;
(iv) avoiding dereferencing and trailing.

Of course, these improvements can be done not only in the case of purely functional programs. Some other optimizations will be mentioned in section 4.

Translation of Flang program into intermediate code of FAM
The set of instructions of Flang Abstract Machine is the modification and extension of instructions of WAM [2]. Additional instructions are incorporated to support new features of Flang and to implement optimizations, some of which are mentioned in this paper. For instance, to return control from functions the compiler uses not only standard instructions *restore, crestore, proceed*, but also *finish* and *return*. As the head of the rule it can use *try, retry, trust, alonetry, trydet, determ, detif, detthen*. Another distinction of FAM is that it does not have *S-stack* to support unification of structures. Instead of it, the compiler uses some new instructions: *strgoto, strsetmark, strgotomark*. FAM has many other new instructions. The thorough decription of this set needs a special publication.

Translation from FAM-code to the native code of a target computer
This step of compilation is less interesting in our compiler. Usually instructions of FAM are treated as macros. But the most expensive in memory instructions are translated as procedures. The main idea of the Flang compiler is to try to do all optimizations on the level of FAM. Thus, on the level of native code there is almost no possibility to do optimizations (maybe, some kind of pipe-line...).

4 Example of compilation

As an example we consider some aspects of the compilation of *quicksort* program:
$partition(X, [], [], []) <= true;$
$partition(X, [Y|Z], [Y|W1], W2) <= X >= Y -> partition(X, Z, W1, W2);$
$partition(X, [Y|Z], W1, [Y|W2]) <= partition(X, Z, W1, W2);$
$qsort([], X) <=> X;$
$qsort([X|Y], Z) <= partition(X, Y, W1, W2) \;\; qsort(W1, [X|qsort(W2, Z)]);$
$?qsort([3, 2, 1, 6, 5, 4, 9, 8, 7], []);$

The first step is the translation of the source program into the standard form. For example, the second rule of *qsort* will be translated into

$qsort([X|Y], Z) <=$
$\quad partition(X, Y, W1, W2)$
$\quad V1 = qsort(W2, Z)$
$\quad V2 = qsort(W1, [X|V1])$
$\quad V2;$

The next step is the type analysis. The main types of arguments are the following
$partition(ground, ground, free, free)$
$qsort(ground, ground)$.

All rules in the program are determinated. So, there is no need for choice points (in FAM to set choice points, a special instruction *setcp* is used).

In the definition of *partition* the second and the third rules have $[Y|Z]$ as the second argument. So, if the system fails to apply the second rule to the goal and tries the third one, the system does not need to use get-instructions for $[Y|Z]$ one more time. This is another useful optimization.

The 3rd and 4th arguments of *partition* have the type *free*. So, the possibility to apply any rule of *partition* does not depend on the contents of the 3rd and the 4th arguments. It depends on the first two arguments of the function (really, only on the second one). The second rule is applied only when $X >= Y$. So, this check should be done before instructions for the 3rd and the 4th arguments (really, these instructions are PUT-instructions, since there is no need for unification!).

Concluding this paper we cite the draft of the FAM-code for *partition*:

```
determ
if (get_nil R2) goto first
get_structure '.' R2
unify_temp_var Y
```

```
    unify_temp_var Z
    if (X >= Y) goto second
    put_structure '.' R3
    build_temp_val Y
    build_temp_var W1
    put_temp_val Z R2
    put_temp_val W1 R3
    restore partition
second:
    put_structure '.' R4
    build_temp_val Y
    build_temp_var W2
    put_temp_val Z R2
    put_temp_val W2 R4
    restore partition
first:
    put_nil R3
    put_il R4
    return
```

References

[1] A.Mantsivoda *Flang: A Functional-Logic Language.* Proc. of Int. conf on Processing Declarative Knowledge.- Kaiserslautern, July 1991.

[2] D.H.D.Warren. *An Abstract Prolog Instruction Set.* Technical Note 309 SRI International, Menlo Park, CA, October 1983.

[3] P. Van Hentenryck. *Constraint Satisfaction in Logic Programming.* The MIT Press, Cambridge, 1989.

[4] H.Boley. A relational/functional Language and its Compilation into the WAM. SEKI Report SR-90-05, University of Kaiserslautern, 1990.

[5] P.L.van Roy *Can Logic Programming Execute as Fast as Imperative Programming.* PhD Th., November 1990.

[6] Taylor A. PhD Th., 1991.

FIDO: Finite Domain Consistency Techniques in Logic Programming

Manfred Meyer*
German Research Center for Artificial Intelligence (DFKI)
P. O. Box 20 80
D-6750 Kaiserslautern
Germany

Hans-Günther Hein & Jörg Müller
Computer Science Department
University of Kaiserslautern
P. O. Box 30 49
D-6750 Kaiserslautern
Germany

Abstract

In this paper we discuss different implementation for FIDO, a logic programming language with finite domain constraints and consistency techniques. These approaches range from meta–interpretation over horizontal compilation (source–to–source transformation) through vertical compilation down to an extended Warren Abstract Machine. We will stress the horizontal compilation approach which already shows promising results, whereas the deeper integration on a lower implementation layer by extending the unification and control instructions of the WAM will finally give much better performance results.

1 Overview

Logic programming offers a convenient and declarative way for stating constrained search problems due to its relational form and nondeterminism. Unfortunately, standard logic programming languages such as PROLOG are also very inefficient for solving these problems. This has led to the definition of several constraint logic programming languages (e.g. [3, 5, 9]) combining the declarative aspects of logic programming with the efficiency of constraint–solving techniques.

One of these languages is FIDO [12, 13], which extends a usual logic programming language by providing **FI**nite **DO**mains together with efficient consistency techniques such as *forward checking* and *lookahead* [14]. This extension leads to a substantial improvement on efficiency for the class of combinatorial problems and contributes to make constraint

*The work of the first author is carried out as part of the ARC-TEC project, supported by BMFT under grant ITW 8902 C4. Please, send correspondence to the first author; e-mail: meyer@dfki.uni-kl.de

logic programming an appropriate tool for expressing and solving constraint satisfaction problems.

In the FIDO laboratory different approaches towards an integration of finite domain consistency techniques in logic programming are investigated. Starting with the implementation of a *meta-interpreter* for FIDO written in PROLOG (cf. [13]), we subsequently investigated the more sophisticated approach of horizontally compiling FIDO programs into SEPIA [10] by making use of the built-in coroutining mechanism. This ***horizontal compilation*** approach (cf. [12]) revealed that the integration and handling of domain variables is the major issue concerning run-time efficiency and implementation complexity. Therefore, we now have started with a ***vertical compilation*** approach (cf. [6]) to compile finite domain constraints down into a WAM architecture by extending the basic WAM data structures and using a *freeze*-like control scheme [2].

Further developments will include the extension of this approach for dealing with hierarchically structured domains as in CONTAX [11], a constraint system currently under development at DFKI as part of the Knowledge Compilation Laboratory COLAB [1].

2 The Meta–Interpretation Approach

When extending PROLOG for handling finite domains by designing a meta–interpreter, one has to implement one's own meta–unification routine working on an explicit representation of meta–variables. In our approach [13], which is similar to that presented in [8], meta–variables are simply represented as PROLOG terms. Domain–variables can then be implemented by extending the terms representing meta–variables introducing additional arguments needed for domain handling and forward checking. In detail, domain–variables are represented as PROLOG terms of the form `meta(Value,Key,Metaterm)` where `Value` gets instantiated only when the value of this domain–variable becomes a singleton. `Key` is used as a key for fast access to variables, and `Metaterm` contains a term holding the current value (`Val`), the domain, the forward–goals depending on this variable, and pointers to variables and constants which have to be checked for inequality with this domain variable. When creating a new meta–variable, `Val` is first unbound. Changing its value because of unification or a call to an $\neq$–constraint or a forward goal is done by instantiating `Val` with a new metaterm of the same form. Therefore, for dereferencing a meta–variable the chain of nested metaterms has to be scanned until reaching an unbound `Val`. This results in an access time depending on the number of unifications the meta–variable took part in.

In FIDO we studied two ways for representing the domains: using bit vectors and ordered linked–lists. Domain representation by using *ordered lists* does not need the domains and their elements to be explicitly declared. Ordinary PROLOG terms may be used as elements, and even the maximum domain size is not limited. The most important operations on domains, represented as lists, have linear time and space complexity, provided that there is some order on the domains for which, if not supplied by the user, the standard term ordering of PROLOG is taken. Using a *bit vector* domain representation makes domain operations more efficient. Even computing the intersection (unification) of two domains becomes a constant–time operation. This will be very important when handling large domains in real–life applications, e.g. in the mechanical engineering domain we are dealing with in the ARC-TEC project at DFKI. On the other side, using a bit vector domain representation requires explicit domain declarations, so that the maximum number of elements as well as the symbolic names of domain elements have to be known to the system. Because of the run-time advantages, we decided to use a bit vector representation for domains in FIDO. Since SEPIA does not support bit vector operations, we have been forced to develop our own bit vector operations and representations on top of SEPIA. Therefore, in the current implementation domains are represented as lists of integers, where each integer represents a substring of the bit vector, so that the usual bit vector operations can be implemented by iteratively using bitwise–AND resp. OR over integers.

With this meta–interpretation approach described so far we have been able to run a lot of benchmarks, mostly taken from [14]. These small applications have already shown a satisfactory run-time behavior of FIDO superior to normal PROLOG execution with backtracking. But, as in general with meta–interpretation, the main disadvantage of this approach remains its lack of good run-time performance resulting from the necessity to do a lot of work on the meta–level which the underlying PROLOG system could do fairly more efficiently.

3 The Horizontal Compilation Approach

Because of the drawbacks of the meta–interpretation approach discussed in the previous section, we went to a next step towards FIDO implementation, which led us to horizontal compilation. The aim of this approach was to build an intermediate step carrying the control regime and the work to be done for domain handling down to lower system layers, finally arriving at the WAM itself. Our claim is that these tasks can be done much more efficiently on a lower implementation layer.

One of the main ideas is to make use of coroutining, which in SEPIA is introduced by using `delay` declarations. By this means which is similar to `wait` declarations in NU-PROLOG more sophisticated control can be gained without introducing much run-time overhead. The very heart of the horizontal compilation model is a preprocessor compiling ordinary PROLOG programs enriched by control declarations (e.g. forward checking, lookahead and domain definitions) into a standard PROLOG program with `delay` clauses, thus performing a source–to–source transformation. This approach is based on ideas presented in [4]. However, there are some interesting differences concerning the domain handling as well as the implementation of consistency techniques. These issues will be discussed in further detail in the following.

3.1 Integrating Finite Domains

A major task of the FIDO system is its ability to handle variables ranging over finite domains. This includes both, the selection of an appropriate representation of the domains and finding methods for efficiently manipulating them. An adequate domain representation scheme for FIDO has to fulfill several requirements. It should

- support sequential and direct access to domain elements,
- guarantee the set property for the domain, and
- reflect an existing order over the domain elements.

The usefulness of having direct access stems from the fact that some of the most important constraints (e.g. the $\neq$ constraint) only eliminate a few elements (or even only one) from the domain. Actually traversing and testing the whole domain for finally deleting only one element seems to be not very economical. The usual list representation scheme is optimal for sequential access whereas it is fatal when direct access to domain elements is required. Thus, representing a domain with N elements as an uninterpreted functor with arity N seems to be more adequate in this case. More elaborate techniques like domain element retrieval using (perfect) hash tables or even using database techniques like B or B^* trees might also be of some interest once the domain sizes become very large.
Shortly, a domain $\{e_1,\ldots,e_n\}$ is internally represented as a term `dom(`$(e_1$`,_),...,(`e_n`,_))`, where the anonymous variables serve as flags. Whenever a domain element is discarded from the domain, its flag is instantiated to `0`. This provides a convenient handling of backtracking—we just leave it to the PROLOG control mechanism. It should be emphasized here that this representation facilitates the direct access to domain elements via the **arg/3** predicate, which—due to the way terms are arranged on the PROLOG heap—is more efficient than a list representation scheme.

Variables ranging over domains described above are called *domain variables.* A domain variable X is internally represented as a 6-tuple $(\&, X_{id}, X_{length}, X_{\#cons}, X_{value}, X_{domain})$. The first argument is a flag used to identify domain variables. X_{id} denotes a unique domain identifier, X_{value} possibly contains the singleton value of X, X_{domain} unifies with an explicit copy of the domain, and X_{length} is the actual length of the domain. This is needed for an efficient implementation of the singleton test and for the usage of first–fail heuristics. The number of constraints, denoted by $X_{\#cons}$ is needed for the first–fail principle as well.

In the following we will show how the user can exploit the domain handling facilities of FIDO. The initialization of the domains in FIDO is different to e.g. CHIP. In FIDO domains have to be declared in the source file. This is done by a declaration of the form **define-domain**(*DomainId, VarSpec, DomSpec*), where *DomainId* is the identifier of the domain, *VarSpec* enumerates the variables from this domain, and *DomSpec* represents the range of that particular domain. FIDO also allows a run-time creation of domains whose cardinality is not known at compile–time. Since the efficiency of handling some built-in constraints such as $>$ or $<$ can greatly benefit from exploiting orders over domains, the user may optionally define an order for the domain. This is contrived by a declaration **domain_order**(*DomainId, Order*), where *DomainId* indicates the domain, and *Order* denotes how the domain is ordered (`asc` for *ascending order* or `desc` for *descending order*). When using standard domain types such as integer subsets, FIDO automatically recognizes the correct ordering.

3.2 Constraints and Consistency Techniques

Consistency techniques studied in FIDO are *forward checking* and *lookahead.* For most applications lookahead is too inefficient as it produces much overhead.[1] This is the reason why, in the horizontal approach, we directed our attention towards forward checking. Besides a general forward checking algorithm for arbitrary constraints, we provide specialized forward checking implementations in order to optimize the access behavior for some frequently used constraints (such as $\neq =$, $<$, and $>$).

Most PROLOG programs written in a natural style show a behavior often referred to as generate–and–test: First, one combination of values for the variables is *generated*, and then it is *tested* against the problem constraints. When a failure occurs, PROLOG responds with backtracking to the previous choice point and generates an alternative value combination. This tends to be a latent source of inefficiency, because many value combinations have to be tested and a lot of intermediate results are unnecessarily recomputed.

It turns out to be a better method, first, to state the problem constraints, and then to successively generate values for the variables. At the same time incompatible domain values can be eliminated. The SEPIA delay mechanism provides an efficient way to realize this method. Stating constraints before instantiating variables leads to a delay of the constraints. If at a later stage of the instantiation process variables become instantiated, their constraints are reactivated and checked again. Thus, all the variable values will be ground and we come up with a definite solution or the constraint system will fail. Also, it is possible that some constraints remain delayed. This can be interpreted as representing a set of solutions satisfying the conditions expressed by these constraints.

In FIDO the user can declare predicates to be executed under a forward checking or lookahead regime. Assume, for example, that the user wants to define a constraint

```
regular(X,N,Y) :- Y =\= X, Y =\= X + N, Y =\= X - N.
```

The functor `=\=` represents the numerical nonequality–constraint. If one likes to force a forward checking execution of the `=\=` constraints used in the definition of `regular/3`, this can be achieved by simply introducing `forward` declarations:

```
regular(X,N,Y) :-
  forward(Y =\= X), forward(Y =\= X+N), forward(Y =\= X-N).
```

During the preprocessing step each call to a `forward`–declared goal will then be replaced by a call to the forward redefinition of the goal. Since it is not known at preprocessing time which arguments may be domain variables, a case distinction has to be done, e.g. `=\=` is compiled into `for_nne_dd`, `for_nne_nd`, `for_nne_dn`, and `for_nne_nn` respectively, which realize variants of the numerical–nonequality–constraint for domain (**d**) and non–domain–variable (**n**) arguments. These specialized constraints finally become delayed and perform the "real" calls to the forward checking resp. lookahead algorithms.

[1] However, *lookahead* may be extremely useful for special kinds of constraints to deal with, e.g. constraints describing a crossword puzzle [14].

3.3 An Example: The N Queens Problem

In the following we will consider the *N Queens Problem* as an example for demonstrating the horizontal transformation approach. Figure 1 shows the initial FIDO source code for the N queens problem.

```
queens(N, L) :-
   define_domain(nqueens, gen_var(N, L),
            gen_int_dom(D, 1, N)).
   safe(L),
   instantiate(L).

safe([]).
safe([X]).
safe([X, Y|Z]) :-
   noattack(X, [Y|Z]),
   safe([Y|Z]).

noattack(X, Y) :- noattack(X, Y, 1).
noattack(X, [], _).
noattack(X, [H|T], N) :-
   regular(X, H, N),
   N1 is N + 1,
   noattack(X, T, N1).

regular(X, Y, N) :-
   forward(Y =\= X),
   forward(Y =\= X + N),
   forward(Y =\= X - N).
```

Figure 1: FIDO source code for the N queens problem

Note the way the `domain` and `forward` declarations are expressed in this program: In contrast to other systems like CHIP which use global declarations, FIDO uses local `forward` and `lookahead` declarations.[2]

```
?- compile('/home/fido/redef/f_nne').

queens(N, L) :-
   queensfidoaux(N, L1),
   transform_dv(L1, L).

queensfidoaux(N, L) :-
   create_domain(queens, N,
            gen_int_dom(D, 1, N), D),
   create_domvars(queens, gen_var(N, L), L),
   safe(L),
   instantiate(L).

safe([]).
safe([X]).

safe([X, Y|Z]) :-
   noattack(X, [Y|Z]),
   safe([Y|Z]).

noattack(X, Y) :- noattack(X, Y, 1).
noattack(X, [], _).
noattack(X, [H|T], N) :-
   regular(X, H, N),
   N1 is N + 1,
   noattack(X, T, N1).

regular(X, Y, N) :-
   for_nne(Y, X),
   for_nne(Y, X + N),
   for_nne(Y, X - N).
```

Figure 2: FIDO code after preprocessing

Figure 2 shows the SEPIA code resulting from the N queens program after preprocessing. A thorough discussion of the way the program works and the problems we had

[2]We have chosen this representation, because in some cases it might be useful to execute one constraint differently in separate environments, e.g. in one procedure call by using *forward checking*, in another call by using the *lookahead* algorithm. This can be done very easily using local declarations.

to overcome during design and implementation would be going too far here. Note, however, the redefined calls to the `forward` constraints, the explicit creation of domains and domain–variables, and the transformation of domain–variables (`transform_dv`) in order to hide implementation details from the user.

3.4 Preliminary Results

Both the FIDO meta–interpreter and the preprocessor are implemented in SEPIA and are running on SUN4 workstations.[3] First run-time results show a significant performance improvement by using horizontal compilation and coroutining: The preprocessed program solves the 16 queens problem about seven times faster (in less than one second) than the meta–interpreted program does with the 8 queens problem.

Finally, since domain variable unification is not covered by the system, it has to be performed by case distinctions in procedure calls. This is the reason why implicit unification of domain variables from different domains, as they appear e.g. in clause head unification done by the PROLOG inference engine, is still an open problem with this approach. In the current implementation the user has to care about not to use this kind of domain unification in FIDO programs.

4 The Vertical Compilation Approach

Although the results for FIDO appear quite reasonable for smaller applications, the limitations become rather clear once we try to solve more complex real–life problems.[4] This elucidates the need for a deeper integration of domain handling, domain variable unification and consistency techniques into the PROLOG system, which leads us to a modification of the WAM.

The two major aspects for integrating finite domain constraints in the WAM are a mechanism for the revised control strategies (*forward checking*, *lookahead*) and an extension of the WAM unification algorithm coping with domain unification together with a set of domain–variable specific WAM instructions. At the moment we have started with the design and implementation of the domain unification routine for handling finite domains. By using the WAM compilation scheme presented in [7], it will then be possible to deduce the new WAM instructions by specializing the basic WAM domain–unification routine. However, the overhead of unnecessary wakeups has to be reduced: expensive groundness checks should not be recomputed when a constraint gets reactivated.

[3]The FIDO system is ftp-able from `minnehaha.rhrk.uni-kl.de` (directory `/pub/languages/fido`).

[4]The run-time for N queens grows significantly for problems of $N > 50$. The CHIP approach providing a deep integration of domains and consistency techniques into the WAM becomes the more superior the more complex the applications are.

Acknowledgements

We would like to thank our colleagues in the constraint group at DFKI for many valuable discussions and lots of comments on earlier versions of this paper. In particular, thanks are due to Stefan Schrödl who implemented the meta-interpreter for FIDO and so contributed to the success of the FIDO project at a very early phase.

References

[1] H. Boley, P. Hanschke, K. Hinkelmann, and M. Meyer. COLAB: A Hybrid Knowledge Compilation Laboratory. *Annals of Operations Research*, 1992.

[2] M. Carlsson. Freeze, Indexing, and Other Implementation Issues in the WAM. In J.-L. Lassez, editor, *Proc. ICLP-87*, Melbourne, Australia, May 1987. MIT Press.

[3] A. Colmerauer. Opening the Prolog III Universe. *BYTE*, pages 177–182, August 1987.

[4] D. de Schreye, D. Pollet, J. Ronsyn, and M. Bruynooghe. Implementing Finite–domain Constraint Logic Programming on Top of a PROLOG–System with Delay–mechanism. In N. Jones, editor, *Proc. of ESOP 90*, pages 106–117, 1990.

[5] M. Dincbas, P. van Hentenryck, H. Simonis, A. Aggoun, T. Graf, and F. Berthier. The Constraint Logic Programming Language CHIP. In *Proc. FGCS'88*, Tokyo, December 1988.

[6] H.-G. Hein. Consistency Techniques in WAM-based Architectures. Master's thesis, Univ. Kaiserslautern, FB Informatik, Postfach 3049, D-6750 Kaiserslautern, 1991. Forthcoming.

[7] H.-G. Hein and M. Meyer. A WAM Compilation Scheme. In *Proceedings of the 2nd Russian Conference on Logic Programming.* Lecture Notes in AI, Springer-Verlag, Heidelberg, 1991.

[8] Ch. Holzbaur. Realization of Forward Checking in Logic Programming through Extended Unification. Technical Report TR–90–11, Austrian Research Institute for Artifical Intelligence, June 1990.

[9] J. Jaffar and J.-L. Lassez. Constraint Logic Programming. In *Proc. POPL-87*, Munich, Germany, 1987.

[10] M. Meier, A. Aggoun, D. Chan, P. Dufresne, R. Enders, D.H. de Villeneuve, A. Herold, P. Kay, B. Perez, E. van Rossum, and J. Schimpf. SEPIA – An Extendible Prolog System. In G. Ritter, editor, *Proc. IFIP 11th World Computer Congress*, pages 1127–1132, 1989.

[11] M. Meyer and C. Jakfeld. CONTAX – A Constraint System over Taxonomies. ARC-TEC Discussion Paper 91-04, DFKI GmbH, P. O. Box 20 80, Kaiserslautern, Germany, 1991.

[12] J. Müller. Design and Implementation of a Finite Domain Constraint Logic Programming System based on PROLOG with Coroutining. Master's thesis, Universität Kaiserslautern, FB Informatik, Postfach 3049, D-6750 Kaiserslautern, 1991. Forthcoming.

[13] S. Schrödl. FIDO: Ein Constraint-Logic-Programming-System mit Finite Domains. ARC-TEC Discussion Paper 91-5, DFKI GmbH, P. O. Box 2080, Kaiserslautern, Germany, 1991.

[14] P. van Hentenryck. *Constraint Satisfaction in Logic Programming.* MIT Press, 1989.

Pierangelo Miglioli, Ugo Moscato, Mario Ornaghi

Department of Information Science - University of Milan

A CONSTRUCTIVE LOGIC APPROACH TO DATABASE THEORY

ABSTRACT

In this paper we propose an approach to database theory based on a constructive logic. The semantics here assumed is a particular one; it is based on the notion of info(K,F) (the information type of F), where K is the set of constants of a first order language L, F is a formula of L and info(K,F) is the set of all the possible pieces of information (within L) on the "truth" of F.
This constructive semantics will be used to treat problems related to relational databases such as disjunctive information and null value.

Introduction

In the recent years there has been a growing interest in the logical approach to database theory. In this frame some authors [e.g., LT1, LT2, Rei] distinguish between a "model theoretic" approach and a "proof-theoretic" one; e.g., the relational approach is viewed as "model theoretic" (where a database represents a model of a set of axioms or constraints), while a "proof-theoretic" approach is related to the so called "deductive databases" [GMN].

If we are interested in semantical and soundness problems, the deductive approach seems to be fruitful, since it allows to reinterpret the model-theoretic one [Rei] and offers a uniform treatment of various aspects [Rei, LT1, LT2] such as disjunctive information, null values, incomplete information and the possibility of incorporating "more world knowledge". Nevertheless, in the duductive approach there is , so to say, an "excess of uniformity", since there is no distinction between the following two levels:

a) the level of the pieces of information which are stored in a database and which can be updated;

b) the level of the general properties of a whole class of databases, e.g., the knowledge of the general laws which always hold and have not to be updated.

To be more precise, such a distinction is not directly evident in the theory representing a deductive database; the distinction between axioms representing facts, constraints, or knowledge is a priori arbitrary and requires suitable metatheoretical considerations.

In this paper we propose a logical approach to database theory based on a particular constructive semantics [MMOU] and on a corresponding constructive deductive apparatus; our approach can be related to the so called BHK explanation of constructivism [Tro2] and there is a strict relation between our notion of a 'piece of information' true in a model and the one of a 'realization number' of a number theoretic formula given in [Kle]; also, even more closely, our notion of constructive validity recalls the notion of 'uniformly solvable problems' given in Medvedev [Med]. Other approaches based on realizability have been given in the field of computer science, e.g. [Vor, HN].

In our attitude, the above levels a) and b) are clearly distinguished within the formal setting itself. As a matter of fact, the considered semantics is based on the interpretation of a first order formula A as a type, where an element of type A represents a PARTICULAR piece of information on the truth of A, related to PARTICULAR models satisfying A (e.g., if $A=B\vee C$, we have models satifying B and models satisfying C).

We will call "information types" the types described by the logical formulas and "pieces of information" the elements of the information types; in this way, we have a formal frame where a logical formula A represents a general property (point b)) of a database, while a piece of information of type A represents a particular content (point a)) of that part of the database related to A.

To be more precise, a database is seen as a triple:

DB = < T , T.info, K >

where T is any set of formulas of a many-sorted first order language, T.info is an element of the information type given by T and K is a finite set of constants.
Intuitively, the triple < T, T.info, K > means that the pieces of information contained in T.info are sufficient to completely answer at least the QUERIES represented by the formulas of T w.r.t. the individuals named by K; this fact allows us to consider a formula both as an information type and as a "query type". E.g., if we consider the following theory T containing only one axiom:

$$T = \{\forall x: \text{employee}\ (\text{works}(x) \lor \neg \text{works}(x))\ \}$$

and K contains the only employee names {John, Mary, Ted}, then a T.info might contain the information "works(John), works(Mary), ¬works(Ted)", another T.info might be "works(John), ¬works(Mary), works(Ted)", and so on; i.e., a T.info contains, for every employee x, the information "x works" or "x doesn't work", which allows to answer any query of the kind "works(x)∨¬works(x)?".

Working with finite databases, K and the information types are intended to be finite; but such a finiteness may be more or less explicit. To be more precise, we may distinguish four cases corresponding to different levels of definiteness:

1) One assumes that the individuals and their relations are completely specified. This corresponds to the possibility of axiomatizing the database as a categorical theory. In this case, the more appropriate logic (and the more powerful) is the classical one.

2) The individuals are assumed to be completely specified (a domain closure axiom is given), but their relations are assumed to be incompletely characterized. This corresponds to a database with incomplete information giving rise to a theory with (a finite number of) non isomorphic (finite) models. Even if a domain closure axiom is given, here the descriptive power of classical logic becomes unadequate in order to discriminate the invariant properties of the database from the ones depending on a complete specification of the involved relations (i.e., on a choice of a particular model among the possible ones). On the contrary, as we will see, our constructive setting, in presence of a domain closure axiom, will enable us to adequately deal with these situations.

3) One assumes the domain closure axiom only implicitly, i.e., one wants to 'uniformly' reason about a database without using this axiom (the characterization of the domain may not yet tbe available, but one may assume that it will be given eventually). Here a deductive database should be seen as a parametric theory where every specification of the parameters gives rise to a theory of the kind discussed in point 2). In our constructive formal setting these theories are adequately treated using Grzegorczyk's principle [Tro1]; this is a logical principle which can be constructively used only in presence of particular axiomatizations (e.g., its addition to intuitionistic arithmetic gives rise to the classical, non constructive one), but it holds in our frame and captures the hypothesis that a domain closure axiom is implicitly given (we remark that from a classical point of view nothing interesting can be said: if the domain closure axiom is implicit we have only that the theory has an unbounded family of models and hence, by general model theoretic results, infinite ones).

4) A domain closure axiom is not even assumed. In this case Grzegorczyk's principle cannot be used (it has no longer a constructive meaning). Also in this case, as we will see, many significant aspects can be captured in our constructive attitude.

The clear distinction between general and particular aspects, TOGETHER with the use of a constructive calculus, allows to distinguish deductions representing GENERAL reasonings on T from deductions of answers to particular queries. As a matter of fact, proofs of our constructive calculus represent formal GENERAL reasonings on T. More precisely, if one constructively proves H from T, he is guaranteed that any instance of the general query of type H can be correctly answered starting from the pieces of information contained in T.info. Moreover, our constructive proofs have a direct computational interpretation, i.e. an algorithm to answer H can be effectively extracted from a constructive proof of H from T.

Finally, there is a class of theories T such that any T.info can be represented in a "traditional" way (e.g., by relations in relational algebras or by Horn-clauses in deductive databases); for this kind of theories the tools developed in traditional approaches can be used in a way compatible with our method. We think that a situation can be reached where different approaches can work together in a sound way under the "control" of a constructive metatheoretical background; our work is in this direction.

In the first section we will discuss some simple examples, which should illustrate how the proposed approach works. The examples have been chosen in order to treat some problems discussed in literature.
For sake of simplicity, the examples will be treated in an informal way, while the formal logical frame upon which they rely will be briefly expounded in the subsequent sections (see also [MMOU, MMOQU]).

In the following we will use the signs $\vdash$ and $\models$ to indicate classical provability and classical validity respectively and will use the signs $\Vdash$ and $|\!\models$ to indicate constructive provability and constructive validity respectively; furthermore, info(K, A) will represent the set of all the pieces of information of type A in a relational language L with constants K and A.info an element of info(K, A).

1. The constructive approach

1.1. The query language and the information types

First of all, we explain the interpretation of first order formulas as queries. The query-interpretation of a formula A allows also to understand which pieces of information are contained in an A.info, since A.info contains what is needed in order to answer the query A. Given a database DB = <T, T.info, K>, we say that:

- a "MODEL" of DB is any relational model of T (in the sense of classical model theory) which also satisfies all the pieces of information in T.info (for a precise def., see sect. 2);

- a query is any closed formula Q of the language of T such that $T \models Q$, where $T \models Q$ means that Q is a logical consequence of T in the sense of classical logic;

- an answer of Q is a piece of information of info(K, Q) which is true in any model of <T, T.info, K>; if not $T \models Q$, Q has no answer.

Before giving the formal treatment we propose some examples, presenting some of the most relevant kinds of query which can be expressed; in the examples of this subsection we will use the following language L (of some DB = <T, T.info, K> here unrelevant):

- individual constants K: John, Mary, teacher, cow-boy, CS200, PH100.

- predicate constants: ILL(x), ISWORKING(x), JOB(x,y), TEACHES(x,y).

Example 1. Disjunctive queries.
A disjunction "A∨B" means that we want the answer A (if A holds) or B (if B holds).
For instance, the query Q = "ISWORKING(John)∨ILL(John)" has two possible answers: namely, the answer "ISWORKING(John)" or the answer "ILL(John)". Correspondingly, the type Q codifies the two pieces of information {ISWORKING(John), ILL(John)}.

Example 2. Existential queries.
An existential "∃xA(x)" means that we want a value c of x, represented by some constant of the considered set of contants K (here a constant of {John,Mary,...}), such that A(c) holds and that we want an answer of A(c).
For instance, a possible answer of the query Q = "∃xJOB(John,x)" is the couple <teacher, JOB(John,teacher)>. Correspondingly, the type Q contains as many elements as the possible jobs are.

Example 3. Universal queries.
An universal query "∀xA(x)" means that we want a method to answer A(c), for every possible instance c for x (i.e., for every constant c of K). In particular, such a method could be provided by suitable stored data.
For instance, an answer of

Q = "∀x(∃j JOB(x,j) ∨ ¬∃j JOB(x,j))"

is a method which, for every instance of x, answers the corresponding query; e.g., for x=John the answer might be: "yes, there is the job teacher such that JOB(John,teacher) holds"; for x=CS200, the answer might be: "no, there is no j: JOB(CS200,j)", and so on.
Correspondingly, a type "∀xA(x)" contains all the functions F : constants c → type A(c).

Example 4. Elementary queries and incomplete information
An elementary formula is any formula E whose info(K, E) contains ONLY ONE element; e.g., info(K, JOB(John,teacher)→JOB(John,teacher)) contains the unique function f : {JOB(John,teacher)} → {JOB(John,teacher)}. Now, the unique knowledge which can be extracted from the unique element of info(K, E) (E elementary) is that E is valid in the theory T of the considered database <T, T.info, K>; in other words, an elementary formula E is unrelevant, when considered as a query, and the corresponding E.info can be omitted. But the elementary formulas are very important, since they can be used to state GENERAL LAWS without requiring to store any T.info; moreover, according to the interpretation of "¬H" as "H → ff" (see sect. 1.3), any formula of the kind "¬ A" is an elementary one. We will use the abbreviation *true* H = ¬¬ H; informally, the operator *true* does not affect the truth value of H, but stops the constructive analysis of H, as shown by the following example.

(A) ∀x(JOB(x,teacher)→*true*(∃cTEACHES(x,c)))

(A) states a general property of being a teacher, i.e., that each teacher teaches at least a course but it doesn't require the knowledge in the database of such a course.
On the other hand, the axiom

(A')∀x(JOB(x,teacher)→(∃cTEACHES(x,c)))

is not elementary and can be read as a general requirement that, for every teacher, the database must contain also the course he teaches.
As we will see in sect. 2.2, this situation can be related with the so called problem of "null values" discussed in literature [Rei]. As a matter of fact, (A) admits the presence of "unknown values" while (A') explicitly requires, for every x, the knowledge of a constant c of K such that TEACHES(x,c) holds.

1.2. The constructive semantics

Our constructive semantics is based on info(K,F) which associates, with every formula F, the set info(K,F) of the pieces of information of type F with respect to the constants of K. Let L be a (possibly many sorted) first order language, K be a non empty set of (possibly sorted) constants and F a formula of L.
We inductively define the set info(K,F):

Def.1.
For a formula F, info(K,F) is so defined:

$\text{info}(K,E) = \{tt\}$ if E is an atomic or negated formula
$\text{info}(K,A\wedge B) = \{\langle a,b\rangle \mid a\in \text{info}(K,A) \text{ and } b\in \text{info}(K,B)\}$
$\text{info}(K,A\vee B) = \{\langle 1,a\rangle \mid a\in \text{info}(K,A)\} \cup \{\langle 2,b\rangle \mid b\in \text{info}(K,B)\}$
$\text{info}(K,A\rightarrow B) = \{f \mid f: \text{info}(K,A) \rightarrow \text{info}(K,B)\}$
$\text{info}(K,\exists xA(x)) = \{\langle c,a\rangle \mid c\in K \text{ and } a\in \text{info}(K,A)\}$
$\text{info}(K,\forall xA(x)) = \{f \mid f: K \rightarrow \text{info}(K,A)\}$

For a set of formulas, we have:

$\text{info}(K,\{A_1,\dots,A_n\}) = \{\langle a_1,\dots,a_n\rangle \ / \ a_i\in \text{info}(K,A_i) \text{ for every index } i\}$.

■

Now, given a (possibly many sorted) structure S with carrier D interpreting the language L and the constants K, we will indicate by

$S \models A\,[\underline{x}|\underline{k}]$

the fact that the formula A is true in S under the assignment of the tuple of free variables $\underline{x}$ with a tuple $\underline{k}$ of values of D. In Def.2 we give the notion of "truth" for info(K,F), where $\mathbf{c}$ indicates the element of D denoted by a constant c:

Def.2.
For a formula we have:

$S \models tt{:}A[\underline{x}|\underline{k}]$	iff $S \models A[\underline{x}|\underline{k}]$
$S \models \langle a,b\rangle{:}A\wedge B[\underline{x}|\underline{k}]$	iff $S \models a{:}A[\underline{x}|\underline{k}]$ and $S \models b{:}B[\underline{x}|\underline{k}]$
$S \models \langle 1,a\rangle{:}A\vee B[\underline{x}|\underline{k}]$	iff $S \models a{:}A[\underline{x}|\underline{k}]$
$S \models \langle 2,b\rangle{:}A\vee B[\underline{x}|\underline{k}]$	iff $S \models b{:}B[\underline{x}|\underline{k}]$
$S \models f{:}A\rightarrow B[\underline{x}|\underline{k}]$	iff $S \models A\rightarrow B[\underline{x}|\underline{k}]$ and, for every $a\in \text{info}(K,A)$, if $S \models a{:}A[\underline{x}|\underline{k}]$ then $S \models f(a){:}B[\underline{x}|\underline{k}]$
$S \models \langle c,a\rangle{:}\exists yA(y)[\underline{x}|\underline{k}]$	iff $S \models a{:}A[\underline{x},y|\underline{k},\mathbf{c}]$
$S \models f{:}\forall yA(y)[\underline{x}|\underline{k}]$	iff $S \models \forall yA(y)[\underline{x}|\underline{k}]$ and, for every $c\in K$, $S \models f(c){:}A[\underline{x},y|\underline{k},\mathbf{c}]$

For a set of formulas, we have:

$S \models \langle a_1,\dots,a_n\rangle{:}\{A_1,\dots,A_n\}$ iff $S \models a_i{:}A_i$ for every index i.

■

The following proposition can be proved:

Prop.1.

i) S ⊨ a: A[x|k] => S ⊨ A[x|k]
ii) If S satisfies the domain closure axiom (DCAx) [Rei], then:

S ⊨ A[x|k] => there is an a such that S ⊨ a:A[x|k]
iii) The analogous of i), ii) hold also for a set of formulas.

■

Now we can give the notion of a model of a database for a theory T = $\{A_1, \ldots, A_n\}$.
Def. 3.
Let DB=<T, T.info, K> be a database for T; we say that S is a MODEL of DB iff S ⊨ T.info:T

■

On the domain closure axiom. By Prop.1 one has that any model of DB is also a model of T; so, the various T.info discriminate various classes of models of T. Also, we have that the domain closure axiom [Rei] (DCAx) plays an important role:

i) if DCAx does not hold in T, then there may be models of T for which there is no true DB;

ii) if DCAx holds in T, then, for every model S of T there is a DB which is true in S.

DB-categorical theories and DCAx. Generally, a DB discriminates a class of models; we say that a theory T is DB-categorical iff every DB for T discriminates a single finite model.
We have that there are theories satisfying DCAx which are not DB-categorical; in these theories we may have the so called [Rei] "null value problem" and/or "disjunctive information problem".

Now we give our notion of *constructive consequence* of an axiomatization T

Def. 4.

T ⊩ H iff for every finite set of constants K and for every T.info∈ info(K, T) there is a h∈ info(K, H) such that for every model S of <T, T.info, K>, S ⊨ h:H

■

The following proposition holds:
Prop. 1.bis.

i) T ⊩ H => T ⊨ H (being ⊨ the usual classical consequence)

■

We remark that T ⊨ H deos not imply, in general, T ⊩ H, i.e., even if T ⊨ H and T contain DCAx, the possibility of devising a h:H such that S ⊨ h:H may heavily depend on the models S of T (which may not be uniquely determined even if a T.info for T is given). This is the **central feature** of the notion of a *constructive consequence* , which can be seen as a *general property* holding in all the databases of T.

In other words, if we know that H is a constructive consequence of T, we know also (by the definition) that, for every DB for T, there is a way to compute a true h:H using only the pieces of information stored in DB; on the contrary, if H is not a constructive consequence of T, then a true h:H may exist, but the pieces of information stored in DB are not sufficient to compute h (i.e. we need some further piece of information on S).

We remark also that here constructiveness is related to the problem of *knowledge, not to the one of recursiveness* (the finite functions are obviously recursive). In this sense we may have a 'Sherlok Holmes situation', where the predicate is_murderer is obviously recursive (given, e.g., 5 suspect people there are 2^5 recursive characteristic functions); but Sherlok Holmes might not have enough *knowledge* to choose the right one.
Let us explain the above insights by an example.

Relation symbols: suspect, murderer ;

Theory T= { ∀x(suspect(x) ∨ ¬ suspect(x)), ∀x *true* (suspect(x) ∨¬ murderer(x)) }

We have that:

T ⊩ ∀x (suspect(x) ∨¬ murderer(x)) but **not** T ⊩ ∀x(murderer(x) ∨¬ murderer(x))

Let us consider, e.g., the database DB1 = <T, < tab1, tt>, {john, mary, ted}>
where tab1 is the function stored in the following table:

x	tab1(x)
john	<1,tt>
mary	<2,tt>
ted	<1,tt>

According to the fact that ∀x (suspect(x) ∨¬ murderer(x)) is a constructive consequence of our T, as one can see, we can extract the following pieces of information from DB1:

suspect(john), ¬ murderer(mary), suspect(ted).

So, we are able to deduce from our knowledge that mary is not the murderer.
But we are not able to answer the question:

murderer(john) ∨ ¬ murderer(john);

to answer this question we need to know pieces of information not contained in tab1. This corresponds to the fact that the above question is not a constructive consequence of our T. This situation is different from the one of *negation as failure* where the answer ¬ murderer(john) would be obtained.

1.3. The logical calculus

Here we give a calculus to prove that a formula is a constructive consequence of a set of axioms. Proofs of our calculus have a **direct interpretation** as programs.
More precisely, we use a variant of Predicate/Transition Petri nets [Bra], in order to represent proofs and their computational interpretation within the *same formal setting*. We call proof nets our nets [MMO3]. We can prove the following:

Soundness theorem. A proof of $T \Vdash A$ is represented by a proof net PN such that, for **every** DB=<T,T.info,K>, one has a computation of PN which extracts from T.info an output $a \in$ info(K,A) such that any model of DB satisfies a:A.

A *derivation* in our net-system is a sequence $N_0, N_1, ..., N_k$ of proof nets, where the proof net N_i ($0 \leq i \leq k$) is built up by (possibly) using proof nets N_j with $j < i$ and by combining proof-figures such as the ones we are going to explain. Not all the possible finite combinations of proof-figures give rise to *sound* proof nets (henceforth simply called proof nets), but only the finite combinations satisfying the reachability property (r) of point 7) below (property (r) can be tested in time which is exponential in the number of proof-figures related to the logical constant $\vee$, but this seems to be reasonable, since a proof net is to be used several times, once it has been built up; alternatively, in order to avoid the reachability test, it is possible to precisely define a set of connection rules, here omitted for sake of conciseness).

In the proof-figures we will use the following formalism:

1) Circles represent *places*, where a place is labelled by a formula and may contain a datum; a datum *must be* an element of info(K,F), where F is the *label* of the considered place.

2) We call *proper parameters* of a net the variables which occur free in at least a label of a place and which are not involved by any $\exists$-elimination figure (see below) of the net; the variables involved by $\exists$-elimination transitions will be called *improper parameters*. Furthemore, improper parameters related to different $\exists$-elimination transitions must be different.

3) A rectangle represents a *transition* and the expression in it represents *'its operation'*. A transition is *enabled* iff all its input places contain a datum and its operation is defined on its input data. Any enabled transition may perform the following *action:* it computes its operation on its input data, assigns the result to its output place and deletes the input data from the input places.

4) A state for a net, called a *marking* M, is a subset of PLACES $\times$ INFO such that, for every $<p,i> \in M$, $i \in$ info(K,label of p); if $<p,i> \in M$ we say that M *assigns* i to p (or, also, that p *contains* i); a computation is a sequence $M_0, M_1, ...,$ where M_{i+1} is obtained from M_i by an action performed by some non deterministically chosen enabled transition. Our nets *do not contain loops*, so that every computation is finite.

5) In our proof nets we have also some mechanisms which are non standard; more precisely:
- the state of a proof net will be a couple $<M,\sigma>$, where M is a marking and σ is a substitution which assigns a constant to every proper parameter and (possibly) to some improper parameter; we call σ the *global assignment*;
- the execution of mkdb transitions (see below) involves executions of other nets.

6) The *input places of a net* are the ones without incoming arcs, while its *output places* are the ones without outcoming arcs; the set of the *assumptions* of a net is the set of the labels of its input places and the set of the *consequences* is the set of the labels of its output places.

7) An *initial state of a net* is a couple $<M,\sigma>$, where M is a marking of its input places and σ is a global assignment assigning all the proper parameters and only them. Our nets must satisfy the following *reachability property*:

(r) for every initial state $<M,\sigma>$, every computation starting with it halts with a final state $<M_t,s_t>$ such that M_t assigns a datum to at least an output place.

8) In the figures, we will use the following conventions:

- X1,X2,... are the input arcs (in the order given in the figures);
- t denotes a term, i.e. a variable or a constant;
- val(t) = t if t is a constant, val(t) = σ(t) if t is a variable assigned by σ, val(t) = undefined otherwise;
- <..,..> constructs a couple;
- P1(<L,R>) = L and P2(<L,R>) = R;
- J1(<1,S>) = S, J1(<2,S>) = undefined and J2(<2,S>) = S, J2(<1,S>) = undefined;
- tt is the operation with constant value tt;
- J(P,<c,S>) = S and one has the *side effect* of adding P/c to the global substitution σ;
- extract X(Y) extracts the piece of information with index Y from the table X; the indexes may be constants (for $\forall$-tables) or elements of an information type (for $\rightarrow$-tables)[1];
- mkdb(H1,..,Hn; A; Nh) constructs a table tab($<a_0,b_0>,...,<a_k,b_k>$), where $\{a_0,...,a_k\}$ = info(K,A) and, for every j ($1 \leq j \leq k$), b_j is a result of a computation of the net Nh, starting from the initial state obtained from the input data X1:H1, .., Xn:Hn, a_j:A and the actual global assignment σ;
- mkdb(H1,..,Hn; P; Nh) constructs a tab($<c_0,b_0>,...,<c_r,b_r>$), where $\{c_0,...,c_r\}$ = K and, for every j ($1 \leq j \leq r$), b_j is a result of a computation of the net Nh starting from the initial state obtained from X1:H1, .., Xn:Hn and from $\sigma \cup \{P/c_j\}$.

We remark that, in general, a mkdb is a nondeterministic operation; one might also compute all the possible tables.

Here we list the proof-figures used in the proof nets; we don't explain here the algorithm to state the reachability property of the above point 7 (it can be tested by a finite set of sample computations).

PROPOSITIONAL FIGURES

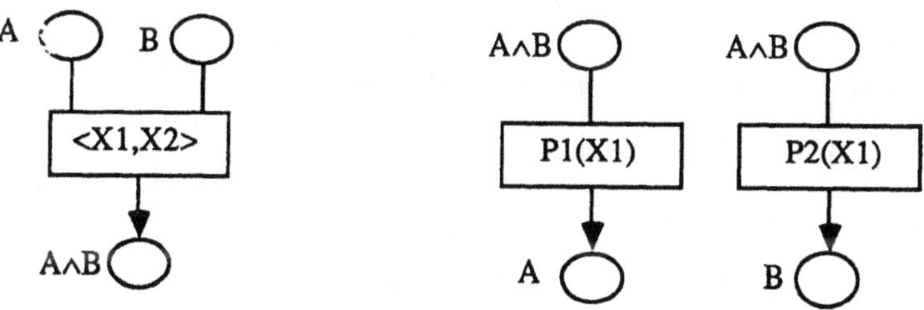

[1] Although from a theoretical point of view we always have tables storing pieces of information, from a practical point of view it is better to consider the possibility of storing programs to extract pieces of information from the basic tables, instead of constructing always new tables.

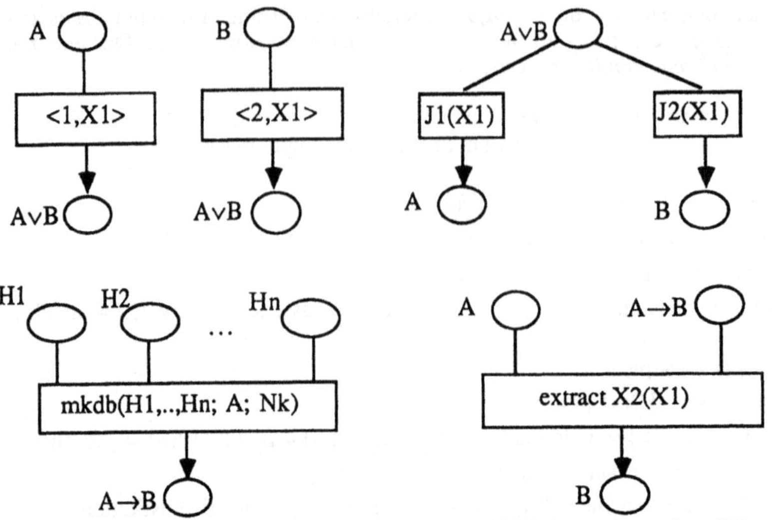

where Nk is a net derived in some previous step and Nk has assumptions H1,...,Hn,A and a unique consequence B

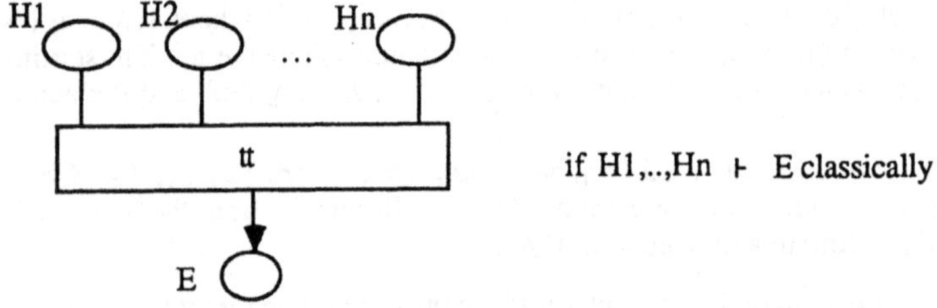

Remark that this rule contains also the intuitionistic rule which says that from absurdum we derive any atomic formula (and, hence, any formula - see [Pra])

PREDICATE FIGURES

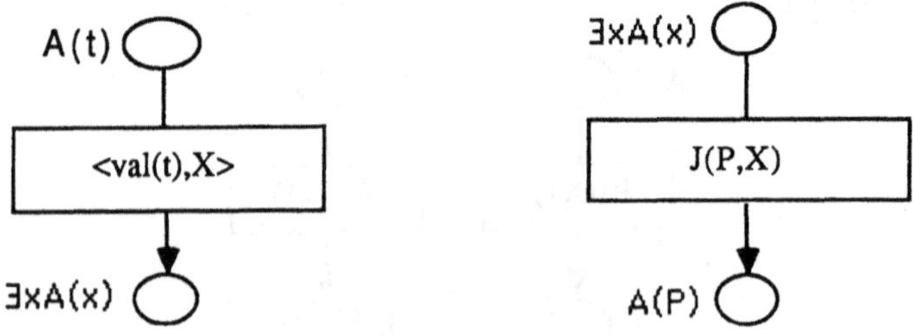

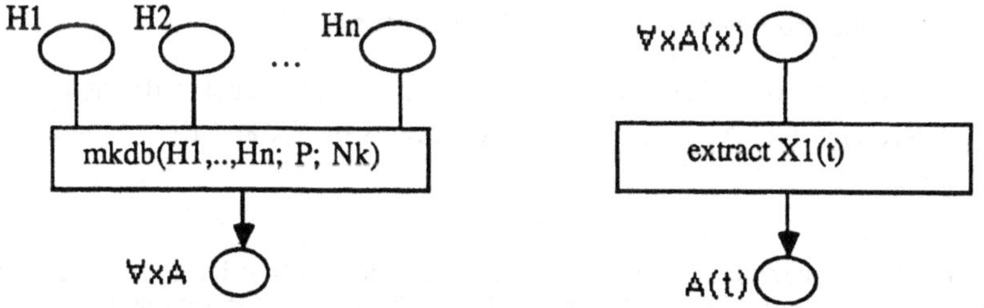

where Nk is a net derived in some previous step and Nk has assumptions H1,...,Hn and a unique consequence A(P), with P not free in the assumptions.

IDENTITY FIGURES:

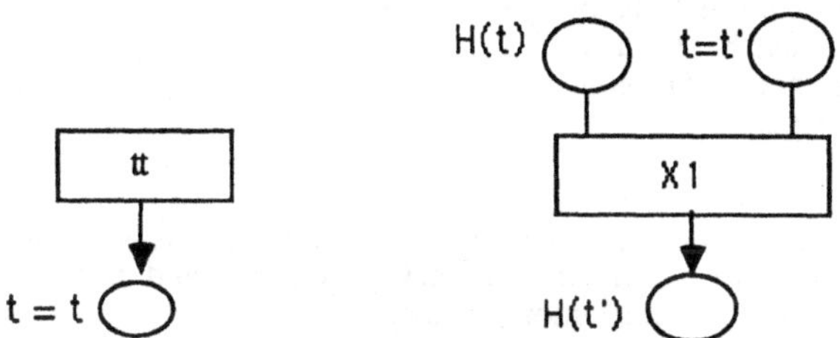

KREISEL AND PUTNAM (KP0, KP1):

KP0) Propositional KP: a figure with one input labelled by $E \rightarrow A \vee B$ (E elementary) and a unique output labelled by $(E{\rightarrow}A){\vee}(E{\rightarrow}B)$; the transition computes the following KP0 operation:
KP0(tab(<tt,<1,a>>)) = <1, tab(<tt,a>)> and KP0(tab(<tt,<2,a>>)) = <2, tab(<tt,a>)>

KP1) Predicative KP: a figure with one input labelled by $E{\rightarrow}\exists xA$ (E elementary) and a unique output labelled by $\exists x(E{\rightarrow}A)$; the transition computes the following KP1 operation:
KP1(tab(<tt,<c,a>>)) = <c, tab(<tt,a>)>

GRZEGORCZYK (G):

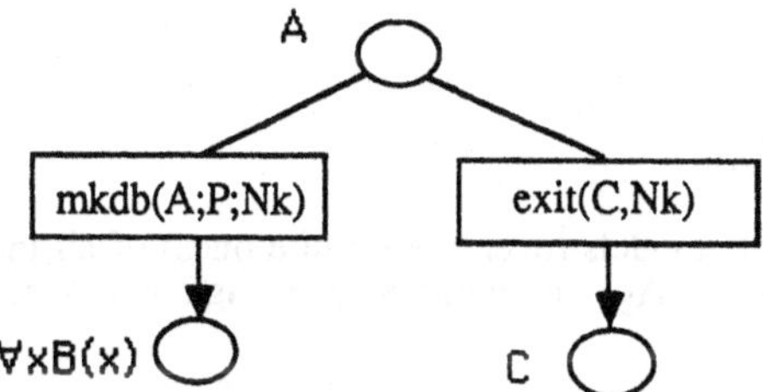

where Nk is a (previous) net with one assumption A and with two consequences B(P) and C and with P not free in A,C.
Explanation of the principle: here mkdb **tries** to construct a table, as in ∀-rule; but, if for some P/c one obtains an exit on C, the construction of the table halts and one obtains a datum for C; on the contrary, if the computation halts with a complete table, the output ∀xB(x) is taken. It is obvious that this procedure is correct whenever the domain closure hypothesis is satisfyed by the current set K of constants.

Def. 5 and Prop.2 characterize the deductive power of our net system.

Def. 5. We say that $A_1,..,A_k \Vdash B_1 \vee .. \vee B_h$ is provable in our net calculus iff there is a derivation $N_1,..,N_m$ such that $A_1,..,A_k$ are the assumptions of N_m and $B_1,..,B_h$ are the consequences of N_m.

■

Prop. 2 . Let INT be any calculus for intuitionistic logic with identity; let us consider the following axioms (K for Kuroda, KP for Kreisel and Putnam, G for Gzregorzyk, see [Tro1]):

at: $\neg\neg at \rightarrow at$ for every atomic at

K: $\forall x \neg\neg A(x) \rightarrow \neg\neg \forall x A(x)$

KP: $(\neg A \rightarrow B \vee C) \rightarrow (\neg A \rightarrow B) \vee (\neg A \rightarrow C)$ and $(\neg A \rightarrow \exists x B(x)) \rightarrow \exists x(\neg A \rightarrow B(x))$

G: $\forall x(A(x) \vee B) \rightarrow (\forall x A(x) \vee B)$

One can prove that the considered calculus is deductively equivalent to:
a) IKA = INT + **at** + **K** if we don't use KP figures;
b) IKA + **KP** if we use KP figures but not G figures;
c) IKA + **KP** + G if we use all the figures.

■

We remark that the rule **at** does not satisfy the uniform substitution principle; this is because the set of the Harrop formulas is not closed under substitution. On the other hand, if we use double negated atomic formulas we have the substitution property.

To state the soundness of our calculus, we need the following definition.

A marking $<M,\sigma>$ is *true* in a structure S iff for every $<p,a> \in M$, if A is the label of p, then $a{:}\sigma' A$ is true in S, for every substitution σ' containing σ.

Then our soundness results can be derived from the following lemma:

Lemma. If we don't use mkdb transitions, every computation is truth preserving.

Using mkdb transitions, one proves that computations are truth preserving by an induction on the mkdb-level (which can be defined in the obvious way).

Starting from the above result, one can prove the following:

Soundness results.

S1) If we don't use G figures, then

$A_1,..,A_k \Vdash B_1 \vee .. \vee B_h \quad \Rightarrow \quad A_1,..,A_k \vDash B_1 \vee .. \vee B_h$

where the validity is related to all the possible models (even the infinite ones) of all the possible databases $<T,T.info,K>$, with $T = \{A_1,..,A_k\}$ and without requiring any domain closure hypothesis.

S2) If we use G figures, soundness is guaranteed considering **only** the **finite models** (of the databases $<T, T.info, K>$) where *each individual is named by at least a constant of* K.

S3) If the considered theory T explicitly contains a domain closure axiom related to a set of constants *fixed by its language* L, then the rule G becomes a *derivable* rule; moreover, we have the following normal form property:

for every formula H there is a disjunctive form $\exists \underline{x}\, (true\, H_1 \vee ... \vee true\, H_n)$ such that:

$T \Vdash H \leftrightarrow \exists \underline{x}\, (true\, H_1 \vee ... \vee true\, H_n)$.

It is to be remarked that, in the proof of S3), the use of a given domain closure axiom $\forall x(x=c1 \vee .. \vee x=cn)$ is essential; this means that we have to know in advance our set of constants.
Now, suppose we have not DCAx because we don't yet know the set of constants. This situation can be related to the one of a DBMS developer; he constructs a DBMS on the basis of a given theory T axiomatizing some applicative domain (e.g., the domain of all the libreries); he has to leave out of consideration the specific databases and the related constants but he can require (e.g. in user's manual) that each database will satisfy DCAx. This corresponds to the correctness of the answers obtained by exaustive searching and this is the meaning of the principle G. See also the discussion on DCAx in the introduction.

2. Some examples

In this section we discuss by examples some of the features of a constructive approach; we give various subsections, related to different aspects.

2.1. Relational theories and reasoning in DB-categorical theories.

Every relational structure M for a language L with constants $K = \{c_0, c_1, \ldots, c_h\}$ and relations $Rel = \{R_0, R_1, \ldots, R_k\}$ is the unique model of a database

$DB_M = \langle Dec(Rel) \cup DCAx(K) \cup UNAx(K), info_M, K \rangle$

where:

$Dec(Rel) = \{ \forall \underline{x}(R_0(\underline{x}) \vee \neg R_0(\underline{x})), \ldots, \forall \underline{x}(R_k(\underline{x}) \vee \neg R_k(\underline{x})) \}$

$DCAx(K) = \forall x \ true(x=c_1 \vee x=c_2 \vee \ldots \vee x=c_h)$

$UNAx(K) = \{\neg c_i=c_k \ / \ i \neq k \}$

In other words, UNAx(K) are the Unique Name Axioms and DCAx(K) is the domain closure axiom. We remark that UNAx(K) and DCAx(K) are elementary formulas; then they don't require any info.

On the contrary, Dec(Rel) requires that, for every relation symbol R_m, $info_M$ contains a table tab_R such that:
for every tuple $\underline{c}$ of constants, $tab_R(c) = \langle 1,tt \rangle$ (which means that "$R(\underline{c})$ holds") or $tab_R(c)=\langle 2,tt \rangle$ (which means that "$\neg R(\underline{c})$ holds").

Now, given a model M, it is easy to see how to construct the tables of $info_M$, in such a way that M is the model of our DB_M.

We have that T = Dec(Rel) ∪ DCAx(K) ∪ UNAx(K) is a DB-categorical theory. But there are many others kinds of DB-categorical theories; in particular, every consistent extension of Dec(Rel) ∪ DCAx(K) ∪ UNAx(K) is DB-categorical. If T is DB-categorical, to know whether or not a query Q can be answered in any DB of T we simply ask whether or not Q is a classical consequence of T; we remark that this holds also if we take K as a parametric set of constants.

2.2. Reasoning in non DB-categorical theories

If a theory T is not DB-categorical, there are classical consequences Q of T which cannot be answered for some (or also for all) databases for T.

But if T $\Vdash$ Q can be proved, by the soundness of our calculus we have that any instance of the query Q can be correctly answered starting **only** from the pieces of information contained in T.info (whichever T.info may be).

Here we give two examples. In the first example we derive the table of a new relation starting from two previously stored relations and from some general knowledge, codified by suitable elementary axioms. In the second one we show how one can use the rule G.

Example 1. Let us consider the following language L:

Relations: ISW (stands for "is working");
HEALTHY
CFB (stands for "coffee break")

Constants: K: parameters

Dec(Rel) : (R1) $\forall x(ISW(x) \vee \neg ISW(x))$

(R2) $\forall x(CFB(x) \vee \neg CFB(x))$

UNAx(K): parametric

General knowledge axioms (which are elementary axioms):

(gk0) $\forall x(ISW(x) \rightarrow HEALTHY(x))$

(gk1) $\forall x(CFB(x) \rightarrow HEALTHY(x))$

(gk2) $\forall x(\neg ISW(x) \wedge \neg CFB(x) \rightarrow \neg HEALTHY(x))$

Now, **without specifying** T.info and K, we can formally prove that the query

(*) $\forall x(HEALTHY(x) \vee \neg HEALTHY(x))$

can be answered; as a matter of fact, we can give the following formal derivation of the "answerability" of (*):

NET1:

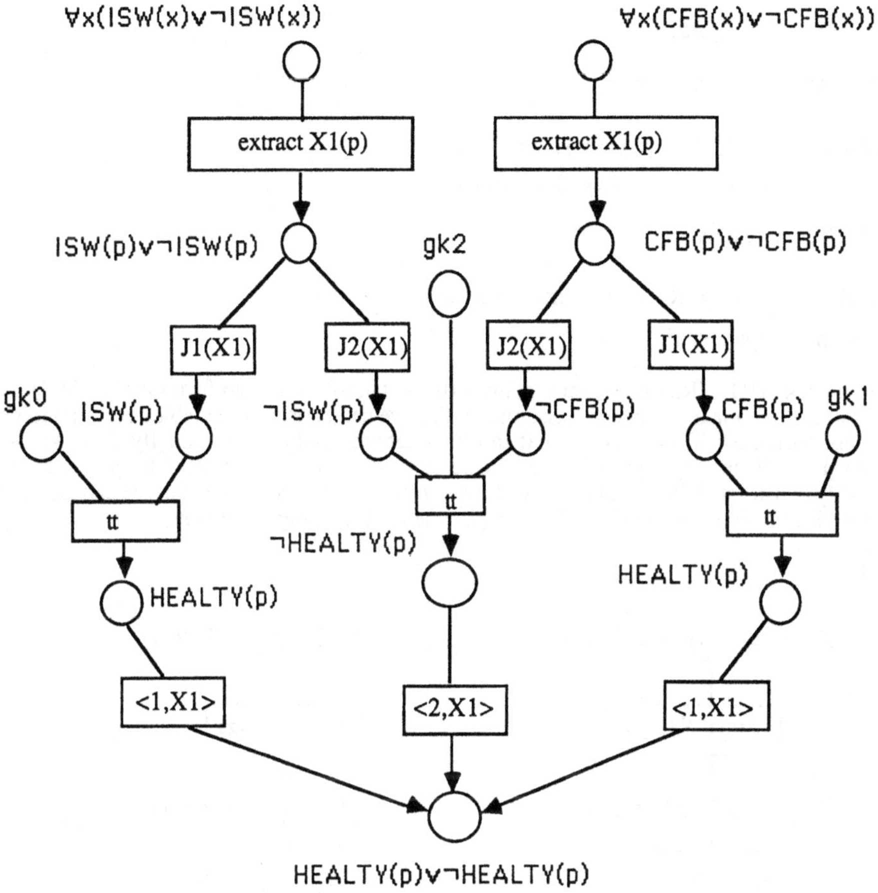

∀x(ISW(x)∨¬ISW(x))
∀x(CFB(x)∨¬CFB(x))
extract X1(p)
extract X1(p)
ISW(p)∨¬ISW(p)
gk2
CFB(p)∨¬CFB(p)
J1(X1)
J2(X1)
J2(X1)
J1(X1)
gk0
ISW(p)
¬ISW(p)
¬CFB(p)
CFB(p)
gk1
tt
tt
tt
¬HEALTY(p)
HEALTY(p)
HEALTY(p)
<1,X1>
<2,X1>
<1,X1>
HEALTY(p)∨¬HEALTY(p)

NET 2

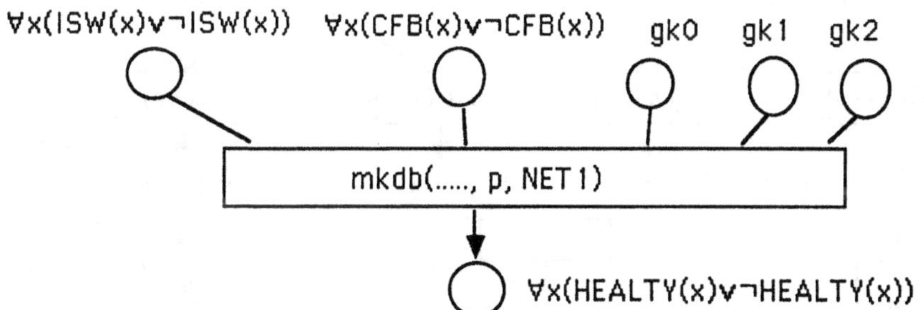

Also, if we drop out [gk_2], then we are no longer able to prove the above (*); but we can give (in a very similar way) a proof of:

(**) $\forall x($ HEALTHY$(x) \vee (\neg$ISW$(x) \wedge \neg$CFB$(x)))$.

Example 2. Let us consider the following language L:

Relations: parametric

Constants: K: parameters

Dec(R,S) : (R1) $\forall x \forall y(R(x,y) \vee \neg R(x,y))$
(R2) $\forall x \forall y(S(x,y) \vee \neg S(x,y))$

DCAx(K): parametric

DefRS: $\forall x \forall y(RS(x,y) \leftrightarrow true(\exists z(R(x,z) \wedge S(z,y))))$

Other axioms: parametric

Here we take (R1), (R2) as axiom schemata (they stand for any two formulas R and S). In other words, we make the generic assumption that we can construct the tables corresponding to these formulas: if we assume that in our databases (which we actually don't know) DCAx(K) will be satisfied, then we can be sure that the table of RS (the join) can be correctly constructed. This can be proved by semantical considerations; but we can also use the rule G (which is valid under DCAx) to provide the following derivation:

net N0

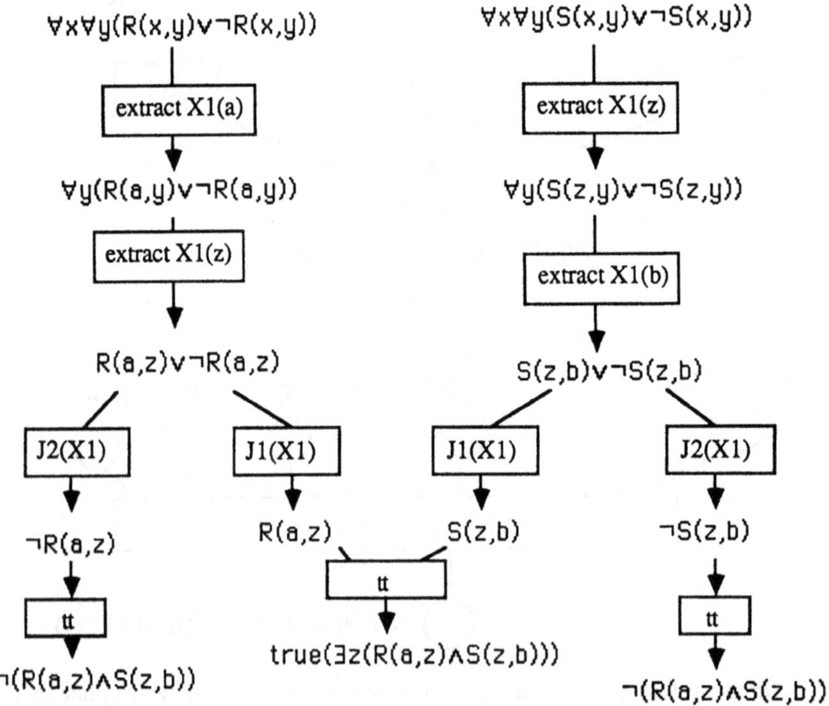

net N1:

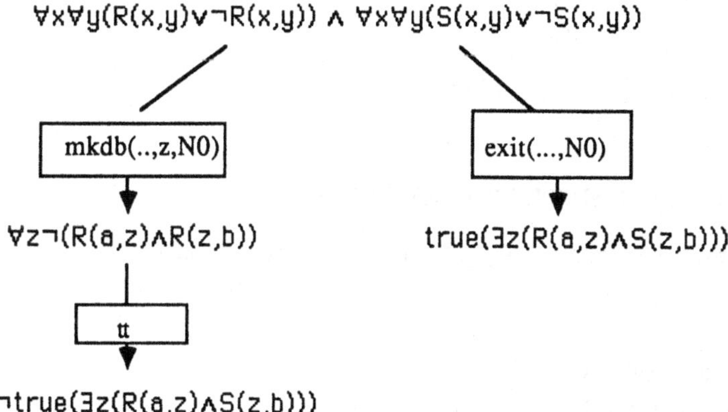

2.3. Reasoning with constraints

We say that a database DB=<T, T.info, K> for a theory T is consistent iff there is at least a model of DB.
We say that a theory T is DB-stable iff every DB for T is consistent.
We say that a theory T is DB-consistent iff there is a consistent DB for T.

An example of DB-stable theory is:

$\forall x(R(x) \vee \neg R(x)), \forall x \ true(H(x) \vee \neg R(x))$

For every DB-stable theory, we can always update our T.info and K without affecting consistency; this is no longer true for DB-consistent but not DB-stable theories. An example of a DB-consistent but not DB-stable theory is:

$\forall x(R(x) \vee \neg R(x)), \forall x \ true(H(x) \vee \neg R(x)), \forall x(H(x) \vee \neg H(x))$

A problem is: can we add to a DB-stable theory a verifiable constraint, i.e., an elementary formula E such that, for every T.info, E is valid with respect to T.info or not (in other words, can we discriminate between the T.info which satisfying E and the others)?

We have that, if we can prove $T \Vdash E \vee \neg E$, then we are in this desired situation; our proof represents also a program P which, for every given T.info, gives us the answer that E holds or $\neg E$ holds. So, if we update T.info, by running P we can verify whether or not the update is consistent with respect to E.

2.4 Elementary axioms for incomplete information and for general laws

In our approach, we use the axioms of T to ***explicitly state*** the pieces of information which are to be stored in a T.info. Here we give an example where we codify a situation of incomplete information on the negative part of a relation.

In this attitude, if we cannot interpret the lack of a tuple $c_1,\ldots,c_n$ in the table of a relation r as the negation of $r(c_1,\ldots,c_n)$, we have to explicitly state this situation in some way.

E.g., we can introduce a new relation symbol r! and an axiom of the following kind:

(r) $\forall x_1,\dots,x_n(r!(x_1,\dots,x_n) \vee\neg\, r!(x_1,\dots,x_n))$

with r! related with r by the elementary axiom:

$\forall x_1,\dots,x_n(r!(x_1,\dots,x_n) \rightarrow r(x_1,\dots,x_n))$.

So, suppose we want to treat the example 4.1 of [Rei] :

Language SUPPL:

- relation symbols: SUPPLIER, SUPPLIES, PART, SUBPART;

- constant symbols: p_1, p_2, p_3, Acme, Foo, ... (possibly, others names).

Stored rel. axioms:

(R1) $\forall x(\mathrm{PART}(x) \vee\neg\, \mathrm{PART}(x))$
(R2) $\forall x(\mathrm{SUPPLIER}(x) \vee\neg\, \mathrm{SUPPLIER}(x))$
(R3) $\forall x,y(\mathrm{SUPPLIES!}(x,y) \vee\neg\, \mathrm{SUPPLIES!}(x,y))$
(R4) $\forall x,y(\mathrm{SUBPART!}(x,y) \vee\neg\, \mathrm{SUBPART!}(x,y))$

Incomplete Information axioms:

(R'3) $\forall x,y(\mathrm{SUPPLIES!}(x,y) \rightarrow \mathrm{SUPPLIES}(x,y))$
(R'4) $\forall x,y(\mathrm{SUBPART!}(x,y) \rightarrow \mathrm{SUBPART}(x,y))$

Unique name axioms.

(R3) and (R4) allow to add any other positive or conditional information on SUPPLIES and on SUBPART *in a consistent way*; e.g., we could add the particular knowledge axiom:

$\mathrm{true}(\mathrm{SUPPLIES}(\mathrm{Foo},p_1) \vee \mathrm{SUPPLIES}(\mathrm{Foo},p_3))$

We remark that this axiom is an elementary one, since the $\vee$ connective is in the scope of the operator "true". It means that we know that Foo supplies p_1 or p_3, but we don't know which.
A similar situation arises in presence of "unknown values" e.g., if we know that there is a supplier of p_1 but we don't know which one it is, we can add the elementary axiom:

$\mathrm{true}(\exists x(\mathrm{SUPPLIER}(x) \wedge \mathrm{SUPPLIES}(x,p_1)))$.

So, in this case, neither "null value omega", nor a three valued logic, nor a K operator is necessary and the distinction between elementary and non elementary axioms is sufficient to represent incomplete and complete information.

Appendix. A comparison with the deductive attitude

In a Prolog-like attitude, we have the following relationship with our formalism:

- Given a goal G, an ***instance*** of G represents the query (in our formalism) $G\vee\neg G$.

E.g., "? healthy(John)" can be seen as the query "healthy(John)$\vee\neg$healthy(John)".

- An open goal G(X) represent the query $\exists XG(X)\vee\neg\exists XG(X)$

E.g., "? healthy(X)." can be seen as the query "$\exists$x healthy(x) $\vee\neg\exists$x healthy(x)".

- If we consider the ***completion*** Comp(P) of a Prolog program P [Lloy] and negation as failure works in a sound way for P, then the answers given by the computations of P are correct. But, if we don't assume Comp(P), then the answers can be wrong or we need to interpret them in some modal way or in some non classical logic. In our approach, we always refer to classical interpretations, but we have a notion of constructive consequence which allows us to discriminate the goals which can be correctly answered from the others even in presence of assumptions weaker than the ones of Comp(P).

- Finally, given a query Q and a theory T, by T $\Vdash$ Q we formally prove that **all** the instance of Q can be correctly answered, *even if we don't know the details* related to the particular T.info or to the particular set K of constants; this means that we are able to study general properties.

Example
To realize the example of sect. 2.2, in Prolog we can write down *database facts* of the kind such as, e.g.:

isw(John).

isw(Mary).

cfb(Ted).

provided that ISW(John), ISW(Mary), CFB(Ted) be in T.info.

(gk_0), (gk_1), (gk_2) can be given by rules such as:

healthy(X) :- isw(X).

healthy(X) :- cfb(X).

As a matter of fact, the completion of the above two clauses is:

$\forall$X (healthy(X) $\leftrightarrow$ isw(X)$\vee$cfb(X))

which is equivalent to (gk_0), (gk_1), (gk_2).

Here the general query: $\forall$x(HEALTHY(x)$\vee\neg$HEALTHY(x))
can be answered since, for every INSTANCE, we have a correct answer "yes" or "no"; e.g.:

? healthy(John) ----> yes

? healthy(Mary) ----> yes

? healthy(Ted). ----> yes

? healthy(Alex). ----> no

By the open goal "? healthy(X)." we obtain

? healthy(X). ----> X=John

which corresponds to an answer to the query "$\exists$x HEALTHY(x)$\vee\neg\exists$x HEALTHY(x)", while, by using "more" we obtain also the answers

X=John, X=Mary, X=Ted

which correspond to the **whole** piece of information required by the query "$\forall$x(HEALTHY(x) $\vee\neg$ HEALTHY(x))" (of course, this piece of information is the table constructed by NET2).

But if (gk_2) cannot be taken as a true general knowledge on our world, then the answer

? healthy(Alex). ----> no

is no longer valid; this corresponds to the fact that we are no longer able to prove $\forall x(HEALTHY(x) \vee \neg HEALTHY(x))$, but only $\forall x(HEALTHY(x) \vee (\neg ISW(x) \wedge \neg CFB(x)))$. So, an automatic theorem prover for constructive logic would be useful to state general and correctness properties for databases.

REFERENCES

[Bra] Brauer W. (ed.) - Net Theory and Applications - LNCS,Springer Verlag,1980.

[GMN] Gallaire H., Minker J., Nicolas J.M. - Logic and Databases: A deductive approach - Computing Surveys 16, 2, 1984.

[HN] Hayashi S., Nakano H. - PX: a computational logic - MIT Press, 1989.

[Kle] Kleene S. - Introduction to metamathematics - Wolters-Noordhoff, 1952.

[Lloy] Lloyd J. W. - Foundations of logic programming - Springer Verlag, 1987.

[LT1] Lloyd J., Topor W. - A Basis for Deductive Database Systems - J. Logic Programming 2, 2, 1985.

[LT2] Lloyd J., Topor W. - A Basis for Deductive Database Systems II - J. Logic Programming 3, 1, 1986.

[Med] Medvedev T. - Finite problems - Sov. Math. Dok. 3, 1962.

[MMO3] Miglioli P., Moscato U., Ornaghi M. - Constructive proofs as programs executable by PrT nets - in Girault C., Reisig W. (ed.) Application and theory of Petri nets, Informatik-Fachberikte, n. 52, Springer Verlag, 1982.

[MMOU] Miglioli P., Moscato U., Ornaghi M., Usberti G. - A costructivism based on classical truth - Notre Dame Journal of Formal Logic, vol. 30, n.1, 1989.

[MMOQU] Miglioli P., Moscato U., Ornaghi M., Quazza S., Usberti G. - Some results on intermediate constructive logics - Notre Dame Journal of Formal Logic, val 30, n.4, 1989.

[Pra] Prawitz D. - Natural deduction: a proof-theoretical study - Almqvist & Wiksell, 1965.

[Rei] Reiter R. - Toward a Logical Reconstruction of Relational Database Theory - in Brodie M., Mylopoulos J., Schmidt J.W. (eds.), On Conceptual Modelling: Perspectives from Artificial Intelligence , Databases and Programming Languages, Springer Verlag, 1984.

[Tro1] Troelstra A.S. - Metamathematical investigations of intuitionistic arithmetic and analysis - LNM, n.344, Springer Verlag, 1973.

[Tro2] Troelstra A.S. - Constructive Mathematics - in Barwise J. (ed.) Handbook of Mathematical Logic, North Holland, 1980.

[Vor] Voronkov A. - Towards the theory of programs in constructive logic - LNCS, n. 432, 1990.

Abstract Syntax and Logic Programming

Dale Miller
Department of Computer and Information Science
University of Pennsylvania
Philadelphia, PA 19104–6389 USA
dale@cis.upenn.edu

Abstract. When writing programs to manipulate structures such as algebraic expressions, logical formulas, proofs, and programs, it is highly desirable to take the linear, human-oriented, concrete syntax of these structures and parse them into a more computation-oriented syntax. For a wide variety of manipulations, concrete syntax contains too much useless information (*e.g.*, keywords and white space) while important information is not explicitly represented (*e.g.*, function-argument relations and the scope of operators). In parse trees, much of the semantically useless information is removed while other relationships, such as between function and argument, are made more explicit. Unfortunately, parse trees do not adequately address important notions of object-level syntax, such as bound and free object-variables, scopes, alphabetic changes of bound variables, and object-level substitution. I will argue here that the *abstract syntax* of such objects should be organized around α-equivalence classes of λ-terms instead of parse trees. Incorporating this notion of abstract syntax into programming languages is an interesting challenge. This paper briefly describes a logic programming language that directly supports this notion of syntax. An example specifications in this programming language is presented to illustrate its approach to handling object-level syntax. A model theoretic semantics for this logic programming language is also presented.

1. Introduction

Consider writing programs in which the data objects to be computed are syntactic structures such as programs, formulas, types, and proofs. A common characteristic of all these structures is that they involve notions of abstractions, scope, bound and free variables, substitution instances, and equality up to alphabetic changes of bound variables. Although the data types available in most computer programming languages are, of course, rich enough to represent all these kinds of structures, such data types do not have direct support for these common characteristics. Instead, "packages" need to be implemented to support such data structures. For example, although it is trivial to represent first-order formulas in Lisp, it is a more complex matter to write Lisp programs that correctly substitute a term into a formula (being careful not to capture bound variables), to test for the equality of formulas up to alphabetic variation, and to determine if a certain variable's occurrence is free or bound. This situation is the same

Implementation	Strings, text (arrays or lists of characters)
Access	Parsers, editors
Good points	1. Readable, publishable. 2. Simple computational models for implementation (arrays, iteration).
Bad points	1. Contains too much information not important for many manipulations: white space, infix/prefix notation, keywords. 2. Important information is not represented explicitly: recursive structure, function–argument relationship, term–subterm relationship.

Figure 1: Characteristics of concrete syntax

when structures like programs or (natural deduction) proofs are to be manipulated and if other programming languages, such as Pascal, Prolog, and ML, replace Lisp.

Before proposing an approach to dealing with representing such syntactic structures in a logic programming language, let us consider current practice in representing syntax in computer programs. Generally, syntax is divided into *concrete* and *abstract* syntax. The first is the linear form of syntax that is readable and typable by a human. Figure 1 characterizes some properties of concrete syntax. The bad points can be overcome by parsing concrete syntax into *parse trees.* Figure 2 characterizes some properties of parse trees. The bad points concerning concrete syntax are now properly addressed, although at significant costs. For example, higher levels of support are required for the programming language and runtime system that encode parse trees. Parse trees, however, are so much more convenient and natural to compute with than strings that these additional costs are outweighed by the advantages to the programmer who must write programs to manipulate syntax. The term *abstract syntax* is often identified with parse trees: we shall reserve the former term for the more "abstract" form of syntax described in the next section.

Parse trees are not without their bad points also. In particular, notions of abstraction within syntax are not supported directly. For example, we have the following unfortunate properties of parse trees for representing syntax containing bound variables.

- Bound variables are, like constants, treated as global objects.
- Concepts such as free and bound occurrences of variables are derivative notions, supported not by programming languages but by programs added on top of the data type for parse trees.
- Although alphabetic variants generally denote the same intended object, the correct choice of such variants is unfortunately very often important.
- Substitution is generally difficult to implement correctly.
- An implementation of substitution for one data structure, say first-order formulas, will not work for another, say functional programs.

There are various computer systems that use a different approach to syntax. They all make use of λ-terms modulo the equations of α, β, and η-conversions and implement

Implementation	first-order terms, linked lists
Access	car/cdr/cons in Lisp, first-order unification in Prolog, or matching in ML.
Good points	1. Recursive structure is immediate. 2. Recursion over syntax is easy to specify. 3. Term–subterm relationship is identified with tree-subtree relationship. 4. Algebra provides a model for many operations on syntax.
Bad points	1. Requires higher-level language support: pointers, linked lists, garbage collection, structure sharing. 2. Notions of scope, abstraction, substitution, and free and bound variables occurrences are not supported.

Figure 2: Characteristics of parse trees

various aspects of $\beta\eta$-unification (often called "higher-order" unification). One of the earliest was designed by Huet and Lang [13]: here, only second-order matching was used to decompose syntax. The generic theorem prover Isabelle uses a fragment of intuitionistic logic with quantification at higher-order types. The Isabelle implementation includes $\beta\eta$-unification at all finite types. The language λProlog [21] is an extension of Prolog that includes, among other things, $\beta\eta$-unification at all finite types. The Elf programming language [23] is a logic programming language implementation of the LF specification language [12] in a style similar to λProlog.

This short paper is organized as follows. In the next section, we shall motivate a notion of abstract syntax that is more "high-level" than parse trees. Section 3 presents the logic programming language $\mathcal{M}$ that incorporates such abstract syntax. In Section 4 an example specification in $\mathcal{M}$ is presented. Finally, a model theory for $\mathcal{M}$ is given in Section 5.

2. Motivating abstract syntax

Consider the recursive structure of first-order terms over the following signature.

$$\Sigma = \{a : i, \quad b : i, \quad f : i \rightarrow i, \quad g : i \rightarrow i \rightarrow i\}$$

Here, i is a primitive type (or sort). These four typed constants can be encoded as the following four inference rules for determining which first-order terms over Σ are correctly constructed.

$$\frac{\Sigma \vdash X : i}{\Sigma \vdash f\, X : i} \qquad \frac{\Sigma \vdash X : i \qquad \Sigma \vdash Y : i}{\Sigma \vdash g\, X\, Y : i}$$

$$\overline{\Sigma \vdash a : i} \qquad \overline{\Sigma \vdash b : i}$$

The following is a proof that the term $g\ (f\ a)\ b$ is a correctly formed first-order term (of type i).

$$\dfrac{\dfrac{\dfrac{}{\Sigma \vdash a : i}}{\Sigma \vdash f\ a : i} \qquad \dfrac{}{\Sigma \vdash b : i}}{\Sigma \vdash g\ (f\ a)\ b : i}$$

Notice that the signature Σ does not change in a proof: it is global and does not need to be written as part of each inference rule.

To consider the structure of λ-terms, let $\Sigma' = \Sigma \cup \{h : (i \to i) \to i\}$ be a signature with the constant h of second-order type. In order to incorporate this new constant into an inference rule, we actually need two rules: one to infer a term with h as its head and one to infer a term of an arrow type (here, $i \to i$). If Γ is a signature that contains Σ', then the two new inference rules are simply

$$\dfrac{\Gamma \vdash U : i \to i}{\Gamma \vdash h\,U : i} \qquad \dfrac{\Gamma,\ x : i \vdash V : i}{\Gamma \vdash \lambda x.V : i \to i}.$$

Here, x is not in Γ. The following is a proof that the term $f\ (h\ (\lambda x(g\ x\ (f\ x))))$ is correctly formed.

$$\dfrac{\dfrac{\dfrac{\dfrac{\dfrac{}{\Sigma',\ x : i \vdash x : i} \qquad \dfrac{\dfrac{}{\Sigma',\ x : i \vdash x : i}}{\Sigma',\ x : i \vdash f\ x : i}}{\Sigma',\ x : i \vdash g\ x\ (f\ x) : i}}{\Sigma' \vdash \lambda x(g\ x\ (f\ x)) : i \to i}}{\Sigma' \vdash h\ (\lambda x(g\ x\ (f\ x))) : i}}{\Sigma' \vdash f\ (h\ (\lambda x(g\ x\ (f\ x)))) : i}$$

(Also replace Σ by Γ in the four rules for a, b, f, and g.) Notice that now, the signatures do not remain constant within proofs: as one moves up through such a proof, signatures can get larger. This suggests that a good notion of bound variable is essentially "scoped constant": it acts like a constant that is not visible from the top of the term, but may become visible when a descent is made through the abstraction. Thus, we state the first of two principles that are needed to support our notion of abstract syntax.

Principle 1. Recursion through syntax containing bound variables requires signatures (contexts) to be dynamically augmented.

The second principle supporting our notion of abstract syntax is rather obvious but produces serious problems for integrating into a programming language.

Principle 2. The equality of syntax should be (at least) α-conversion.

If the equations of α-conversion are assumed then terms are not freely generated and simple destructuring is not a sensible operation. For example, the two terms $\lambda x(fxx)$ and $\lambda y(fyy)$ denote the same syntactic object. If, however, λ-abstraction is treated as a two place constructor, then these equal terms can be decomposed into unequal parts: that is, into x and y and into fxx and fyy.

An approach to solving this problem is to try to decompose syntax using unification modulo α-conversion. For example, consider the following signature over two primitive types i (representing object-level terms) and b (representing object-level formulas):

$$\begin{array}{lll} \forall : (i \to b) \to b & \wedge : b \to b \to b & \supset : b \to b \to b \\ r : i \to b & s : i \to b & t : b. \end{array}$$

Consider attempting to decompose formula

$$\forall \lambda y((ry \supset sy) \wedge t)$$

by unifying it with the formula $\forall \lambda x(P \wedge Q)$, where P and Q are free variables. This pair has no unifiers (modulo α-conversion) since no substitution instance for P will be able to bind the variable x: we are assuming that substitution at the meta-level is the correct declarative substitution that avoids bound variable capture. This example illustrates that unification using purely α-conversion is not able to cope with decomposing syntax involving a bound variable. If we change this example by attempting to match the same formula with $\forall \lambda x(Px \wedge Q)$ we now find that there is exactly one unifier (up to α-conversion), namely,

$$\{P \mapsto \lambda w(rw \supset sw), Q \mapsto t\}.$$

This substitution is a unifier, however, when α *and* β-conversions are assumed since after substituting for P and Q, the resulting term $\forall \lambda x([\lambda w(rw \supset sw)x] \wedge t)$ requires a β-reduction and an α-conversion before it is equal to $\forall \lambda y((ry \supset sy) \wedge t)$.

For some additional matching examples of this kind, consider matching the following pair of open terms (free variables are capital letters) with closed λ-term over the signature Σ.

$$\begin{array}{ll} (1)\ \lambda x \lambda y(f\ (H\ x)) & \lambda u \lambda v(f\ (f\ u)) \\ (2)\ \lambda x \lambda y(f\ (H\ x)) & \lambda u \lambda v(f\ (f\ v)) \\ (3)\ \lambda x \lambda y(g\ (H\ y\ x)\ (f\ (L\ x))) & \lambda u \lambda v(g\ u\ (f\ u)) \\ (4)\ \lambda x \lambda y(g\ (H\ x)\ (L\ x)) & \lambda u \lambda v(g\ (g\ a\ u)\ (g\ u\ u)) \end{array}$$

The second pair cannot be matched for reasons similar to those described above. The other three cases yield unique matches, assuming α and η-conversion.

$$\begin{array}{lll} (1) & H \mapsto \lambda w(f\ w) & \\ (3) & H \mapsto \lambda y \lambda x.x & L \mapsto \lambda x.x \\ (4) & H \mapsto \lambda x(g\ a\ x) & L \mapsto \lambda x(g\ x\ x) \end{array}$$

All of these examples use a very weak form of β-conversion. In particular, they continue to work if β-conversion is replaced by β_0 conversion, which is defined by the equation $(\lambda x.B)x = B$.

In the next section we present a meta-logic $\mathcal{M}$ that supports both of the principles of abstract syntax that we have described above. The language has as its equality theory α, β, and η-conversion for the simply typed λ-terms. It is possible to significantly weaken the logic $\mathcal{M}$ to a logic called L_λ where the equality theory only needs to be a restricted form of α, β_0, and η-conversion. This equality theory is weak enough so that unification in it is decidable and most general unifiers exist when unifiers exist. It is also strong enough to support the two principles of abstract syntax presented above. The logic L_λ is describe in the papers [15, 16]. We shall not be concerned with it further here.

Abstract syntax is characterized in Figure 3: in the rest of this paper, we shall discuss the logic $\mathcal{M}$ and how it supports this notion of syntax. The paper [14] describes an approach to incorporating abstract syntax into the ML programming language [19]. What we are calling abstract syntax in this paper has also been called "higher-order abstract syntax" in [24].

Implementation	α-equivalence classes of $\beta\eta$-normal λ-terms of simple types
Access	$\beta\eta$-unification or a restriction of $\beta_0\eta$-unification (as in L_λ)
Good points	1. Bound variable names are inaccessible so many technical problems regarding them disappear. 2. Substitution is easy to support for every data structure containing abstracted variables. 3. Semantics is provided by proof theory, logical relations, and Kripke models.
Bad points	1. Requires higher-level support: dynamic contexts, extensions to first-order unification, and a richer notion of equality. 2. No robust, well-defined, and generally available programming language supports this notion of syntax (yet).

Figure 3: Characteristics of abstract syntax.

3. A Logic programming language that incorporates abstract syntax

Let S be a set of primitive types. Type expressions are all first-order expressions built from primitive types and the infix, function type constructor $\rightarrow$. This constructor associates to the right: read $\tau_1 \rightarrow \tau_2 \rightarrow \tau_3$ as $\tau_1 \rightarrow (\tau_2 \rightarrow \tau_3)$. Let $\mathcal{S}$ be a finite set of predicate symbols that are sorted using expressions of the form $\langle\tau_1, \ldots, \tau_n\rangle$ for $n \geq 0$, where $\tau_1, \ldots, \tau_n$ are types. Using a primitive type for propositions, say o as in [2], then the sort for predicates could be considered as a type of the form $\tau_1 \rightarrow \cdots \rightarrow \tau_n \rightarrow o$. We shall not, however, give predicates functional types: the expression $\langle\tau_1, \ldots, \tau_n\rangle$ is not a type expression.

Signatures are sets of associations of types to tokens such as

$$\Sigma = \{c_1 : \tau_1, \ldots, c_n : \tau_n, \ldots\}.$$

Signatures can be finite or infinite and are sometimes called *type assignments*. The usual functional requirement holds: if $c : \tau$ and $c : \sigma$ are members of Σ, then τ and σ

are the same type. The expression $\Sigma + c : \tau$ is legal only if c is not assigned by Σ, in which case that expression is equal to

$$\{c : \tau, c_1 : \tau_1, \ldots, c_n : \tau_n, \ldots\}.$$

A *Σ-term of type τ* is a closed λ-term all of whose constants are in Σ and which has type τ. Notice that a given λ-term may be a Σ-term at different types; for example, consider the term $\lambda x.x$. *Σ-formulas* are defined in the following fashion.

- If Q is a predicate in $\mathcal{S}$ that is sorted with $\langle \tau_1, \ldots, \tau_n \rangle$ and t_i is a Σ-term of type τ_i (for $i = 1, \ldots, n$), then $Qt_1 \cdots t_n$ is a Σ-formula. In particular, it is an *atomic* Σ-formula.
- If B and C are Σ-formulas then $B \wedge C$ and $B \supset C$ are Σ-formulas.
- If B is a $\Sigma + x : \tau$-formula then $\forall_\tau x.B$ is a Σ-formula.

Equality of terms and formulas is determined using the usual rules of $\beta\eta$-conversion.

The collection of Σ-formulas over the primitive types in S and the predicates in $\mathcal{S}$ is denoted by $\mathcal{M}(S, \mathcal{S})$, which is written as simply $\mathcal{M}$ if S and $\mathcal{S}$ can be determined from context. A proof system for $\mathcal{M}(S, \mathcal{S})$ is given by the sequent rules in Figure 4. The triple $\Sigma\,;\, \mathcal{P} \longrightarrow B$ is a sequent if $\mathcal{P} \cup \{B\}$ is a set of Σ-formulas. We shall assume that the rules of $\beta\eta$-conversion are used whenever needed to join two inference rules together. The syntax $\mathcal{P}, B$ is short for $\mathcal{P} \cup \{B\}$. The expression $\Sigma; \mathcal{P} \vdash B$ means that the sequent $\Sigma\,;\, \mathcal{P} \longrightarrow B$ is provable (without cut). If there is a Σ-term for all primitive types, then this proof system coincides with the more common notion of intuitionistic sequent calculus. Since $\mathcal{P}$ in the sequent $\Sigma\,;\, \mathcal{P} \longrightarrow B$ is a set, the usual structural rules of thining, contraction, and exchange are not needed.

$$\frac{\Sigma\,;\, \mathcal{P} \longrightarrow B \qquad \Sigma\,;\, \mathcal{P} \longrightarrow C}{\Sigma\,;\, \mathcal{P} \longrightarrow B \wedge C}\ \wedge\text{-R}$$

$$\frac{\Sigma\,;\, \mathcal{P} \longrightarrow B \qquad \Sigma\,;\, C, \mathcal{P} \longrightarrow A}{\Sigma\,;\, B \supset C, \mathcal{P} \longrightarrow A}\ \supset\text{-L}$$

$$\frac{\Sigma\,;\, B, C, \Delta \longrightarrow A}{\Sigma\,;\, B \wedge C, \Delta \longrightarrow A}\ \wedge\text{-L} \qquad \frac{\Sigma\,;\, B, \mathcal{P} \longrightarrow C}{\Sigma\,;\, \mathcal{P} \longrightarrow B \supset C}\ \supset\text{-R}$$

$$\frac{t \text{ is a } \Sigma\text{-term of type } \tau \qquad \Sigma\,;\, \mathcal{P}, B[t/x] \longrightarrow C}{\Sigma\,;\, \mathcal{P}, \forall_\tau x\, B \longrightarrow C}\ \forall\text{-L}$$

$$\frac{\Sigma + c{:}\tau\,;\, \mathcal{P} \longrightarrow B[c/x]}{\Sigma\,;\, \mathcal{P} \longrightarrow \forall_\tau x\, B}\ \forall\text{-R} \qquad \frac{}{\Sigma\,;\, \mathcal{P}, B \longrightarrow B}\ \text{initial}$$

Figure 4: Proof rules for $\mathcal{M}$.

There are two forms of cut rules for this sequent calculus: one works with the signature of the antecedent (called the *subst* rule) and one works with the formulas of the antecedent (called simply the *cut* rule). Both rules are displayed in Figure 5. The following theorem is know as the *cut elimination* theorem.

Theorem 3.1. *A sequent is provable with the two rules of cut and subst (Figure 5) if and only if it is provable without these two inference rules.*

The proof of this fact follows from Gentzen's original result augmented with elementary facts about the meta-theory of the $\beta\eta$-theory of simply typed λ-terms. Notice that since $\mathcal{M}$ does not admit predicate quantification, the cut-elimination result follows the usual line for first-order logics. Sequent proofs in this paper will not contain instances of cut or subst.

$$\frac{\Sigma\,;\,\mathcal{P}, B \longrightarrow C \qquad \Sigma'\,;\,\mathcal{P}' \longrightarrow B}{\Sigma'\,;\,\mathcal{P}' \longrightarrow C}\ \text{cut}$$

$$\frac{\Sigma + x:\tau\,;\,\mathcal{P} \longrightarrow B \qquad t \text{ is a } \Sigma'\text{-term of type } \tau}{\Sigma'\,;\,\mathcal{P}' \longrightarrow B[t/x]}\ \text{subst}$$

Figure 5: Cut and subst rules for $\mathcal{M}$. Here, $\Sigma \subseteq \Sigma'$ and $\mathcal{P} \subseteq \mathcal{P}'$.

The following theorem provides an abstract justification for referring to $\mathcal{M}$ as a logic programming language. This theorem says that *goal-directed* search for proofs in $\mathcal{M}$ is a complete search method.

Theorem 3.2. *A sequent proof is uniform if every occurrence of a sequent in that proof with a non-atomic right-hand side is the conclusion of a right-introduction rule. Then, a sequent is provable in $\mathcal{M}$ if and only if it has a uniform proof.*

This is easily proved by using permutations of inference rules to convert any cut-free proof into a uniform proof. The proof of this result can be found in [18] where a richer logic than $\mathcal{M}$ is considered. A similar proof is given in [3] for a strictly first-order logic (quantification at primitive types only). Both of these papers also motivate why uniform proofs and goal-directed search are useful for characterizing logic programs.

Many examples of using $\mathcal{M}$ as a specification language and using λProlog to implement them have been considered. For example, in the area of theorem proving see the papers [3, 4, 5, 6, 22]; in the area of meta-programming of functional programs see the papers [7, 8, 9, 10, 11, 17]. The logic $\mathcal{M}$ is very similar to the logic hh^ω in [6]. The next section presents one of the example specifications described in [8].

4. Functional program as objects

Let $S_0 = \{i\}$ and let $S = \{eval : \langle i, i\rangle\}$. Let Σ_0 be the signature containing the following constants:

$$true : i,\ false : i,\ 0 : i,\ 1 : i,\ 2 : i, \ldots$$
$$=: i \to i \to i,\ + : i \to i \to i,\ if : i \to i \to i \to i$$
$$app : i \to i \to i,\ abs : (i \to i) \to i,\ fix : (i \to i) \to i$$

The type i denotes object-level functional programs. Obviously, it is possible to add more constants to this signature so that Σ_0-terms of type i denote richer functional programs. The functional program

$$fun\ g\ x\ y = if\ x = y\ then\ x\ else\ g\ y\ y$$

can be represented by the Σ_0-term of type i:

$$(fix\ \lambda g(abs\ \lambda x(abs\ \lambda y(if\ (=\ x\ y)\ x\ (app\ (app\ g\ y)\ y))))).$$

Notice that abstractions in the object-level, functional program are mapped to abstractions in the meta-level term. Using this kind of encoding of functional programs, it is impossible to specify predicates in $\mathcal{M}$ that can make distinctions between two α-convertible functional programs. Thus, this encoding obeys Principle 2.

An evaluator for this object language can be described in $\mathcal{M}(S_0, S_0)$ using some simple formulas: for several examples of such evaluators see [9, 11]. Here, we shall reduce this functional programming language down to it smallest, interesting core. Let Σ_1 be just the signature for application and abstraction, namely, $\{app : i \to i \to i,\ abs : (i \to i) \to i\}$. This kind of representation is derived from the mapping of the untyped λ-terms into simply typed λ-terms. In particular, the pure, untyped λ-terms modulo α-conversion can be identified with Σ_1-terms of type i modulo $\beta\eta$-conversion.

A specification of a call-by-name evaluator for Σ_1-terms in $\mathcal{M}$ is given by the following two formulas.

$\forall_{i\to i}R\ (eval\ (abs\ R)\ (abs\ R))$
$\forall_{i\to i}R\forall_i M, N, V\ (eval\ M\ (abs\ R) \wedge eval\ (R\ N)\ V \supset eval\ (app\ M\ N)\ V)$

Notice there that meta-level β-reduction, in the expression $(R\ N)$, is used to do object-level substitution. Call-by-value evaluation can also be axiomized using the following two formulas.

$\forall_{i\to i}R\ (eval\ (abs\ R)\ (abs\ R))$
$\forall_{i\to i}R\forall_i M, N, V, P\ (eval\ M\ (abs\ R) \wedge eval\ N\ P \wedge eval\ (R\ P)\ V \supset eval\ (app\ M\ N)\ V)$

Let $\mathcal{P}_1$ be the set of two formulas specifying call-by-name evaluation. If $\Sigma_1; \mathcal{P}_1 \vdash eval\ t\ s$ then we say that t evaluates to s.

It is natural to try and extend evaluation so that it can evaluate under abstractions. That is, evaluation could be extended (over the Σ_0 signature) to relate the term

$$(fix\ \lambda f\ (abs\ \lambda x\ (if\ true\ (+\ x\ 1)\ (app\ (app\ f\ x)\ x))))$$

and the term

$$(fix\ \lambda f\ (abs\ \lambda x\ (+\ x\ 1))).$$

That is, evaluation can be pushed through abstractions to reduce redexes that are not at the top-level. Over the signature Σ_1 and formulas $\mathcal{P}_1$ the evaluation predicate only relates the term $(abs\ \lambda x(app\ (abs\ (\lambda y\ y))\ x))$ to itself. It should be possible to "lift" evaluation so that the internal redex $(app\ (abs\ (\lambda y\ y))\ x)$ can be reduced. This is problematic since this internal redex is not a Σ_1-term since it has x free in it. Thus, we need to understand how to evaluate expressions over "mixed" values. This problem is solved by dynamically adding x to the signature. Let $\mathcal{C}_1$ be the set of the following two formulas.

$\forall_{i\to i}R, S\ (\forall_i x, y\ (eval\ x\ y \supset eval\ (Rx)\ (Sy)) \supset eval\ (abs\ R)\ (abs\ S))$
$\forall_i M, N, P, Q\ (eval\ M\ P \wedge eval\ N\ Q \supset eval\ (app\ M\ N)\ (app\ P\ Q))$

The first of these formulas lifts evaluation over object-level abstractions. It can be read operationally as follows: To prove the atomic formula

$$eval\ (abs\ \lambda u.t)\ (abs\ \lambda v.s),$$

try the following steps:

- Introduce two new constants, say $c : i$ and $d : i$, not mentioned in the current signature (corresponds to using $\forall$-R).
- Add the atomic formula $eval\ c\ d$ to the current program (corresponds to using $\supset$-R).
- Attempt to prove $eval\ (t[c/u])\ (s[d/v])$ in the augmented signature and program (corresponds to using β-reduction). Here, c plays the role of the bound variable name when we descend into $\lambda u.t$.

This is an illustration of how Principle 1 is supported in $\mathcal{M}$.

Notice that given $\mathcal{P}_1 \cup \mathcal{C}_1$, the proof rules for *eval* are now more nondeterministic. For every Σ_1-term t, $\Sigma_1; \mathcal{P}_1 \cup \mathcal{C}_1 \vdash\ eval\ t\ t$; that is, the extension of *eval* is reflexive. Given the same context, the atomic formula

$$eval\ (abs\ \lambda x(app\ (abs\ (\lambda y\ y))\ x))\ (abs\ \lambda x\ x)$$

is also provable. See [8] for a discussion about how such syntactic lifting of evaluation is related to the notion of "mixed" or "symbolic" evaluation.

5. A Kripke model semantics

A model theory for $\mathcal{M}(\mathcal{S}, \mathcal{S})$ can be based on the following kind of Kripke models.

Definition 5.1. A *dependent pair* is a pair $\langle\Sigma, \mathcal{P}\rangle$ where Σ is a signature and $\mathcal{P}$ is a set of Σ-formulas. Define $\langle\Sigma, \mathcal{P}\rangle \preceq \langle\Sigma', \mathcal{P}'\rangle$ whenever $\Sigma \subseteq \Sigma'$ and $\mathcal{P} \subseteq \mathcal{P}'$. A *Kripke model*, $[\mathcal{W}, I]$, is the specification of a set of *worlds* $\mathcal{W}$, which is a set of dependent pairs, and a function I, called an *interpretation*, that maps pairs in $\mathcal{W}$ to sets of atomic formulas. The mapping I must satisfy the two conditions:

- $I(\langle\Sigma, \mathcal{P}\rangle)$ is a set of λ-normal, atomic Σ-formulas, and
- for all $w, w' \in \mathcal{W}$ such that $w \preceq w'$, $I(w) \subseteq I(w')$ (*i.e.*, I is order preserving). ∎

Satisfiability (also called *forcing*) in a Kripke model is defined as follows.

Definition 5.2. Let $[\mathcal{W}, I]$ be a Kripke model, let $\langle\Sigma,\mathcal{P}\rangle \in \mathcal{W}$, and let B be a Σ-formula. The three place *satisfaction* relation $I, \langle\Sigma,\mathcal{P}\rangle \Vdash B$ is defined by induction on the structure of B.

- $I, \langle\Sigma,\mathcal{P}\rangle \Vdash B$ if B is atomic and the λ-normal form of B is in $I(\langle\Sigma,\mathcal{P}\rangle)$.
- $I, \langle\Sigma,\mathcal{P}\rangle \Vdash B \wedge B'$ if $I, \langle\Sigma,\mathcal{P}\rangle \Vdash B$ and $I, \langle\Sigma,\mathcal{P}\rangle \Vdash B'$.
- $I, \langle\Sigma,\mathcal{P}\rangle \Vdash B \supset B'$ if for every $\langle\Sigma',\mathcal{P}'\rangle \in \mathcal{W}$ such that $\langle\Sigma,\mathcal{P}\rangle \preceq \langle\Sigma',\mathcal{P}'\rangle$ and $I, \langle\Sigma',\mathcal{P}'\rangle \Vdash B$ then $I, \langle\Sigma',\mathcal{P}'\rangle \Vdash B'$.
- $I, \langle\Sigma,\mathcal{P}\rangle \Vdash \forall_\tau x.B$ if for every $\langle\Sigma',\mathcal{P}'\rangle \in \mathcal{W}$ such that $\langle\Sigma,\mathcal{P}\rangle \preceq \langle\Sigma',\mathcal{P}'\rangle$ and for every Σ'-terms t of type τ, the relation $I, \langle\Sigma',\mathcal{P}'\rangle \Vdash B[t/x]$ holds.

The *signature of an interpretation* I is the largest signature that is contained in all worlds of the partial order underlying I. If Σ_0 is the signature of the interpretation I and B is a Σ_0-formula, then we write $I \Vdash B$ if $I, w \Vdash B$ for all $w \in \mathcal{W}$. ∎

This notion of model is similar to that of *Kripke λ-models* described in [20].

Definition 5.3. Let $\langle\Sigma,\mathcal{P}\rangle$ be a dependent pair. The *canonical model* for $\langle\Sigma,\mathcal{P}\rangle$ is defined as the model with the set of worlds $\{\langle\Sigma',\mathcal{P}'\rangle \mid \langle\Sigma,\mathcal{P}\rangle \preceq \langle\Sigma',\mathcal{P}'\rangle\}$ and where I is defined so that $I(\langle\Sigma',\mathcal{P}'\rangle)$ is the set of all λ-normal, atomic formulas A so that $\Sigma';\mathcal{P}' \vdash A$. ∎

Lemma 5.4. *Cut-elimination (Theorem 3.1) holds for $\mathcal{M}$ if and only if the following holds: for every dependent pair $\langle\Sigma,\mathcal{P}\rangle$ and every Σ-formula B, $\Sigma;\mathcal{P} \vdash B$ if and only if $I \Vdash B$, where I is the canonical model for $\langle\Sigma,\mathcal{P}\rangle$.*

Proof. Assume first that cut-elimination holds for $\mathcal{M}$. We now prove by induction on the structure of B that $\Sigma;\mathcal{P} \vdash B$ if and only if $I, \langle\Sigma,\mathcal{P}\rangle \Vdash B$.

Case: B is atomic. The equivalence is trivial.

Case: B is $B_1 \wedge B_2$. This case is simple and immediate.

Case: B is $B_1 \supset B_2$. Assume first that $\Sigma;\mathcal{P} \vdash B_1 \supset B_2$. By Theorem 3.2, $\Sigma;\mathcal{P}\cup\{B_1\} \vdash B_2$. To show $I, \langle\Sigma,\mathcal{P}\rangle \Vdash B_1 \supset B_2$, let $\langle\Sigma',\mathcal{P}'\rangle \in \mathcal{W}$ be such that $\langle\Sigma,\mathcal{P}\rangle \preceq \langle\Sigma',\mathcal{P}'\rangle$ and $I, \langle\Sigma',\mathcal{P}'\rangle \Vdash B_1$. By the inductive hypothesis, $\Sigma';\mathcal{P}' \vdash B_1$ and by cut-elimination, $\Sigma';\mathcal{P}' \vdash B_2$. By induction again, we have $I, \langle\Sigma',\mathcal{P}'\rangle \Vdash B_2$. Thus, $I, \langle\Sigma,\mathcal{P}\rangle \Vdash B_1 \supset B_2$. For the converse, assume $I, \langle\Sigma,\mathcal{P}\rangle \Vdash B_1 \supset B_2$. Since $\Sigma;\mathcal{P} \cup \{B_1\} \vdash B_1$, the inductive hypothesis yields $I, \langle\Sigma,\mathcal{P} \cup \{B_1\}\rangle \Vdash B_1$. By the definition of satisfaction of implication we must have $I, \langle\Sigma,\mathcal{P} \cup \{B_1\}\rangle \Vdash B_2$. But by the inductive hypothesis again, $\Sigma;\mathcal{P} \cup \{B_1\} \vdash B_2$, and $\Sigma;\mathcal{P} \vdash B_1 \supset B_2$.

Case: B is $\forall_\tau x B_1$. Assume first that $\Sigma;\mathcal{P} \vdash \forall_\tau x B_1$. By Theorem 3.2, $\Sigma \cup \{d\};\mathcal{P} \vdash B_1[d/x]$ for any constant d not in Σ. To show $I, \langle\Sigma,\mathcal{P}\rangle \Vdash \forall_\tau x B_1$, let $\langle\Sigma',\mathcal{P}'\rangle \in \mathcal{W}$ be such that $\langle\Sigma,\mathcal{P}\rangle \preceq \langle\Sigma',\mathcal{P}'\rangle$ and t is a Σ'-term of type τ. By cut-elimination on signatures (the subst rule), we have $\Sigma';\mathcal{P}' \vdash B_1[t/x]$. By induction we have $I, \langle\Sigma',\mathcal{P}'\rangle \Vdash B_1[t/x]$. Thus, $I, \langle\Sigma,\mathcal{P}\rangle \Vdash \forall_\tau x B_1$. For the converse, assume $I, \langle\Sigma,\mathcal{P}\rangle \Vdash \forall_\tau x B_1$. Let d be a constant not a member of Σ. Since d is a $\Sigma\cup\{d\}$-term, $I, \langle\Sigma\cup\{d\},\mathcal{P}\rangle \Vdash B_1[d/x]$ by the definition of satisfaction of universal quantification. But by the inductive hypothesis again, $\Sigma \cup \{d\};\mathcal{P} \vdash B_1[d/x]$ and $\Sigma;\mathcal{P} \vdash \forall_\tau x B_1$.

Now assume the equivalence: for every dependent pair $\langle\Sigma,\mathcal{P}\rangle$ and every Σ-formula B, $\Sigma;\mathcal{P} \vdash B$ if and only if $I \Vdash B$, where I is the canonical model for $\langle\Sigma,\mathcal{P}\rangle$. We now show that any sequent that can be proved using occurrences of the cut and subst rules can be proved without such rules. In particular, we show that if $\langle \Sigma, \mathcal{P} \rangle \preceq \langle \Sigma', \mathcal{P}' \rangle$ then each of the following holds.

(1) If $\Sigma'; \mathcal{P}' \vdash B$ and $\Sigma; \mathcal{P}, B \vdash C$ then $\Sigma'; \mathcal{P}' \vdash C$.

(2) If t is a Σ'-term of type τ and $\Sigma + x : \tau; \mathcal{P} \vdash B$ then $\Sigma'; \mathcal{P}' \vdash B[t/x]$ (of course, x does not occur in Σ).

From these facts, any number of occurrences of the cut and subst rules can be eliminated from a proof containing them.

To prove (1), assume that $\Sigma'; \mathcal{P}' \vdash B$ and $\Sigma; \mathcal{P}, B \vdash C$. Thus, $\Sigma; \mathcal{P} \vdash B \supset C$. By the assumed equivalence, $I, \langle \Sigma', \mathcal{P}' \rangle \Vdash B$ and $I, \langle \Sigma, \mathcal{P} \rangle \Vdash B \supset C$. By the definition of satisfaction for implication, $I, \langle \Sigma', \mathcal{P}' \rangle \Vdash C$. By the assumed equivalence again, this yields $\Sigma'; \mathcal{P}' \vdash C$.

To prove (2), assume that t is a Σ'-term of type τ and that $\Sigma + x : \tau; \mathcal{P} \vdash C$. Thus, $\Sigma; \mathcal{P} \vdash \forall_\tau x.B$. By the assumed equivalence, $I, \langle \Sigma, \mathcal{P} \rangle \Vdash \forall_\tau x.B$. By the definition of satisfaction for universal quantification, we have $I, \langle \Sigma', \mathcal{P}' \rangle \Vdash B[t/x]$. By the assumed equivalence again, this yields $\Sigma'; \mathcal{P}' \vdash B[t/x]$. QED

Given Theorem 3.1, this lemma provides an immediate proof of the following theorem.

Theorem 5.5. *Let $\langle \Sigma, \mathcal{P} \rangle$ be a dependent pair and let I be the canonical model for $\langle \Sigma, \mathcal{P} \rangle$. For all Σ-formulas B, $\Sigma; \mathcal{P} \vdash B$ if and only if $I \Vdash B$. In particular, for every $B \in \mathcal{P}$, $I \Vdash B$.*

This theorem can be sharpened using the following definition of *order* for types and for formulas.

Definition 5.6. The *order of type* τ, written $ord(\tau)$, is 0 if τ is primitive; otherwise τ is of the form $\tau_1 \rightarrow \tau_2$, in which case, the order of τ is $\max(1 + ord(\tau_1), ord(\tau_2))$.

The *order of formula* B, written $\text{ord}(B)$, is 0 if B is atomic; is $\max(ord(B_1), ord(B_2))$ if B is $B_1 \wedge B_2$; is $\max(1 + ord(B_1), ord(B_2))$ if B is $B_1 \supset B_2$; and is $\max(1 + ord(\tau), ord(B_1))$ if B is $\forall_\tau x.B_1$. ∎

Notice that if B has order 1 then B is (modulo weak equivalences) a first-order Horn clause theory.

Next we define the notion of the canonical model at a given order. Such models contain, in a sense, fewer worlds than the canonical models introduced in Definition 5.3.

Definition 5.7. A dependent pair $\langle \Sigma, \mathcal{P} \rangle$ is of order n if all the types in Σ are of order n or less and all the formulas in $\mathcal{P}$ are of order n or less. Let $\langle \Sigma, \mathcal{P} \rangle$ be a dependent pair of order n. The *canonical model of order* n for $\langle \Sigma, \mathcal{P} \rangle$ is $[\mathcal{W}, I]$ where $\mathcal{W}$ is the set of all dependent pairs $\langle \Sigma', \mathcal{P}' \rangle$ of order n such that (*i*) Σ' extends Σ with constants of order at most $n-2$, and (*ii*) $\mathcal{P}'$ extends $\mathcal{P}$ with Σ'-formulas of order at most $n-2$. The mapping I is defined as before, namely, for all $\langle \Sigma', \mathcal{P}' \rangle \in \mathcal{W}$, the set $I(\langle \Sigma', \mathcal{P}' \rangle)$ contains all atomic A so that $\Sigma'; \mathcal{P}' \vdash A$. ∎

Notice that if $\langle \Sigma, \mathcal{P} \rangle$ is of order 1 then Σ is a first-order signature (all constants are of order 0 or 1) and $\mathcal{P}$ is a set of Horn clauses. The canonical model for such a dependent pair contains just one world, namely, the pair $\langle \Sigma, \mathcal{P} \rangle$.

Lemma 5.8. *Cut-elimination (Theorem 3.1) holds for $\mathcal{M}$ if and only if the following holds: Let $n \geq 1$, let $\langle \Sigma, \mathcal{P} \rangle$ be a dependent pair of order n, let I be the canonical model of order n for $\langle \Sigma, \mathcal{P} \rangle$, and let B be a Σ-formula of order $n-1$. Then $\Sigma; \mathcal{P} \vdash B$ if and only if $I \Vdash B$.*

Proof. Assume first that cut-elimination holds for $\mathcal{M}$. We now prove by induction on the structure of B that $\Sigma; \mathcal{P} \vdash B$ if and only if $I, \langle \Sigma, \mathcal{P} \rangle \Vdash B$. The forward part of this equivalence is the same as in the proof of Lemma 5.4. Thus we only show details of the reverse implication for the two interesting cases.
Case: B is $B_1 \supset B_2$. Thus the order of B_1 is $n-2$ or less. Assume $I, \langle \Sigma, \mathcal{P} \rangle \Vdash B_1 \supset B_2$. Since $\Sigma; \mathcal{P} \cup \{B_1\} \vdash B_1$ and $\langle \Sigma, \mathcal{P} \cup \{B_1\} \rangle \in \mathcal{W}$, the inductive hypothesis yields $I, \langle \Sigma, \mathcal{P} \cup \{B_1\} \rangle \Vdash B_1$. By the definition of satisfaction of implication we must have $I, \langle \Sigma, \mathcal{P} \cup \{B_1\} \rangle \Vdash B_2$. But by the inductive hypothesis again, $\Sigma; \mathcal{P} \cup \{B_1\} \vdash B_2$ and $\Sigma; \mathcal{P} \vdash B_1 \supset B_2$.
Case: B is $\forall_\tau x B_1$. Thus the order of τ is $n-2$ or less. Assume $I, \langle \Sigma, \mathcal{P} \rangle \Vdash \forall_\tau x B_1$. Let d be a constant not a member of Σ. Since d is a $\Sigma \cup \{d\}$-term and since $\langle \Sigma \cup \{d\} \rangle$ is a member of $\mathcal{W}$, then we have $I, \langle \Sigma \cup \{d\}, \mathcal{P} \rangle \Vdash B_1[d/x]$ by the definition of satisfaction of universal quantification. But by the inductive hypothesis again, we have $\Sigma \cup \{d\}; \mathcal{P} \vdash B_1[d/x]$ and $\Sigma; \mathcal{P} \vdash \forall_\tau x B_1$.

The fact that cut-elimination holds follows just as in the proof of Lemma 5.4, except here we need to use the equivalence at various different orders. QED

We shall need the following technical result.

Lemma 5.9. *Let $\langle \Sigma, \mathcal{P} \rangle$ be a dependent pair of order $n \geq 1$, and let $[\mathcal{W}, I]$ be the canonical model of order n for $\langle \Sigma, \mathcal{P} \rangle$. Let $\langle \Sigma', \mathcal{P}' \rangle \in \mathcal{W}$, and let $[\mathcal{W}', I']$ be the canonical model of order n for $\langle \Sigma', \mathcal{P}' \rangle$. For all Σ'-formulas B of order n, $I, \langle \Sigma', \mathcal{P}' \rangle \Vdash B$ if and only if $I' \Vdash B$.*

Proof. Simple induction on the structure of B. QED

The next theorem shows that if $\langle \Sigma, \mathcal{P} \rangle$ is a dependent pair of order n then the canonical model for $\langle \Sigma, \mathcal{P} \rangle$ of order n is, in fact, a model for $\mathcal{P}$.

Theorem 5.10. *Let $\langle \Sigma, \mathcal{P} \rangle$ be a dependent pair of order n and let $[\mathcal{W}, I]$ be the canonical model of order n for $\langle \Sigma, \mathcal{P} \rangle$. If B is of order n or less, then $\Sigma; \mathcal{P} \vdash B$ implies $I \Vdash B$.*

Proof. We prove the following by induction on the structure of B: for every $\langle \Sigma', \mathcal{P}' \rangle \in \mathcal{W}$, if $\Sigma'; \mathcal{P}' \vdash B$ then $I, \langle \Sigma', \mathcal{P}' \rangle \Vdash B$.
Cases: B is atomic or B is conjunctive. These cases are simple.
Case: B is $B_1 \supset B_2$ where B_1 is of order $n-1$ or less. Let $\langle \Sigma', \mathcal{P}' \rangle \in \mathcal{W}$ and let $\langle \Sigma'', \mathcal{P}'' \rangle \in \mathcal{W}$ be such that $\langle \Sigma', \mathcal{P}' \rangle \preceq \langle \Sigma'', \mathcal{P}'' \rangle$ and $I, \langle \Sigma'', \mathcal{P}'' \rangle \Vdash B_1$. Let $[\mathcal{W}'', I'']$ be the canonical model of order n for $\langle \Sigma'', \mathcal{P}'' \rangle$. By Lemma 5.9, $I'' \Vdash B_1$. By Lemma 5.8, $\Sigma''; \mathcal{P}'' \vdash B_1$. By cut-elimination, $\Sigma''; \mathcal{P}'' \vdash B_2$. By the inductive hypothesis, we have $I, \langle \Sigma'', \mathcal{P}'' \rangle \Vdash B_2$. By the definition of satisfaction, we have $I, \langle \Sigma', \mathcal{P}' \rangle \Vdash B_1 \supset B_2$.
Case: B is $\forall_\tau x.B_1$ where τ is of order $n-1$ or less. Let $\langle \Sigma', \mathcal{P}' \rangle \in \mathcal{W}$ and let $\langle \Sigma'', \mathcal{P}'' \rangle \in \mathcal{W}$ be such that $\langle \Sigma', \mathcal{P}' \rangle \preceq \langle \Sigma'', \mathcal{P}'' \rangle$ and let t be a Σ''-term of type τ. By cut-elimination, $\Sigma''; \mathcal{P}'' \vdash B_1[t/x]$. By the inductive hypothesis, we have $I, \langle \Sigma'', \mathcal{P}'' \rangle \Vdash B_1[t/x]$. By the definition of satisfaction, we have $I, \langle \Sigma', \mathcal{P}' \rangle \Vdash \forall_\tau x.B_1$. QED

If Theorem 5.10 is specialized to just the case for order 1, it provides the familiar "minimal model" construction for first-order Horn clause theories [1]. Thus, Theorem 5.10 can be seen as a generalization of that model construction to arbitrary orders.

Notice that the converse to Theorem 5.10 is not generally true if the formula B is of order n. For example, let i be the only primitive type, let p and q be the only predicates, each of sort $\langle i \rangle$, let Σ be the signature $\{a : i\}$ and let $\mathcal{P}$ be the set of Σ-formulas

$$\{p\ a,\ \forall_i x\ (p\ x \supset q\ x)\}.$$

Then, the formula of order 1, $\forall_i x\ (q\ x \supset p\ x)$ is valid in the canonical model of order 1 for $\langle \Sigma, \mathcal{P} \rangle$ but it is not provable from Σ and $\mathcal{P}$.

Consider again the problems of evaluation under abstractions within a functional program (Section 4). The pair $\langle \Sigma_1, \mathcal{P}_1 \cup \mathcal{C}_1 \rangle$ is of order 2. The canonical model of order 2 for this pair is built by considering all those pairs $\langle \Sigma', \mathcal{P}' \rangle$ so that Σ' extends Σ_1 with some number of constants of type i and where $\mathcal{P}'$ extends $\mathcal{P}_1 \cup \mathcal{C}_1$ with some number of formulas of the form *eval* t s where t and s are Σ'-terms (conjunctions of such atoms are also allowed). The interpretation mapping is built by the usual provability construction. Thus, an alternative way to view "lifted" evaluation is: t evaluates to s if and only if the atomic formula *eval* t s is true in this model.

It is worth making the following simple observation about how canonical models can be considered minimal. We shall say that a Kripke model $\mathcal{N}$ satisfies $\langle \Sigma, \mathcal{P} \rangle$ if Σ is contained in the signature of $\mathcal{N}$ and if for every $B \in \mathcal{P}$, $\mathcal{N} \Vdash B$.

Theorem 5.11. *Let $\langle \Sigma, \mathcal{P} \rangle$ be a dependent pair, and let $\mathcal{K}$ be the canonical model for $\langle \Sigma, \mathcal{P} \rangle$. If $\mathcal{N}$ is a model of $\langle \Sigma, \mathcal{P} \rangle$ then $\mathcal{K} \Vdash B$ implies $\mathcal{N} \Vdash B$.*

Proof. Since $\mathcal{K} \Vdash B$ then $\Sigma; \mathcal{P} \vdash B$. By the soundness of Kripke models and the fact that $\mathcal{N}$ models $\langle \Sigma, \mathcal{P} \rangle$, we have $\mathcal{N} \Vdash B$. QED

6. Conclusions

Just as concrete syntax inadequately represents the structure of most syntactic objects, parse trees also inadequately represent the structure of syntactic objects containing bound variables. Thus a more high-level notion of syntax, called here ***abstract syntax***, is desirable. A logic $\mathcal{M}$ makes it possible to specify computations that support this notion of abstract syntax. The logic programming language λProlog can be used to provide implementations of such specifications made in $\mathcal{M}$. The semantics of specifications written in this meta-logic can be described using Kripke models.

7. Acknowledgements

During the academic year 1990 – 1991, I was visiting the Universities of Edinburgh and Glasgow. At the University of Glasgow, this work has been supported by a British Science and Engineering Research Council visiting fellowship research grant. At the University of Edinburgh, this work has been supported by SERC Grant No. GR/E 78487 "The Logical Framework" and ESPRIT Basic Research Action No. 3245 "Logical Frameworks: Design, Implementation, and Experiment." At the University of Pennsylvania, this work has been supported by ONR N00014-88-K-0633 and NSF CCR-87-05596. Thanks to Eva Ma for her proofreading help.

8. References

[1] K. R. Apt and M. H. van Emden. Contributions to the theory of logic programming. *Journal of the ACM*, 29(3):841 – 862, 1982.

[2] Alonzo Church. A formulation of the simple theory of types. *Journal of Symbolic Logic*, 5:56–68, 1940.

[3] Amy Felty. *Specifying and Implementing Theorem Provers in a Higher-Order Logic Programming Language.* Ph.D. thesis, University of Pennsylvania, August 1989.

[4] Amy Felty. A logic program for transforming sequent proofs to natural deduction proofs. In Peter Schroeder-Heister, editor, *Extensions of Logic Programming: International Workshop, Tübingen FRG, December 1989*, volume 475 of *Lecture Notes in Artificial Intelligence.* Springer-Verlag, 1991.

[5] Amy Felty and Dale Miller. Specifying theorem provers in a higher-order logic programming language. In *Ninth International Conference on Automated Deduction*, pages 61 – 80, Argonne, IL, May 1988. Springer-Verlag.

[6] Amy Felty and Dale Miller. Encoding a dependent-type λ-calculus in a logic programming language. In Mark Stickel, editor, *Proceedings of the 1990 Conference on Automated Deduction*, volume 449, pages 221–235. Springer Lecture Notes in Artificial Intelligence, 1990.

[7] John Hannan and Dale Miller. Uses of higher-order unification for implementing program transformers. In *Fifth International Logic Programming Conference*, pages 942–959, Seattle, Washington, August 1988. MIT Press.

[8] John Hannan and Dale Miller. Deriving mixed evaluation from standard evaluation for a simple functional programming language. In Jan L. A. van de Snepscheut, editor, *1989 International Conference on Mathematics of Program Construction*, volume 375 of *Lecture Notes in Computer Science*, pages 239–255. Springer-Verlag, 1989.

[9] John Hannan and Dale Miller. A meta-logic for functional programming. In H. Abramson and M. Rogers, editors, *Meta-Programming in Logic Programming*, chapter 24, pages 453–476. MIT Press, 1989.

[10] John Hannan and Dale Miller. From operational semantics to abstract machines: Preliminary results. In M. Wand, editor, *Proceedings of the 1990 ACM Conference on Lisp and Functional Programming*, pages 323–332. ACM Press, 1990.

[11] John J. Hannan. *Investigating a Proof-Theoretic Meta-Language for Functional Programs.* Ph.D. thesis, University of Pennsylvania, August 1990.

[12] Robert Harper, Furio Honsell, and Gordon Plotkin. A framework for defining logics. In *Second Annual Symposium on Logic in Computer Science*, pages 194–204, Ithaca, NY, June 1987.

[13] Gérard Huet and Bernard Lang. Proving and applying program transformations expressed with second-order patterns. *Acta Informatica*, 11:31–55, 1978.

[14] Dale Miller. An extension to ML to handle bound variables in data structures: Preliminary report. In *Informal Proceedings of the Logical Frameworks BRA Workshop*, June 1990. Available as UPenn CIS technical report MS-CIS-90-59.

[15] Dale Miller. A logic programming language with lambda-abstraction, function variables, and simple unification. *Journal of Logic and Computation*, 1991.

[16] Dale Miller. Unification of simply typed lambda-terms as logic programming. In *Eigth International Logic Programming Conference*, Paris, France, June 1991. MIT Press.

[17] Dale Miller and Gopalan Nadathur. A logic programming approach to manipulating formulas and programs. In Seif Haridi, editor, *IEEE Symposium on Logic Programming*, pages 379–388, San Francisco, September 1987.

[18] Dale Miller, Gopalan Nadathur, Frank Pfenning, and Andre Scedrov. Uniform proofs as a foundation for logic programming. *Annals of Pure and Applied Logic*, 51:125–157, 1991.

[19] Robin Milner, Mads Tofte, and Robert Harper. *The Definition of Standard ML*. MIT Press, 1990.

[20] John C. Mitchell and Eugenio Moggi. Kripke-style models for typed lambda calculus. In *Second Annual Symposium on Logic in Computer Science*, pages 303–314, Ithaca, NY, June 1987.

[21] Gopalan Nadathur and Dale Miller. An Overview of λProlog. In *Fifth International Logic Programming Conference*, pages 810–827, Seattle, Washington, August 1988. MIT Press.

[22] Frank Pfenning. Partial polymorphic type inference and higher-order unification. In *Proceedings of the ACM Lisp and Functional Programming Conference*, 1988.

[23] Frank Pfenning. Elf: A language for logic definition and verified metaprogramming. In *Fourth Annual Symposium on Logic in Computer Science*, pages 313–321, Monterey, CA, June 1989.

[24] Frank Pfenning and Conal Elliot. Higher-order abstract syntax. In *Proceedings of the ACM-SIGPLAN Conference on Programming Language Design and Implementation*, 1988.

DEDUCTION SEARCH WITH GENERALIZED TERMS

Vladimir S. Neiman

An efficient method for generating derivable objects in calculuses over terms is proposed. The method is based on the fact that inference rules may require only partial information about their premises. It means that we can apply an inference rule not to single terms (or n-tuples of terms, if the rule has n premises) but to sets of terms (sets of n-tuples) whose elements are equivalent to each other with respect to the rule. This may reduce considerably the running time of deduction search algorithms. In some cases this approach even turns infinite search space into finite one.

The proposed method is applicable to a wide range of calculuses. In particular, this method can be used for optimization of logic program execution.

1. INTRODUCTION

Let a calculus with terms as derivable objects be fixed. We are interested in one of the following problems:

PROBLEM 1. Given a term T, decide whether it is derivable in the calculus.

PROBLEM 2. Decide whether there exists a derivable term which satisfies some property P.

PROBLEM 3. Generate all derivable terms which satisfy some property P.

If the set of derivable terms in the calculus is finite then both of these problems may be solved by the straightforward algorithm which generates sequentially all derivable terms. This algorithm can be used to attack these problems even in case when infinitely many derivable terms exist, though in this case it is not guaranteed to terminate for problems 1 and 2 (and definitely will not terminate for problem 3). Usually the straightforward algorithm is not the best choice even in situations when it is guaranteed to terminate - more efficient calculus-specific methods and strategies are known for many calculuses. These methods may be divided into two classes:

1) A new calculus may be constructed which is better suited for automated deduction. Thus, if problem 1 is approached, it is usually a good idea to try to construct a new calculus where all deductions start with the term T. Deducibility of a special object in the new calculus means that T is deducible in the original calculus (this is the main idea underlying so-called *goal-driven* deduction search methods).

2) Methods of the second class leave the original calculus intact, but either forbid certain applications of inference rules, or define an order in which inference rules must be applied. Such methods are usually called *strategies*.

Methods of both classes are calculus-specific, i.e. each one is designed for a particular calculus. One can hardly conceive a method of the first or the second class which may be used for a number of different calculuses.

This paper presents an approach which can be used in a wide range of calculuses over terms. This approach utilizes the fact that inference rules may require only partial information about their premises, i.e. they may able to generate an answer without full analysis of their arguments. For example, this is usually the case with inference rules based on the unification procedure. Thus, an attempt to unify P(T1) with Q(T2), where P and Q are different predicate symbols and T1 and T2 are some terms, terminates unsuccessfully without examining T1 or T2. On the other hand, an attempt to unify P(X) with P(T), where X is a variable, terminates successfully without examination of T.

We shall call our approach *symbol-based* because it does not process terms as indivisible entities, but rather process only one symbol of a term at one step. On the other hand, each step serves not only for a particular term T, but for all terms which do not differ from T in the symbols processed so far.

Transformation of a usual deduction into a symbol-based deduction may be done in the following way: first, each application of an inference rule is split into several steps consisting in processing of *one symbol* of the rule's premises. Then these small steps are merged again, but in a different way: steps corresponding to processing of arguments which do not differ from each other in symbols processed so far are combined together.

If two terms differ only in subterms which are insignificant to an inference rule, then they may be considered as *equivalent* with respect to the rule. The symbol-based algorithm does not process insignificant symbols at all. So, equivalent terms are not distinguished during applications of the rule.

The symbol-based approach is not dedicated to a particular calculus. Of course, it cannot yield alone the same results as efficient methods for automated deduction designed specially for a particular calculus. But an advantage of the symbol-based approach is that it is usually *compatible* with methods designed for particular calculuses. In fact, methods of the first class consist in shifting to another calculus. If we want to combine such method with the symbol-based approach we must simply apply symbol-based approach to the new calculus rather than to the old one. Methods of the second class (strategies) may be extended to symbol-based inferences in the following way: we can forbid (postpone) an application of an inference rule to a set of n-tuples if the strategy prescribes to forbid (postpone) the application of the rule to each element of the set.

The idea of the symbol-based approach for one particular calculus (goal-driven deductions in Horn fragment of the first-order logic) was described in [1]. Here this approach is proposed in a significantly more general situation.

2. SOME PRELIMINARIES

We assume that a signature (set of symbols together with their arities) is fixed. In a calculus these symbols may be divided into several classes (e.g. predicate symbols, functional symbols, variables, etc.), but this distinction is not significant to us.

DEFINITION. *Term* is an expression of the form F(T1, T2, ..., Tn), where F is an n-place symbol from the signature and T1, T2, ..., Tn are terms. If n=0 then the brackets may be omitted.

DEFINITION. *Generalized term* is either * (asterisk) or an expression of the form F(T1, T2, ..., Tn), where F is an n-place symbol and T1, T2, ..., Tn are generalized terms. The brackets may be omitted if n=0. Each occurrence of the asterisk in a generalized term is called a *parameter*. An *instance* of a generalized term is a result of replacement of all parameters with some terms.

Informally a generalized term is a term some subterm of which are unknown; these subterms are denoted by asterisks. It will be convenient to identify a generalized term with the set of all its instances.

DEFINITION. *Inference rule with n premises* is an algorithm which transforms an n-tuple of terms (these terms are called *premises* of the rule) into a finite set of terms (*conclusions* of the rule). The set of conclusions may be empty. *A calculus* is a pair consisting of a finite set of inference rules and a finite set of terms called *axioms*.

One additional assumption must be made about inference rules in order to make symbol-based approach possible. Namely, the inference rules must be able to work in a situation when some subterms of their premises are unknown. Let us formulate this requirement more precisely.

ADDITIONAL REQUIREMENT TO INFERENCE RULES. Each inference rule with n premises must be applicable to an arbitrary n-tuple of generalized terms (G1, G2, ..., Gn). One of the following answers must be returned:

1) Conclusions cannot be generated until values of some parameters of the premises are specified. In this case a pointer to a desired parameter in the premises must be returned.

2) All unknown subterms are really inessential. In this case a set of generalized terms {C1, C2, ..., Ck} (conclusions of the rule) must be generated. With each parameter in a conclusion the pointer to a parameter in the premises of the rule must be associated. This answer should be sound in the following sense: if we apply this rule to such n-tuple of terms (T1, T2, ..., Tn) that each Ti is an instance of Gi, then the set of conclusions must coincide with the set of terms obtained from {C1, C2, ..., Ck} by replacing all parameters with the values of the corresponding subterms of T1, T2, ..., Tn.

EXAMPLE. Let one inference rule F(a,X) -> G(X,b,X) be given. This rule has one premise and works in the following way: if a premise is unifiable with F(a,X), then one conclusion G(t,b,t) is generated, where t is the term which was substituted instead of X. Otherwise the set of conclusions is empty. This rule produces the following answers when called with generalized terms as an argument:

Argument F(*,*). Answer: conclusions cannot be generated until the value of the first parameter is given.

Argument F(b,*). Answer: the set of conclusions is empty.

Argument F(a,*). Answer: one conclusion G(*/1,b,*/1).

Argument F(a,F(*,*)). Answer: one conclusion G(F(*/1,*/2), b, F(*/1,*/2)) (*/i means that this parameter is associated with the i-th parameter of the premises).

3. GENERAL OVERVIEW OF THE SYMBOL-BASED METHOD

Due to the limited size of this paper we cannot provide the full description of the symbol-based approach; so we confine ourselves to outlining main data structures and main algorithms. An example showing how this approach may be used in a particular calculus is given in the next section.

The main data kept during the inference process are the *list of waiting operations* which contains operations ready to be executed, and *the term tree* which represents the set of derived terms. Deduction is being done in the following way: while the list of waiting operations is not empty, its elements are selected and executed. Each operation may change global data (including the list of waiting operations). If the list becomes empty, it means that the deduction is completed and the term tree represents the set of all deducible terms.

The order in which elements from the list of waiting operations are chosen and executed is not fixed. In other words, the symbol-based approach allows us to use different strategies of computation.

The term tree is a tree whose nodes are generalized terms. The root of the tree is always the generalized term *. The following condition must be satisfied. If a node T has sons T1, T2, ..., Tk, then each Ti must be obtained from T by replacing one parameter of T with a generalized term of the form F(*,*,...,*),where F is some symbol (the asterisks are absent if the arity of F is equal to 0). The parameter of T which is replaced must be the same for all sons; it is called *the point for expansion of T.*

The terms themselves are kept at the leaves of the tree. The above mentioned condition means that when we traverse the tree from the root to a leaf, on each step one new symbol of the destination term becomes known. In order to describe how terms are represented, we need the following definition.

DEFINITION. A *term over the tree* is either a pointer to a node of the term tree or an expression of the form F(T1, T2, ..., Tn), where F is an n-place symbol and T1, T2, ..., Tn are terms over the tree. The brackets may be omitted if n=0. A *refinement* of a generalized term having n parameters is an n-tuple of terms over the tree. In other words, refinement associates a term over the tree with each parameter of T.

A list of refinement is associated with each leaf of the term tree. Each refinement represents one or several (in some cases even infinitely many - see figure 4 below) derived terms. Figure 1 shows an example of the term tree. This tree represents the set of terms {P(A), P(F(A)), P(F(X)), Q(A), Q(B), Q(X)}. To the right of each leaf which has at least one parameter the list of refinement is shown in square brackets (in this example each refinement is simply a term).

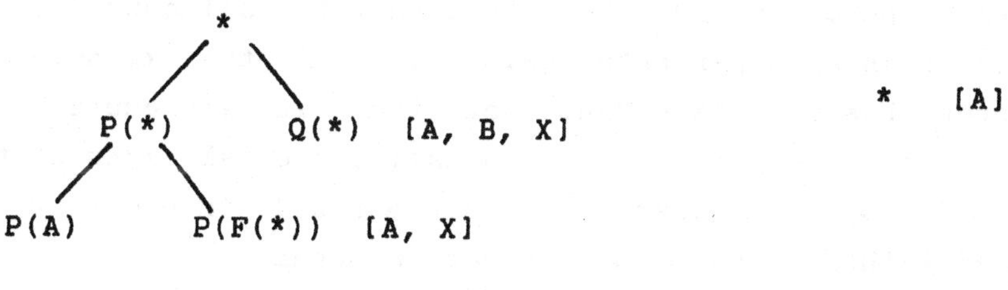

FIGURE 1

FIGURE 2

The arguments of all operations are nodes of the term tree. In other words all operations are executed on generalized terms rather than on ordinary ones. Even if a node of the tree has several refinements associated with it, inference operations would nevertheless work only with the node itself unless values of some parameters of the node are required. In this case *the operation of expansion* of this node is executed. This operation determines the point for expansion, then creates sons of the node and distributes its refinements among its sons. Then the operation which has required the value of a parameter is repeated with all the sons.

For each inference rule of the calculus a corresponding *inference operation* of the symbol-based method exists. If an inference rule has n premises, then the corresponding inference operation has n arguments, each one being a node of the term tree. The operation is executed in the following way. First, it forms the n-tuple of generalized terms from its arguments and passes it to the inference rule. If the inference rule returns the set of conclusions then each conclusion is inserted in the term tree. *The operation of insertion in the term tree* is itself a rather complex operation because a conclusion, being a generalized term, may correspond to several leaves of the tree simultaneously. So during execution of the insertion operation new elements may be added to lists of refinements, and (or) new nodes of the term tree may be created.

Otherwise, if the inference rule called is unable to generate conclusions, it returns the pointer to a parameter in the premises whose value is required. This means that the inference operation must be repeated with all sons of the node to which this parameter belongs. So, if the expansion of this node has not been done yet, then it is executed now. Then the inference operation is called for all sons of the node; besides, this operation is memorized in order to be executed with sons which may appear in future.

In the initial moment the term tree has only one node - the root *. The list of refinements of the root contains one element for each axiom A of the calculus, namely the term A. The list of waiting operations has one element for each inference rule of the calculus. All arguments of the operations are equal to the root of the term tree.

4. EXAMPLE

Let us consider the following calculus. The signature consists of 1-place symbols F and G, 0-place symbols A and B, and the set of 0-place symbols {X, Y, Z, ...} which are marked as variables. There are three inference rules:

X -> F(X)
G(X) -> X
F(G(X)) -> G(F(X))

A rule L -> R has one premise and works in the following way: if the premise is unifiable with L, then the rule returns the answer $\{R\varphi\}$, where φ is the most general unifying substitution. Otherwise the empty set of conclusions is returned.

The calculus has one axiom - the term A. We are interested in deducibility of the term B in this calculus.

It is clear that these inference rules and the axiom A may be regarded as Horn clauses, and our problem is equivalent to that of deducibility of B from these clauses in the standard predicate calculus. The example is chosen in such a way that B is not deducible, but existing inference methods for Horn clauses (e.g. those described in [2]) cannot prove it. Thus, the bottom-up (fact-driven) method would generate the infinite sequence of terms A, F(A), F(F(A)), F(F(F(A))), etc. and would never terminate. On the other hand, the top-down (goal-driven) method would generate the infinite sequence of subgoals B, G(B), G(G(B)), G(G(G(B))), etc. and would also never terminate. The third clause is added in order to make it harder to determine statically the fact that the first and the second clauses cannot interact one with the other in these deductions.

The term tree in the initial moment is shown on figure 2, and the list of waiting operations includes three elements - one for each inference rule. The first inference operation terminates successfully and produces the answer {F(*)}. The second and the third operations are unable to generate conclusions when called with the argument *. It means that the operation of expansion of the root must be executed. The term tree after the expansion is shown on figure 3 (*/1 designates the first parameter of the node *). Notice that only the second and the third operations will be repeated on the new nodes; the first inference operation, being terminated successfully on the root of the term tree, would never be executed anew.

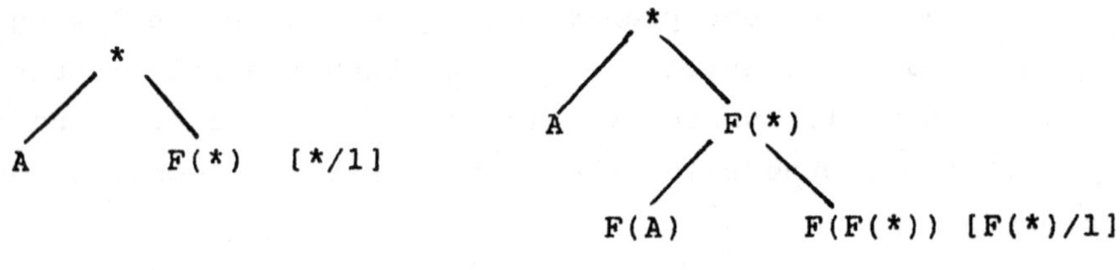

FIGURE 3 FIGURE 4

The second inference operation terminates on both new nodes with the empty sets of conclusions (because G(X) is not unifiable neither with A nor with F(*)). The third operation terminates on the node A with the empty set of conclusions (F(G(X)) is not unifiable with A), but cannot terminate on the node F(*). So the node F(*) also needs to be expanded. The term tree after the expansion is shown on figure 4.

Only the third inference operation must be applied to the new nodes F(A) and F(F(*)). On both nodes it terminates with the empty sets of conclusions. Now the list of waiting operations becomes empty. It means that the current term tree represents the set of all derivable terms. Notice that the infinite set of terms {A, F(A), F(F(A)), ... } is represented by the finite term tree.

Now we can answer the question whether the term B is derivable. We see that it is an instance of the root of the tree, but is not an instance of any son of the root. This indicates that B is non-derivable.

REFERENCES

[1] Neiman, V.S. Using partially defined terms in logic inferences. In: Theory and Applications of Artificial Intelligence (Bulgaria, Sozopol, 1989), pp. 241-248 (in Russian)

[2] Kowalski, R. Logic for problem solving (North-Holland, 1979)

A simple Transformation from Prolog-written Metalevel Interpreters into Compilers and its Implementation

Gustaf Neumann
Vienna University of Economics and Business Administration
Institute of Information Processing
Augasse 2–6
A–1090 Vienna, Austria

neumann@wu-wien.ac.at

Abstract

In this paper we describe a grammar formalism for program transformation and its implementation in Prolog. Whereas Definite Clause Grammars are merely working on a string of tokens the formalism presented here acts on semantic items such as Prolog literals. This grammar will be used to implement a very compact form of a compiler generator that transforms solve-like interpreters into compilers. Finally the compiler generator will be applied on itself to obtain a more efficient version of the compiler generator.

1 Overview

This paper is divided in three main sections: Section two introduces a grammar formalism tailored for the transformation of Prolog Clauses. Section three describes the approach of interpreter directed compilation of logic programs[1]. In the last section a very compact implementation of the compiler generator is given that uses the program transformation grammars of section two.

2 A Grammar Formalism for the Transformation of Prolog Clauses

The formalism of definite clause grammars (DCG, [PW80]) is a very popular tool for writing parsers in Prolog. DCG rules are just a syntactically different appearance of Prolog clauses where the two arguments (used for the processing of the list of tokens) are implicitly added to predicates. DCG allow to build structures (such as parse trees) during parsing. Several extensions to DCG have been proposed, but most of them are developed in the context of natural language analyses [Per81, Abr84, McC87] or they provide means to add more variables to predicates as DCG do [vR88]. In our experience DCG have two main disadvantages when they are applied to program transformation:

1. While DCG process a list of Prolog atoms, we want to be able to work directly on the program clauses which should be seen as literals and logical connectors.
2. In order to keep the grammar simple we want to have a formalism where the routine work of processing the input and generating the output literals causes as little syntactical overhead as possible.

The program transformation grammar consists of grammar rules describing the input clauses and literal (local) transformations. The grammar consists of rules of the structure

$$\langle \textit{nonterminal} \rangle ::= \langle \textit{body} \rangle .$$

where ⟨*nonterminal*⟩ is a nonterminal symbol and ⟨*body*⟩ is a list of items separated by commas. Each item is either

- a nonterminal symbol, or

[1]A more detailed discussion of the topics of this section can be found in [Neu90].

- a terminal symbol [⟨*syntactic-type*⟩ : ⟨*semantic-type*⟩]. ⟨*syntactic-type*⟩ is either a head literal "`hl`" or a body literal "`bl`". ⟨*semantic-type*⟩ is an arbitrary Prolog term used for the literal transformation;
- Prolog goals can be used in the grammar between curly brackets (exactly like in definite clause grammars).

The specifications of the literal transformations are of the form

t_lit(⟨semantic-type⟩,⟨input-literal⟩,⟨output-literal⟩,⟨context⟩*)*

where ⟨*semantic-type*⟩ refers to the semantic types used in the grammar rules (see the examples below). ⟨*input-literal*⟩ is a literal G which is transformed into an ⟨*output-literal*⟩ G'. ⟨*context*⟩ specifies the type of the transformation wanted.

Parsing a terminal symbol using [`hl` : ⟨*semantic-type*⟩] or [`bl` : ⟨*semantic-type*⟩] causes an implicit generation of a translated output literal. In cases where explicit control over the output literals is wanted, a construction [`ibl` : ⟨*semantic-type*⟩ = `X`] is used, where "`X`" is bound with the translated literal. This construction will be used only in the last section and will be omitted in the discussion of this section. The full implementation of the grammar interpreter is given in the appendix.

```
% Grammar describing Clauses
clause ::= [hl:any], body.
body ::= [bl:any], body.
body ::= {true}.

% Literal Transformations
t_lit(any, L,L, ident).
t_lit(any, L,i(L), wrap).
t_lit(any, i(L),L, unwrap).
```

Figure 1: A program transformation grammar for isomorphic transformations

The first example illustrates isomorphic transformations of Prolog clauses, where the input and output clauses have exactly the same structure, but each literal G of the input is transformed into a literal G' in the output according to a context variable (see Figure 1). The only semantic type used in this example is "`any`".

```
% Interpreter for the Grammar Rules
% pg(GrammarGoal, InClauses0,InClauses1, Context, OutClauses0,OutClauses1)

pg((A,B), In0,In2, Context, Out0,Out2) :- !,
      pg(A,In0,In1, Context, Out0,Out1),
      pg(B,In1,In2, Context, Out1,Out2).
pg({Goal}, In,In, Context, Out,Out) :- !, call(Goal).
pg([bl:Type], In0,In1, Context, Out0,Out1) :- !,
      get_literal(i,In0,Literal0,In1),
      t_lit(Type, Literal0,Literal1, Context),
      get_literal(o,Out0,Literal1,Out1).
pg([hl:Type], In0,In1, Context, Out0,Out1) :- !,
      get_head_literal(i,In0,Literal0,In1),
      t_lit(Type, Literal0,Literal1, Context),
      get_head_literal(o,Out0,Literal1,Out1).
pg(Head,In0,In1, Context, Out0,Out1) :-
      (Head ::= Body),
      pg(Body, In0,In1, Context, Out0,Out1).
```

Figure 2: Interpreter for the program transformation grammar rules

A straightforward Prolog implementation of the program transformation grammar is given in Figure 2. The predicate "`get_literal/4`" (or "`get_head_literal/4`") is used to parse a literal from the body (or the head) of an input clause and to construct a body (or head) literal in the output clause. A full definition of these predicates is given in the appendix. Sample goals to the grammar interpreter and the resulting transformed clauses are presented in Figure 3.

```
:-   pg(clause, (a :- b, c, d),true, ident, A,true), portray_clause(A), nl,
     pg(clause, (a :- b, c, d),true, wrap, B,true), portray_clause(B), nl,
     pg(clause, B,true, unwrap, C,true), portray_clause(C).
```

```
a :-                    i(a) :-                 a :-
      b,                      i(b),                   b,
      c,                      i(c),                   c,
      d.                      i(d).                   d.
```

Figure 3: Output of the isomorphic transformations

Isomorphic transformations could be used in an intelligent Prolog environment to implement search and replace functions over Prolog goals. Much more interesting are, however, non-isomorphic transformations, where the structure of the input and output clauses differs.

The second example (Figure 4) illustrates how arguments in the grammar rules could be used to pass information between grammar goals. The grammar rules define a schema for single recursive clauses. Mutual recursion is not covered by these grammar rules. The first rule of the grammar describes the base case of the recursive definition, the second rule is used for simple single recursive clauses. The variable "`Sig`" in "`srec(Sig)`" is used for the functor/arity pair defined by the instances of this schema. The semantic types used in the grammar are "`base(Sig)`" (the head of the base definition), "`head(Sig,Args)`" (the head of the recursive definition), "`rec(Sig,Args)`" (the recursive call) and "`nrec(Sig)`" (a non-recursive call).

These grammar rules will be used to guide the transformation of instances of this schema into rules with enhanced functionality. This functionality will be specified together with a programming technique (that should be used) in the initial goal to the program transformator. As programming techniques we use here *upward propagation* and *accumulator* technique. [Lak89] uses meta level interpreters to achieve similar results.

For each programming technique a set of literal transformations is given (see Figure 4). Each set covers all semantic types used in the grammar. The function definition consists of a name and a term "`f/5`", where an initial value, input and output value, a goal to access values in the head literal, and finally a goal performing the actual function are given. The predicate "`enhance/4`" is the top level predicate, which receives as first argument the functor/arity pair of the clause to be processed, followed by the name of the function that should be added, followed by the name of the programming technique to be used and the functor of the output predicates. Sample calls to "`enhance/4`" and the resulting output are given in Figure 5.

3 Interpreter directed Compilation of Logic Programs

Both compilers and interpreters are means to execute programs written in a programming language L_s which is not directly executable in the given environment. A compiler translates the source program into a target program in an executable language L_m that will be run at some later time after the compilation. Very often L_m is the machine language. An interpreter on the other hand is an emulator of a hypothetical machine that is able to perform computations according to the source program.

Compilers and interpreters are metaprograms written in an implementation language L_i that use programs in the source language L_s as input data. In both cases the source program is analyzed and transformed into an internal representation, which is some kind of data structure. While an interpreter performs computations according to this internal representation, a compiler generates essentially the same instructions as output in some target language L_m. Both types of metaprograms are very similar up to the point where the intended computational steps according to the source program are identified. We will see in the sequel how interpreter directed compilation exploits these similarities.

In order to perform computations a program P, a goal G and some means to execute the program (say the interpreter I) are needed. The program P models the problem domain. In the goal G the user specifies which computations should be performed. The specification of an execution contains a definition part P and a goal part G:

$$I(P,G)$$

The execution of $I(P,G)$ leads to results R_I that consist of success/failure/runtime error, variable substi-

% Grammar describing single recursive clauses with a base case

```
srec(Sig)  ::= [hl:base(Sig)], others(Sig).
srec(Sig)  ::= [hl:head(Sig,Args)], others(Sig), [bl:rec(Sig,Args)], others(Sig).

others(Sig)  ::= [bl:nrec(Sig)], others(Sig).
others(Sig)  ::= {true}.
```

% Literal Transformations for upward propagation technique

```
t_lit(base(Sig), L,L1, up-f(S,_,_,_)-N) :- addarg(L,Sig,[S],N,L1).
t_lit(head(Sig,A), L,L1, up-f(_,_,_,Goal/L=V,_)-N) :- Goal, addarg(L,Sig,[A],N,L1).
t_lit(rec(Sig,X1), L,(L1,P), up-f(S,X0,X1,_,P)-N) :- addarg(L,Sig,[X0],N,L1).

t_lit(nrec(F/A),L,L,_) :- \+ functor(L,F,A).
```

% Literal Transformations for accumulator technique

```
t_lit(base(Sig), L,(L1:-L2), accu-f(S,_,_,_,_)-N) :-
        addarg(L,Sig,[X],N,L1),
        addarg(L,Sig,[S,X],N,L2).
t_lit(base(Sig), L,L1, accu-f(S,_,_,_,_)-N) :- addarg(L,Sig,[X,X],N,L1).
t_lit(head(Sig,A), L,L1, accu-f(_,X0,_,Goal/L=V,_)-N) :- Goal,addarg(L,Sig,[X0,A],N,L1).
t_lit(rec(Sig,A), L,(P,L1), accu-f(_,_,X1,_,P)-N) :- addarg(L,Sig,[X1,A],N,L1).

addarg(L,F/A,Arg,NewName,L1) :-
        functor(L,F,A),
        L =..  [F|Any], append(Any,Arg,Args), L1 =..  [NewName|Args].
```

% function(FunctionName, f(StartValue,From,To,GoalForHeadVariable/Head=HeadVariable,GoalToBeAdded))

```
function(add1,f(0,X0,X1,true/_=[],sum(X0,1,X1))).
function(list_add,f(0,X0,X1,(Head=..[_,[V|_]|_])/Head=V,sum(X0,V,X1))).


enhance(Sig,FunctionName,Technique,NewName) :-
        function(FunctionName,Function),
        myclauses(Clause),
        pg(srec(Sig), Clause,true, Technique-Function-NewName, A,true),
        portray_clause(A),
        fail.
```

Figure 4: Grammar rules for single recursive clauses, literal transformations for the upward propagation and the accumulator technique, definition for the functions "add1" and "list_add"

tutions in G and side effects[2]. Let us assume I is an interpreter running on the underlying machine M. I consists of clauses that describe how to deal with program clauses of P. Therefore the more detailed execution formula revealing the interpreter is

$$M(I \cup P, I \circ G)$$

where the definition part contains clauses for I and P; the goal part $I \circ G$ specifies that goal G should be run on I. The results R_M of the execution now contain besides the variable substitutions of G the variable substitutions of I. We are interested in a transformation program τ that will be applied on P and G leading to P' and G' so that the execution of

$$M(P', G')$$

leads to results $R_M{}'$ that are identical to R_M. Goals $I \circ G$ and G' will contain the same variables. Figure 6 shows the homomorphism of the three execution models.

In the literature several approaches to achieve similar results can be found. Most of these approaches are based on partial evaluation [JSS85, FF88, TF86] where an interpreter I is evaluated with respect to a

[2]A more detailed definition of the results of a computation is given in [Neu86]

```
:-      enhance(a/0,add1,up,add_up).
:-      enhance(a/0,add1,accu,add_accu).
```

```
a.                      add_up(0).              add_accu(A) :- add_accu(0,A).
a :-                    add_up(A) :-            add_accu(A,A).
        b,                      b,              add_accu(A,B) :-
        a,                      add_up(B),              b,
        d.                      sum(B,1,A),             sum(A,1,C),
                                d.                      add_accu(C,B),
                                                        d.
```

```
:-      enhance(list/1,list_add,up,sum_add).
:-      enhance(list/1,list_add,accu,sum_accu).
```

```
list([]).               sum_add([],0).          sum_accu([],A) :- sum_accu([],0,A).
list([A|B]) :-          sum_add([A|B],C) :-     sum_accu([],A,A).
        list(B).                sum_add(B,D),   sum_accu([A|B],C,D) :-
                                sum(D,A,C).             sum(C,A,E),
                                                        sum_accu(B,E,D).
```

Figure 5: Sample goals to "enhance/4" using the upward propagation or accumulator technique, followed by the source clauses and the achieved output

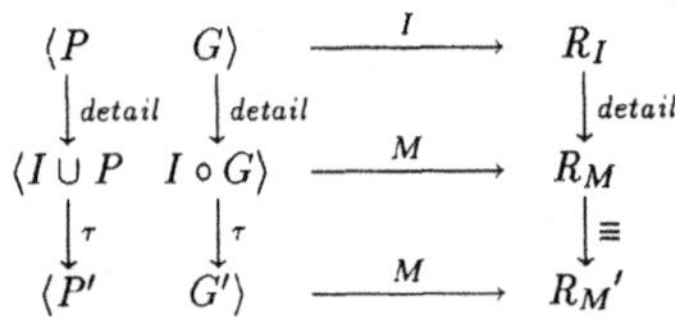

Figure 6: Homomorphism between execution models

given program P; the specialized, residual program is seen as a compiled program P'. Interpreter directed compilation is an analytical approach where a compiler τ is derived from a given interpreter I by classifying the goals in I. In a separate step τ is applied on P resulting in P'.

The base assumption of interpreter directed compilation is that an interpreter and a compiler perform the same program decomposition and analysis. If it is possible to factor out the goals of an interpreter that are needed for this task, it is also possible to obtain a compiler that outputs (modified) goals of the source program P and the remaining instructions of the interpreter as the compiled program.

The key element of the transformation is the analysis of the goals used in the interpreter. The goals of the interpreter are classified according to the following categories:

I: *Interpreter calls* are goals in the right-hand-sides of the interpreter clauses, where either the interpreter I is called recursively or another interpreter is called. A common use for recursive interpreter calls is to decompose compound statements of the source program (eg. conjunctive goals) in Prolog by unification in the head of the clause followed by recursive interpreter calls to process the constituents. More generally it is necessary to determine calls of interpreters; these are calls of predicates that process (part of) the program structure further.

C: *Checking predicates* are used in the interpreter to figure out certain properties of the predicates in the source program. During the interpretation or compilation the nature of each predicate in P is determined either by matching in the clause heads (of the I or τ) or by checking predicates. According to this classification different rules of the interpreter will be applied. Checking predicates are used when the matching in the head of the interpreter clause will not suffice to determine the type of the goal. A typical example for a checking predicate in a Prolog metainterpreter is the "sys(X)" predicate that is used to test whether its argument is a system predicate. Note that a checking predicate is always side effect free and never instantiates its argument further. In the interpreter for the grammar rules of the previous section all checking is done in the heads of the clauses of the interpreter; no explicit checking predicates are necessary. Very often a cut symbol is used after a checking predicate to commit to the chosen clause.

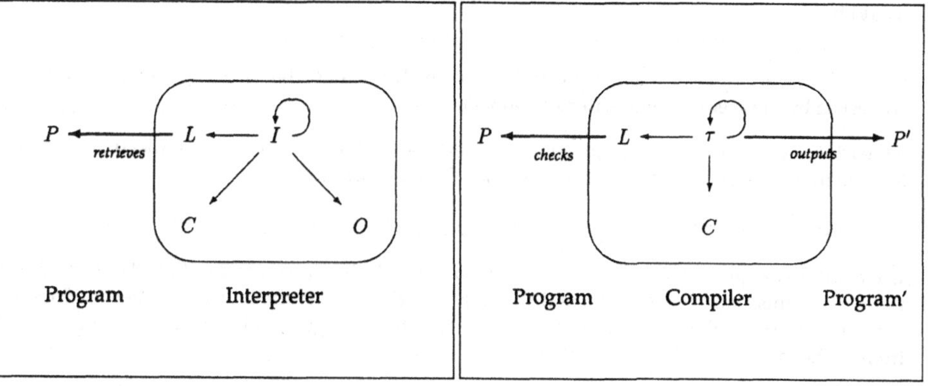

Figure 7: Classification of goals in the interpreter and the compiler: C are checking predicates, L are link predicates, and O are other predicates

L: *Link predicates* are needed to access the source program P. A link predicate binds its argument with a program structure. Link predicates always terminate and typically refer to facts or a system predicate to retrieve clauses.

O: *Others:* Goals that do not belong to the categories I, C and L.

This classification guides the generation of the compiler. Calls of the interpreter I (commonly recursive) are transformed into calls of the corresponding translator τ. Checking predicates C (and cut symbols used to determine the applicable rule of the interpreter) are performed identically in the interpreter and the translator. Link predicates L are the fulcrum of the structural inversion where a goal processing program is converted into a clause processing program. All other goals O occurring in the interpreter will be moved into the clauses of P' in the positions relative to the goals of P where they would be executed in I (see the abstract interpreter below).

The translator τ has one argument more than the interpreter I where the compiled program will be returned. The goals $I \circ G$ will be transformed into goals G'. The goals G' differ in two ways from $I \circ G$:

- An atomic goal from the source program corresponds to a conjunctive goal G' when class O goals are used in the interpreter.
- The arguments of the interpreter that are not used for passing the program structure are moved into the goals G' (argument pushing).

The clauses of the interpreter I containing the link predicates is split into a clause of τ that transforms a clause of P into a clause of P' and a second clause where corresponding goals (which might unify with the heads of P or P' respectively) are translated. The generated clause processing rule of τ performs the argument pushing for heads of clauses in P', the goal processing clause performs argument pushing for goals. Note that an interpreter without a link predicate has neither a program P, nor a program P' and no argument pushing will be performed.

The classification is done in this paper in part by user declarations and in part automatically. The following information is needed for the compiler generator of the next section:

- The *interpreter description* is a signature of an interpreter used in two ways: First as a pointer to a certain interpreter (see following items) and second, to define which argument of the interpreter I is used for argument pushing. An interpreter description for the grammar interpreter I_{pg} of the previous section is "`pg(p,s,s,s,s,s)`", where the constant "p" refers to the argument of the interpreter in which the program goals are expected. The "p" argument will appear in P' as predicate symbol. The constants "s" denote the additional arguments that will be pushed into the program literals in P'.
- For each *checking predicate* of the interpreter a fact is defined stating that a certain predicate is used as a checking predicate in a given interpreter. A hypothetical definition for a checking predicate in I_{pg}

would be

```
checking_pred(pg(p,s,s,s,s,s),Goal,sys(Goal))
```

where the first argument is the interpreter description. The second argument contains the goal G that is checked by the predicate in the third argument.[3]

- For each *link predicate* in the interpreter a fact is defined stating that a certain predicate is used as a link predicate in a given interpreter. A definition for I_{pg} would be

```
link_pred(pg(p,s,s,s,s,s),Head,Body,(Head::=Body),(Head::=Body))
```

  where the first argument is the interpreter description of the interpreter in question. The second and third arguments stand for the head and the body of a clause in P, the fourth argument defines the syntactical form in which clauses are passed to τ and the last argument is a goal to retrieve a clause from P; the fourth and fifth argument in I_{pg} are identical.

- Calls to interpreters can be determined automatically by comparing all used interpreter descriptions with the goal in question. Mutual calls of interpreters and recursive calls of interpreters are treated the same way.

- In an environment where several interpreters and compilers are used naming problems easily occur. An easy solution to avoid name clashes is to use the same functor for each derived τ and add one additional argument containing the interpreter description of the interpreter from where τ was derived (the functors "`tauG`" and "`tauP`" are used here for the goal and clause compiler respectively).

4 An Implementation of the Compiler Generator using the Program Transformation Grammars

The transformation of an interpreter into a compiler can be done by a clausewise transformation [Neu90]. Each clause of the interpreter corresponds to one "`tauG`"-clause in the part of the compiler that compiles goals. For interpreter clauses containing link predicates a "`tauP`" clause is generated in addition which accepts input clauses from P and outputs the compiled clauses of P'. The different treatment of link predicates is reflected in the structure of the program transformation grammar.

Some of the goals of the interpreter will appear (maybe in modified form) in the clauses of the compiler τ, some goals will be moved into the clauses of the compiled program P'. In the compiler generator presented in Figure 8 the generation of clauses for "`tauG`" is performed implicitly, while the generation of clauses for "`tauP`" and the treatment of goals for P' is handled explicitly using [`ibl` : ⟨*semantic-type*⟩ = `X`]. The resulting "`tauP`" clauses are returned in the variable "`TauP`" of the top level grammar rule "`interpreter_clause`". In some places in the grammar the predicate "`simplify_body`" is used which removes "`true`" literals from constructed goals.

The semantic types used in the grammar are "`i_call/2`" (for category I), "`check`" (category C), "`linkp/6`" (category L) and "`other`" (category O). The cut symbol "`!`" is treated like a category. We do not provide full treatment cut symbols occuring in the interpreter. In the grammar of Figure 8 cut symbols are only allowed in front or immediately after checking predicates.

An "`interpreter_clause`" has a head literal of type I and a (potentially empty) body. In its argument either a clause processing rule "`tauP`" or the literal "`true`" is returned. The grammar rules for "`i_body`" express the different treatment of clauses containing a link predicate. The first grammar rule is for interpreter clauses without link predicates, which are built up by a sequence of literals belonging to I, O, C and the cut symbol. The constant "`true`" in the last argument of "`i_body`" expresses that for this grammar rule a "`tauP`" clause is never generated.

The second grammar rule for "`i_body`" treats only interpreter rules containing link predicates, where a certain ammount of arguments is needed to express the dependencies of the link predicate. As noted above the grammar does not allow cut symbols right of a I, O or L call, where in the general case new predicate symbols in P' would be needed.

[3]This definition is meaningless for I_{pg}, because it contains no goal matching with "`sys(Goal)`".

% The grammar for the compiler generator

```
interpreter_clause(TauP) ::=
        [hl:i_call(Body, Args)],
        i_body(Body0, Args, TauP),
        {simplify_body(Body0, Body)}.

i_body(Body, _, true) ::= cond_check, simple_body(Body).
i_body(BodyG, AH, TauP) ::=
        cond_check(TP_Goal), other(A),
        [bl:linkp(TP_Goal, BodyG, TauP, TauP0, AH, Body)],
        linkp_rule(C, TauP0),
        other(B),
        {simplify_body((B,C), Body)}.

simple_body(true) ::= {true}.
simple_body((A,Body)) ::= [bl:i_call(A,_)], simple_body(Body).
simple_body((A,Body)) ::= [ibl:other=A], simple_body(Body).

linkp_rule((G1,C,Rest), (TC,TCs)) ::=
        [ibl:i_call(G1,_)=TC],
        other(C),
        linkp_rule(Rest,TCs).
linkp_rule(true, true) ::= {true}.

other(true) ::= {true}.
other((A,B)) ::= [ibl:other=A], other(B).

cond_check(true) ::= {true}.
cond_check((A,B)) ::= [ibl:!=A], cond_check(B).
cond_check((A,C)) ::= [ibl:check=A], cond_check(C).

cond_check ::= {true}.
cond_check ::= [bl:!], cond_check.
cond_check ::= [bl:check], cond_check.
```

% Literal transformations for compiler generator

```
t_lit(_,V,_,_) :- var(V), !, fail.
t_lit(i_call((Tc,G1),A1),exec_interpreter(Ic),
            (is_interpreter(Ic,Id),i_to_tau(Id,Ic,Tc,Gg,G1,A1)),I) :-
        var(Ic), !.
t_lit(i_call((Tc,G1),A1),exec_interpreter(Ic),true,I) :-
        is_interpreter(Ic,Id), !,
        i_to_tau(Id,Ic,Tc,G,G1,A1).
t_lit(i_call(\+G1,A1),\+Ic,Tc,I) :- !, i_to_tau(I,Ic,Tc,G,G1,A1).
t_lit(i_call(G1,A1),Ic,Tc,I) :- !, i_to_tau(I,Ic,Tc,G,G1,A1).
t_lit(check,G,G,I) :- !, is_checkp(I,G).
t_lit(linkp(Check,G1,TauP,TC, P/A,Body),L,
            (au(H,P,A,G1), \+ \+ L, !),I) :- !,
        simplify_body((Check,au(H,P,A,G2),TC),TauP_Body),
        TauP=(tauP(I,IC,(G2:-Body)):-TauP_Body),
        linkp(I,IC,L,H,B).
t_lit(other,call(G),G,I) :- !.
t_lit(other,G,G,I) :- !, is_other(I,G).
t_lit(!,!,!,I).
```

Figure 8: The grammar and the literal transformations for the compiler generator

The literal transformations do not permit variable goals in the interpreter (first clause). The second and third clause of the literal transformations are for variable goals in the object program P. The next two clauses are for positive and negated calls of interpreters. In the clause for "`linkp/6`" the goal for the compiler and the "`tauP`" clause are generated. Finally we should note that a "`call(X)`" in the interpreter is treated as O predicate (typical use: "`sys(X), call(X)`").

The full program listing of the grammar interpreter and all utility predicates used is given in the appendix of this paper. The compiler generator produces identical output for all examples of [Neu86]. The version using the program transformation grammars is much shorter (between a third and a half of the original version), much easier to understand and to modify. The presented version is a simplified version of the grammar the author uses. The full version allows a more general treatment of the cut and performs the variable renaming needed in one example in [Neu88]. The restrictions of [Neu90] apply to this compiler generator as well.

4.1 Self Applicability

Self applicability of compiler generators is a big challenge since [Fu71] (see [FF88, JSS85]). Partial evaluation is an instrument to evaluate a program with respect to some of its data; the result is a specialized program. Furthermore partial evaluation can be used for the compilation of programs. Futamura [Fu71] suggests three projections using a partial evaluator:

1. Partial evaluation of an interpreter with respect to a program; the result is a compiled program.
2. Partial evaluation of a partial evaluator with respect to an interpreter; the result is a compiler.
3. Partial evaluation of a partial evaluator with respect to a partial evaluator; the result is a compiler-generator.

There are some similarities between the futamura projections and our approach of interpreter directed compilation. The key transformation in our approach is the transformation of an interpreter into a compiler performed by the compiler generator $\mathcal{T}$:

$$\mathcal{T}(I) = \tau$$

This transformation corresponds to the second Futamura projection. Compiling a program with the obtained compiler corresponds to the first projection:

$$\tau(P) = P'$$

In the sequel we investigate for the counterpart of the third futamura projection in the framework of interpreter directed compilation.

Let us view the compiler generator $\mathcal{T}$ as a grammar ϕ interpreted by an interpreter I_{pg}

$$\mathcal{T} = I_{pg} \circ \phi$$

If we apply the compiler generator $\mathcal{T}$ on the interpreter I_{pg} we will obtain a compiler for the grammar:

$$\mathcal{T}(I_{pg}) = \tau_{pg}$$

The resulting compiler applied to the grammar ϕ will lead to a Prolog program ϕ' which is equivalent to $I_{pg} \circ \phi$, which is just a more detailed notation of $\mathcal{T}$.

$$\tau_{pg}(\phi) = \phi' \equiv I_{pg} \circ \phi \equiv \mathcal{T}$$

Therefore, if the grammar interpreter I_{pg} is transformed by ϕ' (instead of $\mathcal{T}$ as above), the identical grammar compiler τ_{pg} is obtained:

$$\phi'(I_{pg}) = \tau_{pg}$$

The predicates "`interpreter_clause`" and "`i_body`" of ϕ' are presented in Figure 9.

ϕ' is much more efficient than the original program $\mathcal{T}$ (consisting of the grammar rules and the grammar interpreter), but it has the drawback that it is much harder to understand and to maintain. The part of $\mathcal{T}$ that is more likely to change is the grammar. In order to modify ϕ' one has to care about details of the grammar and the details of the interpreter.

Exactly like $\mathcal{T}$, the compiler generator ϕ' produces identical output for all examples of [Neu86]. These examples are about 10 different applications of metalevel interpreters that range from explanation generation, uncertain reasoning, turtle graphics or goal freezing to forward chaining.

```
interpreter_clause(A,(B :- C),D,E,F,G) :-
        t_lit(i_call(H,I),B,J,E),
        get_head_literal(o,F,J,K),
        i_body(L,I,A,C,D,E,K,G),
        simplify_body(L,H).
i_body(A,B,true,C,D,E,F,G) :-
        cond_check(C,H,E,F,I),
        simple_body(A,H,D,E,I,G).
i_body(A,B,C,D,E,F,G,H) :-
        cond_check(I,D,J,F,G,K),
        other(L,J,M,F,K,N),
        get_literal(i,M,O,P),
        t_lit(linkp(I,A,C,Q,B,R),O,S,F),
        get_literal(o,N,S,T),
        other(U,P,V,F,T,W),
        linkp_rule(X,Q,V,E,F,W,H),
        simplify_body((L,U,X),R).
```

Figure 9: The first three rules of ϕ'

5 Conclusion

We have shown that in order to obtain an efficient version of the compiler generator ϕ', we can apply it on (central parts of) itself. We see the interesting aspect of this self applicability mainly in the software engineering benefits of bootstrapping like in classical compiler construction (see for instance [Wir84]).

We do not believe that this type of self applicability is that Futamura had in mind with his third projection. In Futamura's framework of partial evaluation a single partial evaluator is capable to specialize programs, to generate compiled programs, to generate compilers and to generate even a compiler-compiler. As indicated by the slow progress it seems to be very hard to develop such a powerful partial evaluator for a reasonable language subset.

Another observation is interesting: The T program presented in [Neu86] was written as a plain Prolog program on the same detail level as ϕ'. The program was much more complicated in its overall structure. As stated earlier the size of the program was more than twice the size of the approach using program transformation grammars. We believe that the use of several interpretation layers does indeed lead to systems where it is easier to express the intended semantics of a problem. Furthermore, such systems can be compiled automatically to efficient programs using compiler generators like the one presented here.

References

[Abr84] H. Abramson: *"Definite Clause Translation Grammars"*, in: Proceedings of the 1984 International Symposium on Logic Programming, 1984, pp 233–240.

[FF88] H. Fujita, K. Furukawa: *"A Self-Applicable Partial Evaluator and its Use in Incremental Compilation"*, New Generation Computing, Vol. 6, Nos. 2 and 3, pp. 91–118, 1988.

[Fu71] Y. Futamura: *"Partial Evaluation of Computation Process – an Approach to a Compiler-Compiler"*, Systems, Computers, Controls, Vol. 2, No. 5, 1971, pp. 45–50.

[Gal86] J. Gallagher: *"Transforming Logic Programs by Specializing Interpreters"*, Proceedings of ECAI 86, pp. 109-122, Brighton 1986.

[JSS85] N.D. Jones, P. Sestoft, H. Søndergaard: *"The Generation of a Compiler Generator"*, 1st International Conference on Rewriting Techniques and Applications, Dijon 1985, LNCS 202, pp. 120–140.

[Lak89] A.. Lakhotia: *"Incorporating 'Programming Techniques' into Prolog Programs"*, in: Proceedings of the North American Logic Programming Conference 1989, MIT Press, Cambridge 1989.

[McC87] M. McCord: *"Natural Language Processing and Prolog"*, in: A. Walker et al.: *"Knowledge Systems and Prolog"*, Addison Wesley, Reading 1987.

[Neu86] G. Neumann: *"Meta-Interpreter Directed Compilation of Logic Programs into Prolog"*, IBM-Research Report RC 12113 (#54357), Yorktown Heights, New York 1986.

[Neu88] G. Neumann: *"Metaprogrammierung und Prolog"*, Addison-Wesley, Bonn 1988.

[Neu90] G. Neumann: *"Transforming Interpreters into Compilers by Goal Classification"*, in: Proceedings of the Meta90, 4-6 April 1990, Leuven, Belgium 1990.

[Pag88] F.G. Pagan: *"Converting Interpreters into Compilers"*, Software-Practice and Experience, Vol. 18, No. 6, pp. 509–527, June 1988.

[PW80] F.C.N. Pereira, D.H.D. Warren: *"Definite Clause Grammars for Language Analysis—A Survey of the Formalism and a Comparison with Augmented Transition Networks"*, Artificial Intelligence, Vol. 13, Nr. 3, pp. 231-278, May 1980.

[Per81] F.C.N. Pereira: *"Extraposition Grammars"*, American Journal of Computational Linguistics, 1981, vol. 7, no. 4, pp 243–255.

[TF86] A. Takeuchi, K. Furukawa: *"Partial Evaluation of Prolog Programs and its Application to Meta Programming"*, IFIP 1986.

[vR88] P. Van Roy: *"A Useful Extension to Prolog's Definite Clause Grammar Notation"*, SIGPLAN Notices, Volume 24, Number 11, November 1989.

[Wir84] N. Wirth: *"Compilerbau"*, Teubner Verlag, Stuttart 1984.

6 Appendix

```
:-      op(1200,xfx,'::='),
        op(900,xfy,(:)).

ptg(A, In0,In1, I, Out0,Out1) :- var(A), !,
        exec_interpreter(ptg(A, In0,In1, I, Out0,Out1)).
ptg(\+ A, In0,In0,I, Out0,Out0) :- !, \+ ptg(A, In0,In1, I, Out0,Out1).
ptg(true, In,In, _, Out,Out) :- !.
ptg((A,B), In0,In2, I, Out0,Out2) :- !,
        ptg(A, In0,In1, I, Out0,Out1),
        ptg(B, In1,In2, I, Out1,Out2).
ptg([ibl:Type=Literal1], In0,In1, I, Out,Out) :- !,
        get_literal(i,In0,Literal0,In1),
        t_lit(Type,Literal0,Literal1,I).
ptg({Goal}, In,In, I, Out,Out) :- !, call(Goal).
ptg([bl:Type], In0,In1, I, Out0,Out1) :- !,
        get_literal(i,In0,Literal0,In1),
        t_lit(Type,Literal0,Literal1,I),
        get_literal(o,Out0,Literal1,Out1).
ptg([hl:Type], In0,In1, I, Out0,Out1) :- !,
        get_head_literal(i,In0,Literal0,In1),
        t_lit(Type,Literal0,Literal1,I),
        get_head_literal(o,Out0,Literal1,Out1).
ptg(A, In0,In1, I, Out0,Out1) :- (A ::= B), ptg(B, In0,In1, I, Out0,Out1).


i_to_tau(I,Goal,tauG(Id,G,G1,Ss),G,G1,S/Ss) :-
        is_interpreter(Goal,Id),!,
        get_prg(Id,Goal,_,G),
        get_states(Id,Goal,S,Ss).

is_checkp(I,Pred)      :-  nonvar(Pred), \+ \+ checking_pred(I,_,Pred).
is_linkp(I,Pred)       :-  nonvar(Pred), \+ \+ link_pred(I,_,_,_,Pred).
linkp(I,IC,Pred,H,B)   :-  nonvar(Pred), link_pred(I,H,B,IC,Pred).
is_other(I,Pred) :-
        \+ Pred = (_,_),
        \+ is_linkp(I,Pred),
        \+ is_checkp(I,Pred),
```

```
        \+ Pred = !,
        \+ Pred = true,
        \+ Pred = exec_interpreter(_),
        \+ is_interpreter(Pred,_),
        \+ (Pred = (\+ P), is_interpreter(P)).

is_interpreter(P) :- nonvar(P), is_interpreter(P,_).
is_interpreter(P,Id) :- mgt(P,Id), link_pred(Id,_,_,_,_),!.


simplify_body(A,B) :- flatten(A,A1), simplify_true(A1,B).

flatten(A,A) :- literal(A), !.
flatten((A,B),(A,C)) :- literal(A), !, flatten(B,C).
flatten(((A,B),C),D) :- flatten((A,B,C),D).

simplify_true(A,A) :- literal(A), !.
simplify_true((A,B),C) :- A==true, !, simplify_true(B,C).
simplify_true((A,B),A) :- simplify_true(B,C), C==true, !.
simplify_true((A,B),(A,C)) :- simplify_true(B,C).

literal(A) :- var(A), !.
literal(A) :- \+ A = (_,_).


get_literal(i,A,A,true) :- literal(A), \+ A==true, !.
get_literal(o,A,C,A) :- C==true, !.
get_literal(_,AB,A,B) :- AB=(A,B), !.

get_head_literal(_,(H:-B),X,Y) :- !, X=H, Y=B.
get_head_literal(_,A,A,true) :- !.


get_states(I,P,States,Vars) :-
   locate(s,I,States),
   get_states(P,States,Vars).

get_states(P,[],[]) :-!.
get_states(P,[S|T],[State|States]) :- arg(S,P,State), get_states(P,T,States).

get_prg(I,P,Pos,Prg) :-
   locate(p,I,Pos),
   get_prg_(I,P,Pos,Prg).

get_prg_(I,P,[],Functor) :- !,
   functor(I,Functor,_) .
get_prg_(_,P,[Po],Prg) :- nonvar(P), arg(Po,P,Prg).


locate(Atom,Term,Pos) :-
   Term =..  [M|List],
   locate_(Atom,List,1,Pos).

locate_(A,[A|T],Pos,[Pos|PosT]) :- !, NPos is Pos+1, locate_(A,T,NPos,PosT).
locate_(A,[_|T],Pos,      PosT)  :- !, NPos is Pos+1, locate_(A,T,NPos,PosT).
locate_(_,[],_,[]).


au(P,[],_,P) :- !.
au(P,[H|T],[V|Vs],P2) :- !,
   P =..  [F|Args], append(Args,[V|Vs],Args1), P2=..  [F|Args1].
```

FREE DEDUCTION:
AN ANALYSIS OF "COMPUTATIONS" IN CLASSICAL LOGIC

Michel Parigot
Equipe de logique – CNRS UA 753
45–55 5ème étage, Université Paris 7
2 place jussieu, 75251 PARIS Cedex 05, FRANCE
e–mail: parigot@logique.jussieu.fr

Abstract: Cut–elimination is a central tool in proof–theory, but also a way of computing with proofs used for constructing new functional languages. As such it depends on the properties of the deduction system in which proofs are written.

For intuitionistic logic, natural deduction allows a cut–elimination procedure which effectively provides a computation mechanism with deep theoretical properties such as confluence and strong normalisation. For classical logic, on the contrary, neither natural deduction nor sequent calculus provide a suitable cut–elimination, and, in fact, the computational meaning of classical proofs is an open problem.

In this paper, a new deduction sytem is introduced: free deduction. Free deduction is an adequate system for classical logic, allowing a global cut–elimination procedure in the style of intuitionistic natural deduction. We prove that, provided a choice of inputs (which corresponds to a fundamental non–determinism of classical logic), the cut–elimination procedure in free deduction provides a computation mechanism for classical logic which satisfies confluence and strong normalisation.

1. INTRODUCTION

Cut–elimination is a central tool in proof–theory, but also a way of computing with proofs. As such it depends on the properties of the deduction system in which proofs are written.

In the case of intuitionistic logic, natural deduction allows a cut–elimination procedure which effectively provides a computation mechanism – widely used as a theoretical (and even implementational) base for functional programming languages. Via the so–called Curry–Howard isomorphism, it corresponds to normalisation in typed lambda–calculus. As such it has deep theoretical properties which ensure its computational pertinence: (i) confluence (or Church–Rosser property), which says that the computation mechanism is deterministic, and (ii) strong normalisation, which says that every computation terminates.

But if we turn to classical logic, the situation radically changes. Classical natural deduction is no more satisfactory from the viewpoint of cut–elimination and even the notion of cut–free proof becomes problematic. The addition of the absurdity rule conflicts with the definition of a suitable cut–elimination. Sequent calculus, on the other hand, is a deduction system for classical logic in which cut–elimination is a powerfull tool for proof–theory. But it seems difficult to assign a computational meaning to cut–elimination in sequent calculus. This is not only because the computation process fails to have good properties, but more deeply because it is not defined in a canonical way.

What are the reasons for that situation? One could simply give an a priori answer such as: this is just that classical logic is not constructive. Because I find this answer unsatisfactory, I have tried to get a more technical answer, which could explain in what precise sense it is not constructive. The first step was to separate the problems due to classical logic from the ones due to the deduction systems in which classical proofs are written. This led me to the definition of a new deduction system for classical logic: free deduction (FD).

Free deduction is a fully symetric system (in the style of sequent calculus) in which formulas and sequents are the usual ones, and the notion of cut internalised (as in natural deduction). It enjoys the cut–elimination theorem, and could be used as an alternative system for proof–theoretic studies. Sequent calculus and natural deduction are obtained from free deduction using a very natural and uniform restriction: it suffices to kill certain premises of the rules (they are not mentionned, but have the effect of axioms). In this respect, FD appears both as a system and a meta–system in which one can transform proofs from one system to another.

But the main distinctive properties of free deduction concern the computational aspects. Two cut–elimination procedures are available in FD: a local procedure (as in sequent calculus) which proceeds by permutation of rules and a global one (as in natural deduction) which proceeds by composition of proofs. The existence of global cut–elimination procedure is a crucial property of FD: it effectively provides a computation mechanism for classical logic, for which confluence and strong normalisation hold. A confluence result may seem strange, because of the non–deterministic nature of classical logic. But in fact, non–determinism is entirely included in the choice of the inputs: one has to choose whether the input of a formula is to the right or to the left of the sequents.

Looking from the point of view of FD, one obtains a good explanation of the computational differences between natural deduction and sequent calculus. In these systems, the choice of the inputs is syntactically imposed by the choice of the killed premisses of the rules. In (intuitionistic) natural deduction inputs are systematically to the left (in a sequent $A_1,\ldots,A_n \vdash B$, the formulas $A_1,\ldots,A_n$ are inputs and B is an output) and this leads to a deterministic computation mechanism (the usual functional interpretation of intuitionistic proofs). In sequent calculus, inputs are not always available, and this prevents to define a suitable computational mechanism.

In free deduction, all the inputs are available. As a consequence cut–elimination provides a non–deterministic computation mechanism for classical logic. But we prove that the fundamental non–determinism of classical logic amounts to a choice of the inputs: each such choice give birth to a deterministic computation mechanism satisfying confluence and strong normalisation (note that for us, determinism refers to the result and not to the computation process, otherwise confluence would have no meaning).

In what follows we only discuss propositional free deduction, but the system and the results obviously generalise to the predicate case.

Notations: latin letters A,B,C,.. denote formulas and greek letters Γ,Δ,Π,Σ denote sets of formulas;

2. THE SYSTEM OF FREE DEDUCTION

2.1 Rules of free deduction

Axiom

$A \vdash A$

Conjunction

$$\frac{\Gamma, A\wedge B \vdash \Delta \quad \Pi \vdash A, \Sigma \quad \Pi' \vdash B, \Sigma'}{\Gamma, \Pi, \Pi' \vdash \Delta, \Sigma, \Sigma'} \qquad \frac{\Gamma \vdash A\wedge B, \Delta \quad \Pi, A \vdash \Sigma \ \{\Pi, B \vdash \Sigma\}}{\Gamma, \Pi \vdash \Delta, \Sigma}$$

Disjunction

$$\frac{\Gamma, A\vee B \vdash \Delta \quad \Pi \vdash A, \Sigma \ \{\Pi \vdash B, \Sigma\}}{\Gamma, \Pi \vdash \Delta, \Sigma} \qquad \frac{\Gamma \vdash A\vee B, \Delta \quad \Pi, A \vdash \Sigma \quad \Pi', B \vdash \Sigma'}{\Gamma, \Pi, \Pi' \vdash \Delta, \Sigma, \Sigma'}$$

Negation

$$\frac{\Gamma, \neg A \vdash \Delta \quad \Pi, A \vdash \Sigma}{\Gamma, \Pi \vdash \Delta, \Sigma} \qquad \frac{\Gamma \vdash \neg A, \Delta \quad \Pi \vdash A, \Sigma}{\Gamma, \Pi \vdash \Delta, \Sigma}$$

Implication

$$\frac{\Gamma, A\to B \vdash \Delta \quad \Pi, A \vdash \Sigma \ \{\Pi \vdash B, \Sigma\}}{\Gamma, \Pi \vdash \Delta, \Sigma} \qquad \frac{\Gamma \vdash A\to B, \Delta \quad \Pi \vdash A, \Sigma \quad \Pi', B \vdash \Sigma'}{\Gamma, \Pi, \Pi' \vdash \Delta, \Sigma, \Sigma'}$$

A sequent between brackets beside a rule means that there are in fact two rules, the one which is written and the one formed by replacing the right premise by the sequent between brackets.

The left premise of a rule is called the main premise and the other ones, secondary premises; the formula mentionned in the premise is called the active formula of the premise and the other ones form the context; a rule whose main premise has A as active formula is called an elimination of A.

Meaning of the rules

Sequents are interpreted as usual: the sequent $A_1,\ldots,A_n \vdash B_1,\ldots,B_m$ is true if and only if the formula $A_1 \wedge\ldots\wedge A_n \to B_1 \vee\ldots\vee B_m$ is true; in particular $\vdash A$ (resp. $A \vdash$) is true if and only if A is true (resp. false).

The intuitive meaning of the rules is that the premises give the conditions to derive a contradiction; for instance for the connective and:

- one derives a contradiction from "A ∧ B false" and "A true and B true";
- one derives a contradiction from "A ∧ B true" and "A false" (or "B false").

Structural rules

There are two ways of adding structural rules to FD :

(i) Explicit (or local) structural rules: weakening and contraction are governed by their usual rules.

(ii) Implicit (or global) structural rules:

- weakening is obtained (as in natural deduction) by the convention that the active formula of a premise does not necessary occur; when it occurs the premise is called strict;
- contraction is obtained by the convention that the contexts in the premises of a rule are always contracted in the conclusion of the rule.

As usual, a more complex structure is in fact needed from an algorithmic viewpoint: one needs indexed formulas A^x (or x:A in the notation of typed lambda–calculus) and not only formulas.

Explicit structural rules are more convenient for the local cut–elimination procedure and implicit structural rules are more convenient for the global one.

Proposition (completeness) FD is complete for provability in classical logic. Moreover,

(i) there is a "direct" translation of (cut–free) proofs of sequent calculus into (cut–free) proofs in this system (which don't use additional structural rules);

(ii) there is a "direct" translation of proofs of natural deduction into proofs in this system, (which respects the structure of cuts).

Remark. Like natural deduction, free deduction can be presented both as a deduction system of sequents and a deduction system of formulas.

Variant. A important variant FD' of the system FD (which is not considered in this paper) is obtained by "unifying" the pairs of rules for $\wedge$d, $\vee$g and $\neg$g using secondary premises with two active formulas:

$$\frac{\Gamma \vdash A \wedge B,\ \Delta \qquad \Pi,\ A,\ B \vdash \Sigma}{\Gamma,\ \Pi \vdash \Delta,\ \Sigma}$$

$$\frac{\Gamma,\ A \vee B \vdash \Delta \qquad \Pi \vdash A,\ B,\ \Sigma}{\Gamma,\ \Pi \vdash \Delta,\ \Sigma}$$

$$\frac{\Gamma,\ A \rightarrow B \vdash \Delta \qquad \Pi,\ A \vdash B,\ \Sigma}{\Gamma,\ \Pi \vdash \Delta,\ \Sigma}$$

In FD' binary connectives become symetric, from the algorithmic point of view.

2.2 Formula property

The formula property is a very simple property, which is in fact the base of the global cut–elimination procedure and therefore of the main computational properties of free deduction. It says that each formula occuring in the conclusion of a proof comes from an axiom (in the case of explicit strutural rules one needs to add "or from a weakening"). This has to be compared with natural deduction where we have only a left formula property: each formula occuring to the left of the sequent conclusion of a proof comes from an axiom.

2.3 Permutability property

The permutability property is the base of the local cut-elimination property: it is a systematization of a partly existing property of sequent calculus.

Permutability property between logical rules

The rules of FD are permutable in the following sense: suppose that we have a rule R' whose conclusion is a premise of a rule R and C_1 is the active formula of this premise; then we can permute the two rules in the following way: first apply R to the premise of R' containing C_1 and then R' to the conclusion; the conclusion of the proof is unchanged (in fact the only sequent which is changed is the intermediate sequent between the two rules) and no structural rule is added. Moreover the permutation process is reversible: if one permutes R and R' and then R' and R, we get the original proof.

Here is the general scheme of permutation (in order to cover the general situation we represent sequents by lists of formulas, indicate active formulas in boldface and consider the premises of a rule as unordered):

$$\dfrac{\dfrac{\Gamma'_1, C_1, \mathbf{C'_1} \qquad \Gamma'_2, \mathbf{C'_2} \qquad \Gamma'_3, \mathbf{C'_3}}{\Gamma'_1, \Gamma'_2, \Gamma'_3, \mathbf{C_1}}\,R' \qquad \Gamma_2, \mathbf{C_2} \qquad \Gamma_3, \mathbf{C_3}}{\Gamma'_1, \Gamma'_2, \Gamma'_3, \Gamma_2, \Gamma_3}\,R$$

becomes

$$\dfrac{\dfrac{\Gamma'_1, C'_1, \mathbf{C_1} \qquad \Gamma_2, \mathbf{C_2} \qquad \Gamma_3, \mathbf{C_3}}{\Gamma'_1, \Gamma_2, \Gamma_3, \mathbf{C'_1}}\,R \qquad \Gamma'_2, \mathbf{C'_2} \qquad \Gamma'_3, \mathbf{C'_3}}{\Gamma'_1, \Gamma_2, \Gamma_3, \Gamma'_2, \Gamma'_3}\,R'$$

Permutability property for contractions

The permutation of a contraction and a logical rule is not reversible: the permutation is only possible if the contraction precedes the logical rule (otherwise the contracted occurences can come from different premises of the logical rule, and the permutation is impossible). There are two different cases: if the contracted formula is not an active formula of the logical rule, then the permutation is obvious and doesn't change the structure of the proof; if the contracted formula is an active formula of the logical rule, then the permutation produces a duplication of the logical rule:

$$\dfrac{\dfrac{\Gamma_1, C_1, C_1}{\Gamma_1, \mathbf{C_1}}\,Ct \qquad \Gamma_2, \mathbf{C_2} \qquad \Gamma_3, \mathbf{C_3}}{\Gamma_1, \Gamma_2, \Gamma_3}\,R$$

becomes

$$\dfrac{\dfrac{\Gamma_1, C_1, C_1 \qquad \Gamma_2, C_2 \qquad \Gamma_3, C_3}{\Gamma_1, C_1, \Gamma_2, \Gamma_3}\,R \qquad \Gamma_2, C_2 \qquad \Gamma_3, C_3}{\dfrac{\Gamma_1, \Gamma_2, \Gamma_3, \Gamma_2, \Gamma_3}{\overline{\overline{\Gamma_1, \Gamma_2, \Gamma_3}}}\,Ct}\,R$$

<u>**Permutability property for weakenings**</u>

The permutation of a weakening and a logical rule is interesting only if <u>the weakening precedes the logical rule</u> (otherwise the permutation is possible but arbitrary: one has to choose an arbitrary premise of the logical rule). There are two different cases: if the weakened formula is not an active formula of the logical rule, then the permutation is obvious and doesn't change the structure of the proof; if the weakened formula is an active formula of the logical rule, then the permutation produces an <u>erasing</u> of the logical rule:

$$\dfrac{\dfrac{\Gamma_1}{\Gamma_1, C_1}\,Wk \qquad \Gamma_2, C_2 \qquad \Gamma_3, C_3}{\Gamma_1, \Gamma_2, \Gamma_3}\,R$$

becomes

$$\dfrac{\Gamma_1}{\Gamma_1, \Gamma_2, \Gamma_3}\,Wk \quad \text{(double line)}$$

3. LOGICAL PROPERTIES OF FREE DEDUCTION

<u>3.1 Cuts and their elimination</u>

A <u>cut</u> in a proof is either (i) an elimination R of a weakened formula A, or (ii) a right (resp. left) elimination R of a formula A, such that the proof of the main premise of R contains a left (resp. right) elimination of the same formula A ("the same" means that the two occurences of A are related by an axiom); in this case, A is called a <u>cut–formula</u>.

<u>**Example**</u>. A∧B is a cut–formula in the following proof:

$$\dfrac{\dfrac{\begin{matrix} d_0 \\ \vdots \\ \Gamma, A\wedge B \vdash A\wedge B, \Delta\end{matrix} \qquad \begin{matrix} d_1 \\ \vdots \\ \Pi \vdash A, \Sigma\end{matrix} \qquad \begin{matrix} d_2 \\ \vdots \\ \Pi' \vdash B, \Sigma'\end{matrix}}{\Gamma, \Pi, \Pi' \vdash A\wedge B, \Delta, \Sigma, \Sigma'}\,\wedge l \qquad \begin{matrix} d_3 \\ \vdots \\ \Pi'', A \vdash \Sigma''\end{matrix}}{\Gamma, \Pi, \Pi', \Pi'' \vdash \Delta, \Sigma, \Sigma', \Sigma''}\,\wedge r$$

This notion of cut enjoys the usual proof–theoretic properties.

<u>Subformula property</u>: each formula which appears in a proof without cut is a subformula of a formula of the conclusion of the proof.

<u>Cut elimination theorem</u>: each proof can be effectively transformed into a proof without cut.

The cut–elimination theorem is proved using a local procedure inspired by Gentzen procedure for sequent calculus (cf [2]). Because of the permutation property, it can be presented more uniformly.

Here is an example of a proof without cuts which is neither in natural deduction, nor in sequent calculus:

$$\dfrac{\dfrac{A\vee\neg A \vdash A\vee\neg A \qquad \neg A \vdash \neg A}{\neg A \vdash A\vee\neg A}\vee l \qquad \dfrac{A\vee\neg A \vdash A\vee\neg A \qquad A \vdash A}{A \vdash A\vee\neg A}\vee l}{\vdash A\vee\neg A}\neg r$$

3.2 Interpretation of sequent calculus

There is a direct interpretation of (cut–free) sequent calculus into free deduction. Axioms (and structural rules) are unchanged and logical rules of sequent calculus appear as particular cases of the corresponding rules of free deduction: it suffices to <u>kill the main premise</u> (i.e. the main premise do not appear in the rule, but has the effect of an axiom). Note that one would obtain a different version of sequent calculus starting from the alternative system FD' instead of FD.
The killed premises are written inside brackets.

$$\dfrac{[A\wedge B \vdash A\wedge B] \qquad \Pi \vdash A,\ \Sigma \qquad \Pi' \vdash B,\ \Sigma'}{\Pi,\ \Pi' \vdash A\wedge B,\ \Sigma,\ \Sigma'} \qquad\qquad \dfrac{[A\wedge B \vdash A\wedge B] \qquad \Pi,\ A \vdash \Sigma \quad \{\Pi,\ A \vdash \Sigma\}}{\Pi,\ A\wedge B \vdash \Sigma}$$

$$\dfrac{[A\vee B \vdash A\vee B] \qquad \Pi \vdash A,\ \Sigma \quad \{\Pi \vdash B,\ \Sigma\}}{\Pi \vdash A\vee B,\ \Sigma} \qquad\qquad \dfrac{[A\vee B \vdash A\vee B] \qquad \Pi,\ A \vdash \Sigma \qquad \Pi',\ B \vdash \Sigma'}{\Pi,\ \Pi',\ A\vee B \vdash \Sigma,\ \Sigma'}$$

$$\dfrac{[\neg A \vdash \neg A] \qquad \Pi,\ A \vdash \Sigma}{\Pi \vdash \neg A,\ \Sigma} \qquad\qquad \dfrac{[\neg A \vdash \neg A] \qquad \Pi \vdash A,\ \Sigma}{\Pi,\ \neg A \vdash \Sigma}$$

$$\dfrac{[A\to B \vdash A\to B] \qquad \Pi,\ A \vdash \Sigma \quad \{\Pi \vdash B,\ \Sigma\}}{\Pi \vdash A\to B,\ \Sigma} \qquad\qquad \dfrac{[A\to B \vdash A\to B] \qquad \Pi \vdash A,\ \Sigma \qquad \Pi',\ B \vdash \Sigma'}{\Pi,\ \Pi',\ A\to B \vdash \Sigma,\ \Sigma'}$$

The following proposition shows that free deduction can be seen as a meta–system in which proofs can be transformed internally (by permutation of rules) from one system to another.

Proposition. There is a procedure which transform each (cut–free) proof of free deduction into a (cut–free) proof of sequent calculus without cut.

The procedure is based on the permutation of rules whose main premise is not an axiom; another possibility is the global elimination of structural cuts of section 4.3.

3.3 Interpretation of natural deduction

Rules of natural deduction also appear as particular cases of rules of free deduction. One crucial property of natural deduction is to have inputs to the left. It can be expressed by a killing of (i) the main premise, for left rules, and (ii) the secondary premises with a left active formula, for right rules. The resulting system is a natural deduction system with multiple conclusions.

$$\frac{[A\wedge B \vdash A\wedge B] \quad \Pi \vdash A,\ \Sigma \quad \Pi' \vdash B,\ \Sigma'}{\Pi,\ \Pi' \vdash A\wedge B,\ \Sigma,\ \Sigma'} \qquad \frac{\Gamma \vdash A\wedge B,\ \Delta \quad [A \vdash A] \quad \{[B \vdash B]\}}{\Gamma \vdash A,\ \Delta}$$

$$\frac{[A\vee B \vdash A\vee B] \quad \Pi \vdash A,\ \Sigma \quad \{\Pi \vdash B,\ \Sigma\}}{\Pi \vdash A\vee B,\ \Sigma} \qquad \frac{\Gamma \vdash A\vee B,\ \Delta \quad [A \vdash A] \quad [B \vdash B]}{\Gamma \vdash A,\ B,\ \Delta}$$

$$\frac{[\neg A \vdash \neg A] \quad \Pi,\ A \vdash \Sigma}{\Pi \vdash \neg A,\ \Sigma} \qquad \frac{\Gamma \vdash \neg A,\ \Delta \quad \Pi \vdash A,\ \Sigma}{\Gamma,\ \Pi \vdash \Delta,\ \Sigma}$$

$$\frac{[A\to B \vdash A\to B] \quad \Pi,\ A \vdash \Sigma \quad \{\Pi \vdash B,\ \Sigma\}}{\Pi \vdash A\to B,\ \Sigma} \qquad \frac{\Gamma \vdash A\to B,\ \Delta \quad \Pi \vdash A,\ \Sigma \quad [B \vdash B]}{\Gamma,\ \Pi \vdash B,\ \Delta,\ \Sigma}$$

For the usual natural deduction system, there is one more requirement, which corresponds to the functional interpretation of natural deduction: the ouput must be unique or, in other words, sequents must have at most one formula to the right. It can be also obtained by a killing of premises, starting from a variant of FD with at most one formula to the right (the restriction of FD to sequents with at most one formula to the right is complete for classical logic). Moreover in order to keep completeness with respect to classical logic one additional rule is needed: the absurdity rule. It is obtained from left negation rule by killing the secondary premise, instead of the main premise (thus destroying the symetry).

4. COMPUTATIONAL PROPERTIES OF FREE DEDUCTION

In this section we define and study a global procedure for cut–elimination in free deduction. In fact, this procedure does more than cut elimination. Because free deduction contains cut–free subsystems of different natures (like sequent calculus and natural deduction) which are complete for classical logic, one has the possibility to compute not only cut–free proofs, but cut–free proofs of a particular form. For doing this, one has to define generalised notions of cuts including not only logical cuts, but also structural cuts depending on the particular form choosen.

In what follows we investigate the case where the particular form is a cut–free proof of sequent calculus. In this case a generalised cut is a rule whose main premise is not a strict axiom.

4.1 Composition of proofs

The global cut–elimination procedure uses implicit structural rules. We therefore deal with indexed formulas A^x, A^y, B^x, B^y... (indexed are mentionned only when necessary). In sequents, left and right occurences receive distinct indexes. A bounded occurence of A^x in a proof d is an occurence of A^x in a subproof of a premise of a rule where A^x is active; a free occurence of A^x in d, is an occurence which is not bounded. We suppose for simplicity that an active formula A^x of a sequent in a proof never occurs outside the subproof of this sequent. A formula is free (resp. bounded) in a proof, if it has a free (resp. bounded) occurence in this proof.

The basic mechanism of the global cut–elimination procedure is the composition of proofs, as for intuitionistic natural deduction or typed lambda–calculus: one replaces in a proof d, occurences of a hypothesis A, by a proof e of A. In natural deduction inputs are necessary to the left of sequents and therefore this is the only way of composing proofs; in free deduction one has the freedom to decide, for each formula, whether the input is to the right or to the left and this gives two possible ways of composing proofs:

input to the left: $d[[\vdash A^z]e/A^x\vdash]$

input to the right: $d[[A^z\vdash]e/\vdash A^x]$.

For A^x not bounded in d, the proof $d[[\vdash A^z]e/A^x\vdash]$ (resp. $d[[A^z\vdash]e/\vdash A^x]$) is defined as the result of replacing in d, each axiom $A^x \vdash A^y$ (resp. $A^y \vdash A^x$) by the proof $e[A^y/A^z]$, where $e[A^y/A^z]$ is defined, as usual, as the result of replacing the formula A^z by the formula A^y in the proof e (a formal definition of substitution is given in § 5.1). Note that because of our conventions on indexes one could avoid to mention explicitely left and right occurences, but we would loose clarity.

4.2 Logical cuts

A logical cut is a right (resp. left) elimination of a non contracted formula A^x, whose main premise is the conclusion of a left (resp. right) elimination of A^y, whose main premise is the axiom $A^y \vdash A^x$ (resp $A^x \vdash A^y$).

Each connective has an associated cut. Let us examine the cases of negation and conjunction.

negation

$$\dfrac{\dfrac{\neg A \vdash \neg A \qquad \overset{d_1}{\Pi, A^x \vdash \Sigma}}{\Pi \vdash \neg A, \Sigma}\neg l \qquad \overset{d_2}{\Pi' \vdash A^y, \Sigma'}}{\Pi, \Pi' \vdash \Sigma, \Sigma'}\neg r$$

The natural way to eliminate this cut is to replace the proof by its reducts, which is obtained, as in natural deduction, by choosing a **left input** for A:

$$\begin{array}{c} d_1[[\vdash A^y]d_2/A^x\vdash] \\ \cdots\cdots \\ \Pi, \Pi'' \vdash \Sigma, \Sigma'' \end{array}$$

where $\Pi'' = \Pi'$ and $\Sigma'' = \Sigma'$ if $A^x$is not weakened in d_1, or $\Pi'' = \Sigma'' = \emptyset$ otherwise.
But there is another possible reducts, corresponding to the choice of a **right input** for A:

$$\begin{array}{c} d_2[[A^x\vdash]d_1/\vdash A^y] \\ \cdots\cdots \\ \Pi, \Pi'' \vdash \Sigma, \Sigma'' \end{array}$$

where $\Pi'' = \Pi$ and $\Sigma'' = \Sigma$ if A^y is not weakened in d_2, or $\Pi'' = \Sigma'' = \emptyset$ otherwise.
Because of the formula property, these reducts are well defined, but they are in general completely different. In the existence of these two possibilities, lies the fundamental computational non-determinism of classical logic.

conjunction

$$\dfrac{\dfrac{A\wedge B \vdash A\wedge B \qquad \overset{d_1}{\Pi \vdash A^y, \Sigma} \qquad \overset{d_2}{\Pi' \vdash B, \Sigma'}}{\Pi, \Pi' \vdash A\wedge B, \Sigma, \Sigma'}\wedge l \qquad \overset{d_3}{\Pi'', A^x \vdash \Sigma''}}{\Pi, \Pi', \Pi'' \vdash \Sigma, \Sigma', \Sigma''}\wedge r$$

One has again two possible reductions:

$$\begin{array}{c} d_3[[\vdash A^y]d_1/A^x\vdash] \\ \cdots\cdots \\ \Pi_1, \Pi'' \vdash \Sigma_1, \Sigma'' \end{array} \quad \text{or} \quad \begin{array}{c} d_1[[A^x\vdash]d_3/\vdash A^y] \\ \cdots\cdots \\ \Pi, \Pi_2 \vdash \Sigma, \Sigma_2 \end{array}$$

where Π_1,Σ_1 are respectively Π,Σ or $\emptyset$, and Π_2,Σ_2 are respectively Π'',Σ'' or $\emptyset$.

4.3 Strutural cuts

A structural cut is a generalised cut which is not a logical cut. The elimination of a structural cut R is obtained by reversing the order of the rules: one applies the rule R directly to the axioms containing occurences of the main formula of R. This elimination can be defined as a composition of proofs (suppose for example that the main formula is to the right):

$$\frac{\begin{matrix} d_1 \\ \cdots\cdots \\ \Gamma \vdash A^y, \Delta \end{matrix} \quad \cdots\cdots}{\Gamma, \Gamma' \vdash \Delta, \Delta'} R$$

reduces to

$$\begin{matrix} d_1[[A^w \vdash]e/\vdash A^y] \\ \cdots\cdots\cdots\cdots \\ \Gamma, \Gamma'' \vdash \Delta, \Delta'' \end{matrix}$$

with e := $\dfrac{A^w \vdash A^y \quad \cdots\cdots}{\Gamma', A^w \vdash \Delta'}$, w is a new index and Γ'',Δ'' are respectively Γ',Δ' or $\emptyset$.

Remark. The procedure of elimination of structural cuts allows to transform a proof in free deduction into a proof which is "essentially" in sequent calculus: one has only to write the cuts as aplications of the cut–rule of sequent calculus.

4.4 Confluence and strong normalisation

We have seen in 4.2 that there is a computational non–determinism in classical logic, which corresponds to the freedom for the choice of the inputs. From now on we will fix the inputs. For this we consider signed formulas A_l or A_r, where l and r are signs which indicate where the input is (the sign is considered as a part of the formula), and we suppose that the elimination of logical cuts is done according to the signs; for example

$$\frac{\dfrac{A_r \wedge B \vdash A_r \wedge B \quad \begin{matrix} d_1 \\ \cdots\cdots \\ \Pi \vdash A_r{}^y, \Sigma \end{matrix} \quad \begin{matrix} d_2 \\ \cdots\cdots \\ \Pi' \vdash B, \Sigma' \end{matrix}}{\Pi, \Pi' \vdash A_r \wedge B, \Sigma, \Sigma'} \wedge l \qquad \begin{matrix} d_3 \\ \cdots\cdots \\ \Pi'', A_r{}^x \vdash \Sigma'' \end{matrix}}{\Pi, \Pi', \Pi'' \vdash \Sigma, \Sigma', \Sigma''} \wedge r$$

reduces to

$$\begin{matrix} d_1[[A_r{}^x \vdash]d_3/\vdash A_r{}^y] \\ \cdots\cdots\cdots\cdots \\ \Pi, \Pi_2 \vdash \Sigma, \Sigma_2 \end{matrix}$$

where Π_2,Σ_2 are respectively Π'',Σ'' or $\emptyset$.

Remark The semantic of formulas is not changed by the addition of signs: signs are just external annotations for cut–elimination without any effect on the proofs themselves.

Once the inputs are fixed, cut–elimination in free deduction becomes a deterministic computation mechanism for classical logic which is very like the one provided by natural deduction for intuitionistic logic (where the inputs are fixed by the syntax). In particular it enjoys the confluence and strong normalisation properties. Moreover these properties hold for the structural and logical cuts separately. Here is a summary of the results proved in the next section.

<u>Theorem A</u>. The procedure of elimination of structural cuts enjoys:
- the confluence property, up to the order of logical cuts, and
- the strong normalisation property.

<u>Theorem B</u>. The procedure of elimination of logical cuts enjoys:
- the confluence property, and
- the strong normalisation property.

<u>Theorem C</u>. The procedure of elimination of generalised cuts enjoys:
- the confluence property, and
- the strong normalisation property.

<u>Remark</u>. These results apply in particular to proofs of sequent calculus and allow to compute corresponding cut–free proofs in sequent calculus, but the computation cannot be done inside sequent calculus! (i.e. in general, intermediate proofs are not proofs of sequent calculus). In fact, having free deduction in mind, we can define inside sequent calculus cut–elimination procedures giving the same results.

<u>Remark</u>. For the "unified" rules of FD' the situation is a little bit more complex. The choice of the inputs doesn't completely determine the situation in one case: if the secundary active formulas of a unified rule are both outputs, then the order between the two corresponding inputs is not determined by the signs; here is the case of the conjunction:

$$\dfrac{\dfrac{A_r \wedge B_r \vdash A_r \wedge B_r \qquad \overset{d_1}{\Pi \vdash A_r{}^{y}, \Sigma} \qquad \overset{d_2}{\Pi' \vdash B_r{}^{z}, \Sigma'}}{\Pi, \Pi' \vdash A_r \wedge B, \Sigma, \Sigma'}\wedge l \qquad \overset{d_3}{\Pi'', A_r{}^{x}, B_r{}^{t} \vdash \Sigma''}}{\Pi, \Pi', \Pi'' \vdash \Sigma, \Sigma', \Sigma''}\wedge r$$

In this case we have two possible reductions which correspond to two ways of simulating the unified rule of FD' by the corresponding rules of FD (in this respect, unified rules seem not primitive).
There are many possible determinizations of the calculus: binary connectives bearing a priority between their components, or restrictions on the notion of proof.

5. PROOFS OF CONFLUENCE AND STRONG NORMALISATION

Because of lack of place we cannot give full proofs of the results. We will give relatively detailed proofs for local confluence – which is the heart of the problem – and only sketch the rest. Confluence is deduced from local confluence and finiteness of developments (see [1]). We have choosen combinatorial proofs which seem to me more explicative than abstract ones.

In this section proofs are represented linearly as terms (as in typed lambda–calculus). A term is either an axiom or of the form $(\llbracket X_1\rrbracket d_1\ \llbracket X_2\rrbracket d_2\ \llbracket X_3\rrbracket d_3)$ or $(\llbracket X_1\rrbracket d_1\ \llbracket X_2\rrbracket d_2)$ where d_i are terms, X_i are of the form $\vdash A$ or $A\vdash$, and $\llbracket\,\rrbracket$ is a binding; this notation allows to represent the application of a rule to previous proofs d_i with an explicit mention of the active formulas (considered as bounded) together with their position in the sequent; for example,

$$\dfrac{\begin{array}{ccc} d_1 & d_2 & d_3 \\ \cdots\cdots & \cdots\cdots & \cdots\cdots \\ \Gamma,\ \mathbf{A\wedge B}\vdash\Delta & \Pi\vdash\mathbf{A},\ \Sigma & \Pi'\vdash\mathbf{B},\ \Sigma' \end{array}}{\Gamma,\ \Pi,\ \Pi'\ \vdash\Delta,\ \Sigma,\ \Sigma'}$$

is linearly represented as

$$(\llbracket A\wedge B\vdash\rrbracket d_1\ \ \llbracket\vdash A\rrbracket d_2\ \llbracket\vdash B\rrbracket d_3).$$

For convenience, we often consider the intermediate expression $\llbracket X\rrbracket d$ as a term. A proof ending with a (structural, logical, generalised) cut is called a (structural, logical, generalised) redex, and cut–elimination becomes reduction of redexes.
Convention: to simplify exposition we will often consider only the case of binary rules and state the properties up to obvious symetries.

5.1 Substitution

Definition. For A^y not bounded in d, $d[\llbracket A^x\vdash\rrbracket e/\vdash A^y]$ is defined inductively as follows

$$\{A^z\vdash A^y\}[\llbracket A^x\vdash\rrbracket e/\vdash A^y] = e[A^z/A^x]$$
$$\{C^u\vdash C^v\}[\llbracket A^x\vdash\rrbracket e/\vdash A^y] = C^u\vdash C^v \text{ if } C^v\neq A^y \text{ (i.e. } C\neq A \text{ or } v\neq y)$$
$$\{\llbracket X\rrbracket d\}[\llbracket A^x\vdash\rrbracket e/\vdash A^y] = \llbracket X\rrbracket d[\llbracket A^x\vdash\rrbracket e/\vdash A^y]$$
$$(u\ v)[\llbracket A^x\vdash\rrbracket e/\vdash A^y] = (u[\llbracket A^x\vdash\rrbracket e/\vdash A^y]\ v[\llbracket A^x\vdash\rrbracket e/\vdash A^y])$$

Lemma 1 If $A^u\neq A^z$, then $e[\llbracket A^z\vdash\rrbracket f/\vdash A^x][A^u/A^y] = e[A^u/A^y][\llbracket A^z\vdash\rrbracket\{f[A^u/A^y]\}/\vdash A^x]$; in particular, if A^y is not free in f, then $e[\llbracket A^z\vdash\rrbracket f/\vdash A^x][A^u/A^y] = e[A^u/A^y][\llbracket A^z\vdash\rrbracket f/\vdash A^x]$

Lemma 2
(i) If A^y and A^x are not free in f and $A^y\neq B^z$, then
$d[\llbracket A^x\vdash\rrbracket e/\vdash A^y][\llbracket B^t\vdash\rrbracket f/\vdash B^z] = d[\llbracket B^t\vdash\rrbracket f/\vdash B^z][\llbracket A^x\vdash\rrbracket\{e[\llbracket B^t\vdash\rrbracket f/\vdash B^z]\}/\vdash A^y]$.
(ii) If A^y and A^x are not free in f and A^y,B^z are not related by an axiom in d, then
$d[\llbracket A^x\vdash\rrbracket e/\vdash A^y][\llbracket\vdash B^t\rrbracket f/B^z\vdash] = d[\llbracket\vdash B^t\rrbracket f/B^z\vdash][\llbracket A^x\vdash\rrbracket\{e[\llbracket\vdash B^t\rrbracket f/B^z\vdash]\}/\vdash A^y]$.

5.2 Structural reduction

We will have to consider proofs up to the order of the logical cuts; this equivalence relation is denoted by $\sim$. If r is redex $(\llbracket A^x\vdash\rrbracket d_1\ d_2)$ or $(\llbracket\vdash A^x\rrbracket d_1\ d_2)$, then $\llbracket A^x\vdash\rrbracket d_1$ is called the function of r, d_2 the argument of r and A^x the binder of r.

Definition. The contractum $d_1\langle d_2/A^x\vdash\rangle$ of a structural redex $(\llbracket A^x\vdash\rrbracket d_1\, d_2)$ is $d_1[\beta/A^x\vdash]$ with $\beta = \llbracket\vdash A^u\rrbracket(\llbracket A^v\vdash\rrbracket\{A^v\vdash A^u\}\, d_2)$ for A^u, A^v not occuring in d_2. Structural reduction $\overset{s}{\triangleright}$ is the reflexive and transitive closure of the one–step structural reduction $\overset{s}{\triangleright}_1$ defined by:

$(\llbracket A^x\vdash\rrbracket d_1\, d_2) \overset{s}{\triangleright}_1 d_1\langle d_2/A^x\vdash\rangle$, if $(\llbracket A^x\vdash\rrbracket d_1\, d_2)$ is a strutural redex;
$(d_1\, d_2) \overset{s}{\triangleright}_1 (d_1\, d_2')$, if $d_2 \overset{s}{\triangleright}_1 d_2'$;
$(d_1\, d_2) \overset{s}{\triangleright}_1 (d_1'\, d_2)$, if $d_1 \overset{s}{\triangleright}_1 d_1'$;
$\llbracket X\rrbracket d \overset{s}{\triangleright}_1 \llbracket X\rrbracket d'$, if $d \overset{s}{\triangleright}_1 d'$.

Remark 3 The contractum is defined for structural redexes, but more generally, if $u = (\llbracket A^x\vdash\rrbracket d_1\, d_2)$ and $v = d_1\langle d_2/A^x\vdash\rangle$, then $u \overset{s}{\triangleright} v$ or $v \sim u$; in fact one of the following situations hold:
– u is a structural redex and $u \overset{s}{\triangleright}_1 v$;
– u is a logical redex and $u \sim v$;
– u is not a redex and $u = v$.

Lemma 4 If $u \overset{s}{\triangleright}_1 v$ then $u[A^x/A^y] \overset{s}{\triangleright}_1 v[A^x/A^y]$.

Lemma 5
(i) If $u \overset{s}{\triangleright}_1 v$ then there exist u' and v' such that $u' \sim v'$, $u[\llbracket A^y\vdash\rrbracket d/\vdash A^x] \overset{s}{\triangleright} u'$ and $v[\llbracket A^y\vdash\rrbracket d/\vdash A^x] \overset{s}{\triangleright} v'$ (in the case where the binder of the reduced redex of u is not related to A^x in u, we have in fact $u[\llbracket A^y\vdash\rrbracket d/\vdash A^x] \overset{s}{\triangleright}_1 v[\llbracket A^y\vdash\rrbracket d/\vdash A^x]$).
(ii) If $u \overset{s}{\triangleright}_1 v$, then there exists u' such that $u\langle e/\vdash A^x\rangle \overset{s}{\triangleright}_1 u'$ and $v\langle e/\vdash A^x\rangle \sim u'$.

Proof. (i) We prove the result by induction on $\overset{s}{\triangleright}_1$. Let $u_1 = u[\llbracket A^y\vdash\rrbracket d/\vdash A^x]$ and $v_1 = v[\llbracket A^y\vdash\rrbracket d/\vdash A^x]$. The only interesting case is when v is the contractum of u.
1. $u = (\llbracket\vdash B^z\rrbracket d_1\, d_2)$ and $B^z \neq A^x$.
We have

$v = d_1[\beta/\vdash B^z]$ with $\beta = \llbracket B^u\vdash\rrbracket(\llbracket\vdash B^v\rrbracket\{B^u\vdash B^v\}\, d_2)$
$u_1 = (\llbracket\vdash B^z\rrbracket\{d_1[\llbracket A^y\vdash\rrbracket d/\vdash A^x]\}\, d_2[\llbracket A^y\vdash\rrbracket d/\vdash A^x])$;
$v_1 = d_1[\beta/\vdash B^z][\llbracket A^y\vdash\rrbracket d/\vdash A^x]$;

and u_1 reduces to u' where

$u' = d_1[\llbracket A^y\vdash\rrbracket d/\vdash A^x][\llbracket B^u\vdash\rrbracket(\llbracket\vdash B^v\rrbracket\{B^u\vdash B^v\}\, d_2[\llbracket A^y\vdash\rrbracket d/\vdash A^x])/\vdash B^z]$
$= d_1[\llbracket A^y\vdash\rrbracket d/\vdash A^x][\beta[\llbracket A^y\vdash\rrbracket d/\vdash A^x]/\vdash B^z]$.

Because B^z is not free in d and $B^z \neq A^x$, we have $u' = v_1$ by lemma 2.(i).
2. $u = (\llbracket B^z\vdash\rrbracket d_1\, d_2)$ with $B \neq A$ (and more generally with B^z not related to A^x in d_1).
We have

$v = d_1[\beta/B^z\vdash]$ with $\beta = \llbracket\vdash B^u\rrbracket(\llbracket B^v\vdash\rrbracket\{B^v\vdash B^u\}\, d_2)$
$u_1 = (\llbracket B^z\vdash\rrbracket\{d_1[\llbracket A^y\vdash\rrbracket d/\vdash A^x]\}\, d_2[\llbracket A^y\vdash\rrbracket d/\vdash A^x])$;
$v_1 = d_1[\beta/B^z\vdash][\llbracket A^y\vdash\rrbracket d/\vdash A^x]$;

and u_1 reduces to u' where

$u' = d_1[\llbracket A^y\vdash\rrbracket d/\vdash A^x][\llbracket\vdash B^u\rrbracket(\llbracket B^v\vdash\rrbracket\{B^v\vdash B^u\}\, d_2[\llbracket A^y\vdash\rrbracket d/\vdash A^x])/B^z\vdash]$
$= d_1[\llbracket A^y\vdash\rrbracket d/\vdash A^x][\beta[\llbracket A^y\vdash\rrbracket d/\vdash A^x]/B^z\vdash]$.

Because B^z is not free in d and not related to A^x in d_1, we have $u' = v_1$ by lemma 2.(ii).

3. $u = ([\![A^z\vdash]\!]d_1\ d_2)$.

We have

$$v = d_1[\beta/A^z\vdash] \text{ with } \beta = [\![\vdash A^u]\!]([\![A^v\vdash]\!]\{A^v\vdash A^u\}\ d_2)$$

$$u_1 = ([\![A^z\vdash]\!]\{d_1[[\![A^y\vdash]\!]d/\vdash A^x]\}\ d_2[[\![A^y\vdash]\!]d/\vdash A^x]);$$

$$v_1 = d_1[\beta/A^z\vdash][[\![A^y\vdash]\!]d/\vdash A^x];$$

and u_1 reduces to u' where

$$u' = d_1[[\![A^y\vdash]\!]d/\vdash A^x][[\![\vdash A^u]\!]([\![A^v\vdash]\!]\{A^v\vdash A^u\}\ d_2[[\![A^y\vdash]\!]d/\vdash A^x])/A^z\vdash]$$
$$= d_1[[\![A^y\vdash]\!]d/\vdash A^x][\beta[[\![A^y\vdash]\!]d/\vdash A^x]/A^z\vdash].$$

We prove by induction on d_1 that there exists v' such that $v_1 \overset{s}{\triangleright} v'$ and $v' \sim u'$. The only interesting case is when d_1 is an axiom. If d_1 is an axiom other than $A^z \vdash A^x$, then A^z and A^x are not related in d_1 and we have in fact $u' = v_1$ by lemma 2.(ii)(of course A^z is not free in d). Suppose that d_1 is the axiom $A^z \vdash A^x$. In this case we have

$$v_1 = ([\![A^v\vdash]\!]\{A^v\vdash A^x\}\ d_2)[[\![A^y\vdash]\!]d/\vdash A^x]$$
$$= ([\![A^v\vdash]\!]d[A^v/A^y]\ d_2[[\![A^y\vdash]\!]d/\vdash A^x])$$

$$u' = d[A^z/A^y][\beta[[\![A^y\vdash]\!]d/\vdash A^x]/A^z\vdash]$$
$$= d[A^z/A^y][[\![\vdash A^u]\!]([\![A^v\vdash]\!]\{A^v\vdash A^u\}\ d_2[[\![A^y\vdash]\!]d/\vdash A^x])/A^z\vdash].$$

Let $w = d[A^v/A^y][\gamma/A^v\vdash]$, where $\gamma = [\![\vdash A^u]\!]([\![A^w\vdash]\!]\{A^w\vdash A^u\}\ d_2[[\![A^y\vdash]\!]d/\vdash A^x])$. By remark 3, we have $v_1 \overset{s}{\triangleright} w$ or $v_1 \sim w$; because A^v and A^z are not free in d, we also have $u' = w$; therefore we have $v_1 \overset{s}{\triangleright} u'$ or $v_1 \sim u'$: in the first case we take $v' = u'$ and in the second $v' = v_1$.

(ii) We just have to check that in case 3 of the previous proof we have $u' \sim v_1$ if d is $([\![\vdash A^w]\!]\{A^y \vdash A^w\}\ e)$. The only interesting case is when d_1 is the axiom $A^z \vdash A^x$. In this case we have

$$v_1 = ([\![A^v\vdash]\!]\{A^v\vdash A^x\}\ d_2)[[\![A^y\vdash]\!]d/\vdash A^x]$$
$$= ([\![A^v\vdash]\!]([\![\vdash A^w]\!]\{A^v \vdash A^w\}\ e)\ d_2[[\![A^y\vdash]\!]d/\vdash A^x])$$

$$u' = ([\![\vdash A^w]\!]\{A^z \vdash A^w\}\ e)[\beta[[\![A^y\vdash]\!]d/\vdash A^x]/A^z\vdash]$$
$$= ([\![\vdash A^w]\!]\{A^z \vdash A^w\}\ e)[[\![\vdash A^u]\!]([\![A^v\vdash]\!]\{A^v\vdash A^u\}\ d_2[[\![A^y\vdash]\!]d/\vdash A^x])/A^z\vdash]$$
$$= ([\![\vdash A^w]\!]([\![A^v\vdash]\!]\{A^v\vdash A^w\}\ d_2[[\![A^y\vdash]\!]d/\vdash A^x])\ e).$$

Therefore $v_1 \sim u'$.

Lemma 6 If $u \overset{s}{\triangleright}_1 v$ then $w[[\![A^y\vdash]\!]u/\vdash A^x] \overset{s}{\triangleright} w[[\![A^y\vdash]\!]v/\vdash A^x]$.

Proof. The result is proved by induction on w. The only interesting case is when w is an axiom. Let $w_1 = w[[\![A^y\vdash]\!]u/\vdash A^x]$ and $w_2 = w[[\![A^y\vdash]\!]v/\vdash A^x]$. If $w = C^u \vdash C^v$ with $C^v \neq A^x$, then $w_1 = w_2 = w$. If $w = A^z \vdash A^x$, then we have $w_1 = u[A^z/A^y]$ and $w_2 = v[A^z/A^y]$, and therefore $w_1 \overset{s}{\triangleright} w_2$ by lemma 4.

Lemma 7 If $u \overset{s}{\triangleright}_1 v$ and $u \sim u'$, then there exists v' such that $v \sim v'$ and $u' \overset{s}{\triangleright}_1 v'$.

Theorem 1 Structural reduction is locally confluent, up to the order of the logical redexes, i.e. if $u \overset{s}{\triangleright}_1 u_1$ and $u \overset{s}{\triangleright}_1 u_2$, then there exist v_1 and v_2 such that $v_1 \sim v_2$, $u_1 \overset{s}{\triangleright} v_1$ and $u_2 \overset{s}{\triangleright} v_2$.

Proof. The proof is by induction on u. The only interesting case is when u is one of the redexes which are reduced, say $u = ([\![A^x\vdash]\!]d_1\ d_2)$ and $u_1 = d_1\langle d_2/A^x\vdash\rangle$. We consider the different possibilities for u_2.

1. $u_2 = ([A^x\vdash]d_1\ d_2')$ with $d_2 \overset{s}{\triangleright}_1 d_2'$.
In this case u_2 is a structural redex and $u_2 \overset{s}{\triangleright}_1 v = d_1\langle d_2'/A^x\vdash\rangle$; by lemma 6, we have also $u_1 \overset{s}{\triangleright} v$.
2. $u_2 = ([A^x\vdash]d_1'\ d_2)$ with $d_1 \overset{s}{\triangleright}_1 d_1'$.
Let $u_2' = d_1'\langle d_2/A^x\vdash\rangle$. Because $d_1 \overset{s}{\triangleright}_1 d_1'$, there exist (by lemma 5) v_1 such that $u_1 \overset{s}{\triangleright} v_1$ and $v_1 \sim u_2'$. By remark 3, we also have $u_2 \overset{s}{\triangleright} u_2'$ or $u_2 \sim u_2'$: in the first case we take $v_2 = u_2'$; in the second case we take $v_2 = u_2$; in each case we have $v_1 \sim v_2$, $u_1 \overset{s}{\triangleright} v_1$ and $u_2 \overset{s}{\triangleright} v_2$.

<u>Theorem 2</u> Structural reduction enjoys the strong normalisation property.

<u>Proof</u> (sketch of). We prove that each structural reduction sequence of a term t is finite, by induction on the number of structural redexes of t. Suppose that t has at least one redex (otherwise the result is trivial). Consider the redex $r = ([X]u_1\ u_2)$ or $r = ([X]u_1\ u_2\ u_3)$ whose first argument u_2 is the rightmost one. We label t as follows: the binder of r receives the label 1, and the others a label $\neq 1$. Let t' obtained from t by reducing r; t' has less structural redexes than t and therefore by induction hypothesis, each structural reduction sequence of t' is finite.
Suppose now that t has an infinite structural reduction sequence (t_i). We consider a function φ between terms which reduces every structural redex whose binder is labelled 1. Using the fact the sequences of reduction of structural redexes whose binder is labelled 1 are finite, we can extract an infinite subsequence (s_j) of (t_i) such that $s_0 = t$ and for each j, there exists v_j such that $\varphi s_j \overset{s}{\triangleright}_1 v_j$ and $v_j \sim \varphi s_{j+1}$. Using lemma 7, we can transform (s_j) into an infinite structural reduction sequence of t'; this is impossible and therefore each structural reduction sequence of t is finite.

<u>Corollary 3</u> Structural reduction is confluent up to the order of the logical redexes, i.e. if $u \overset{s}{\triangleright} u_1$ and $u \overset{s}{\triangleright} u_2$, then there exist v_1 and v_2 such that $v_1 \sim v_2$, $u_1 \overset{s}{\triangleright} v_1$ and $u_2 \overset{s}{\triangleright} v_2$.

5.3 Logical reduction

One defines notions of function, argument and binder for logical redexes. For example, if r is the redex $([\vdash C_r \wedge D]([C_r \wedge D\vdash]\{C_r \wedge D \vdash C_r \wedge D\}\ [\vdash C_r^u]d_1\ [\vdash D]d_2)\ [C_r^v\vdash]d_3)$, then $[\vdash C_r^u]d_1$ is called the <u>function</u> of r, $[C_r^v\vdash]d_3$ the <u>argument</u> and C_r^u the <u>binder</u>; the <u>contractum</u> of r is $d_1[[C_r^v\vdash]d_3/\vdash C_r^u]$.

<u>Definition</u>. <u>Logical reduction</u> $\overset{1}{\triangleright}$ is the reflexive and transitive closure of the one-step logical reduction $\overset{1}{\triangleright}_1$ defined by:

$d \overset{1}{\triangleright}_1 d'$, if d a logical redex and d' its contractum;
$(d_1\ d_2) \overset{1}{\triangleright}_1 (d_1\ d_2')$, if $d_2 \overset{1}{\triangleright}_1 d_2'$;
$(d_1\ d_2) \overset{1}{\triangleright}_1 (d_1'\ d_2)$, if $d_1 \overset{1}{\triangleright}_1 d_1'$;
$[X]d \overset{1}{\triangleright}_1 [X]d'$, if $d \overset{1}{\triangleright}_1 d'$.

<u>Lemma 8</u> If $u \overset{1}{\triangleright}_1 v$ then $u[A^x/A^y] \overset{1}{\triangleright}_1 v[A^x/A^y]$.

Lemma 9
If $u \overset{1}{\rhd}_1 v$ then $u[\llbracket A_r{}^y\vdash\rrbracket d/\vdash A_r{}^x] \overset{1}{\rhd}_1 v[\llbracket A_r{}^y\vdash\rrbracket d/\vdash A_r{}^x]$ and $u[\llbracket\vdash A_1{}^y\rrbracket d/A_1{}^x\vdash] \overset{1}{\rhd}_1 v[\llbracket\vdash A_1{}^y\rrbracket d/A_1{}^x\vdash]$.

Proof. We prove the result by induction on $\overset{1}{\rhd}_1$ for a right input (the proof is the same for a left one). Let $u_1 = u[\llbracket A_r{}^y\vdash\rrbracket d/\vdash A_r{}^x]$ and $v_1 = v[\llbracket A_r{}^y\vdash\rrbracket d/\vdash A_r{}^x]$. The only interesting case is when u is the logical redex which is reduced; suppose for example it is a redex for a conjunction.
1. $u = (\llbracket\vdash C_r{\wedge}D\rrbracket(\llbracket C_r{\wedge}D\vdash\rrbracket\{C_r{\wedge}D \vdash C_r{\wedge}D\}\ \llbracket\vdash C_r{}^u\rrbracket d_1\ \llbracket\vdash D\rrbracket d_2)\ \llbracket C_r{}^v\vdash\rrbracket d_3)$.
We have

$$v = d_1[\llbracket C_r{}^v\vdash\rrbracket d_3/\vdash C_r{}^u];$$
$$u_1 = (\llbracket\vdash C_r{\wedge}D\rrbracket(\llbracket C_r{\wedge}D\vdash\rrbracket\{C_r{\wedge}D \vdash C_r{\wedge}D\}\ \llbracket\vdash C_r{}^u\rrbracket d_1'\ \llbracket\vdash D\rrbracket d_2')\ \llbracket C_r{}^v\vdash\rrbracket d_3')$$

with $d_i' = d_i[\llbracket A_r{}^y\vdash\rrbracket d/\vdash A_r{}^x]$ for i=1,2,3;

$$v_1 = d_1[\llbracket C_r{}^v\vdash\rrbracket d_3/\vdash C_r{}^u][\llbracket A_r{}^y\vdash\rrbracket d/\vdash A_r{}^x];$$

and $u_1 \overset{1}{\rhd}_1 w$ where

$$w = d_1'[\llbracket C_r{}^v\vdash\rrbracket d_3'/\vdash C_r{}^u]$$
$$= d_1[\llbracket A_r{}^y\vdash\rrbracket d/\vdash A_r{}^x][\llbracket C_r{}^v\vdash\rrbracket\{d_3[\llbracket A_r{}^y\vdash\rrbracket d/\vdash A_r{}^x]\}/\vdash C_r{}^u].$$

Because $C_r{}^u \neq A_r{}^x$ and $C_r{}^u$ is not free in d, we have by lemma 2.(i) $w = v_1$ and therefore $u_1 \overset{1}{\rhd}_1 v_1$.
2. $u = (\llbracket\vdash C_1{\wedge}D\rrbracket(\llbracket C_1{\wedge}D\vdash\rrbracket\{C_1{\wedge}D \vdash C_1{\wedge}D\}\ \llbracket\vdash C_1{}^u\rrbracket d_1\ \llbracket\vdash D\rrbracket d_2)\ \llbracket C_1{}^v\vdash\rrbracket d_3)$.
We have

$$v = d_3[\llbracket\vdash C_1{}^u\rrbracket d_1/C_1{}^v\vdash];$$
$$u_1 = (\llbracket\vdash C_1{\wedge}D\rrbracket(\llbracket C_1{\wedge}D\vdash\rrbracket\{C_1{\wedge}D \vdash C_1{\wedge}D\}\ \llbracket\vdash C_1{}^u\rrbracket d_1'\ \llbracket\vdash D\rrbracket d_2')\ \llbracket C_1{}^v\vdash\rrbracket d_3')$$

with $d_i' = d_i[\llbracket A_r{}^y\vdash\rrbracket d/\vdash A_r{}^x]$ for i=1,2,3;

$$v_1 = d_3[\llbracket\vdash C_1{}^u\rrbracket d_1/C_1{}^v\vdash][\llbracket A_r{}^y\vdash\rrbracket d/\vdash A_r{}^x];$$

and $u_1 \overset{1}{\rhd}_1 w$ where

$$w = d_3'[\llbracket\vdash C_1{}^u\rrbracket d_1'/C_1{}^v\vdash]$$
$$= d_3[\llbracket A_r{}^y\vdash\rrbracket d/\vdash A_r{}^x][\llbracket\vdash C_1{}^u\rrbracket\{d_1[\llbracket A_r{}^y\vdash\rrbracket d/\vdash A_r{}^x]\}/C_1{}^v\vdash].$$

By choice of the inputs $C_1 \neq A_r$ and therefore $C_1{}^v$ is not related to $A_r{}^x$ in d_3; moreover $C_1{}^v$ is not free in d, and therefore we have by lemma 2.(ii) $w = v_1$ and $u_1 \overset{1}{\rhd}_1 v_1$.

Lemma 10 If $u \overset{1}{\rhd}_1 v$ then $w[\llbracket A^y\vdash\rrbracket u/\vdash A^x] \overset{1}{\rhd} w[\llbracket A^y\vdash\rrbracket v/\vdash A^x]$.

Theorem 4 Logical reduction is locally confluent, i.e. if $u \overset{1}{\rhd}_1 u_1$ and $u \overset{1}{\rhd}_1 u_2$, then there exists v such that $u_1 \overset{1}{\rhd} v$ and $u_2 \overset{1}{\rhd} v$.

Proof. The proof is by induction on u. The only interesting case is when u is one of the redexes which are reduced. Suppose for instance that

$$u = (\llbracket\vdash C_r{\wedge}D\rrbracket(\llbracket C_r{\wedge}D\vdash\rrbracket\{C_r{\wedge}D \vdash C_r{\wedge}D\}\ \llbracket\vdash C_r{}^u\rrbracket d_1\ \llbracket\vdash D\rrbracket d_2)\ \llbracket C_r{}^v\vdash\rrbracket d_3);$$
$$u_1 = d_1[\llbracket C_r{}^v\vdash\rrbracket d_3/\vdash C_r{}^u].$$

We consider the different possibilities for u_2.
1. $u_2 = (\llbracket\vdash C_r{\wedge}D\rrbracket(\llbracket C_r{\wedge}D\vdash\rrbracket\{C_r{\wedge}D \vdash C_r{\wedge}D\}\ \llbracket\vdash C_r{}^u\rrbracket d_1'\ \llbracket\vdash D\rrbracket d_2)\ \llbracket C_r{}^v\vdash\rrbracket d_3)$ with $d_1 \overset{1}{\rhd}_1 d_1'$.
Let v be $d_1'[\llbracket C_r{}^v\vdash\rrbracket d_3/\vdash C_r{}^u]$. We have $u_2 \overset{1}{\rhd}_1 v$ and by lemma 9, $u_1 \overset{1}{\rhd}_1 v$.
2. $u_2 = (\llbracket\vdash C_r{\wedge}D\rrbracket(\llbracket C_r{\wedge}D\vdash\rrbracket\{C_r{\wedge}D \vdash C_r{\wedge}D\}\ \llbracket\vdash C_r{}^u\rrbracket d_1\ \llbracket\vdash D\rrbracket d_2')\ \llbracket C_r{}^v\vdash\rrbracket d_3)$ with $d_2 \overset{1}{\rhd}_1 d_2'$.
Let v be $d_1[\llbracket C_r{}^v\vdash\rrbracket d_3/\vdash C_r{}^u]$. We have $u_2 \overset{1}{\rhd}_1 v$ and $u_1 = v$.

3. $u_2 = ([\vdash C_r \wedge D]([C_r \wedge D \vdash]\{C_r \wedge D \vdash C_r \wedge D\}\ [\vdash C_r^{u}]d_1\ [\vdash D]d_2)\ [C_r^{v} \vdash]d_3')$ with $d_3 \overset{1}{\rhd}_1 d_3'$.
Let v be $d_1[[C_r^{v} \vdash]d_3'/\vdash C_r^{u}]$. We have $u_2 \overset{1}{\rhd}_1 v$ and by lemma 10, $u_1 \overset{1}{\rhd} v$.

<u>Theorem 5</u> Logical reduction enjoys the finiteness of developments property.

<u>Proof</u> (sketch of). Consider a term t labelled as follows: the binders of the logical redexes are labelled 1 or 2 and the other ones 0. We call i–redex (resp. $\bar{i}$–redex) a logical redex whose binder is labelled i (resp. ≠i). We prove that each reduction sequence of $\bar{0}$–redexes of t is finite, by induction on the number of $\bar{0}$–redexes of t. Suppose that t has at least one $\bar{0}$–redex (otherwise the result is trivial). Consider the $\bar{0}$–redex r whose argument is the rightmost one (we suppose for simplicity that the argument is always to the right of the function). We change the labels of the $\bar{0}$–redexes as follows: the binder of r receives the label 1, and the binders of the other logical redexes, the label 2. Let t' obtained from t by reducing r; t' has less $\bar{0}$–redexes than t and therefore by induction hypothesis, each reduction sequence of $\bar{0}$–redexes of t' is finite.
Suppose now that t has an infinite sequence (t_i) of reduction of $\bar{0}$–redexes. We get a contradiction in the same way as for theorem 2 using a function ψ between terms which reduces every 1–redex

<u>Corollary 6</u> Logical reduction is confluent, i.e. if $u \overset{1}{\rhd} u_1$ and $u \overset{1}{\rhd} u_2$, then there exist v such $u_1 \overset{1}{\rhd} v$ and $u_2 \overset{1}{\rhd} v$.

<u>5.4 Generalised reduction</u>

<u>Definition</u>. <u>Generalised reduction</u> $\rhd$ is the reflexive and transitive closure of the one–step generalised reduction $\rhd_1$ defined by: $u \rhd_1 v$ if and only if $u \overset{s}{\rhd}_1 v$ or $u \overset{1}{\rhd}_1 v$.

<u>Lemma 11</u> If $u \overset{1}{\rhd}_1 v$, then there exists v' such that $u\langle d/\vdash A^x\rangle \rhd v'$ and $v' \sim \overset{1}{\rhd}_1 v\langle d/\vdash A^x\rangle$.

<u>Theorem 7</u> Generalised reduction is locally confluent, up to the order of the logical redexes, i.e. if $u \rhd_1 u_1$ and $u \rhd_1 u_2$, then there exist v_1 and v_2 such that $v_1 \sim v_2$, $u_1 \rhd v_1$ and $u_2 \rhd v_2$.

<u>Proof</u>. The proof is by induction on u. The only interesting case is when u_1 (or u_2) is the contractum of u. If the reduced redexes are both structural ones or both logical ones we apply the theorems 1 and 4. We consider the remaining cases.
1. $u \overset{1}{\rhd}_1 u_1$ and $u \overset{s}{\rhd}_1 u_2$.
Suppose for instance that

$$u = ([\vdash C_r \wedge D]([C_r \wedge D \vdash]\{C_r \wedge D \vdash C_r \wedge D\}\ [\vdash C_r^{u}]d_1\ [\vdash D]d_2)\ [C_r^{v} \vdash]d_3);$$
$$u_1 = d_1[[C_r^{v} \vdash]d_3/\vdash C_r^{u}].$$

We consider the different possibilities for u_2.
1.1. $u_2 = ([\vdash C_r \wedge D]([C_r \wedge D \vdash]\{C_r \wedge D \vdash C_r \wedge D\}\ [\vdash C_r^{u}]d_1'\ [\vdash D]d_2)\ [C_r^{v} \vdash]d_3)$ with $d_1 \overset{s}{\rhd}_1 d_1'$.
Let v be $d_1'[[C_r^{v} \vdash]d_3/\vdash C_r^{u}]$. Because $d_1 \overset{s}{\rhd}_1 d_1'$, there exist (by lemma 5) v_1 and v_2 such that $v_1 \sim v_2$,

$u_1 \overset{s}{\triangleright} v_1$ and $v \overset{s}{\triangleright} v_2$; because $u_2 \overset{1}{\triangleright}_1 v$, we have in fact $u_1 \triangleright v_1$ and $u_2 \triangleright v_2$.

1.2. $u_2 = ([\vdash C_r\wedge D]([C_r\wedge D\vdash]\{C_r\wedge D \vdash C_r\wedge D\}\ [\vdash C_r^{u}]d_1\ [\vdash D]d_2')\ [C_r^{v}\vdash]d_3)$ with $d_2 \overset{s}{\triangleright}_1 d_2'$.

Let v be $d_1[[C_r^{v}\vdash]d_3/\vdash C_r^{u}]$. We have $u_2 \overset{1}{\triangleright}_1 v$ and $u_1 = v$.

1.3. $u_2 = ([\vdash C_r\wedge D]([C_r\wedge D\vdash]\{C_r\wedge D \vdash C_r\wedge D\}\ [\vdash C_r^{u}]d_1\ [\vdash D]d_2)\ [C_r^{v}\vdash]d_3')$ with $d_3 \overset{s}{\triangleright}_1 d_3'$.

Let v be $d_1[[C_r^{v}\vdash]d_3'/\vdash C_r^{u}]$. We have $u_2 \overset{1}{\triangleright}_1 v$ and by lemma 6, $u_1 \overset{s}{\triangleright} v$.

2. $u \overset{s}{\triangleright}_1 u_1$ and $u \overset{1}{\triangleright}_1 u_2$.

Suppose that $u = ([A^{x}\vdash]d_1\ d_2)$ and $u_1 = d_1\langle d_2/A^{x}\vdash\rangle$. We consider the different possibilities for u_2.

2.1. $u_2 = ([A^{x}\vdash]d_1\ d_2')$ with $d_2 \overset{1}{\triangleright}_1 d_2'$.

In this case u_2 is a structural redex and $u_2 \overset{s}{\triangleright}_1 v = d_1\langle d_2'/A^{x}\vdash\rangle$; by lemma 10, we have also $u_1 \overset{1}{\triangleright} v$.

2.2. $u_2 = ([A^{x}\vdash]d_1'\ d_2)$ with $d_1 \overset{1}{\triangleright}_1 d_1'$.

Let $v = d_1'\langle d_2/A^{x}\vdash\rangle$. By lemma 11, there exists v_1 such that $u_1 \triangleright v_1$ and $v_1 \sim v$. By remark 3, we also have $u_2 \overset{s}{\triangleright} v$ or $u_2 \sim v$: in the first case we take $v_2 = v$ and in the second one $v_2 = u_2$; in each case we have $v_1 \sim v_2$, $u_1 \triangleright v_1$ and $u_2 \triangleright v_2$.

__Theorem 8__ Generalised reduction enjoys the finiteness of developments property.

__Proof__. The proof is essentially a combination of the proofs of theorem 2 and 5.

__Corollary 9__ Generalised reduction is confluent, i.e. if $u \triangleright u_1$ and $u \triangleright u_2$, then there exists v such that $u_1 \triangleright v$ and $u_2 \triangleright v$.

__Theorem 10__ Generalised reduction enjoys strong normalisation property.

__Proof__ (sketch of). We call strict a term in which every binder has at least one occurence. The first step is the reduction of the problem to strict terms, by interpreting arbitrary terms into strict terms in a way which preserves the reduction sequences. In a second step we prove the strong normalisation for strict terms in the following combinatorial way. We prove that each reduction sequence of a strict term t is finite, by induction on $\Phi(t) = (d(t),n(t))$ with lexicographic ordering, where d(t) is the maximum degree of the redexes of t (the degree of a redex is the length of the main formula of the corresponding cut) and n(t) is the number of redexes of degree d(t). Let r be a redex of maximal degree whose first argument is the rightmost one. We label t as follows: the binder of r receives label 1, and the other binders, a label $\neq 1$. Let t' obtained from t by reducing r; we have $\Phi(t') < \Phi(t)$ and by induction hypothesis, each reduction sequence of t' is finite.

Suppose now that t has an infinite reduction sequence (t_i). We get a contradiction in the same way as for theorem 2 using a function Θ between terms which reduces every 1–redex (in order to "transform" (t_i) into an infinite reduction sequence of t', we use the fact that the terms are strict and the finiteness of developments).

RELATED WORKS / FURTHER WORKS

There is a growing interest in the extraction of programs from classical proofs in particular through the works of T. Griffin and C. Murthy. Though not directly related, our work has been partly motivated by this idea which is in the air at the moment.

In a parallel non communicating work, J.Y. Girard has built a new computational system for classical logic LC, based on an operational semantic. It could be interesting to see whether FD and LC share some properties.

Among expected applications of FD are: the extraction of programs from classical proofs written in FD; the study of proof search strategies from the viewpoint of FD; the use of FD as a metasystem for studying usual logical systems.

BIBLIOGRAPHY

[1] BARENDREGT, The lambda calculus, North–Holland, 1985.

[2] GIRARD, Proof theory and logical complexity, Bibliopolis, 1987.

[3] HOWARD, The formulae–as–types notion of construction, in "To HB Curry...", Academic Press, 1980.

[4] PRAWITZ, Natural deduction, Almqvist&Wiksell, 1965.

GENTZEN-TYPE CALCULI FOR MODAL LOGIC S4 WITH BARCAN FORMULA

Aida Pliuškevičienė

Institute of Mathematics and Informatics,
Lithuanian Academy of Sciences,
Akademijos 4, Vilnius, 232600, Lithuania
email: logica @ ma – mii.lt.su

Introduction

Logic programming systems based on modal logic are presented for example in [1, 2]. It is well-known that resolution-type systems serve as a basis for the interpreters of logic programming systems. In [3, 7] it was shown that soundness and completeness of various resolution procedures can be proved in an evident way through explicit translations between resolution refutations and cut-free sequential calculus. In [8] the first order calculi without Barcan formula (i.e. the formula of the type $\forall x \Box A(x) \supset \Box \forall x A(x)$) for various modal logics (both in Gentzen-type and in resolution-type) are presented. In [5] it was shown how to construct cut-free calculus for propositional modal logic $S4$. It is also known that an extension of this result for the first order modal logic $S4$ without Barcan formula is rather traditional. The purpose of this paper is to construct a cut-free sequential calculus for the first order modal logic $S4$ with Barcan formula. To obtain the cut-free calculus we extend the index method, introduced for the modal logic S5 in [5]. Using the index method the non-clausal resolution for the modal logic S5 was described in [9]. The results of this paper may be extended to other modal logics with Barcan formula. The proof of completeness for axiom systems including Barcan formula in terms of constant domain models is presented in [4].

Description of the calculi for the considered modal logic S4

Let H be the Hilbert-type calculus for the first order logic without equality (see e.g. [6]). The language of the first order modal logic $S4$ is obtained from the language of the calculus H by adding modality $\Box$ (necessity) which is interpreted as a reflexive and transitive binary relation. The formulas are determined with the help of logic symbols and modality $\Box$, as usual. Let $HS4^B$ (Hilbert-type calculus for the first order modal logic $S4$ with Barcan formula) be the calculus obtained from H by adding the following postulates.

Axioms: (1) $\Box(A \supset B) \supset (\Box A \supset \Box B)$; (2) $\Box A \supset A$; (3) $\Box A \supset \Box\Box A$; (4) $\forall x \Box A(x) \supset \Box \forall x A(x)$. The rule of inference: $A / \Box A$.

Let us construct the sequential calculus for the logic under consideration. A sequent is an expression of the form $\Gamma \to \Delta$, where Γ, Δ are arbitrary multisets of the formulas (i.e. the order of formulas in Γ, Δ is disregarded). Let us consider only such sequents in which no variable occurs free and bound at the same time. The formula $\mathfrak{A} = A_1 \wedge \ldots \wedge A_n \supset B_1 \vee \ldots \vee B_m$ is called the image formula of the sequent $A_1, \ldots, A_n \rightarrow B_1, \ldots, B_m$. When $n = 0$ $\mathfrak{A} = \bigvee_{i=1}^{m} B_i$ and if $m = 0$ then $\mathfrak{A} = \bigwedge_{i=1}^{n} A_i \supset F$ where F is false.

Let $GS4^B$ be the calculus defined by the following postulates.
Axiom: $\Gamma, A \rightarrow \Delta, A$
Rules of inference:

$$\frac{\Gamma, A \rightarrow \Delta, B}{\Gamma \rightarrow \Delta,\ A \supset B}\ (\rightarrow\supset) \qquad \frac{\Gamma \rightarrow \Delta, A;\ B, \Gamma \rightarrow \Delta}{A \supset B,\ \Gamma \rightarrow \Delta}\ (\supset\rightarrow)$$

$$\frac{\Gamma \rightarrow \Delta, A, B}{\Gamma \rightarrow \Delta,\ A \vee B}\ (\rightarrow \vee) \qquad \frac{A, \Gamma \rightarrow \Delta;\ B, \Gamma \rightarrow \Delta}{A \vee B, \Gamma \rightarrow \Delta}\ (\vee \rightarrow)$$

$$\frac{\Gamma \rightarrow \Delta, A;\ \Gamma \rightarrow \Delta, B}{\Gamma \rightarrow \Delta,\ A \wedge B}\ (\rightarrow \wedge) \qquad \frac{A, B, \Gamma \rightarrow \Delta}{A \wedge B,\ \Gamma \rightarrow \Delta}\ (\wedge \rightarrow)$$

$$\frac{\Gamma, A \rightarrow \Delta}{\Gamma \rightarrow \Delta, \urcorner A}\ (\rightarrow \urcorner) \qquad \frac{\Gamma \rightarrow \Delta, A}{\urcorner A, \Gamma \rightarrow \Delta}\ (\urcorner \rightarrow)$$

$$\frac{\Box\Gamma \rightarrow A}{\Sigma, \Box\Gamma \rightarrow \Delta, \Box A}\ (\rightarrow \Box) \qquad \frac{A, \Box A, \Gamma \rightarrow \Delta}{\Box A, \Gamma \rightarrow \Delta}\ (\Box \rightarrow)$$

$$\frac{\Gamma \rightarrow \Delta, A(b)}{\Gamma \rightarrow \Delta, \forall x A(x)}\ (\rightarrow \forall) \qquad \frac{A(t), \forall x A(x), \Gamma \rightarrow \Delta}{\forall x A(x), \Gamma \rightarrow \Delta}\ (\forall \rightarrow)$$

$$\frac{\Gamma \rightarrow \Delta, A(t), \exists x Ax}{\Gamma \rightarrow \Delta, \exists x A(x)}\ (\rightarrow \exists) \qquad \frac{A(b), \Gamma \rightarrow \Delta}{\exists x A(x), \Gamma \rightarrow \Delta}\ (\exists \rightarrow)$$

$$\frac{\Gamma \rightarrow \Delta,\ \forall x \Box A(x)}{\Gamma \rightarrow \Delta, \Box\forall x A(x)}\ (\rightarrow \Box\forall) \qquad \frac{\Gamma \rightarrow \Delta, A;\ A, \Gamma \rightarrow \Delta}{\Gamma \rightarrow \Delta}\ (\text{cut})$$

The rules of inference for quantifiers satisfy conventional conditions.

Theorem 1. $HS4^B \vdash A \Leftrightarrow GS4^B \vdash \rightarrow A$.

Proof: analogously as in [6, §77].

Let G_1S4^B be the calculus obtained from $GS4^B$ replacing the rule of inference (cut) by the rule of inference (cut$^\Box$) obtained from (cut) replacing the formula A by the formula of the shape $\Box C$.

To obtain a cut-free calculus for the logic under consideration let us introduce formulas with the indices. An ordered sequence (k-tuple) of natural numbers $< i_1, \ldots, i_k >$ will be called an index (differently from [5], where the index is a natural number); $\alpha, \beta, \gamma, \alpha_1, \beta_1, \gamma_1, \ldots$ are variables for the index. Every predicate variable P having its own index γ is denoted by P^γ. A predicate variable without an index is regarded as a variable with the index zero. Two identical predicate variables with different indices are considered as different variables. A formula A with the index γ is denoted by $(A)^\gamma$ and is defined in the following way:

1. $(E)^\gamma = E^\gamma$, where E is an elementary formula;
2. $(A \odot B)^\gamma = (A)^\gamma \odot (B)^\gamma$, $\odot \in \{\supset, \wedge, \vee\}$;
3. $(\sigma A)^\gamma = \sigma(A)^\gamma$, $\sigma \in \{\urcorner, \Box\}$;
4. $\big(QxA(x)\big)^\gamma = Qx\big(A(x)\big)^\gamma$, $Q \in \{\forall, \exists\}$.

Let $\alpha = < i_1, \dots, i_m >, \beta = < j_1, \dots, j_n >$ be arbitrary indices and i be any natural number. Then the notation α, β means an index $< i_1, \dots, i_m, j_1, \dots, j_n >$ and the notation α, i means an index $< i_1, \dots, i_m, i >$.

Let us define the relation $\alpha \leqslant \beta$ for two arbitrary indices $\alpha = < i_1, \dots, i_m >$ and $\beta = < j_1, \dots, j_n >$ in a usual way: $\alpha = \beta$ if $m = n$ and $\forall k (1 \leqslant k \leqslant m)(i_k = j_k)$. If $\alpha \neq \beta$, then $\alpha < \beta$ if $m < n$ or $m = n$ and for some $l \leqslant m$ $\quad \forall k(1 \leqslant k < l)(i_k = j_k) \wedge (i_l < j_l)$.

Let $\alpha = < i_1, \dots, i_k >$, then the length of α denoted by $|\alpha|$ is k.

Let $\gamma_1, \dots, \gamma_m$ be all the indices entering a sequent $\Gamma \to \Delta$ as indices of the formulas from Γ, Δ. Let us divide these indices into classes. For any k let us select from the sequence of indices $\gamma_1, \dots, \gamma_m$ all the indices having the same length k and denote this sequence by $\beta_1, \dots, \beta_{n_k}$. So for any k $|\beta_1| = \dots = |\beta_{n_k}| = k$ and $\{\beta_1, \dots, \beta_{n_k}\} \subseteq \{\gamma_1, \dots, \gamma_m\}$. For any index $\beta_i (1 \leqslant i \leqslant n_k)$ the set $\{\beta_1, \dots, \beta_{n_k}\}$ will be called β_i-index class.

Let $KS4^B$ be the calculus obtained from $GS4^B$ replacing the rules of inference $(\to \Box)$, $(\Box \to)$, $(\to \Box\forall)$ by the following ones in which the formula $\Box A$ in the conclusion of the rule has the index α, i and $|\alpha, i| = k$, in other words $\alpha, i = \beta_i \in \{\beta_1, \dots, \beta_i, \dots, \beta_{n_k}\}$. Besides if $k = 1$ then $|\alpha| = 0$ and i may be zero or an empty word (which is identified with zero).

$$\frac{\Gamma \to \Delta, (A)^{\alpha, \alpha_1}}{\Gamma \to \Delta, \Box(A)^{\alpha, i}} \; (\to \Box^\alpha)$$

where $\alpha_1 = i + l$ (l is any natural number, $l \geqslant 1$) if $\alpha, i = \max(\beta_1, \dots, \beta_i, \dots, \beta_{n_k})$. Otherwise α_1 is k-tuple such that $|\alpha, \alpha_1| > \max(|\gamma_1|, \dots, |\gamma_m|)$. The index α, α_1 will be called eigen index of the rule $(\to \Box^\alpha)$. It is obvious that eigen index does not enter the conclusion of the rule $(\to \Box^\alpha)$.

$$\frac{(A)^{\alpha, i+l}, \Box(A)^{\alpha, i}, \Gamma \to \Delta}{\Box(A)^{\alpha, i}, \Gamma \to \Delta} \; (\Box^\alpha \to)$$

l is zero or arbitrary natural number such that $\alpha, i + l \in \{\beta_1, \dots, \beta_i, \dots, \beta_{n_k}\}$.

Below in the paper for simplicity we will omit parentheses in the notation of formulas with the index, i.e. we use A^α instead of $(A)^\alpha$. The notation Π^α will mean $A_1^\alpha, \dots, A_n^\alpha$ if $\Pi = A_1, \dots, A_n$. In the examples we will also omit parentheses when writing the index as k-tuple.

Example 1. Let us demonstrate the application of the rule of inference $(\to \Box^\alpha)$ reflecting the choice of eigen index in an application of $(\to \Box^\alpha)$.

Let the conclusion of $(\to \Box^\alpha)$ is $C^{1,1}$, A^1, $B^2 \to \Box C^0$, $\Box D^2$, A^0. Then $\gamma_1 = 1, 1$, $\gamma_2 = 1$, $\gamma_3 = 2$, $\gamma_4 = 0$. If we apply $(\to \Box^\alpha)$ bottom-up with respect to the formula $\Box D^2$, then $\alpha, i = 2$. So $|\alpha, i| = 1$, α, i-index class is $\{0, 1, 2\}$ and $2 = \max(0, 1, 2)$. Thus, the application of $(\rightarrow \Box^\alpha)$ has the form

$$\frac{C^{1,1}, A^1, B^2 \rightarrow \Box C^{\prime 0}, D^3, A^0}{C^{1,1}, A^1, B^2 \rightarrow \Box C^0, \Box D^2, A^0}$$

If we apply $(\rightarrow \Box^\alpha)$ bottom-up with respect to the formula $\Box C^0$, then $\alpha, i = 0$, $|\alpha, i| = 1$, α, i-index class is $\{0, 1, 2\}$ and $0 < \max(0, 1, 2)$. Because $\max(|\gamma_1|, |\gamma_2|, |\gamma_3|, |\gamma_4|) = 2$, $|\alpha| = 0$, $|\alpha_1|$ must be greater than 2. Thus, the application of $(\rightarrow \Box^\alpha)$ has the form

$$\frac{C^{1,1}, A^1, B^2 \rightarrow C^{0,0,1}, \Box D^2, A^0}{C^{1,1}, A^1, B^2 \rightarrow \Box C^0, \Box D^2, A^0}$$

Let K_1S4^B be the calculus obtained from $KS4^B$ by removing (cut).

Cut elimination theorem

The derivation D in $GS4^B$ (in $KS4^B$) will be called pure if
(1) each variable does not enter free and bound at the same time;
(2) all eigen variables of $(\rightarrow \forall)$, $(\exists \rightarrow)$ (all eigen variables of $(\rightarrow \forall)$, $(\exists \rightarrow)$ and eigen index of $(\rightarrow \Box^\alpha)$) are distinct from one another;
(3) if a variable (index) occurs as an eigen variable (eigen index) in a sequent S from D, then this variable (index, respectively) occurs only in sequents above S.

Lemma 1. An arbitrary derivation in the calculus $I \in \{GS4^B, KS4^B\}$ may be translated into a pure derivation with the same end sequent.

Proof: follows from the possibility to rename eigen variables (eigen index) of an application of the rules of inference $(\rightarrow \forall)$, $(\exists \rightarrow)((\rightarrow \Box^\alpha)$, respectively.)

Lemma 2. In the calculus $I_1 \in \{G_1S4^B, K_1S4^B\}$ the following structural rules of inference are derivable

$$\frac{\Gamma \rightarrow \Delta}{\Gamma \rightarrow \Delta, A} (\rightarrow W) \qquad \frac{\Gamma \rightarrow \Delta}{A, \Gamma \rightarrow \Delta} (W \rightarrow)$$

besides in case of K_1S4^B if α is any index from Γ, Δ and the formula A has some index β then β is any index such that $|\beta| \neq |\alpha|$ or if $|\beta| = |\alpha|$ then $\beta \leqslant \gamma$ where γ is maximum index from α-index class.

Proof: let us rename eigen variables and eigen index of an application of the rules of inference $(\rightarrow \forall)$, $(\exists \rightarrow)$, $(\rightarrow \Box^\alpha)$ in such a way that they do not enter the formula A indicated in the conclusion of $(\rightarrow W)$, $(W \rightarrow)$. After that the proof is carried out by induction on height of the given derivation.

A derivation in the calculus $I_1 \in \{G_1S4^B, K_1S4^B\}$ will be called atomic, if all the axioms are of the form $\Gamma, E \rightarrow \Delta, E$, where E is an elementary formula.

Lemma 3. An arbitrary derivation in the calculus $I_1 \in \{G_1S4^B, K_1S4^B\}$ may be reconstructed into an atomic one.

Proof: by induction on the complexity of the formula A in $\Gamma, A \rightarrow \Delta, A$.

Lemma 4. (1) All the rules of inference of K_1S4^B (except for $(\to \Box^\alpha)$ when the conclusion of this rule contains more than one positive occurrence of the formula of the form $\Box A$) are invertible in K_1S4^B. (2) All the rules of inference of G_1S4^B except for $(\to \Box)$, $(\to \Box\forall)$, $(\text{cut}^\Box)$ are invertible in G_1S4^B, besides $h(V_1) \leqslant h(V)$, where $h(V)$ is the height of the given derivation, $h(V_1)$ is the height of resulting one.

Proof: by induction on the height of the given derivation using Lemma 3.

Lemma 5. In the calculus $I_1 \in \{G_1S4^B,\ K_1S4^B\}$ the following structural rules of inference are derivable

$$\frac{\Gamma \to \Delta, C, C}{\Gamma \to \Delta, C}\ (\to C) \qquad\qquad \frac{C, C, \Gamma \to \Delta}{C, \Gamma \to \Delta}\ (C \to)$$

Proof: by induction on $w|C| + h(V)$, where $|C|$ is the complexity of the formula C, $h(V)$ is the height of the given derivation, using Lemma 4.

Lemma 6. Let $K_1S4^B \vdash \Gamma, \Pi^\alpha \to \Theta^\alpha, \Delta$ where $|\alpha| > \max(|\gamma_1|, \dots, |\gamma_n|)$, $\gamma_1, \dots, \gamma_n$ are all the indices from Γ, Δ, then either $K_1S4^B \vdash \Gamma \to \Delta$, or $K_1S4^B \vdash \Pi^\delta \to \Theta^\delta$, where δ is any index.

Proof: by induction on the height of the given derivation and using restriction for the choice of index in $(\to \Box^\alpha)$, $(\Box^\alpha \to)$.

Lemma 7. Let $K_1S4^B \vdash \Gamma, \Pi^{\alpha,i+l} \to \Delta^{\alpha,i+l}, \Theta$, where $l \geqslant 1$ and $\alpha, i+l \notin \Gamma, \Theta$; $\alpha, i = \max(\beta_1, \dots, \beta_n)$ where β_j belongs to α, i–index class from $\Gamma \to \Theta$, then $K_1S4^B \vdash \Gamma, \Pi^{\alpha,i+k} \to \Delta^{\alpha,i+k}, \Theta$, where $k \geqslant 0$.

Proof: in the given derivation let us replace all occurrences of the index $\alpha, i+l$ $(l \geqslant 1)$ by $\alpha, i+k (k \geqslant 0)$. It is not difficult to get convinced that performing such a replacement all the axioms and the application of any rule of inference (r) (except the case analyzed below in point 3) are transformed to the same axioms and to the applications of the same rule of inference (r). Let us consider only characteristic cases.

1. Let $(r) = (\Box^\alpha \to)$ then the application has the form:

$$\frac{\Gamma, B^{\alpha,j+h}, \Box B^{\alpha,j}, \Pi^{\alpha,i+l} \to \Delta^{\alpha,i+l}, \Theta}{\Gamma, \Box B^{\alpha,j}, \Pi^{\alpha,i+l} \to \Delta^{\alpha,i+l}, \Theta} (\Box^\alpha \to)$$

From the conditions of the Lemma and restriction of $(\Box^\alpha \to)$ we get that $j = i+l$ or $j \leqslant i \leqslant i+k$. Only the case when $j+h = i+l$ is interesting. Performing the replacement mentioned above in the premise of $(\Box^\alpha \to)$ under consideration we have that in the first case (when $j = i+l$) $j+h = i+l$, so $h = 0$, $K_1S4^B \vdash S_1 = \Gamma, B^{\alpha,i+k}, \Box B^{\alpha,i+k}, \Pi^{\alpha,i+k} \to \Delta^{\alpha,i+k}, \Theta$; in the second case (when $j \leqslant i \leqslant i+k$) $K_1S4^B \vdash S_1 = \Gamma, B^{\alpha,i+k}, \Box B^{\alpha,j}, \Pi^{\alpha,i+k} \to \Delta^{\alpha,i+k}, \Theta$. In both cases applying $(\Box^\alpha \to)$ to S_1 we get the desired derivation.

2. Let $(r) = (\to \Box^\alpha)$ and the application have the form:

$$\frac{\Gamma, \Pi^{\alpha,i+l} \to \Delta^{\alpha,i+l}, B^{\alpha,i+l+n}, \Theta}{\Gamma, \Pi^{\alpha,i+l} \to \Delta^{\alpha,i+l}, \Box B^{\alpha,i+l}, \Theta} (\to \Box^\alpha),$$

besides $n \geqslant 1$. Performing the replacement mentioned above, we have $K_1S4^B \vdash S_1 = \Gamma, \Pi^{\alpha,i+k} \to \Delta^{\alpha,i+k}, B^{\alpha,i+k+n}, \Theta$. As $n \geqslant 1$, then applying $(\to \square^\alpha)$ to the S_1 we get the desired derivation.

3. Let $(r) = (\to \square^\alpha)$ and the application have the form:

$$\frac{\Gamma, \Pi^{\alpha,i+l} \to \Delta^{\alpha,i+l}, B^{\alpha,\alpha_1}, \Theta}{\Gamma, \Pi^{\alpha,i+l} \to \Delta^{\alpha,i+l}, \square B^{\alpha,j}, \Theta}(\to \square^\alpha),$$

where $j \leqslant i < i+l$ and α_1 is k-tuple such that $|\alpha, \alpha_1| > \max(|\gamma_1|, \ldots, |\gamma_n|)$, $\gamma_1, \ldots, \gamma_n$ are all the indices entering the conclusion of the application of the rule of inference under consideration. Only the case when $j = i$ and $k = 0$ is interesting (in all other cases the replacement of all the indices $\alpha, i+l$ by $\alpha, i+k$ in this application does not disturb the restriction of the rule $(\to \square^\alpha)$). Having performed the replacement in the given derivation we can get some applications of the rule of inference $(\to \square^\alpha)$ where the restriction fails. In that case let us consider the upper one. Then $K_1S4^B \vdash \Gamma, \Pi^{\alpha,i} \to \Delta^{\alpha,i}, B^{\alpha,\alpha_1}, \Theta$. By Lemma 6 either (1) $K_1S4^B \vdash \Gamma, \Pi^{\alpha,i} \to \Delta^{\alpha,i}, \Theta$ or (2) $K_1S4^B \vdash S_1 =\to B^\delta$, where δ is any index. Let us take the index $i+1$ for δ. Then applying $(\to \square^\alpha)$ to S_1 and by Lemma 2 we get $K_1S4^B \vdash \Gamma, \Pi^{\alpha,i} \to \Delta^{\alpha,i}, \square B^i, \Theta$. In the case (1) by Lemma 2 we get the desired derivation.

Theorem 2. Let $\mathcal{D}$ be a pure derivation of a sequent S in $I = GS4^B$ $(KS4^B)$, then $I_1 \vdash S$, where $I_1 = G_1S4^B(K_1S4^B)$.

Proof: by induction on $\omega^2 n + \omega|A| + h(V_1) + h(V_2)$, where n is the number of (cut) in the given derivation $\mathcal{D}$, $|A|$ is the complexity of the cut-formula of the receding application of (cut) in $\mathcal{D}$, $h(V_1)$, $h(V_2)$ are the heights of premises of the receding application of (cut). In the proof of this theorem Lemmas 1, 2, 6, 7 are used and every time the upper application of (cut) rule is considered. Let us consider only the case when the upper application of (cut) in $\mathcal{D}$ has the shape:

$$\frac{\mathcal{D}_1\left\{\dfrac{\mathcal{D}_1'\{\Gamma \to \Delta, A^{\alpha,\gamma}}{\Gamma \to \Delta, \square A^{\alpha,i}}(\to \square^\alpha)\right. \qquad \dfrac{\mathcal{D}_2'\{A^{\alpha,i+k}, \square A^{\alpha,i}, \Gamma \to \Delta}{\square A^{\alpha,i}, \Gamma \to \Delta}(\square^\alpha \to)}{\Gamma \to \Delta}\text{(cut)}$$

Applying (cut) to $\mathcal{D}_1$, $\mathcal{D}_2'$ before having used Lemma 2 and using the induction hypothesis we get that $K_1S4^B \vdash S_3 = \Gamma,\ A^{\alpha,i+k} \to \Delta$. Now let us consider two cases. (A) indices α, γ and α, i belongs to the same index class, i.e. $\gamma = i + l$ $(l \geqslant 1)$. Relying on Lemma 7, let us replace in $\mathcal{D}_1'$ α, γ by $\alpha, i+k$. We get $K_1S4^B \vdash S_1' = \Gamma \to \Delta,\ A^{\alpha,i+k}$. Applying (cut) to S_1', S_3 and using the induction hypothesis we get $K_1S4^B \vdash \Gamma \to \Delta$. Now let us consider the case (B) when α, γ and α, i do not belong to the same index class, i.e. $|\alpha, \gamma| > \max(|\gamma_1|, \ldots, |\gamma_n|)$, where $\gamma_1, \ldots, \gamma_n$ are all the indices from Γ, Δ. In this case by Lemma 6 we have either

(1) $K_1S4^B \vdash \Gamma \to \Delta$, or

(2) $K_1S4^B \vdash\to A^\delta$. In the case (1) we have got the desired derivation. In the case (2) having taken $\alpha, i+l$ $(l \geqslant 1)$ for arbitrary δ and by applying Lemma 2 we proceed as in the case (A).

Remark. A necessity of the rule of inference $(\text{cut}^\square)$ in G_1S4^B can be illustrated by the derivation of the sequent $\forall x \square A(x) \to \square(\forall x A(x) \lor B)$.

Example 2. We shall show that in K_1S4^B the sequent $S = \rightarrow \Box A^0, \Box\neg\Box A^0$ is not derivable. Indeed, applying bottom-up to S the rules $(\rightarrow \Box^\alpha)$, $(\rightarrow \neg)$ (with respect to $\Box\neg\Box A^0$) we get the sequent $S_1 = \Box A^1 \rightarrow \Box A^0$. Applying bottom-up to S_1 the rule $(\rightarrow \Box^\alpha)$ we get the sequent $S_2 = \Box A^1 \rightarrow A^{0,1}$ which is not derivable relying on the restriction of the rule $(\Box^\alpha \rightarrow)$. We get the same result applying the rule $(\rightarrow \Box^\alpha)$ with respect to $\Box A^0$ at first.

Example 3

Let us consider a predicate version of the example from [7, page 225] i.e. the sequent $S = \rightarrow \Box\neg\Box\neg B$, where $B = (\Box\forall x A(x) \vee \Box\neg\forall x\Box A(x))$. The sequent is not derivable in $S4$ without Barcan axiom and derivable with Barcan axiom. Let us demonstrate two versions of the derivation in K_1S4^B of the sequent S. The second version is longer than the first one but it demonstrates that the application of the rule of inference $(\rightarrow \Box^\alpha)$ sometimes is invertible even if the conclusion contains more then one positive occurrence of the formula of the form $\Box A$. Let us denote $\forall x A(x)$ by D; $\forall x\Box A(x)$ by C.

First version:

$$\Box\neg B^1, C^2, \Box A^2(b), A^3(b) \rightarrow \Box D^1, A^3(b), \Box\neg C^2 \quad (\Box^\alpha \rightarrow)$$
$$\Box\neg B^1, C^2, \Box A^2(b) \rightarrow \Box D^1, A^3(b), \Box\neg C^2 \quad (\forall \rightarrow)$$
$$\Box\neg B^1, C^2 \rightarrow \Box D^1, A^3(b), \Box\neg C^2 \quad (\rightarrow \Box^\alpha), (\rightarrow \forall)$$
$$\Box\neg B^1, C^2 \rightarrow \Box D^1, \Box D^2, \Box\neg C^2 \quad (\Box^\alpha \rightarrow), (\neg \rightarrow), (\rightarrow \vee)$$
$$\Box\neg B^1, C^2 \rightarrow \Box D^1 \quad (\rightarrow \neg)$$
$$\Box\neg B^1 \rightarrow \Box D^1, \neg C^2 \quad (\rightarrow \Box^\alpha)$$
$$\Box\neg B^1 \rightarrow \Box D^1, \Box\neg C^1 \quad (\Box^\alpha \rightarrow), (\neg \rightarrow), (\rightarrow \vee)$$
$$\Box\neg B^1 \rightarrow \quad (\rightarrow \Box^\alpha), (\rightarrow \neg)$$
$$\rightarrow \Box\neg\Box\neg B^0$$

Second version:

$$\Box\neg B^1, C^{1,0}, C^3, A^4(b_1) \rightarrow A^2(b), \Box D^2, A^4(b_1), \Box\neg C^3 \quad (\forall \rightarrow), (\Box^\alpha \rightarrow)$$
$$\Box\neg B^1, C^{1,0}, C^3 \rightarrow A^2(b), \Box D^2, A^4(b_1), \Box\neg C^3 \quad (\rightarrow \Box^\alpha), (\rightarrow \forall)$$
$$\Box\neg B^1, C^{1,0}, C^3 \rightarrow A^2(b), \Box D^2, \Box D^3, \Box\neg C^3 \quad (\Box^\alpha \rightarrow), (\neg \rightarrow), (\vee \rightarrow)$$
$$\Box\neg B^1, C^{1,0}, C^3 \rightarrow A^2(b), \Box D^2 \quad (\rightarrow \Box^\alpha)(\rightarrow \neg)$$
$$\Box\neg B^1, C^{1,0} \rightarrow A^2(b), \Box D^2, \Box\neg C^2 \quad (\Box^\alpha \rightarrow), (\neg \rightarrow), (\rightarrow \vee)$$
$$\Box\neg B^1, C^{1,0} \rightarrow A^2(b) \quad (\rightarrow \Box^\alpha), (\rightarrow \neg)$$
$$\Box\neg B^1 \rightarrow A^2(b), \Box\neg C^1 \quad (\rightarrow \Box^\alpha), (\rightarrow \forall)$$
$$\Box\neg B^1 \rightarrow \Box D^1, \Box\neg C^1 \quad (\Box^\alpha \rightarrow), (\neg \rightarrow), (\rightarrow \vee)$$
$$\Box\neg B^1 \rightarrow \quad (\rightarrow \Box^\alpha), (\rightarrow \neg)$$
$$\rightarrow \Box\neg\Box\neg B^0$$

The equivalence index and index-free calculi

We shall prove that the calculi $KS4^B$ and $GS4^B$ are equivalent for the index-free sequents.

Lemma 8. $GS4^B \vdash S \Rightarrow KS4^B \vdash S$.

Proof: relying on Theorem 1 for proving Lemma it is sufficient to prove the axioms (1), (2), (3), (4) of $HS4^B$. Let us remind that the formula without index is considered as the formula with the index zero.

1)
$$\cfrac{\cfrac{\cfrac{A^1 \to A^1; \quad B^1 \to B^1}{\Box(A \supset B), \Box A \to B^1}\,(\Box^\alpha \to),(\supset^\alpha\to),(\supset\to),(W\to)}{\Box(A \supset B), \Box A \to \Box B}\,(\to \Box^\alpha)}{\to \Box(A \supset B) \supset (\Box A \supset \Box B)}\,(\to\supset),(\to\supset)$$

2)
$$\cfrac{\cfrac{A \to A}{A, \Box A \to A}\,(W\to)}{\Box A \to A}\,(\Box^\alpha \to)$$

3)
$$\cfrac{\cfrac{\cfrac{A^2 \to A^2}{\Box A \to A^2}\,(\Box^\alpha \to),(W\to)}{\Box A \to \Box\Box A}\,(\to \Box^\alpha),(\to \Box^\alpha)}{\to \Box A \supset \Box\Box A}\,(\to\supset)$$

4)
$$\cfrac{\cfrac{\cfrac{\cfrac{A^1(b) \to A^1(b)}{A^1(b), \Box A(b), \forall x \Box A(x) \to A^1(b)}\,(W\to)}{\Box A(b), \forall x \Box A(x) \to A^1(b)}\,(\Box^\alpha \to)}{\forall x \Box A(x) \to A^1(b)}\,(\forall\to)}{\to \forall x \Box A(x) \supset \Box \forall x A(x)}\,(\to\supset),(\to\Box^\alpha),(\to\forall)$$

To prove the statement inverse to Lemma 8 let us prove some auxiliary Lemmas.

Let K^*S4^B be the calculus obtained from K_1S4^B by eliminating $(\to \Box^\alpha)$ and adding $(\to \Box), (\Box \to), (\to \Box\forall), (\text{cut}^{\Box})$.

Let K^+S4^B be the calculus obtained from K_1S4^B by adding $(\to \Box)$, $(\Box \to), (\to \Box\forall), (\text{cut}^{\Box})$.

Lemma 9. Let $K^*S4^B \vdash \Gamma, \Pi^{\alpha,\gamma} \to \Delta^{\alpha,\gamma}, \Theta$, then $K^*S4^B \vdash \Gamma \to \Box(\Pi^{\alpha,i} \supset \Delta^{\alpha,i})$, Θ, where α, i is some index such that $\alpha,\gamma > \alpha, i$ and α,γ is greater than any index from Γ, Θ and there are no formulas with the index α,γ in Γ, Θ; $\Pi \supset \Delta$ is an image formula of the sequent $\Pi \to \Delta$.

Proof: by induction on the height of the given derivation. The basis is considered in an obvious way. Let (r) be the rule of inference which is applied last in the given derivation. Let us consider only some characteristic cases.

1. $(r) = (\Box^\alpha \to)$ and the application of $(\Box^\alpha \to)$ has the form:

$$\frac{\Gamma, A^{\alpha,\beta,j+l}, \Box A^{\alpha,\beta,j}, \Pi^{\alpha,\beta,j} \to \Delta^{\alpha,\beta,j}, \Theta}{\Gamma, \Box A^{\alpha,\beta,j}, \Pi^{\alpha,\beta,j} \to \Delta^{\alpha,\beta,j}, \Theta}\,(\Box^\alpha \to),$$

i.e. $\gamma = \beta, j$. Relying on the formulation of Lemma 9 and the rule $(\Box^\alpha \to)$ we have that $l = 0$. Therefore applying the induction hypothesis to the premise of $(\Box^\alpha \to)$ we have $K^*S4^B \vdash S_1 = \Gamma \to \Box(A^{\alpha,i} \wedge \Box A^{\alpha,i} \wedge \Pi^{\alpha,i} \supset \Delta^{\alpha,i}),\ \Theta$. It is easy to verify that $K^*S4^B \vdash S_2 = \Box(A^{\alpha,i} \wedge \Box A^{\alpha,i} \wedge \Pi^{\alpha,i} \supset \Delta^{\alpha,i}) \to \Box(\Box A^{\alpha,i} \wedge \Pi^{\alpha,i} \supset \Delta^{\alpha,i})$. Applying $(\mathrm{cut}^{\Box})$ to $S_1,\ S_2$ we obtain the desired derivation.

2. $(r) = (\Box^\alpha \to)$ and the application of $(\Box^\alpha \to)$ has the form:

$$\frac{\Gamma, \Box A^{\alpha,j}, A^{\alpha,i+l}, \Pi^{\alpha,i+l} \to \Delta^{\alpha,i+l}, \Theta}{\Gamma, \Box A^{\alpha,j}, \Pi^{\alpha,i+l} \to \Delta^{\alpha,i+l}, \Theta}\ (\Box^\alpha \to),$$

i.e. $\gamma = i + l,\ j \leqslant i$. Applying the induction hypothesis we have $K^*S4^B \vdash S_1 = \Gamma, \Box A^{\alpha,j} \to \Box(A^{\alpha,i} \wedge \Pi^{\alpha,i} \supset \Delta^{\alpha,i}), \Theta$. It is easy to verify that $K^*S4^B \vdash S_2 = \Box(A^{\alpha,i} \wedge \Pi^{\alpha,i} \supset \Delta^{\alpha,i}),\ \Box A^{\alpha,i} \to \Box(\Pi^{\alpha,i} \supset \Delta^{\alpha,i})$. Applying $(\mathrm{cut}^{\Box})$ to $S_1,\ S_2$ we get $K^*S4^B \vdash S_3 = \Gamma, \Box A^{\alpha,j}, \Box A^{\alpha,i} \to \Box(\Pi^{\alpha,i} \supset \Delta^{\alpha,i}),\ \Theta$. It is easy to verify that $K^*S4^B \vdash S_4 = \Box A^{\alpha,j} \to \Box A^{\alpha,i}$. Applying $(\mathrm{cut}^{\Box})$ to S_4, S_3 we get the desired derivation.

3. $(r) = (\to \Box)$ and the application of $(\to \Box)$ has the form

$$\frac{\Box\Pi^{\alpha,\gamma} \to A^{\alpha,\gamma}}{\Gamma, \Box\Pi^{\alpha,\gamma} \to \Box A^{\alpha,\gamma}, \Delta^{\alpha,\gamma}, \Theta}(\to \Box)$$

By the induction hypothesis we have $K^*S4^B \vdash S_1 = \to \Box(\Box\Pi^{\alpha,i} \supset A^{\alpha,i})$. It is easy to verify that $K^*S4^B \vdash S_2 = \Box(\Box\Pi^{\alpha,i} \supset A^{\alpha,i}) \to \Box(\Box\Pi^{\alpha,i} \supset \Box A^{\alpha,i} \vee \Delta^{\alpha,i})$. Applying $(\mathrm{cut}^{\Box})$ to S_1, S_2 we obtain the desired derivation.

4. $(r) = (\to \forall)$ and the application of $(\to \forall)$ has the form

$$\frac{\Gamma, \Pi^{\alpha,\gamma} \to \Delta^{\alpha,\gamma}, A^{\alpha,\gamma}(b), \Theta}{\Gamma, \Pi^{\alpha,\gamma} \to \Delta^{\alpha,\gamma}, \forall x A^{\alpha,\gamma}(x), \Theta}(\to \forall)$$

By the induction hypothesis we have $K^*S4^B \vdash S_1 = \Gamma \to \Box(\Pi^{\alpha,i} \supset \Delta^{\alpha,i} \vee A^{\alpha,i}(b)), \Theta$, where $b \notin \Gamma, \Pi, \Delta, \Theta$. Applying $(\to \forall)$, $(\to \Box\forall)$, to S_1, we have $K^*S4^B \vdash S_1' = \Gamma \to \Box\forall x(\Pi^{\alpha,i} \supset \Delta^{\alpha,i} \vee A^{\alpha,i}(x)), \Theta$. We can verify that $K^*S4^B \vdash S_2 = \Box\forall x(\Pi^{\alpha,i} \supset \Delta^{\alpha,i} \vee A^{\alpha,i}(x)) \to \Box(\Pi^{\alpha,i} \supset \Delta^{\alpha,i} \vee \forall x A^{\alpha,i}(x))$. Applying $(\mathrm{cut}^{\Box})$ to S_1', S_2 we obtain the desired derivation.

Lemma 10. $K^+S4^B \vdash S \Rightarrow K^*S4^B \vdash S$.

Proof: by induction on the number of applications of $(\to \Box^\alpha)$ in the given derivation using Lemma 9.

Lemma 11. Let S is index-free sequent, then $K^*S4^B \vdash S \Rightarrow G_1S4^B \vdash S$.

Proof: as S is index-free sequent then relying on the condition on $(\Box^\alpha \to)$ we can verify (starting from the sequent S) that all the applications of $(\Box^\alpha \to)$ coincide with the applications of $(\Box \to)$.

Theorem 3. Let S be an index-free sequent then $GS4^B \vdash S \Leftrightarrow KS4^B \vdash S$.

Proof: follows from Lemmas 8, 10, 11 and Theorem 2.

REFERENCES

1. Farinas del Cerro C.L., MOLOG: A system that extends PROLOG with modal logic. New Generation Computing, 4, (1986), 35–51.
2. Gabbay D., Modal and Temporal Logic programming, in book "Temporal logics and their applications" ed. A.Galton, Academic Press, London, 1987, 197–237.
3. Gallier J.H., Logic for computer science: foundations of automatic theorem proving, Harper and Row, New York, 1986.
4. Hughes G.E., Cresswell M.J., An Introduction to modal logic, Methuen and Co., London, 1968.
5. Kanger S.G., Provability in logic, Acta Universiatis Stockholmiensis, Stockholm Studies in Philosophy 1, 1957.
6. Kleene S.C., Introduction to metamathematics. North-Holland. Amsterdam. 1952.
7. Mints G., Gentzen-type systems and resolution rules. Part I. Propositional logic. LNCS, 417, (1990), 198–231.
8. Mints G., Gentzen-type systems and resolution rules. Part II. Predicate logic. Manuscript 1990.
9. Pliuškevičienė A., Nondisjunctive resolution for modal logics. COLOG-88, Part 2. Tallinn. 1988, 93–95 (in Russian).

LOGICAL FOUNDATION FOR LOGIC PROGRAMMING BASED ON FIRST ORDER LINEAR TEMPORAL LOGIC

Regimantas PLIUŠKEVIČIUS
Institute of Mathematics and Informatics,
Lithuanian Academy of Sciences,
232600 Vilnius, Akademijos 4, Lithuania
email: logica @ ma – mii.lt.su

Introduction

It is well-known (see e.g. [8, 15, 16]) that a cut-free sequential calculus serves as a foundation for logic programming. In [2, 6, 10, 18] several logic programming systems are presented. The logic programming systems from [10, 18] are based not on the proof but on constructing a model to satisfy a formula, i.e. the logic programming system works only in the finite state space case. The logic programming systems from [2, 6] are based on proofs but the language of these systems does not include clauses connected with intricate inductive proofs. More precisely: for these proofs there is no need to search for the invariant formula, i.e. such a formula $\mathcal{R}$ that $\models \mathcal{R} \supset \circ\mathcal{R}$. It must be stressed that to find invariant formulas is the key element in the finitary derivations in temporal logics, substitution and logical connectives play a secondary and simpler role. The purpose of this paper is to present some foundation of constructing a sequential variant with the efficient proof – theoretical properties for first order linear temporal logic with $\circ$ ("next") and $\Box$ ("always") without equality and flexible functional symbols. It is well-known that in temporal logic predicate and function symbols may be either flexible (time-dependent) or rigid (time-independed). Some computer science works use the words "local" and "global" for "flexible" and "rigid", respectively. Thus, if "busy" is a flexible unary predicate symbol and "computer" is a rigid constant symbol (i.e. a nullary rigid function symbol), then "busy (computer) $\wedge\neg\circ\Box$ busy (computer)" express that the computer is busy in the initial state and not busy from there on. Variables are rigid, free variables have an implicit universal quantitication. In [1, 4, 5, 12, 22] it was proved that if we consider linear temporal logic with flexible functional symbols, then such logic is incomplete. But it will become complete (see [1, 4, 5]) under nonstandard semantics and will allow new rules to define auxiliary predicates, i.e. in fact allowing the deduction tools of second order logic. In [7] it was proved that first order temporal logic with unary predicate symbols and with equality but without flexible function symbols is incomplete, because the Peano arithmetic can be imbedded in this logic. In [11] it was proved that first order linear temporal logic (with $\circ$, $\Box$) without equality and function symbols and without flexible constants becomes complete under standard Kripke-style semantic upon the addition of the infinitary rule: if A, $\circ A$, $\circ\circ A, \ldots$, then $\Box A$. (This infinitary rule is similar to the familiar ω-induction rule). In [19, 20] it was proposed to use the method of reduction of restricted first order infinitary linear temporal logic described in [11] to the finitary one. The main result of this paper can be stated as follows: let a sequent S be provable in infinitary first order linear logic (with $\bigcirc, \Box$) without equality and flexible function symbols then S^* is provable in finitary cut-free calculus, where S^* is the sequent applied to S standard skolemization operation (see e.g. [8, 17]). This transformation involves some direct method for finding the invariant formula. The proposed method of constructing a cut-free finitary calculus by means of reduction of infinitary calculus may be extended to temporal logic with other temporal operators and also to a nonlinear temporal model.

1. Description of the infinitary calculus $G_{L\omega}$

The formulas are determined with the help of logical symbols $\supset, \wedge, \vee, \urcorner, \exists, \forall$ and temporal operators $\bigcirc$ and $\Box$ as usual. The predicate symbols may be either flexible or rigid, the functional symbols may be only rigid. A sequent is an expression of the form $\Gamma \to \Delta$, where Γ, Δ are arbitrary finite sets (not sequences or multisets) of formulas. As in [19] we introduce formulas with indices (denoted by A^i and certifying the truth value of A in the i-th moment of time):

1) $(E^i)^k := E^{i+k}$, where E is an atomic formula, i is zero (which is identified with an empty word) or any natural number, k is any natural number;
2) $(A \odot B)^k := A^k \odot B^k, \quad \odot \in \{\supset, \wedge, \vee, \}$;
3) $(\sigma A)^k := \sigma A^k, \quad \sigma \in \{\urcorner, \bigcirc, \Box, \exists x, \forall x\}$.

Any formula with an index will be also called a formula.

The calculus $G_{L\omega}$ is defined by the following postulates.

Axiom: $\Gamma, A \to \Delta, A$.

Rules of inference:

1) temporal ones:

$$\Gamma \to \Delta, A^1 / \Gamma \to \Delta, \bigcirc A \qquad (\to \bigcirc)$$

$$A^1, \Gamma \to \Delta / \bigcirc A, \Gamma \to \Delta \qquad (\bigcirc \to)$$

$$A, \Box A^1, \Gamma \to \Delta / \Box A, \Gamma \to \Delta \qquad (\Box \to)$$

$$\{\Gamma \to \Delta, A^k\}_{k \in w} / \Gamma \to \Delta, \Box A \qquad (\to \Box_w),$$

where $k \in w$ means that $k \in \{0, 1, \ldots\}$; here and below Γ^1 means $A_1^{k_1+1}, \ldots, A_n^{k_n+1}$, if $\Gamma = A_1^{k_1}, \ldots, A_n^{k_n}$, $n \geqslant 1$, $k_i \geqslant 0$, $1 \leqslant i \leqslant n$;

2) logical rules of inference consist of traditional invertible rules of inference (for the considered logical operators) (see, e.g. [19]), except for the rules of inference $(\to \exists)$, $(\forall \to)$. For example, $(\to \exists)$: $\Gamma \to, \Delta\ A(t),\ \exists x A(x) / \Gamma \to \Delta,\ \exists x A(x)$ satisfies the following condition of "minus-normality": the eigen-term t belongs to the conslusion of $(\to \exists)$ if the conslusion of $(\to \exists)$ has not any terms, then $t = a$, where a is some fixed free variable; analogously for the $(\forall \to)$.

Remark 1.1. Let $G^{=}_{L\omega}$ be the calculus obtained from $G_{L\omega}$ by adding the postulates to the equality. Let us consider the sequent $S = \to \exists x \Box (f(x) = a \supset g(f(x)) = x))$, which is "temporal" versus of intuitionistic logic example from [14]. It is easy to verify that S is derivable in $G^{=}_{L\omega}$ without the condition of "minus-normality" and non derivable in $G^{=}_{L\omega}$ with this condition.

Derivations in the calculus $G_{L\omega}$ are built in an usual way (for the calculi with ω-rule), i.e. in the form of an infinite tree (with finite branches); the height of a derivation D is an ordinal (defined in a traditional way) denoted by $O(D)$. Let I be some calculus, then notation $I \vdash S$ means that the sequent S is derivable in S.

Lemma 1.1. In $G_{L\omega}$ the structural rule $\Gamma \to \Delta/\Gamma, \Pi \to \Delta, \Omega$ (W) is admissible, besides $O(D) = O(D_1)$, where D is the given derivation, D_1 is the resulting one.

Proof. Let us rename the eigen-variables of the applications of $(\to \forall), (\exists \to)$ in such a way that they do not enter Π, Ω. Afterwards the proof is carried out by induction on $O(D)$.

A derivation in the calculus $G_{L\omega}$ will be called atomic if all axioms have the form $\Gamma, E \to \Delta, E$, where E is an atomic formula.

Lemma 1.2. An arbitrary derivation in $G_{L\omega}$ may be transformed into an atomic one.

Proof. Let A be from the axiom $\Gamma, A \to A, \Delta$; $\Box(A)$ be the number of occurrences of the temporal operator $\Box$ in A; $l(A)$ be the number of occurrences of all operators in A. The proof of Lemma carried out by induction on $w.\ \Box(A) + l(A)$ (i.e. by double induction $< \Box(A), l(A) >$).

Lemma 1.3. Let (i) be the rule of inference of $G_{L\omega}$; S be the conclusion, S_1 be any premise of (i). Then $G_{L\omega} \vdash S \Rightarrow G_{L\omega} \vdash S_1$, besides $O(D_1) \leqslant O(D)$, where D is the given atomic derivation, D_1 is the resulting one.

Proof: by induction on $O(D)$ using Lemma 1.2 (when $(i) = \{(\to \exists), (\forall \to)\}$ Lemma 1.1 is applied).

A model M over which a formula of temporal logic under consideration is interpreted as a pair $< N, V >$, where N is a triple $< D, \omega, \leqslant >$ (called frame), V is a valuation function, defined in a traditional way; (see, e.g. [8]), ω is a set of all natural numbers, $\leqslant$ is the usual order relation on ω.

The concept "A is valid in $M =< N, V >$ at time point $k \in \omega$" (in symbols $M, k \models A$) is defined as follows:

1) $M, k \models P^l(t_1, \dots, t_m) \iff < V(t_1), \dots, V(t_m), k + l > \in V(P)$;
2) $M, k \models \circ A \iff M, k+1 \models A$;
3) $M, k \models \Box A \iff \forall l (l \in k) M, k + l \models A$.

Other cases are defined as in first order logic (see, e.g. [8]). Using this concept we can define (in a traditional way) the concept of universally valid sequent.

Theorem 1.1 (soundness of $G_{L\omega}$). If $G_{L\omega} \vdash S$, then S is universally valid.

Proof. The theorem is easily proved by induction on $O(D)$, where D is the given derivation.

Theorem 1.2 (completeness of $G_{L\omega}$). If a sequent S is universally valid, then $G_{L\omega} \vdash S$.

Proof: analogously as in [8,19].

Remark 1.2. From Theorem 1.2 there follows the admissibility of cut in $G_{L\omega}$.

2. Obtaining the quantifier-free sequences in $G_{L\omega}$

In this section we shall prove that $G_{L\omega} \vdash S \Rightarrow G_{L\omega} \vdash S^{cf}$, where S^{cf} is the quantifier-free sequent obtained from S in the way described below.

Let A be a formula then $\widetilde{\forall} A$ denotes the closure of free variables in A.

Lemma 2.1 (monotone replacement of formulas). Let A, B, C be arbitrary formulas and the formula C^* be obtained as a result of the replacement of the formula B instead of some positive (negative) occurrence of A in C, then

$$G_{L\omega} \vdash \widetilde{\forall}\Box(A \supset B) \to \widetilde{\forall}\Box(C \supset C^*)$$
$$(G_{L\omega} \vdash \widetilde{\forall}\Box(A \supset B) \to \widetilde{\forall}\Box(C^* \supset C),$$

respectively).

Proof. Let d be the number of occurrences of logical and temporal symbols which enter the scope of the replaceable occurrence of A in C. The proof of Lemma 2.1 is carried out by induction on d using the following facts:

$$G_{L\omega} \vdash \widetilde{\forall}\Box(D \supset E) \to \widetilde{\forall}\Box(\urcorner E \supset \urcorner D);$$
$$G_{L\omega} \vdash \widetilde{\forall}\Box(D \supset E) \to \widetilde{\forall}\Box(\sigma D \supset \sigma E),\ \sigma \in \{\circ, \Box, \forall x, \exists x\};$$
$$G_{L\omega} \vdash \widetilde{\forall}\Box(D \supset E) \to \widetilde{\forall}\Box((D \odot M) \supset (E \odot M));$$
$$G_{L\omega} \vdash \widetilde{\forall}\Box(D \supset E) \to \widetilde{\forall}\Box((M \odot D) \supset (M \odot E)), \odot \in \{\wedge, \vee\};$$
$$G_{L\omega} \vdash \widetilde{\forall}\Box(D \supset E) \to \widetilde{\forall}\Box((M \supset D) \supset (M \supset E));$$
$$G_{L\omega} \vdash \widetilde{\forall}\Box(D \supset E) \to \widetilde{\forall}\Box((E \supset M) \supset (D \supset M)).$$

Lemma 2.2 (equivalent replacement of formulas). $G_{L\omega} \vdash \widetilde{\forall}\Box(A \equiv B) \to \widetilde{\forall}\Box(C \equiv C^*)$, where A, B, C, C^* is the same as in Lemma 2.1. $A \equiv B$ means $(A \supset B)\wedge(B \supset A)$.

Proof: follows immediately from Lemma 2.1.

Let $S = A_1, \ldots, A_n \to B_1, \ldots, B_m$ then the formula $S^f = \bigwedge_{i=1}^{n} A_i \supset \bigvee_{j=1}^{m} B_j$ is called the image formula of S.

Lemma 2.3. $G_{L\omega} \vdash S \iff G_{L\omega} \vdash\to S^f$.

Proof: traditionally (see, e.g. [8]), using Remark 1.2.

Lemma 2.4 ("next" – elimination). Let S be a sequent, S° be the sequent obtained from S by replacing all occurrences of the formulas of the form $\circ A^i$ $(i \geqslant 0)$ by the formula A^{i+1}, then $G_{L\omega} \vdash S \iff G_{L\omega} \vdash S^\circ$.

Proof. Since $G_{L\omega} \vdash\to \circ A^i \equiv\to A^{i+1}$, the proof of the Lemma is carried out using Lemmas 2.2, 2.3.

Let S be a sequent, S^* be the sequent, obtained from S with the help of standard skolemization (in short: st-skolemization) operation (see, e.g. [8, 17]), S^* will be called st-skolemized variant of the S.

Remark 2.1. We use the notion "standard skolemization" to differ from "flexible skolemization" used [3] (which is in fact skolemization operation for higher order modal logic introduced in [9]), which introduces the flexible functional symbols, therefore leads to second order logic. Different from st-skolemization operation which yields only deductive equivalent (i.e. leads to a conservative extension of a calculus under consideration), the flexible skolemizations operation applied to the sequent S yields the sequent $S^{\#}$ such that $I \vdash\rightarrow (S^f \equiv S^{\#f})$, where I is second order logic-like calculus.

Lemma 2.5. Let S be a sequent, S^* be the st-skolemized variant of S then $G_{L\omega} \vdash S \Rightarrow G_{L\omega} \vdash S^*$ $(*)$.

Proof. The proof is carried out using Lemmas 2.3, 2.1, and following the facts: $G_{L\omega} \vdash \widetilde{\forall}\Box(A(t) \supset \exists x A(x))$, $G_{L\omega} \vdash \widetilde{\forall}\Box(\forall x A(x) \supset A(t))$ and admissibility of cut in $G_{L\omega}$.

Remark 2.2. Let us consider the inversion of the implication $(*)$ from Lemma 2.5, i.e. let us consider the implication $G_{L\omega} \vdash S^* \Rightarrow G_{L\omega} \vdash S$ $(**)$. It is clear that the implication $(**)$ is not true for an arbitrary sequent S. For example $G_{L\omega} \vdash \Box A(b) \rightarrow \exists x \Box A(x)$ (b is 0-place skolem-function), but $G_{L\omega} \nvdash \Box \exists x A(x) \rightarrow \exists x \Box A(x)$. Let us state some condition under which the implication $(**)$ is valid. Let us call the application of $(\rightarrow \exists), (\forall \rightarrow)$ incorrect if the main formula of the application contains the positive occurrence of $\Box$ and eigen-term of the application contain the skolem-function entering in the parametric formula (of the application) of the type $\Box A$. The derivation of S^* in $G_{L\omega}$ will be called correct if it does not contain the incorrect applications of $(\rightarrow \exists)$, $(\forall \rightarrow)$. Now we can state a new implication $(**')$: if the given derivation of S^* in $G_{L\omega}$ is correct, then $(**)$ is valid, moreover we can prove $(*')$, obtained from $(*)$ by adding the condition that the resulting derivation is correct. The justification of $(*')$, $(**')$ is beyond the scope of the paper.

Let $S' \cdot B(QxA(x))$ mean the st-skolemized sequent $S = \Gamma_1 \rightarrow \Gamma_2$, where $\Gamma_i = B(QxA(x))$, Γ'_i ($i = 1$ or $i = 2$, $Q \in \{\forall, \exists\}$); $\forall x A(x)$ occurs negatively ($\exists x A(x)$ occurs positively) in S; let an isolated occurrence of $QxA(x)$ do not enter the scope of quantifiers; let $G_{L\omega} \vdash S$; let $S^h = S' \cdot B(\bigwedge_{i=1}^{n} A(t_i))$ ($S^h = S' \cdot B(\bigvee_{i=1}^{n} A(t_i))$) mean the sequent obtained from S replacing the isolated occurrence of the formula $\forall x A(x)$ (the formula $\exists x A(x)$) by the formula $\bigwedge_{i=1}^{n} A(t_i)$ (formula $\bigvee_{i=1}^{n} A(t_i)$) where $t_1, \ldots, t_n$ ($n \geqslant 1$) is the finite sequence of eigen-terms of applications of $(\forall \rightarrow)$ (of $(\rightarrow \exists)$, respectively) the main formulas of which are ancestors of the isolated occurrence of the formula $\forall x A$ (formula $\exists x A$) in the given atomic derivation of S. The described operation will be denoted by (h).

Lemma 2.6. Let $S = S' \cdot B(QxA(x))$ be the st-skolemized sequent described above, let S^h be the sequent obtained from S with the help of the operation (h), then $G_{L\omega} \vdash S \Rightarrow G_{L\omega} \vdash S^h$ for some terms $t_1, \ldots, t_n$ ($n \geqslant 1$).

Proof. Since the given derivation is atomic, in the given derivation there are applications of $(\forall \rightarrow)$, $(\rightarrow \exists)$ the main formulas of which are ancestors of the isolated occurrence of the formula $QxA(x)$. The proof is carried out on $O(D)$ (where D is given atomic derivation) using the condition of "minus-normality" on the $(\forall \rightarrow,)$ $(\rightarrow \exists)$. Let us consider only the case when the last step of the given derivation D has the shape:

$$\frac{D_i\{\Gamma, B(\forall x A(x)) \rightarrow \Delta,\ A^i\}_{i \in \omega}}{\Gamma,\ B(\forall x A(x)) \rightarrow \Delta,\ \Box A}\ (\rightarrow \Box_\omega)$$

Applying to D_i the induction hypothesis, we have that $G_{L\omega} \vdash S'_i = \Gamma, B(\bigwedge_{j=1}^{n_i} A(t_j)) \rightarrow \Delta, A^i$ $(i \in \omega)$. Relying on the condition of "minus-normality" we get that $\forall j$ $(1 \leqslant j \leqslant n_i)$ n_j is a finite natural number. Applying Lemma 2.1 and $(\rightarrow \Box_\omega)$, we get that $G_{L\omega} \vdash \Gamma, B(\bigwedge_{j=1}^{l} A(t_j)) \rightarrow \Delta, \Box A$, where $l = \sum_{k=1}^{n_i} n_k$.

Remark 2.3. The condition of "minus-normality" in $(\forall \rightarrow)$, $(\rightarrow \exists)$ plays a crucial role in justification of finiteness of the sequence of eigen-terms of applications of $(\forall \rightarrow)$, $(\rightarrow \exists)$ which may be applied in an infinitary way.

Lemma 2.7. Let S, S^h be the same as in Lemma 2.6 then $G_{L\omega} \vdash S^h \Rightarrow G_{L\omega} \vdash S$.

Proof. Let $Q = \forall$ (the case $Q = \exists$ is considered in a dual way). Let us replace all the ancestors of the isolated occurrence of the formula $\bigwedge_{i=1}^{n} A(t_i)$ by the formula $\forall x A(x)$. The axiom $A(t_i) \rightarrow A(t_i)$ will be replaced by the application of (W) (which is admissible in $G_{L\omega}$) and $(\forall \rightarrow)$, applications of $(\wedge \rightarrow)$ (with the main formula $\bigwedge_{i=1}^{n} A(t_i)$) will be replaced by the applications of $(\forall \rightarrow)$.

Let S be a st-skolemized sequent, S^H be the sequent obtained from S by eliminating all negative (positive) occurrences of $\forall$(of $\exists$, respectively) with the help of operation (h).

Theorem 2.1 (analogue of Herbrand's theorem for $G_{L\omega}$). Let S be a st-skolemized sequent, then $G_{L\omega} \vdash S \iff G_{L\omega} \vdash S^H$.

Proof: by means of Lemmas 2.6, 2.7.

Remark 2.4. Instead of choosing the finite sequence of terms from the given derivation of the st-skolemized sequent, it is possible to present the working algorithm to generate the terms starting from the sequent S itself (not from the derivation of S); of course it can not be an effective way, in general, to formulate the criterion of termination of the algorithm (we can justify only the existence of termination of the algorithm). In these terms one can obtain more traditional formulation of Herbrand's theorem for $G_{L\omega}$ than Theorem 2.1. However, the justification of the algorithm is rather laborious and it is beyond of the scope of the paper.

Let S be a sequent, the sequent obtained from S by eliminating the operator $\bigcirc$(next) and eliminating the quantifiers will be called reduced variant of S (shortly: reduced sequent).

Theorem 2.2. $G_{L\omega} \vdash S \Rightarrow G_{L\omega} \vdash S^r$, where S^r is the reduced variant of the S.

Proof: follows from Lemmas 2.4, 2.5 and Theorem 2.1.

3. Some preparative lemmas for reduction of infinitary calculus $G_{L\omega}$ to finitary calculus G_L

In this section we shall consider only reduced sequences. A sequent S will called singular if (1) S contains positive occurrences of $\Box$; (2) all negative occurrences of $\Box$ may enter only the scope of positive occurrences of $\Box$. A sequent S will be called ordinary if S contains both positive and negative occurrences of $\Box$ (which do not enter the scope of positive ones). A sequent S will be called simple if S does not contain any positive occurrences of $\Box$. Let us define the notion of the reduction $R\{i\}$ in a calculus J of a sequent S to the sequences $S_1, \ldots, S_n$, where of $\{i\}$ is the set of rules of inference invertible in J. $R\{i\}$ is a tree of sequences; the lowest sequent of $R\{i\}$ is the sequent S; every sequent in $R\{i\}$ except S is an upper sequent of the rule of inference $(k) \in \{i\}$ whose lower sequent is also in $R\{i\}$, i.e. $R\{i\}$ consists of the bottom-up applications of the rules of inference from $\{i\}$. The topmost sequences of $R\{i\}$ are the sequences $S_1, \ldots, S_n$ and any axiom of J. The notation $S\{i\} \Rightarrow \{S_1, \ldots, S_n\}$ will mean that it is possible to construct the reduction $R\{i\}$ of the sequent S to the sequences $S_1, \ldots, S_n$. From the definition of $R\{i\}$ we have that if $S\{i\} \Rightarrow \{S_1, \ldots, S_n\}$, then $J \vdash S \Rightarrow J \vdash S_j (1 \leqslant j \leqslant n)$. We shall say that the sequent S is absorbed by some sequent S' if S can be obtained from S' with the help of the structural rule of inference $(\mathcal{W})$. Let us consider a reduction $R\{i\}$ of the sequent S to the sequents $S_1, \ldots, S_n$. The sequent $S^* \in R\{i\}$ will be called saturated if $\exists S' \in R\{i\}$ which is below S^* and $S^* = S'$. The sequent $S^* \in R\{i\}$ will be called closed if S^* is either saturated or absorbed by a saturated sequent which is in $R\{i\}$. The reduction $R\{i\}$ of S to the sequents $S_1, \ldots, S_n$ will be called closed if $\forall j (1 \leqslant j \leqslant n)$ S_j is closed. Now we shall define the set $\{i\}$ of rules of inference invertible in $G_{L\omega}$ and some tactics with the help of which we shall construct a closed reduction in $G_{L\omega}$ of the given sequent S to the sequents $S_1, \ldots, S_n$. At first let us introduce the following rules of inference:

$$\Gamma \to \Delta, \Box A / \Gamma \to \Delta, \Box A^1 \quad (\to \Box^{-1})$$
$$\Gamma \to \Delta, \Box A^1 / \Gamma \to \Delta, \Box A \quad (\to \Box^1).$$

Lemma 3.1. The rule of inference $(\to \Box^{-1})$ is admissible in $G_{L\omega}$, i.e. the rule of inference $(\to \Box^1)$ is invertible in $G_{L\omega}$.

Proof: by using cut and its admissibility in $G_{L\omega}$.

Now we shall set a canonical form of a sequent S (recall that we consider only reduced sequents). The sequent S is in a canonical form and will be called primary if $S = \Sigma_1, \Pi_1^1, \Box(\Delta_1^1) \to \Sigma_2, \Pi_2^1, \Box(\Delta_2^1)$, where $\Sigma_i = \varnothing (i = 1, 2)$ or consists of formulas which do not contain $\Box$ and indices (such formulas will be called logical formulas); $\Pi_i = \varnothing (i = 1, 2)$ or consists of atomic formulas (possible with indices), $\Delta_i = \varnothing (i = 1, 2)$ or consists of arbitrary quantifier – free formulas; here and below $\Box(\Delta)$ stands for $\Box B_1, \ldots, \Box B_n$ if $\Delta = B_1, \ldots, B_n$; let us recall that Γ^1 means $A_1^{k_1+1}, \ldots, A_n^{k_n+1}$, if $\Gamma = A_1^{k_1}, \ldots, A_n^{k_n} (k_i \geqslant 0)$; if S is singular, then $\Delta_1 = \varnothing$; if S is ordinary, then $\Delta_1 \neq \varnothing$ and $\Delta_2 \neq \varnothing$. A sequent S will be called quasiprimary if $S = \Sigma_1, \Pi_1^1, \Box(\theta_1) \to \Sigma_2, \Pi_2^1, \Box(\theta_2)$, i.e. θ_1, θ_2 have no indices. Let $G_{L\omega} \vdash S$, let $S\{i\} \Rightarrow \{S_1, \ldots, S_n\}$, where $S_j (1 \leqslant j \leqslant n)$ is the primary (quasiprimary) sequent, $\{i\}$ consists of bottom-up applications of $(\to \Box^1)$ and the rules of inference of $G_{L\omega}$, except of $(\to \circ), (\circ \to), (\to \Box_\omega)$ and rules of inference for quantifiers (consist of bottom-up applications of the rules of inference of $G_{L\omega}$, except $(\to \circ)$, $(\circ \to), (\to \Box_\omega), (\Box \to)$ and the rules for quantifiers, respectively); $\{S_1, \ldots, S_n\}$ will be called a primary (quasiprimary) expansion of S and will be denoted by $P(S)$ (by $QP(S)$, respectively), besides each antecedent (succedent) member of the form $\Box A^1$ containing in a vertex of $P(S)$ arises from the corresponding application of $(\Box \to)((\to \Box^1)$, respectively).

Let us introduce the following rules of inference $S/S_i(A)$; $S_i/S(\bar{A})$, where $i = 1$ or $i = 2$, S is a primary sequent, i.e. $S = \Sigma_1, \Omega_1^1 \to \Sigma_2, \Omega_2^1$ (where Σ_1, Σ_2 consist of logical formulas); $S_1 = \Sigma_1 \to \Sigma_2$; $S_2 = \Omega_1 \to \Omega_2$; S_1 will be called a logical upper sequent of the rule of inference $(\bar{A})$, S_2 will be called the temporal one.

Lemma 3.2. The rule of inference (A) is admissible in $G_{L\omega}$, i.e. the rule of inference $(\bar{A})$ is invertible in $G_{L\omega}$.

Proof: by induction on $O(D)$ where D is the given atomic derivation of the sequent S in $G_{L\omega}$.

Lemma 3.3. Let $G_{L\omega} \vdash S$ then it is possible to construct $P(S)$ and $QP(S)$.

Proof: using the invertibility the rules of inference of $G_{L\omega}$ using Lemma 3.1, 3.2.

Let $S_i \in P(S)(1 \leqslant i \leqslant n)$, let $S'_j(1 \leqslant j \leqslant m \leqslant n)$ be the temporal upper sequent of the application of the rule of inference $(\bar{A})$ whose lower sequent is the sequent S_i, then the set $\{S'_1, \ldots, S'_m\}$ will be called a temporal expansion of S and will be denoted by $T(S)$.

Lemma 3.4. Let $G_{L\omega} \vdash S$ then it is possible to construct $T(S)$.

Proof: follows from Lemmas 3.2, 3.3.

Let S be a quasiprimary ordinary sequent. Let us define the notion of subformula and invariant subformula of an arbitrary formula from reduced sequent. These notions are obtained from the notion of subformula of an arbitrary formula the propositional logic by adding the following two points: (1) the subformulas (invariant subformulas) of A^1 are the subformulas (invariant subformulas) of A and A^1 itself; (2) the subformulas (invariant subformulas) of $\Box B$ are subformulas of B and the formula $\Box B^k$, $(k \in \omega)$ (invariant subformulas of B and the formula $\Box B$ itself). Therefore the set of subformulas of the formula A containing $\Box$ is infinite opposite to the set of invariant subformulas of A which is finite. We shall show that with the help of the rule of inference $(\bar{A})$ we can construct the reduction R of the given sequent S to the quasiprimary sequences containing only invariant subformulas of formulas from S. By the finiteness of the set of invariant subformulas of formulas from S we can justify that the reduction R is closed. Let us introduce the notion of n-th $(n \geqslant 0)$ resolvent of the sequent S (denoted by $\mathcal{R}^n(S)$). $\mathcal{R}^n(S)$ is a set of quasiprimary sequences $S_1, \ldots, S_n$. The reduction of S to the quasiprimary sequences $S_1, \ldots, S_n$ can be described as follows: (1) let us construct $P(S)$ and then with the help of $(\bar{A})$, $T(S)$ is obtained; for each $S' \in T(S)$ let us construct $QP(S')$ (each sequent from $QP(S')$ consists of some invariant subformulas of formulas from S); (2) let $S^* \in QP(S')$, if S^* is closed, then let us consider another sequent from $QP(S')$ otherwise let us set $S^* = S$ and returns to (1). The steps (1),

(2) are repeated until closed reduction is obtained. Let us describe $\mathcal{R}^n(S)$ in a more formal way. Let $n = 0$ then $\mathcal{R}^n(S) = S$. Let $n > 0$, let $S_i \in \mathcal{R}^n(S)$, the sequent S_i will be called saturated if $\exists S' \in \mathcal{R}^k(S)$ $(k = 0, 1, \ldots, n-1)$ such that $S_i = S'$; the sequent S_i will be called closed if either S_i is saturated or S_i absorbed by some saturated sequent $S' \in \mathcal{R}^k(S)$ $(k = 1, 2, \ldots, n-1)$, otherwise S_i will be called nonclosed. Let $S_i \in \mathcal{R}^n(S)$ and let S_i be nonclosed then $\mathcal{R}^{n+1}(S)$ is defined in the following way. The set $\bigcup_i \bigcup_j QP(S_{ij})$ where $S_{ij} \in T(S_i)$ will be called a resolvent of S_i and will be denoted by $\mathcal{R}(S_i)$; $\mathcal{R}^{n+1}(S) = \bigcup_i \mathcal{R}(S_i)$. The reduction of S to the sequences $S_i \in \mathcal{R}^n(S)$ will be called n-th resolvent reduction of S and denoted by $\mathcal{R}^n R(S)$. If all the vertexes of $\mathcal{R}^n R(S)$ are closed, then such $\mathcal{R}^n R(S)$ will be called closed and will be denoted by $C\mathcal{R}^n R(S)$.

Lemma 3.5. Let S be an ordinary quasiprimary sequent, let $G_{L\omega} \vdash S$, then it is possible to construct $C\mathcal{R}^n R(S)$, besides all saturated vertexes of $C\mathcal{R}^n R(S)$ are the sequences of the form $S_i = \Pi_i, \Box(\Gamma_1) \to \Delta_i, \Box(\Gamma_2)$ $(1 \leqslant i \leqslant m)$, where $\Gamma_1, \Gamma_2 \neq \varnothing$, Π_i, Δ_i are invariant subformulas of Γ_1; Π_i consists of atomic formulas, Δ_i consists of either atomic formulas or the formulas of the type $\Box A$.

Proof: using the invertibility of the rules of inference of $G_{L\omega}$ (except of $(\to \bigcirc), (\bigcirc \to), (\to \Box_\omega)$ and the rules for quantifiers) and $(\to \Box^1)$, $(\bar{A})$, also the finiteness of the set of invariant subformulas of formulas from S and the construction of $C\mathcal{R}^n R(S)$.

Example 3.1. Let us consider the sequent from [13, page 17]: $S = \Box(\Omega),\ P \to Q, \Box\neg R$, where $\Box(\Omega) = \Box(P \leftrightarrow \neg P^1),\ \Box(R \supset P),\ \Box(Q \leftrightarrow \neg Q^1),\ \Box(R \supset Q)$ ($A \leftrightarrow B$ means $(A \supset B) \wedge (B \supset A)$). Applying to S the algorithm for finding $C\mathcal{R}^n R(S)$ we get that $C\mathcal{R}^n R(S) = \{\Box(\Omega),\ P \to Q, \Box\neg R;\ \Box(\Omega), Q \to P,\ \Box\neg R\}$.

Lemma 3.6 (main property of the saturated vertex of $C\mathcal{R}^n R(S)$). Let $G_{L\omega} \vdash S$, where S is an ordinary quasiprimary sequent, let S' be a saturated vertex of $C\mathcal{R}^n R(S)$, let $\mathcal{R}(S') = \{S_1, \ldots, S_n\}$. Then $\forall i (1 \leqslant i \leqslant n)$ S_i is saturated or is absorbed by saturated vertex of $C\mathcal{R}^n R(S)$ different from S'.

Proof: follows from the construction of $C\mathcal{R}^n R(S)$ and the finiteness of the set of vertexes of $C\mathcal{R}^n R(S)$.

Example 3.2. Let $S = P, \Box\Omega \to \Box A$ (where $\Omega = ((P \vee \neg\Box Q) \supset (P^1 \vee \neg\Box Q^1))$; $A = (P \vee \neg\Box Q)$. Then $C\mathcal{R}^n R(S) = \{S_1; S_2\}$, where $S_1 = P, \Box\Omega \to \Box A$; $S_2 = \Box\Omega \to \Box Q, \Box A$; $\mathcal{R}(S_1) = \{S_1;\ S_2\}$; $\mathcal{R}(S_2) = \{\Box\Omega, P \to \Box Q, \Box A;\ \Box\Omega \to \Box Q, \Box Q, \Box A\}$, the sequent $\Box\Omega, P \to \Box Q, \Box A$ is absorbed by S_1 and $\Box\Omega \to \Box Q, \Box Q, \Box A = S_2$.

Lemma 3.7 (main separation Lemma). Let $G_{L\omega} \vdash S$, where S is an ordinary quasiprimary sequent, let $S_i = \Pi_i, \Box(\Gamma_1) \to \Delta_i, \Box(\Gamma_2)$ be any saturated vertex of $C\mathcal{R}^n R(S)$, then either (1) $\forall i (1 \leqslant i \leqslant n)$ there exists a formula $\Box A \in \Box(\Gamma_2)$ such that $G_{L\omega} \vdash S'_i = \Pi_i, \Box(\Gamma_1) \to \Delta_i, \Box A$, or (2) $G_{L\omega} \vdash S^0 = \Box(\Gamma_1) \to \Box(\Gamma_2)$.

Proof: using Lemmas 3.5, 3.6.

Remark 3.1. The set of sequences $\{S'_1 = \Pi_1, \Box(\Gamma_1) \to \Delta_1; \ldots; S'_n = \Pi_n, \Box(\Gamma_1) \to \Delta_n\}$ corresponding to the set of sequents S'_i $(1 \leqslant i \leqslant n)$ from Lemma 3.7 will be called an invariant set of the sequent S and will be denote by INV(S).

Let us consider the case when S is singular. Let us consider a more specific form of the primary sequent for a singular sequent S. A singular sequent S will be called strong primary if $S = \Sigma_1 \to \Sigma_2, \Box B_1^{k_1}, \ldots, \Box B_n^{k_n} (n \geqslant 0)$, $k_l > t \geqslant 0$ $(1 \leqslant l \leqslant n)$, t is the greatest index in Σ_1, Σ_2; $\Box \notin \Sigma_1, \Sigma_2$.

Lemma 3.8. Let $G_{L\omega} \vdash S$, where S a singular sequent, then $S\{i\} \Rightarrow \{S_1, \ldots, S_n\}$, where $S_j (1 \leqslant j \leqslant n)$ is a strong primary sequent, $\{i\}$ consists of bottom-up applications of $(\to \Box^1)$ and the rules of inference of $G_{L\omega}$, except $(\to \circ), (\circ \to), (\to \Box_\omega), (\Box \to)$ and rules of inference for quantifiers.

Proof: using the invertibility of the rules of inference from $\{i\}$.

Let us introduce the following rules of inference $S/S_i(A^*)$; $S_i/S(\bar{A}^*)$, where $i = 1$ or $i = 2$, S is a strong primary (see above) $S_1 = \Sigma_1 \to \Sigma_2$; $S_2 = \to \Box(\Omega)$.

Lemma 3.9. The rule of inference (A^*) is admissible in G_{L_ω}, i.e. the rule of inference $(\bar{A}^*)$ is invertible in $G_{L\omega}$.

Proof: by a repeated application (starting from S) the rule of inference (A).

4. Description of the finitary calculus $\mathbf{G_L}$ and reduction of $\mathbf{G_L}$ to $\mathbf{G_{L\omega}}$

The postulates of G_L are the same as in $G_{L\omega}$ except the rule of inference $(\to \Box_\omega)$ which is replaced by the following ones:

$$\Gamma \to \Delta/\Pi, \Gamma^1 \to \Delta^1, \theta \qquad (+1)$$

$$\Gamma, \Box(\Omega) \to \Delta, \mathcal{R}; \; \mathcal{R} \to \mathcal{R}^1; \; \mathcal{R} \to A/$$
$$\Pi, \Gamma, \Box(\Omega) \to \theta, \Delta, \Box A \qquad (\to \Box_1)$$

$$\Gamma \to \Delta, A; \; \Gamma \to \Delta, \Box A^1/\Gamma \to \Delta, \Box A \qquad (\to \Box_2)$$

$$S_1; \ldots; \; S_n/\Gamma, \Box(\Omega) \to \Delta, \Box A_1, \ldots, \Box A_n \qquad (\to \Box_3),$$

where $S_1 = \Box(\Omega) \to A_1, \Box A_2, \ldots, \Box A_n; \ldots; S_n = \Box(\Omega) \to \Box A_1, \ldots, \Box, A_{n-1}, A_n$.

The rule of inference $(\to \Box_1)$ satisfies the following conditions. Let the conclusion of the $(\to \Box_1)$ be a quantifier-free sequent, then $\Gamma, \Box(\Omega) \to \Delta, \Box A$ is a quasiprimary sequent; the formula R (called an invariant formula) has the form $\bigvee_{i=1}^{n} (\Pi_i^{\wedge} \wedge \urcorner \Delta_i^{\vee}) \wedge (\Box(\Omega))^{\wedge}$, where $\{\Pi_1, \Box(\Omega) \to \Delta_1, \ldots, \Pi_n, \Box(\Omega) \to \Delta_n\} = \text{INV}(S^0)$ (see the end of section 3), where S^0 is a quasiprimary sequent such that $\Gamma, \Box(\Omega) \to \Delta, \Box A$ is a vertex of $\mathcal{CR}^n R(S^0)$ (condition *); the notation $\Gamma^{\wedge}(\Gamma^{\vee})$ means conjunction (disjunction, respectively) of the formulas from Γ. If the conclusion of $(\to \Box_1)$ contains quantifiers, then the method of finding the invariant formula is based on reducing the conclusion S to the quantifier-free sequent S^{qf} and returning back to S (see Example 4.1 (d), below). Another way for finding $\mathcal{R}$ is based on describing the method for searching $\mathcal{CR}^n R(S)$ for arbitrary sequents. The foundations of this method are beyond the scope of the paper.

Example 4.1

(a) Let S be the same as in Example 3.1, then $\mathcal{R} = ((P \wedge \urcorner Q) \vee (Q \wedge \urcorner P)) \wedge (\Box(\Omega))^{\wedge}$. It is easy to verify that $I \vdash \Box(\Omega), P \to Q, \mathcal{R}$ (1), where I is the calculus, obtained from G_L by the dropping rules of inference $(+1)$ and $(\to \Box_i)$ $(i =\in \{1, 2, 3\})$, $IN \vdash \mathcal{R} \to \mathcal{R}^1$ (2), where IN is the calculus obtained from G_L by dropping the rules of inference $(\to \Box_1), (\to \Box_3)$; $I \vdash \mathcal{R} \to \urcorner R(3)$. Applying to (1), (2), (3) $(\to \Box_1)$, we get $G_L \vdash S$.

(b) Let $S = \Box\Omega \to A, \Box\neg\Box A$, where $\Omega = \neg(\neg A \wedge \Box(A \supset \Box A))$ (see axiom 5 from [21, page 176] and from [13, page 15]). Applying to S the algorithm for finding $\mathcal{CR}^n R(S)$ we get that $\mathcal{CR}^n R(S) = \{S^0 = \Box\Omega \to \Box(A \supset \Box A), \Box\neg\Box A\}$. Let us consider the case (2) of Lemma 3.7. Bottom-up applying to S^0 $(\to \Box_3)$, we get $S_1 = \Box\Omega, A \to \Box A, \Box\neg\Box A$ and $S_2 = \Box\Omega, \Box A \to \Box(A \supset \Box A)$. Bottom-up applying to S_2 $(\to \Box_3)$, we get $G_L \vdash S_2$. Applying to S_1 the algorithm for finding $\mathcal{CR}^n R(S_1)$, we get that $\mathcal{CR}^n R(S_1) = \{S_1^+ = \Box\Omega \to \Box A, \Box\neg\Box A\}$ (because the sequent $\Box\Omega \to \Box A, \Box(A \supset \Box A), \Box\neg\Box A$ is absorbed by S_1^+). If we consider the case (2) of Lemma 3.7, then we get the initial sequent S. Therefore we consider the case (1) from Lemma 3.7 and get $R = \neg\Box A \wedge \Box\Omega$. It is easy to verify that $I \vdash \Box\Omega \to \Box A, \mathcal{R}(1)$; $IN \vdash \mathcal{R} \to \mathcal{R}^1(2)$; $I \vdash \mathcal{R} \to \neg\Box A(3)$. Applying to (1), (2), (3) $(\to \Box_1)$, we get $G_L \vdash S_1 = \Box\Omega, A \to \Box A, \Box\neg\Box A$. Applying $(\to \Box_3)$ to S_1, S_2, we get $G_L \vdash S$. Now we shall show that we get the same if we consider the case (1) of Lemma 3.7 to S^0. In this case $\mathcal{R} = \neg\Box(A \supset \Box A) \wedge \Box\Omega$. It is easy to verify that $I \vdash \Box\Omega \to \mathcal{R}(1)$ and $I_1 \vdash \mathcal{R} \to \neg\Box A(3)$, where $I_1 = I + (\to \Box_3)$. Let us show that $IN \vdash \mathcal{R} \to \mathcal{R}^1$ (2). Bottom-up applying to $\mathcal{R} \to \mathcal{R}^1$ $(\to \neg), (\neg \to), (\Box \to)$, $(\to \Box_2), (+1)$ we get the sequent $S^* = \Box(A \supset \Box A), \Box\Omega \to \Box A$. Bottom-up applying to S^* $(\Box \to)$ (with respect to $\Box\Omega$ and $\Box(A \supset \Box A)$) we get that $IN \vdash S^*$, therefore $IN \vdash \mathcal{R} \to \mathcal{R}^1$. Applying to (1), (2), (3) $(\to \Box_1)$ we get $G_L \vdash S^0$. Afterwards, applying (+1) and successively applying $(\to \Box_2)$ we get $G_L \vdash S$. Let us show that $G_L^* \vdash S$, where G_L^* is more efficient then G_L and is described in the following way. Let $(\to \Box_1^*)$ be the rule of inference obtained from $(\to \Box_1)$ by (1) changing the condition $(*)$ by the following one: $R = \bigvee_{i=1}^{n} (\Pi_i^\wedge \wedge \neg\Delta_i^\vee)^k \wedge (\Box(\Omega^k))^\wedge$ ($k \in \{0,1\}$ and depends on the construction of $\mathcal{CR}^n R(S)$); $\{\Pi_1, \Box\Omega \to \Delta_1, \ldots, \Pi_n, \Box\Omega \to \Delta_n\} = \mathrm{INV}(S)$, S is the conclusion of $(\to \Box_1^*)$; (2) adding the premise of the form $\Gamma \to \Delta, A$, and changing the right premise by the following one $R \to A^1$, if $k = 1$. Let G_L^* be the calculus obtained from G_L by dropping (+1) and by replacing $(\to \Box_1)$ by $(\to \Box_1^*)$. It is easy to show that $G_L^* \vdash S = \Box\Omega \to A, \Box\neg\Box A$ $(\Omega = \neg(\neg A \wedge \Box(A \supset \Box A)))$. Indeed, $G_L^* \vdash \Box\Omega \to A, R$ (1) (where $R = \neg\Box(A \supset \Box A) \wedge \Box\Omega$); $G_L^* \vdash R \to R^1$ (2); $G_L^* \vdash R \to \neg\Box A$ (3). Applying $(\to \Box_1^*)$ to (1), (2), (3) we get $G_L^* \vdash S$.

(c) Let $S = \Box(\Omega) \to \Box\neg A, \Box\neg B$, where $\Box(\Omega) = \Box\neg(A \wedge B), \Box\neg(A \wedge \neg\Box\neg B), \Box\neg(B \wedge \neg\Box\neg A)$ (see axiom 4 from [20, page 176] and from [13, page 15]). Bottom-up applying to S $(\to \Box_3)$ we get $S_1 = \Box(\Omega), A \to \Box\neg B$ and $S_2 = \Box(\Omega), B \to \Box\neg A$. Bottom-up applying to S_1 $(\Box \to)$ (with respect to $\Box\neg(A \wedge \neg\Box\neg B)$) we get $I \vdash S_1(1)$. Bottom-up applying to S_2 $(\Box \to)$ (with respect to $\Box\neg(B \wedge \neg\Box\neg A)$) we get $I \vdash S_2(2)$. Applying to (1), (2), $(\to \neg), (\to \Box_3)$ we get $I_1 \vdash S$.

(d) Let $S = \forall x P(x, f_1(x)), \Box(\exists y P(b_1, y) \supset \forall x P^1(x, f_2(x))) \to \Box\exists y P(b_2, y)$. Let $G_{L\omega} \vdash S$, as $S = S^*$, then applying Theorem 2.1 to S we can verify that $S^{qf} = A, \Box\Omega \to \Box B$, where $A = P(b_1, f_1(b_1)) \wedge P(b_2, f_1(b_2))$; $B = P(b_2, f_1(b_2)) \vee P(b_2, f_2(b_2))$; $\Omega = (P(b_1, f_2(b_1)) \vee P(b_1, f_1(b_1)) \supset (P^1(b_1, f_2(b_1)) \wedge P^1(b_2, f_2(b_2)))$. Applying to S^{qf} the algorithm for finding $\mathcal{CR}^n R(S^{qf})$ we get that $\mathcal{CR}^n R(S^{qf}) = \{P(b_1, f_2(b_1)), P(b_2, f_2(b_2)), \Box\Omega \to \Box B\}$. Therefore $\mathcal{R} = P(b_1, f_2(b_1)) \wedge P(b_2, f_2(b_2)) \wedge \Box\Omega$. It is easy to verify that $I \vdash P(b_1, f_2(b_1))$, $P(b_2, f_2(b_2)), \Box\Omega \to \mathcal{R}(1)$; $IN \vdash \mathcal{R} \to \mathcal{R}^1(2)$; $I \vdash \mathcal{R} \to B$ (3). Applying to (1), (2), (3) $(\to \Box_1)$ we get $G_L \vdash S_1 = P(b_1, f_2(b_1)) \wedge P(b_2, f_2(b_2)), \Box\Omega \to \Box B$. Applying to S_1 $(+1), (\wedge \to)$ we get $G_L \vdash S_1' = P^1(b_1, f_2(b_1)) \wedge P^1(b_2, f_2(b_2)), \Box\Omega^1 \to \Box B^1(4)$. As $I \vdash S_1'' = A \to P(b_1, f_2(b_1)) \vee P(b_1, f_1(b_1))(5)$, then applying to S_1', S_1'' $(\supset\to), (\Box \to)$

we get $G_L \vdash S_2 = A, \Box\Omega \to \Box B^1(6)$. As $I \vdash S_2' = A \to B(7)$, then applying to S_2', S_2 the $(\to \Box_2)$ we get $G_L \vdash A, \Box\Omega \to \Box B(8)$. It is easy to verify that replacing in the derivations (1)÷(8) the formula $\Box\Omega$ by $\Box\Omega^*$, where $\Omega^* = (\exists y P(b_1, y) \supset \forall x P^1(x, f_2(x)))$; the formula A by the $\forall x P(x, f_1(x))$, the formula $P(b_1, f_2(b_1)) \vee P(b_1, f_1(b_1))$ by the formula $\exists y P(b_1, y)$; the formula $P(b_1, f_2(b_1)) \wedge P(b_2, f_2(b_2))$ by the formula $\forall x P(x, f(x))$; the formula B by the formula $\exists y P(b_2, y)$ we get $G_L \vdash S$. It is easy to verify that $G_L^* \vdash S^{qf} = A, \Box\Omega \to \Box B$. Indeed, $G_L^* \vdash A, \Omega \to R_1$ (1) (where $R_1 = R^1, R$ is the same as above); $G_L^* \vdash R_1 \to R_1^1$ (2); $G_L^* \vdash R_1 \to B^1$ (3), $G_L^* \vdash A, \Box\Omega \to B$ (4). Applying $(\to \Box_1^*)$ to (1), (2), (3), (4) we get $G_L^* \vdash S^{qf}$.
(e) Let $S = E_1, E_2, \Box(\Omega) \to \Box B$, where $\Box(\Omega) = \Box((E_3^1 \vee E_1) \supset (E_3^2 \wedge E_4^2))$; $B = (E_2 \vee E_4^1)$. It is easy to verify that $G_L \vdash S$, but $G_L^* \nvdash S$. Let $(\to \Box_1^{**})$ be the rule of inference obtained from $(\to \Box_1^*)$ by changing the rule of inference $(\bar{A})$ by the new one $(\bar{A}_1)$ (which enambles us to reject some nonessential members in constructing INV(S)) S_i/S, where $i \in \{1,2,3\}$, S is a primary sequent, i.e. $S = \Sigma_1, \Pi_1^1, \Box(\Delta_1^1) \to \Sigma_2, \Pi_2^1, \Box(\Delta_2^1)$ (where Σ_1, Σ_2 consist of logical formulas); $S_1 = \Sigma_1 \to \Sigma_2$; $S_2 = \Pi_{11} \to \Pi_{12}$; $S_3 = \Pi_{21}, \Box(\Delta_1) \to \Pi_{22}, \Box(\Delta_{22})$, where $\{\Pi_{1j}\} \subseteq \{\Pi_j\}$ and $\{\Pi_{1j}\} = \{\Pi_j\} \backslash \{\Pi_{2j}\}$; $\{\Pi_{2j}\} \subseteq \{\Pi_j\}$ and Π_{2j} ($j \in \{1,2\}$) consists of the propositional subformulas (which are defined as the notion of the subformula of a propositional formula) of the formulas from $\Box(\Delta_1), \Box(\Delta_2)$; S_1 S_2 will be called a logical upper sequent of the rule of inference $(\bar{A}_1)$; S_3 will be called the temporal one. Let G_L^{**} be the calculus obtained from G_L^* by changing $(\to \Box_1^*)$ by $(\to \Box_1^{**})$. It is easy to verify that $G_L^{**} \vdash E_1, E_2, \Box(\Omega) \to R(1)$ (where $R = E_3^2 \wedge E_4^2 \wedge \Box(\Omega^1))$; $G_L^{**} \vdash R \to R^1$ (2); $G_L^{**} \vdash R^1 \to A^1$ (3); $E_1, E_2, \Box(\Omega) \to B$ (4). Applying $(\to \Box_1^{**})$ to (1), (2), (3), (4), we get $G_L^{**} \vdash S$.

Lemma 4.1. The structural rule (W) is admissible in G_L.

Proof: by induction on the height of a given derivation.

Lemma 4.2. An arbitrary derivation in G_L may be reconstructed into an atomic one, i.e. all the axioms are of the form $\Gamma, E \to \Delta, E$, where E is an atomic formula.

Proof: analogously as in Lemma 1.2. In the case of the axiom $\Box B \to \Box B$ the rule of inference $(\to \Box_3)$ is applied.

Lemma 4.3. The rules of inference of the calculus G_L, except (+1), $(\to \Box_1)$, $(\to \Box_3)$ are invertible in G_L.

Proof: by induction on the height of a given derivation, using Lemma 4.2.

Lemma 4.4. In $G_{L\omega}$ the rule of inference (+1) is admissible.

Proof: by induction on $O(D)$, where D is the derivation of the upper sequent of (+1).

Lemma 4.5. In $G_{L\omega}$ the following rule of inference is admissible: $\Gamma \to \Delta, R$; $R \to R^1$; $R \to A/\Gamma \to \Delta, \Box A$ $(\to \Box)$.

Proof: using the premises of $(\to \Box)$, the admissibility of cut in $G_{L\omega}$ and by induction on k we get $G_{L\omega} \vdash \Gamma \to \Delta, A^k$ $(k \in \omega)$. Hence, by $(\to \Box_\omega)$ we get that $G_{L\omega} \vdash \Gamma \to \Delta, \Box A$.

Lemma 4.6. In $G_{L\omega}$ the rule of inference $(\to \Box_2)$ is admissible.

Proof: follows from Lemma 4.5, taking $R = (A \wedge \Box A^1)$.

Lemma 4.7. In $G_{L\omega}$ the rule of inference $(\rightarrow \Box_3)$ is admissible.

Proof: applying $(\rightarrow \Box_2), (\Box \rightarrow)$.

Theorem 4.1. $G_L \vdash S \Rightarrow G_{L\omega} \vdash S$.

Proof: follows from Lemmas 4.4, 4.5, 4.6, 4.7.

5. Reduction of $\mathbf{G_{L\omega}}$ to $\mathbf{G_L}$

At first we shall prove that $G_{L\omega} \vdash S \Rightarrow G_L \vdash S$ for any reduced sequent S.

Lemma 5.1. Let $G_{L\omega} \vdash S$, let $R\{i\}$ be a reduction of S to the sequents $S_1, \ldots, S_n$, let $\{i\}$ consist of bottom-up applications of the rules of inference $(\rightarrow \Box^1), (\bar{A}), (\bar{A}^*)$ and the rules of inference of $G_{L\omega}$, except of $(\rightarrow \Box_\omega), (\rightarrow \bigcirc), (\bigcirc \rightarrow)$ and rules of inference for quantifiers; let $G_L \vdash S_j$ $(1 \leqslant j \leqslant n)$ then $G_L \vdash S$.

Proof. Let us consider any two applications of $(\rightarrow \Box_\omega)$ in the given atomic derivation D of S. These applications will be called different if the principal formulas are either different (formulas distinguished by indices are considered as coincident) or coincide but have different descendants. Therefore, different applications of $(\rightarrow \Box_\omega)$ in D are finite and let $(\rightarrow \Box_\omega)[D]$ mean this number. The proof of the Lemma is carried out by induction on $(\rightarrow \Box_\omega)[D]$. To prove the Lemma it is sufficient to show that any bottom-up applications of $(\rightarrow \Box^1)$, $(\bar{A}), (\bar{A}^*)$ in $R\{i\}$ can be replaced by the applications of the rules of inference of G_L. A bottom-up application of $(\rightarrow \Box^1)$ can be replaced (using the "atomicity" of D and induction hypothesis) by application of $(\rightarrow \Box_2)$. A bottom-up application of $(\bar{A})$ can be replaced (using the "atomicity" of D, induction hypothesis, the admissibility of structural rule $(\mathcal{W})$ in G_L) by the applications of $(+1)$, $(\rightarrow \Box_2)$. A bottom-up application of $(\bar{A}^*)$ can be replaced by the applications of $(+1)$ and $(\mathcal{W})$ (which is admissible in G_L).

Lemma 5.2. Let $G_{L\omega} \vdash S$, where S is a quasiprimary ordinary sequent and $C\mathcal{R}^n R(S)$ be such that all saturated vertexes have the form $S_i = \Pi_i, \Box(\Gamma) \rightarrow \Delta_i, \Box A$ $(1 \leqslant i \leqslant n)$, where Π_i, Δ_i consist of invariant subformulas of Γ, then $IN \vdash \mathcal{R} \rightarrow \mathcal{R}^1$, where IN is the calculus obtained from G_L by dropping the rules of inference for quantifiers, rules of inference $(\rightarrow \Box_1), (\rightarrow \Box_3)$; $\mathcal{R} = \bigvee_{i=1}^{n} (\Pi_i^{\wedge} \wedge \neg\Delta_i^{\vee}) \wedge (\Box\Gamma)^{\wedge}$.

Proof. Let us consider any vertex S_i of $C\mathcal{R}^n R(S)$. Let us consider $\mathcal{R}(S_i)$ (i.e. the resolvent of S_i), let $\mathcal{R}(S_i) = \{S_{i1}, \ldots, S_{in}\}$. Let us consider the reduction R_i of S_i $(1 \leqslant i \leqslant n)$ to the sequents $S_{i1}, \ldots, S_{in}$. To prove the Lemma we shall transform the reductions $R_i (1 \leqslant i \leqslant n)$ to the derivations in IN of the sequents $S_i' = \Pi_i, \Box(\Gamma) \rightarrow \Delta_i, R^1$ $(1 \leqslant i \leqslant n)$, from which with the help of $(\wedge \rightarrow), (\neg \rightarrow), (\vee \rightarrow)$ we shall get a derivation in IN of $R \rightarrow R^1$. To show this let us consider the bottom-up application of the rule of inference $(\bar{A})$ in R_i:

$$\frac{S_l'}{S_{ij}^* = \Sigma_{ij}', \Pi_{ij}^1, \Box(\Gamma^1) \rightarrow \Sigma_{ij}'', \Delta_{ij}^1, \Box A^1}\ (\bar{A}),$$

where $l = 1$ or $l = 2$; $S'_1 = \Sigma'_{ij} \to \Sigma''_{ij}$; $S'_2 = \Pi_{ij}, \Box(\Gamma) \to \Delta_{ij}, \Box A$. If $l = 1$, then $IN \vdash S'_1$ (because $\Box \notin \Sigma'_{ij}, \Sigma''_{ij}$). Using the admissibility of $(\mathcal{W})$ in IN (proof is carried out as in the case of G_L) we get $IN \vdash S'^{*}_{ij} = \Sigma'_{ij}, \Pi^1_{ij}, \Box(\Gamma^1) \to \Sigma''_{ij}, \Delta^1_{ij}, \mathcal{R}^1$. Let us consider the case when $l = 2$, then in R_i we substitute the formula $\mathcal{R}$ for selected occurrence of the formula $\Box A$. Using the main property of the saturated vertex of $C\mathcal{R}^n R(S)$ (Lemma 3.6) and $(\to \vee)$, $(\to \wedge)$, $(\urcorner \to)$ we get $IN \vdash S''_2 = \Pi_{ij}, \Box(\Gamma) \to \Delta_{ij}, \mathcal{R}$. Thus, instead of $(\underline{A})$, we get $(+1)$. Other bottom-up applications of the rules of inference in R_i do not change and therefore instead of the reduction R_i we get $IN \vdash S'_i = \Pi_i, \Box(\Gamma) \to \Delta_i, \mathcal{R}^1$ $(1 \leqslant i \leqslant n)$. Applying $(\wedge \to)$, $(\urcorner \to)$, $(\vee \to)$ to all $S'_i (1 \leqslant i \leqslant n)$ we get $IN \vdash \mathcal{R} \to \mathcal{R}^1$.

Lemma 5.3. Let $G_{L\omega} \vdash S$, where S is a quasiprimary ordinary sequent and $C\mathcal{R}^n R(S)$ be such that all saturated vertexes have the form $S_i = \Pi_i, \Box(\Gamma) \to \Delta_i, \Box A$ $(1 \leqslant i \leqslant n)$, where Π_i, Δ_i consist of invariant subformulas of Γ, then $G_L \vdash S$.

Proof: by induction on $(\to \Box_\omega)[D]$, where D is a given atomic derivation. As follows from Lemma 5.1, to prove the Lemma it is sufficient to prove that $G_L \vdash S_i$. With the help of $(\to \vee)$, $(\to \wedge)$, $(\to \urcorner)$ we have $I \quad \vdash \Pi_i, \Box(\Gamma) \to \Delta_i, \mathcal{R}$ (1) $(1 \leqslant i \leqslant n)$, where $\mathcal{R}$ is the same as in Lemma 5.2. By Lemma 5.2 we have $IN \vdash \mathcal{R} \to \mathcal{R}^1$ (2). Using the "atomicity" of D and induction hypothesis we have $G_L \vdash S'_i = \Pi_i, \Box(\Gamma) \to \Delta_i, A$ $(1 \leqslant i \leqslant n)$. With the help of $(\wedge \to)$, $(\urcorner \to)$, $(\vee \to)$ from S'_i $(1 \leqslant i \leqslant n)$ we get $G_L \vdash \mathcal{R} \to A$ (3). Applying $(\to \Box_1)$ to (1), (2), (3) we get $G_L \vdash S_i$ and therefore $G_L \vdash S$.

Remark 5.2. In the proof of Lemma 5.3 the transformation $G_{L\omega} \vdash S \Rightarrow G_L \vdash S$ $(*)$ was justified in the case when a conclusion of $(\to \Box_1)$ coincides with a vertex of $C\mathcal{R}^n R(S_1)$ where S_1 is a quasiprimary sequent (what is possible by Lemma 5.1). Transformation $(*)$ (without using Lemma 5.1 and $(+1)$) in the case when the conclusion of $(\to \Box_1)$ coincides with the root of $C\mathcal{R}^n R(S)$ (see Example 4.1 (*b*), (*d*), (*e*)) is more efficient but needs some additional arguments and is beyond the scope of the paper.

Lemma 5.4. Let $G_{L\omega} \vdash S$, where S is a quasiprimary ordinary sequent and $C\mathcal{R}^n R(S)$ be such that all saturated vertexes have the form $\Box(\Gamma) \to \Box(\Delta)$, then $G_L \vdash S$.

Proof: by induction on $\to (\Box_\omega)[D]$, using $(\to \Box_3)$.

Lemma 5.5. Let $G_{L\omega} \vdash S$, where S is ordinary sequent, then $G_L \vdash S$.

Proof: using Lemmas 3.3, 3.5, 3.7, 5.3, 5.4.

Lemma 5.6. Let $G_{L\omega} \vdash S$, where S is a singular sequent, then $G_L \vdash S$.

Proof: by induction on $(\to \Box_\omega)[D]$, using Lemmas 3.8, 3.9, the rule of inference $(\to \Box_3)$ and Lemma 5.1.

Lemma 5.7. Let $G_{L\omega} \vdash S$, where is S is a reduced sequent, then $G_L \vdash S$.

Proof. Let S be simple (i.e. S does not contain a positive occurrence of $\Box$) then $G_L \vdash S$. Let S be ordinary (singular), then the Lemma follows from Lemma 5.5 (Lemma 5.6, respectively).

Let S^* a be st-skolemized sequent, S^{*H} be the sequent, obtained from S^* by eliminating negative occurrences of quantifiers (see Lemma 2.6, Theorem 2.1).

Lemma 5.8. $G_L \vdash S^{*H} \Rightarrow G_L \vdash S^*$

Proof: almost the same as in Lemma 2.7 (see Example 4.1 (d)).

Theorem 5.1. Let S be a sequent, S^* is the st-skolemized variant of S then $G_{L\omega} \vdash S \Rightarrow G_L \vdash S^*$.

Proof: follows from Theorem 2.2, Lemmas 5.7, 5.8.

Theorem 5.2. (1) If S is universally valid, then $G_L \vdash S^*$; (2) $G_L + (\text{cut}) \vdash S^* \Rightarrow G_L \vdash S^*$.

Proof: follows from Theorems 4.1, 1.1, 1.2, 5.1.

Remark 5.3. We state that Theorems 5.1, 5.2 remain true replacing S^* by S (see Remark 2.2).

REFERENCES

1. M.Abadi, The power of temporal proofs, Theoret. Comput. Sci., Vol. 64, (1989), 35–84.
2. M.Abadi, Z.Manna, Temporal logic programming, Journal Symbolic Computation, Vol. 8, (1989), 277–295.
3. M.Abadi, Z.Manna, A timely resolution, First annual symposium on logic in Computer Science, (1987), 123–130.
4. H.Andreka, J.Nemeti, J.Sain, Henkin-type semantics for program schemes to turn negative results to positive, In: FCT–79, ed.: Budach, Akademia Verlag, Berlin, Band 2 (1979), 18–24.
5. H.Andreka, J.Nemeti, J.Sain, On the strength of temporal proofs, LNCS, Vol. 379, (1989), 135–144.
6. H.Barringer, M.Fisher, D.Gabbay, G.Gough, R.Owens, METATEM: A framework for programming in temporal logic, Technical Report Series, UMCS 89–10–4, Department of Computer Science, University of Manchester, 1989.
7. D.Gabbay, Decidability of some intuitionistic predicate theories, JSL, Vol. 37, (1972), 579–587.
8. J.H.Gallier, Logic for computer Science: Foundations of Automatic Theorem Proving, Harper and Row, New York, 1986.
9. D.Gallin, Intensional and higher-order modal logic, North–Holland, Math. Studies 19, 1975.
10. R.Hale, Temporal logic programming, in book:, Temporal logics and their applications, ed. A.Galton, Academic Press, London, 1987.
11. H.Kawai, Sequential calculus for a first order infinitary temporal logic, Zeitshr. für Math. Logic und Grundlagen der Math., Vol.33 (1987), 423–432.
12. F.Kroger, On the interpretability of arithmetic in temporal logic, Theoret. Comput Sci., Vol. 73 (1990), 47–60.
13. L.L.Maksimova, Interpolation, Beth's property and temporal logic "tomorrow", (in Russian), Preprint No. 90 of Math. Institute of Sibirian Division of the USSR Academy of Sciences, Novosibirsk, 1989.

14. S.J.Maslov, Invertible sequential calculus for predicate constructive logic, (in Russian), Zap. Naučn. Sem. Leningrad. Otdel. Mat. Inst. Steklov (LOMJ), Vol. 4, (1967), 96–111.
15. D.Miller, G.Nadathur, A.Scedrov, Uniform proofs as foundation for logic programming, Annals of Pure and Applied Logic, Vol. 51, (1991), 125–157.
16. G.Mints, Gentzen-type systems and resolution rules. Part 1. Propositional logic. LNCS, Vol. 417, 198–231.
17. G.Mints, Skolem's method of elimination of positive quantifiers in sequential calculi, Soviet Math. Dokl. Vol. 7, (1966), 861–864.
18. B.Moszkowski, Executing temporal programs, Technical Report No. 55, Computer Laboratory, University of Cambridge, 1984.
19. R.Pliuškevičius, Investigation of finitary calculi for temporal logics by means of infinitary calculi, LNCS, Vol. 452, (1990), 464–469.
20. R.Pliuškevičius, Investigation of finitary calculus for a discrete linear time logic by means of finitary calculus, LNCS, Vol. 502, (1991), 504–528.
21. K.Segerberg, Temporal logic of von Wright, (in Russian), Logical derivation. Moscow, "Nauka", (1979), 173–205.
22. A.Szalas, L.Holenderski, Incompleteness of first-order temporal logic with UNTIL, Theoret. Comput. Sci, Vol.57, (1988), 317–325.

LOGIC PROGRAMMING WITH PSEUDO-RESOLUTION

David M. W. Powers[1]
Department of Computer Science
University of Kaiserslautern
D-6750 KAISERSLAUTERN FRG

Abstract

This paper presents a new proof technique for Automated Reasoning and Logic Programming which based on a generalization of the original Connection Graph paradigm of Kowalski and provides a methodology for Logic Programming in this framework.

We show how execution of a logic program can be executed in the logarithm of the number of steps taken by PROLOG and standard resolution theorem provers, or better.

This paper deals primarily with recursion, both in relation its exploitation and its explication. In dealing with explicit recursion, a modified "compartmentalized" connection graph framework emerges. In dealing with implicit recursion, a filter for the compartmentalized connection graph emerges which forces recursion to be explicitly represented.

The method is demonstrated on standard PROLOG examples.

Introduction

This work has arisen from the author's experience in the programming of a number of complex PROLOG applications and his dissatisfaction with the strong control of the SLD procedure. In particular, applications (such as language learning) which need to explore search spaces full of unknowns are unsuited to a strongly deterministic strategy. Programming this flexibility in PROLOG can be done, but involves needless complexity, the use of complex guards, multiple passes, and carefully considered use of guiding cuts.

Here we start from the opposite extreme, a completely general clausal theorem prover without control, and seeks to understand the behaviour of logic programs expressed in such an environment, including how to implement the environment efficiently and whether it is possible to use general search heuristics rather than an explicit control paradigm. The CONG system [Powe88] for CONcurrent logic programming is based on a CONnection Graph theorem prover [Kowa79] and can accept pure PROLOG programs (cutless Horn clauses without builtin predicates) as well as general clause form logic program (again pure without builtins).

---

[1] The work reported here was in the main undertaken while the author was at Macquarie University NSW 2009 AUSTRALIA, and was supported in part by IMPACT Ltd, PETERSHAM NSW 2049 AUSTRALIA, the Australian Telecommunications and Electronics Research Board, and the Australian Research Council (Grant No A48615954). The author is currently supported under ESPRIT BRA 3012: COMPULOG.

Although, in a sense, CONG is a general theorem prover, it has been built for logic programming and with design giving priority to efficiency for the restricted classes of theorems which might be regarded as logic programs. Our techniques both provide an effective proof procedure for theorem proving with full clauses in the connection graph framework and show how standard PROLOG logic programs can be executed efficiently and more completely in this framework.

Model

Standard Connection Graph

Links

A set of clauses to be proven inconsistent are linked into a graph by connecting clauses with a *link* whenever they have unifiable complementary literals. The links may at times most conveniently be regarded as connecting clauses. However, they actually indicate potential resolutions, or equally well potential resolvents. The substitution giving rise to the most general unifier of the linked literals is associated with the link.

Under this definition a link is formally defined only between distinct clauses, although loosely used it may, when the context permits, include pseudolinks. In certain syntactic contexts all forms of links, including, e.g. factor links, may be intended. When we want be absolutely clear we can refer to these links between complementary literals of distinct clauses as resolution links.

Pseudolinks

The connections between unifiable literals of opposite sign within a single clause are termed *pseudolinks*. These links represent the potential for *copies* of the clause to resolve, but are themselves never actually resolved on. They are restricted to inheritance and can inherit to standard resolution links or to new pseudolinks. The substitution giving rise to the most general unifier of the linked literals is associated with the pseudolink.

Resolution rule

Upon resolving on a link, the *resolvent* is the clause containing copies of all literals other than those linked by the selected link substituted by the substitution associated with the resolved link. The linked clauses (resp. terms) are termed the *positive* and *negative parents* according to the sign of the respective linked term. Once it has been resolved upon, a link is deleted, as there is no need ever to repeat this resolution step. There is a sense in which a link already represents the resolvent and in an implementation the same structure may actually be used with just the complementing of a bit to indicate the difference: the resolvent literally replacing the link.

Inheritance rule

Following resolution on a link, the copies of the literals need to be connected in to the rest of the graph. Rather than trying unification with all possible complements, the potential of the connection graph comes from the inheritance of new links from those which have not yet been resolved on (or otherwise deleted) and remain extant as links. The links represent work remaining to be done.

Upon copying a literal, all links associated with that literal are also copied, but the type will change to that appropriate to the new link. Thus links (and pseudolinks) between other terms of the parents will inherit as pseudolinks, and pseudolinks impinging on one of the parent terms will inherit to links from the parent term to the copy of the other term.

Factoring rule; Purity, Tautology and Subsumption deletion

As we are primarily interested in Logic Programs here, we ignore the problem of factoring noting only that it is soundly and completely implemented in CONG [Powe90a].

An important pruning of the graph may be achieved by observing that certain clauses are redundant or cannot contribute to a derivation of the empty clause. While the deletion of such clauses, along with associated links, is possible if we are careful, and considerable savings through prunings can result, the deletions are not germane to the treatment here and introduce complications which lead us to define the connection graph procedure without deletion rules.

Proofs

Logic Programming

The clauses in a connection graph are inconsistent if and only if there exists a sequence of resolutions on links (inheriting and factoring as required) such that the empty clause (containing no literals) is derived. The execution of straightforward cutless PROLOG programs without library predicates is straightforward in a connection graph theorem prover, a PROLOG proof corresponding to one such sequence of links in the connection graph.. However, our aim is to improve on this PROLOG proof by the use of better selections of links through CONG's control and heuristics.

At this point, moreover, we can already observe some of the advantages of programming in a control free (Horn) theorem prover. In particular, the macro effect means that in some sequences of resolutions performed toward the solution of one goal may actually have produced a *lemmas* as resolvent which is useful in the solution of other goals.

This is a very practical effect demonstrated in the very first implementation of CONG [Powe88]. In a small program with multiple 'calls' to the standard PROLOG `append` predicate, we can observe a very clear advantage of CONG over PROLOG. By resolving the recursive clause against the unit clause, a new unit clause with an extra exposed head is produced. By continuing this process on the new unit clauses, the `append` predicate is expanded to a unit clause of the appropriate size to match any given `append` call. Any smaller `append` call will match one of the intermediate resolvents. Any `append` involving a larger list will be reached by continuing the expansion of the predicate, whilst PROLOG would involve starting from scratch. Thus the number of resolutions steps in CONG equals the size of the the largest list to be appended to plus the number of distinct calls to it, whilst in PROLOG it is the sum of the sizes of the first arguments in all calls to `append`.

Extended Connection Graph

The extended connection graph is a graph with the same clauses, links and pseudolinks as the standard form. It may however have additional types of links, or additional subclassification of the standard clauses, terms and links. In this section we mention the new types of links and introduce the concept of *filters* , which governs the use of subclassifications of links.

Factor-links and Subsumption links

The extended connection graph properly includes links between literals of the same sign as well [Eisi88]. But as we have excluded the operations of factoring and subsumption from consideration for the present purpose, we ignore these links too.

Orderings & Restrictions

Within the scope of the extended connection graph procedure there is the possibility of imposing additional conditions to filter out the less helpful choices of next link to resolve (or factor) on. The procedure has a property called *confluence* which says that whichever non-deterministic path through proof-space you take there is always a common successor graph. Soundness says the procedure is valid, completeness says there is a way of finding a proof if there is one, but there is nothing said about getting you there.

Strong completeness adds the condition that the procedure guarantees to get you there. Even strong completeness however may not even be strong enough. Ideally we want our oracle or heuristics to tell us the fastest way to get our proof. Strong completeness is discussed in detail in [Eisi88] and in Part III of this series.

For now we note that there are two types of filters which can be distinguished. Those which say *never* are said to be *restrictions*, and those which say *first* are called *orderings*. The first are quite dangerous – they do not preserve confluence and may actually exclude the part of the search space which contains the proof, affecting completeness. The second are more advisory and preserve the properties of soundness, completeness and confluence. Whether they guarantee termination, finding the proof, is another question.

Ordering restrictions aim to force towards the proof. An ordering which systematically avoids any possibility of the proof escaping is said to be *exhaustive*. An exhaustive ordering leads automatically to strong completeness. Unfortunately none are known for either the standard or the extended connection graph. There may be none!

An example of a restriction is the *unit restriction*. Unit clauses are clauses with exactly one literal. The unit restriction says all resolution steps undertaken must involve.

An example of an ordering strategy is predicate elimination: you resolve only on terms with a particular functor, resolving away occurence by occurence, until none are left. That it can't help is obvious if you consider that any program or problem could be rewritten with a single unary predicate `p(..)` wrapped around the original literals. In this case the ordering gives no help.

Compartmentalized Connection Graph

We turn now to a modification of the connection graph algorithm which reverses some of the basic principles and changes some of its properties dramatically.

Pseudolinked clauses

Recall that in the standard and extended definitions we are permitted only to resolve on links and pseudolinks may only be inherited from. Here we change this around and specify that in the case of clauses with pseudolinks the only permitted operation is resolution on the pseudolink and links may only be inherited from.

Self-resolving clauses with pseudolinks are restricted to self-resolution on these pseudolinks, and links between clauses containing unresolved pseudolinks are declared *illegal*. This ensures that pseudolinks cannot inherit and allows the distinction between two types of operation, resolution on normal links and resolution on pseudolinks, to be extended to a differentiation of distinct phases of the algorithm – a *static* phase in which legal normal links are resolved, and a *dynamic* phase in which legal pseudolinks are resolved.

Static purity

A clause is of no further use in the current static phase once all legal links from a single term have been deleted. The clause is then said to be *statically pure*.

Corridors – links connecting compartments

Clauses with and without pseudolinks are said to be respectively in the dynamic and static compartments. The links between clauses in different compartments are illegal and are said to define a corridor between the two compartments.

Clearly, if the corridor is empty, the algorithm will terminate at the end of the current resp. next static phase, and there is no need to continue work in any current dynamic phase. This is the way termination is achieved in satisfiable clause sets, although it cannot be guaranteed in general, but is usual for logic programs.

Self-resolution

As was implicit in our definition of pseudolink originally, a *self-resolvent* is the result of a resolution of two distinct copies of the clause with variables renamed apart. A consequence of this is that *self-resolution* seems to introduce additional variables. In fact every resolvent always has a brand new (but possibly empty) set of variables. However we tend to reuse the names. In the case of self-resolution we are forced to think up new ones! In the CONG and PROLOG examples it will be noted that the systems always distinguish variables uniquely with numbers.

Positive-to-parent inheritance

The definition of pseudo-resolution allowed in the extended connection graph has the property that the pseudolink inherits to only one normal link after its resolution. There is a sense in which it is only half there to inherit because it is currently selected for resolution and deletion, and if we allowed both descendants of the parent literals in the resolvent to inherit links to the parent literals, resolving on these links would produce identical clauses (mutually subsumable with links to same literals).

We call the original recursive clause the order 1 form, the first resolvent is the order 2 form. Resolving on either of these normal links inherited from the pseudolink would produce an order 3 form, while resolving on the pseudolink inherited from the original pseudolink would produce an order 4 form. The order is in fact the count of the number of copies of other literals of the original clause which are found in a descendant in the absence of factoring.

In fact, we have defined self-resolution in terms of copies. [Eisi88,p128] shows that the self-resolution operation can be added to the connection graph and is then equivalent to making a copy and inheriting links, performing a resolution on one of the two descendants of the pseudolink, and then subsuming away the copy (although it could equally well have been the original). The arbitrary choice of which descendant to resolve on (or equally which copy to resolve away) affects whether the pseudolink inherits from the positive term of the resolvent to the negative term of the original parent, or vice-versa.

Our positive-to-parent inheritance rule arbitrarily disambiguates this in a consistent fashion by requiring that the link to inherit goes from the positive term of the resolvent to the negative term of the parent. This has, in a sense, the side effect of biasing the links in the corridor toward goal direction.

Interestingly [Eisi88] only introduces self-resolution to simplify proofs of formal properties, and elsewhere [Eisi89] shows that it is an unnecessary operation.

Delayed composition

For various reasons pertinent to both efficiency and strong completeness, we note that the composing of new substitutions need not be performed immediately. The link may simply be inherited and retain information about the substitutions which compose to give the associated substitution.

In the case of links which may not actually inherit with a valid substitution this can mean *false* links are present in the graph. False links may delay recognition of purity of a clause. On the other hand, false pseudolinks may accelerate termination of the current static phase and thus tighten the ordering conditions relating to pseudolinks.

```
>cat append
append([],W,W).
append([H|X], Y, [H|Z]) :- append(X, Y, Z).
append([a,b,c,d,e,f,g,h],[i,j,k,l],L)/?

>cong append.cong.2
CONG MQU/UKL$Revision: 1.3.1.3 $$State: Exp $
Copyright (C) DMWP,MQU 1983-84,87-89,90 Version$Date: 88/12/15 15:27:17 $
Cong> pseudores, hyper, minterm, hyperfac, unit, trace, go!
:- append([a,b,c,d,e,f,g,h],[i,j,k,l],L'1)/?
   append([H'1|X'2],Y'3,[H'1|Z'4]):- append(X'2,Y'3,Z'4).
   append([],W'1,W'1).
   append([H'1,H'5|X'6],Y'3,[H'1,H'5|Z'8]):- append(X'6,Y'3,Z'8).
   append([H'1,H'5,H'9,H'13|X'14],Y'3,[H'1,H'5,H'9,H'13|Z'16]):- append(X'14,Y'3,Z'16).
   append([H'1,H'5,H'9,H'13,H'17,H'21,H'25,H'29|X'30],Y'3,[H'1,H'5,H'9,H'13,H'17,H'21,H'25,H
   append([H'1,H'5,H'9,H'13,H'17,H'21,H'25,H'29,H'33,H'37,H'41,H'45,H'49,H'53,H'57,H'61|X'62]
:- append([],[i,j,k,l],Z'33)?
** yes **
L = [a,b,c,d,e,f,g,h,i,j,k,l]
```

Figure 1. *Self-resolving append executed in* CONG.

In this context the orderings are regarded as more important than purity, which is in any case largely superceeded by the concept of static purity.

We specify that composition for links is done only in the static phase, and possibly *lazily* (that is a false link may actually be resolved on), and that composition for pseudolinks is done only in the dynamic phase, and always *eagerly* (that is a false pseudolink may never actually be resolved on).

Note that in proof and search type problems this heuristic could be counter-productive – and demand the use of at least some quick trial unification technique [Wise84,Powe90b]. But in other kinds of logic programs it can act as a form of compilation and be very efficient.

Example of Doubling

We again use append as our example to show the speed at which append can now get up to big lists – the number of resolutions being logarithmic in the length of the largest list appended to and linear in the number of distinct calls (as opposed to recursive calls). In Figure 1, we illustrate this with output from a run: the number of lines and the general flavour is more important than the details and lines have therefore been truncated for compactness and to allow the exponential growth of the (underlined) first argument to be clearly seen.

Example of Explication

The big advantage of compartmentatilzation over any other logic programming effect is the doubling effect which is realizable on explicitly recursive clauses as just shown. Now we want to show how multiply recursively defined PROLOG predicates can be reduced to a canonical form to reduce the amount of explicit recursion. The canonical form has only a single singly recursive clause and doubling produces only one form of a given size.

But multiply recursively defined predicates also hide implicit recursion, where one clause can call the other and vice versa (in PROLOG terms). Thus not only do we have a family of forms generated from each one, but we can have an exponential number of such families. Even once the pseudolinks have been resolved on, the clauses are in general cross linked, and resolution on any one of those links inherits the others as a pseudolink.

We illustrate with a set of Horn clauses which is beyond the power of PROLOG and which has this property.

```
q(g(f(g(f(g(f(a))))))).
q(X)  :- q(f(X)).
q(Y)  :- q(g(Y)).
:- q(a).
```

Note that if the positive and negative terms were interchanged, so that the query was complex and the unit clause simple, it would run under PROLOG. But as it stands PROLOG will only search for unit clauses of the form `q(f(f(..f(a)..)))`, which no `g` functors.

Resolving on the pseudolinks produces the first of a family of recursive clauses containing exclusively one functor or the other. But resolving on a link between the two recursive clauses produce the zero form of one of two families with alternating functors. Confluence and completeness guarantees that CONG can succeed in finding the proof in this way. But the explosion in the number of families is exponential.

We noted that the problem was the implicit recursion which was not already expressed by pseudolinks. Such recursion can also occur in clauses which are not directly recursive, but indirectly or *implicitly*. If we make a slight modification to the above algorithm we will see such an example.

```
q(g(f(g(f(g(f(a))))))).
q(X)  :- p(f(X)).
p(Y)  :- q(g(Y)).
:- q(a).
```

We would like to make this recursion *explicit* so that we may deal with it efficiently. Careful *ordering* of our choice of links in the compartmentalized connection graph can *explicate* such recursion.

Consider what happens if we use straight forward goal directed search as PROLOG does. We generate a sequence of goals: `p(f(a). q(g(f(a)). p(f(g(f(a)))...` This will eventually find the unit clause in this case. If we used unit resolution in a data driven way we would produce a similar but reducing set of positive unit clauses. In both cases the process is linear in the size of the complex term. If we could *explicate* the recursion and then use *doubling* it could be done in a logarithmic number of steps.

We can achieve explication very simply in the static phase of the algorithm: we introduce a rule, technically an *ordering strategy*, which prohibits resolving on an inherited link before its parent link (that is the link it was inherited from) is resolved upon. We call this the *orphaning* filter as it stipulates that a child may not resolve while its parent is alive. This stops the above unit resolution series after the first step. The generated unit clause has a link only because it was inherited from somewhere – one of the cross links. This forces resolving on the parent of the new link first, on a cross link, and forces generation of an explicitly recursive clause containing a pseudolink. The *explication* of recursion achieved by the first static phase can be achieved independent of goal as a "compile-time" program transformation [Powe90b].

Conclusions

We have reviewed how lemmatization can avoid duplicating work in the connection graph, and how taking advantage of the possibility of resolving on self-recursive pseudolinks as permitted in the extended connection graph can allow reduction of the length of proofs and traces logarithmically. We have further shown how filters on the extended connection graph procedure give rise to the possibility of making implicit recursion explicit. All of these techniques and advantages are implemented and achieved in our implementation of CONG.

It remains to note how much of this is necessarily tied to the connection graph representation. The answer is that probably none is. They all arise very simply and naturally in the connection graph paradigm, but it may be appropriate to attempt to apply them as preprocessing techniques feeding a conventional PROLOG compiler – at the moment our interpreted connection graph technique is not competitive with a good compiler, but we believe there is considerable scope for improvement. But we consider again the three enhancements in this light.

Lemmatization is known from other theorem proving techniques, but is perhaps hardest for PROLOG techology – but with fast indexing and recognition of previously solved goals, some progress can be made. However, the the lemmatization (arising e.g. through bottom-up execution) is necessary for instant solving of analogous but not identical goals. This reversal is also a conceivable optimization.

The exploitation of explicit recursion is really a form of middle-out execution and similarly achievable after appropriate analysis.

The explication of implicit recursion is also a middle-out operation and achievable as a compile-time preprocessing step using the algorithm executed by the first static phase of the compartmentalized algorithm with the orphaning filter.

Acknowledgements

I wish to acknowledge the participation of Graham Wrightson, Debbie Meagher, Laz Davila, David Menzies, Martin Wheeler, Graham Epps, Richard Buckland and Philip Nettleton in the MARPIA project at Macquarie University, and their varying contributions to the development of CONG. Laz Davila wrote the first version of the present incarnation of CONG. Debbie Meagher has been responsible for its further development including the addition of compartmentalization.

References

[Eisi88] Norbert Eisinger, "Completeness, Confluence and Related Properties of Clause Graph Resolution", Doctoral Dissertation, SEKI Report SR-88-07, FB Informatik, University of Kaiserslautern FRG (1988)

[Eisi89] Norbert Eisinger, "A Note on the completeness of resolution without self-resolution.", Information Processing Letters **31**, pp323-326 (1989)

[Kowa79] Robert Kowalski, "Logic for Problem Solving", North Holland (1979)

[Powe88] David M. W. Powers, Lazaro Davila and Graham Wrightson, "Implementing Connectiong Graphs for Logic Programming", Cybernetics and Systems '88 (R. Trappl, Ed), Kluwer (April 1988)

[Powe90a] David M. W. Powers, "Compartmentalized Connection Graphs for Concurrent Logic Programming I: Compartmentalization, Transformation and Examples", SEKI Report SR-90-16, University of Kaiserslautern FRG (1990).

[Powe90b] David M. W. Powers, "Compartmentalized Connection Graphs for Concurrent Logic Programming II: Parallelism, Indexing and Unification", SEKI Report SR-90-17, University of Kaiserslautern FRG (1990).

[Wise84] Michael J. Wise, David M. W. Powers, "Indexing PROLOG Clauses via Superimposed Code Words and Field Encoded Words", *Proc. International Symposium on Logic Programming*, IEEE Computer Society, pp.203-210 (February 1984).

BRAVE: An OR-Parallel dialect of Prolog and its application to Artificial Intelligence

T.J.Reynolds and P.Kefalas
Department of Computer Science
University of Essex
Colchester CO4 3SQ, UK
e-mail reynt,kefap@uk.ac.essex

Abstract

We present the Brave system which consists of the simple Horn-clause language Brave, a dialect of Prolog designed for OR-parallel execution, plus meta-Brave, which contains the extra-logical features and executes sequentially. Pure Brave has been stripped of Prolog's **assert** and **retract,** control features like **cut** and side-effect predicates like **write.** Meta-Brave features enable compatability with existing Prolog. Brave allows programmers to exploit an available parallel machine by writing programs in a declarative style which enables easy parallelisation. Meta-Brave also features a number of directives which allow algorithmic knowledge about a domain to be stated independently from the Brave application program. These include informing as to what sort of partial results should be remembered (lemmas), instructing on which goals can be pruned as they must fail and modifying the selection rule from depth-first to breadth-first or best-first. We show how this combination of features makes the Brave system specially suited to writing clear programs for search problems in Artificial Intelligence. Results are presented for the performance of well-known algorithms including parallel branch-and-bound and game-tree search with alpha-beta pruning, obtained using a multi-processor simulator written in C.

1 Introduction

We present the Brave system which consists of the simple Horn-clause language Brave, a dialect of Prolog designed for OR-parallel execution, plus meta-Brave, which contains the extra-logical features and executes sequentially. Pure Brave has been stripped of Prolog's **assert** and **retract,** control features like **cut** and side-effect predicates like **write.** Brave allows programmers to exploit an available parallel machine by writing programs in a declarative style which enables easy parallelisation. We have removed the 'dirty' features because they compromise parallel execution. However, in common with other workers [2] [7], we find we need replacements to code parallel versions of certain algorithms. The need is highlighted in search problems where guidance of the search in one subtree requires information from another subtree. This is necessarily meta-control of the proof process, and cannot be expressed with Horn clauses. We thus introduce a meta-language to provide this control. Meta-Brave also handles I/O and enables compatability with existing Prolog. Meta-Brave further features a number of directives which allow algorithmic knowledge about a domain to be stated independently from the Brave application program. These include informing as to what sort of partial results should be remembered (lemmas), instructing on which goals can be pruned as they must fail and modifying the selection rule from depth-first to breadth-first or best-first. We show how this combination of features makes the Brave system specially suited to writing clear programs for search problems in Artificial Intelligence. In Prolog one use of **assert** is to retain information over backtracking and **cut** can be used to prune undesired subproofs. In a parallel context there can be both synchronous and asynchronous versions of these operators. The term **synchronous** is used to indicate that the control operator, **assert** or **cut**, causes an ordering between an event on one OR-branch of a proof and events on others. This requires global synchronisation of processors and possible compromise of parallel execution. The preservation of existing Prolog semantics requires synchronous operators. The Andorra [2] and Aurora [7] OR-Parallel models explore the possibilities of both asynchronous and synchronous operators. We have decided to abandon precise Prolog semantics for Brave and explore the asynchronous versions of this control. This may hamper the conversion of existing Prolog programs (and programmers), but has the advantage of allowing freer exploitation of parallelism. We shall show how meta-directives stated separately from the program can effectively replace the use of synchronous assert and cut. We will then show how these operators are used in the parallelisation of standard AI search algorithms such as **branch and bound, minimax** and **alpha-beta** pruning.

2 Pure Brave

The language of pure Brave is that of Horn-clauses written with the same syntax as Prolog. The semantics of pure Brave is of Horn-clause logic, modified by the operators detailed in this section. We add to pure Brave a number of features which are not Horn-clause logic, in an attempt to provide the minimal constructs to allow efficient programming. We are attempting to avoid the problems, both semantic and practical, of adding **cut** to the base language. The operational reading is that of SLD-resolution, but with a default selection rule which amounts to a multiple depth-first search strategy, giving rise to a subset of all the refutations for a query. Brave's search strategy is sound but not complete. With such a search strategy a programmer can no longer use clause ordering as a means of controlling proof.

2.1 Replacing uses of cut - the if_then_else construction

A common use of **cut**, which enhances considerably the efficiency of Prolog, is to implement conditional control. For example

```
p :- q, !, r.
p :- s.
```

is coded in Prolog using **cut**. There is a syntax for this sort of **if_then_else** construction in Prolog since we can write:

```
p :- q -> r ; s.
```

We allow this latter syntax in pure-Brave, to allow this specific type of proof-tree pruning. There is, however, a problem with this construction if there is more than one solution for **q**. If **q** is ground, there is no problem as it either succeeds or fails. If however there is at least one set of bindings which cause **q** to succeed, and other sets of bindings that cause **q** to fail, then in a parallel implementation it will be indeterminate to which solution we commit. We will be forced to consider them all to give consistent results. This construction has a clear logical reading if **q** has only one solution. This determinacy may be detectable at compile time if **q** is a built-in or user predicate which is obviously determinate. Otherwise we must continue the execution of **q** and report a run-time error if another solution is found. Example: The program for checking for membership of a list becomes in Brave:

```
member(X,[Y|L]):- X=Y -> true ; member(X,L).
```

which avoids checking all the list every time member is called.

2.2 The colon (:) and dot (.) terminators

In Brave we can use dot (.) to terminate OR-cases to be tried sequentially or colon (:), to terminate OR-cases to be tried in parallel. For pure Horn-clause programs these terminators are semantically equivalent, only influencing the order in which solutions are produced. This is true when the program does not use higher order predicates. For example

```
g :- a:
g :- b:
```

would be tried in parallel, while

```
g :- a.
g :- b.
```

would be tried sequentially.Here is a Brave program which searches a tree of constant branching factor 3:

```
search( leaf(N), N ).
search( tree(Left,_,_), Solution ):-
                search( Left, Solution).
search( tree(_,Middle,_), Solution ):-
                search( Middle, Solution):
search( tree(_,_,Right), Solution ):-
                search( Right, Solution):
```

The search is performed at the leftmost subtree first. As soon as a Solution is found both **Middle** and **Right** subtrees can be searched in parallel, as shown in figure 1 . This is a useful technique for certain guided search problems, like alpha-beta, where the **Middle** and the **Right** subtree could benefit from the first solution found in the **Left** subtree.

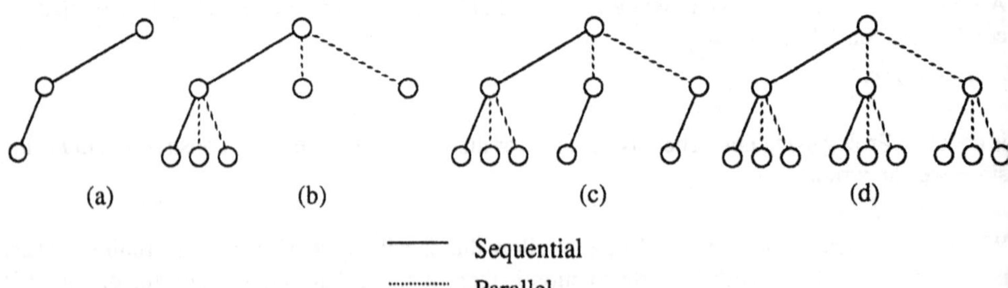

Figure 1 : The execution OR-tree for the search/2 program. Each node corresponds to a call to search/2. Execution is sequential in (a). After a leaf is found, execution proceeds in parallel (b) and switches back to sequential (c) for the leftmost children of the new generated nodes.

2.3 The single predicate

There are occasions when one solution for a goal is required and finding extra solutions is over-computation. Brave is provided with a built-in predicate

single(g1,g2,..)

to allow a programmer to indicate that the goal conjunction contained need be proved at most once. The implementation of **single** requires that a single solution for **g1,g2..** be found, and that any further, non-identical, solutions cause a run-time error. It might seem that it is difficult to obtain any performance benefit from **single** if execution must proceed in this fashion to detect further solutions. The benefit accrues when **single** is part of a clause body such as

.... , single(a), b, ...

and multiple,identical, solutions for **a** would cause repeated computation of **b**.

2.4 Collecting and Expanding Lists of Solutions

Frequently it is desirable to collect all the solutions for a particular goal into a list, and conversely to allow all members of a list to be examined in turn. In Brave this process is crucial and both jobs are performed by built-in predicates. To collect solutions into a list there is

bagof(term, goal, L)

which has the same meaning as **findall** in Prolog. However the resolution strategy of Brave is mutable by Brave constructs, as well as by Meta-Brave. This implies that **bagof** collects a subset of the set of solutions for **goal**.

To expand a list we provide the predicate

expand(t, L)

whose logical reading is

$\exists$ **t1 (t1** $\in$ **L & t1 unifies with t)**

The operational reading is that **expand** unifies **t** successively with each item of **L** and succeeds for each unification that succeeds. This is the same as the common cut-less Prolog coding of member which is used to generate all members of a list. The use of **expand** generates OR-parallelism over the members of a list. A common use of **expand** is to produce operational parallelism from data parallelism. For example a typical style of Prolog coding is to recurse over a list in order to perform some operation on all items in the list.

```
process([Item|In],[New|Out]):-
        perform(Item,New),
        process(In,Out).
```

This is data parallelism which might be performed in another language as a vector operation, or using a higher-order operation like map or apply-to-all functional. In Brave we can use **expand** and **bagof** in combination, to produce as much potential OR-parallelism as the number of items in the list.

```
process(In,Out):-
        bagof(New, (expand(Item,In), perform (Item,New)), Out).
```

4 Meta Brave and search problems

In the domain of Artificial Intelligence applications search plays a large part. A robot seeking to perform a physical assembly task may have the equivalent of a travelling salesman optimisation problem to solve. Determining motion in vision requires solution of the correspondence problem between areas in two succesive visual frames, giving rise to a combinatorial search through plausible matches. In work on natural language ambiguity implies that many grammar rules may be applicable at any point in a parsing algorithm. This applies at all depths in sentence structure giving rise to a large tree of possible parses to be searched. Good algorithms for these types of problems use knowledge about the nature of the tree being searched to reduce the combinatorial explosion. The parsing problem is highly repetitative, as the same sub-parse is required over and over again in different sub-trees. The commonly used chart parsing algorithm builds up a table of completed sub-parses in order to avoid repeating them. Heuristic information can be used to prune the search to avoid whole sub-trees which are (almost) guaranteed not to contain a solution to the problem. A number of classic AI search algorithms make use of knowledge retained from exploration of one subtree to prune useless search in another.

When Prolog is used for AI work it is possible to employ **assert** into the database to hold partial results and **cut** to prune sub-trees. These features are problematical for parallel execution. If Prolog semantics is to be maintained then an **assert** on one OR-branch prevents parallel execution of the others. The same applies to **cut**. Work in the Aurora project [3] has managed to identify various compromise positions, for example **asynchronous asserts** within the body of programs, to be executed immediately wherever they are encountered. In Brave we propose specialised meta-control syntax to separate the control from the logic to aid clarity and help programming discipline. Programmers manipulate a global lemma-base where processors may learn of information found by others, and provide conditions under which goals can be failed early. This meta-control of the search is intended to save unnecessary work, possibly delivering a changed subset of the solution set implied by the pure Brave program. Thus we aim to retain the declarative reading of programs byretaining soundness while avoiding synchronisation problems. The hitch is that we cannot publish everything we prove, nor keep checking to see if someone else has proved something before, on efficiency grounds. We need to provide additional syntax so that a programmer can indicate which goal results to publish as lemmas. The command

:- lemmapattern(predicate,lifetime,type) if condition.

where

type ∈ { **both, success, failure** },
lifetime ∈ { **update, permanent, garbage-collect** },
condition is a conjunction of goals **g1,g2,..**

is a meta-directive which asks any result for **predicate** to be retained as a lemma, provided the **condition** is satisfied. The lemma only retains information about the arguments of **predicate** with a name. Arguments declared as "don't care" (_) are discarded during lemma creation. A **lifetime** flag of **update**, indicates that new lemmas replace the old. The flag **permanent**, means all lemmas are remembered and **garbage-collect** signals that lemmas are to be removed when the subtree with which they are associated has been explored and discarded. The **type** flag indicates whether lemmas are to be retained for **success, failure** or **both**. The information is held in a special database and may be accessed through use of the predicate **lemma/1**. A call to **lemma/1** may not be made from a pure Brave program. This is to discourage 'dirty' programming by not allowing control flow of the object program to depend on bound variables from existing lemmas. There are also a number of specialised directives which cause implicit consultation of the lemma-base.

The command

:- lookup(predicate).

causes the lemma-base instances of **predicate** to be consulted whenever **predicate** is called. This can cause immediate failure or success for as many results as are in the lemma-base. If no lemmas exist for **predicate** then execution proceeds as normal.

From the programmer's point of view this is a simple directive to improve efficiency without altering results. The predicate named will not have to be executed if results for it are already in the lemma-base. The implementation however, presents a number of problems since successive calls of a predicate may not be identical:

(a) the first call may be the most general, i.e. subsume the subsequent, thus the appropriate results must be looked up for each case

(b) a subsequent call may be more general or disjoint, and the new call must be performed, and any duplication with the previous case avoided.

A further complication occurs when one execution of a predicate is current when another starts. We can choose either to synchronise with the first, awaiting each solution as it comes, or to continue to execute **predicate** totally independently, ignoring the first execution. We are experimenting with the latter course. The database of lemmas may become large, and efficient access is crucial. This is particularily true of a multiprocessor implementation which must allow distributed global access. Though lemmas may well be useful on a purely local basis we expect most benefit to come from global access. Methods that are commonly used for indexing a Prolog database are easily adapted for the implementation of lemma-base access.

The command

:- prune(goal) if condition.

is a meta-directive to fail **goal** if the conjunction of goals in **condition** can be satisfied at least once. The **condition** is tried immediately after **goal** is called. The **condition** is proved by the **single** mechanism, as only the existence of one solution is required for pruning. It is important to realise that the **prune** meta-directive does not return any bindings for the variables in **goal**. Control over the search can only be achieved via the meta-level.

The next set of features in Meta Brave are concerned with control over the resolution strategy adopted by pure Brave. The meta-directive:

:-deschedule(goal) if condition.

suggests to the implementation that the process whose immediate task is to solve **goal** be de-scheduled i.e. placed to the back of a goal-queue for later consideration. The goal may eventually pass the test and continue. The suspension refers to the whole subcomputation that follows goal. In contrast, the meta-directive:

:-delay(goal) if condition.

refers to **goal** only, suggesting that the process may continue trying to solve subgoals that follow **goal**. This is equivalent to re-ordering subgoals within the body of a clause. If a goal cannot delay any more (because it is the last goal) then Brave reports a run-time error.

The **deschedule** meta-directive is useful for implementing best-first search strategies. It replaces the need to maintain an explicit frontier set, as the currently best nodes are active whereas the rest remain idle. Once the condition for descheduling is stated, it becomes the responsibility of the underlying implementation to awake nodes as they become more promising. The **delay** directive is useful for relaxing the criteria of groundness in a goal. **Single** and **if-then-else**, as well as arithmetic operations can delay until they become fully ground, as in MU-Prolog [8].

To define further proof strategies, the programmer needs to supply meta-directives which specify scheduling action for particular predicates. A breadth-first strategy is imposed by:

:- strategy(predicate, breadth-first).

This indicates that any goal which unifies with **predicate** is to be scheduled differently from the default strategy. All calls to **predicate** suspend until there are no more active branches in the execution tree. At that point execution is resumed, thus achieving a breadth-first search as far as the named predicate is concerned. Notice that other predicates, ie. those not annotated by the strategy meta-directive, are resolved according to the default multiple depth-first strategy. Figure 2 illustrates the execution of the search program listed in 2.2, where the execution strategy of the predicate **search/2** is breadth-first.

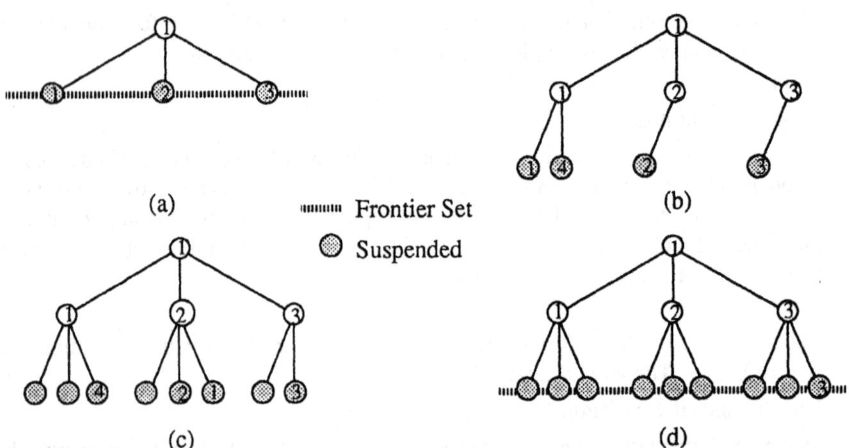

Figure 2 : Execution OR-tree for the search program in 2.2 by 4 procesors. Execution proceeds as normal until an all-suspended situation arises (a). Work is then released for all processors (b), until the next calls to search/2 are encountered (c), which eventually lead to a new all-suspended situation (d).

A best-first strategy is directed by:

:- strategy(predicate(X,Y,....), best-first(X,N)).

which indicates an argument , **X**, in **predicate** which should be bound to integer when **predicate** is called. This integer is used as a measure to indicate priority for execution. Its effect on the scheduler is similar to breadth-first. The only difference is on waking up the processes after no more work is left to be done in the execution tree. All suspended goals are sorted according to the specified argument, **X**, and the best are distributed as active to all processors, so that they get N each. The parameter N is tunable to the application but the number of re-distributed processes is equal to the number of processors available in the machine. So, N*P processes are allowed to continue, where P is the number of processors. The rest of the processes remain suspended. Let us rewrite the search program in 2.2 so that it performs a heuristic search. The heuristic function is **h/2,** where the first argument is the current subtree and the second argument is the heuristic value:

```
:-strategy(search(_,H,_), best-first(H,1)).

search( leaf(N), _, N ).
search( tree(Left,_,_),H,Solution):-
            h(Left,HV),  search(Left,HV,Solution):
search( tree(_,Middle,_),H,Solution):-
            h(Middle,HV), search(Middle,HV,Solution):
search( tree(_,_,Right),H,Solution):-
            h(Right,HV),  search(Right,HV,Solution):
```

As shown in figure 3 the best node is allowed to continue, where best is estimated by the heuristic procedure **h/2.**

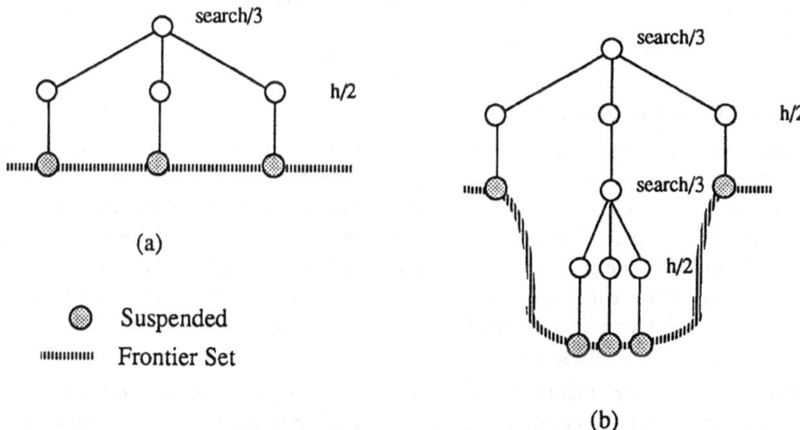

Figure 3 : Execution OR-tree for the heuristic search program executed by one processor. If the heuristic value of the middle node in (a) is the best, then work is released on that node until the next all-suspended situation arises in (b). This behaviour exactly corresponds to Best-First search on sequential execution mode.

5 Programming Examples

In this section we shall show how the features of Brave introduced above can be used to produce parallel versions of classic AI search algorithms. The programs have been tried out on an implementation of Brave for SUN workstations which consists of a compiler from Brave to Brave Abstract Machine (BAM) code, written in Prolog, and a BAM parallel simulator, written in C. The example programs we present were run on a SUN sparcstation with 16 Mbytes of memory, which enables up to 32 processors to be simulated. The Brave Abstract Machine is based on the Warren Abstract Machine [13], and the extentions for OR-parallelism described in the binding array model of Warren [14]. Our machine differs in detail from the Warren model, having evolved separately since 1986, but its main differences are in its provision for the implementation of the Brave language features detailed in this paper.

We shall present the speed-ups obtained for each one of the search problems. The speed-up from a given number of processors is estimated as the number of simulator cycles required for one processor divided by the number of simulator cycles for parallel execution. One simulator cycle represents the time taken to fetch-execute a BAM instruction or remain idle. In order to take into account the overhead of publishing and retrieving a lemma in the lemma base, we have attached a cost for both operations which is the relative cost of each operation compared to one simulator cycle. It is estimated that a lemma creation is as costly as 1-2 simulator cycles whereas a lemma retrieval as 1.5-2.5 simulator cycles. We also take into account the overhead of task switching which is as costly as 40-60 simulator cycles. However, we do NOT consider the overheads due to interprocessor communication in the real parallel machine. The speed-up is defined by the ratio:

$$\mathbf{Speedup} = \frac{\mathbf{Cycles_1}}{\dfrac{\mathbf{Cycles_N + 1.5 \times LC_N + 2 \times LR_N + 50 \times TS_N}}{\mathbf{N}}}$$

where $Cycles_N$, LC_N, LR_N, TS_N are the number of simulator cycles, lemma creation, lemma retrievals and task switching operations for N processors

5.1 The Branch and Bound Algorithm.

Our first programming example is a type of tree search problem. We formulate the problem as searching a state-space tree for the minimum cost leaf, where the cost is known at each node in the tree and is guaranteed to increase towards the leaves. These conditions are plausible for problems such as the Travelling Salesman Problem (TSP), when possible tours are expanded into a tree. The cost then becomes the distance to each node, which clearly accumulates in the required fashion This accumulation property means that once a node is reached by a path which exceeds in distance that of the best tour so far, then all tours which continue from there can be pruned. Thus large sub-trees of the state-space may not be searched at all.

This algorithm searches a tree maintaining a current minimum cost, which it updates as it finds better leaves. Subtrees are pruned below a node if the cost at the node exceeds the current minimum cost. A sequential branch and bound (b&b) is easy to implement by expanding a selection from a frontier set of nodes, at each recursion. The selection strategy could be depth-first, breadth-first or best-first. The current best bound is passed around as an argument. The natural parallel algorithm would independently search subtrees. This makes it impossible however, in a purely Horn-clause context, for one searching agent to benefit from a better bound found by another. Each agent may benefit from bounds it finds itself, but this is less than optimal as the entire subtree being searched by one agent may be useless. The optimal solution is to have a global address space for bounds that can be altered by any processor reaching a solution. Other processors can then prune subtrees according to that value. Brave allows this approach by enabling the publishing of bounds as lemmas. Here is a b&b solution to the TSP in Brave:

```
:-lemmapattern(tour_complete(SolutionPath,NewBound),update,success) if
        lemma(tour_complete(_,CurrentMinBound)),
        NewBound < CurrentMinBound.
:-prune(tsp(_,_,_,CostSoFar,_)) if
        lemma(tour_complete(_,CurrentMinBound)),
        CostSoFar > CurrentMinBound.
:-prune(tsp(Node,[_|Path],_,_,_)) if
        member(Node,Path).

tsp(Node, Path, Path, MinBound, MinBound):-
        tour_complete(Path,MinBound).
tsp(Node, Path, SolutionPath, CostSoFar, MinBound):-
        arc(Node, NextNodes),
        expand(NewNode, NextNodes),
        distance(Node, NewNode, Distance),
        NewCost is CostSoFar + Distance,
        tsp(NewNode, [NewNode|Path],SolutionPath,NewCost,MinBound).

tour_complete(Path,Bound):- nodes(N),length(Path,N).
```

Without the meta-directives this program is an exhaustive search, which is simple to program but expensive to execute. It is a parallel search which potentially forks new processes using **expand** with an arity equal to the number of children. The **lemmapattern** directive instructs the implementation to remember the cost of any complete tour. Notice that the predicate **tour_complete** is chosen as lemma instead of the predicate **tsp** in order to avoid recording success for all recursive calls of **tsp** up to the root. The **prune** directive instructs the implementation to fail any **tsp** goal which has a current cost which exceeds the current minimum cost in the lemma-base. The second **prune** directive is a loop check which could be implemented within the program algorithm, but is conveniently added as a meta-condition.

Notice that by simply adding the meta-directive:

```
:- strategy(tsp(_,_,_,C,_),best-first(C,1)).
```

the search proceeds in a best-first manner instead of the default depth-first.

We have run the TSP program with the default multiple depth-first strategy on our parallel Brave simulator. The test problem is to find the shortest tour of a 8 node graph with randomly generated distances between each pair of nodes. The results shown in figure 4 were chosen as the most representative of the parallel b&b behaviour. The vertical axis is the nominal speedup obtained from the simulator shown against rising numbers of processors. Each diagram corresponds to a different random distribution of nodes. Figure 4.a corresponds to a search tree where the best solution is among the leftmost branches whereas in figure 4.b the best solution is among the

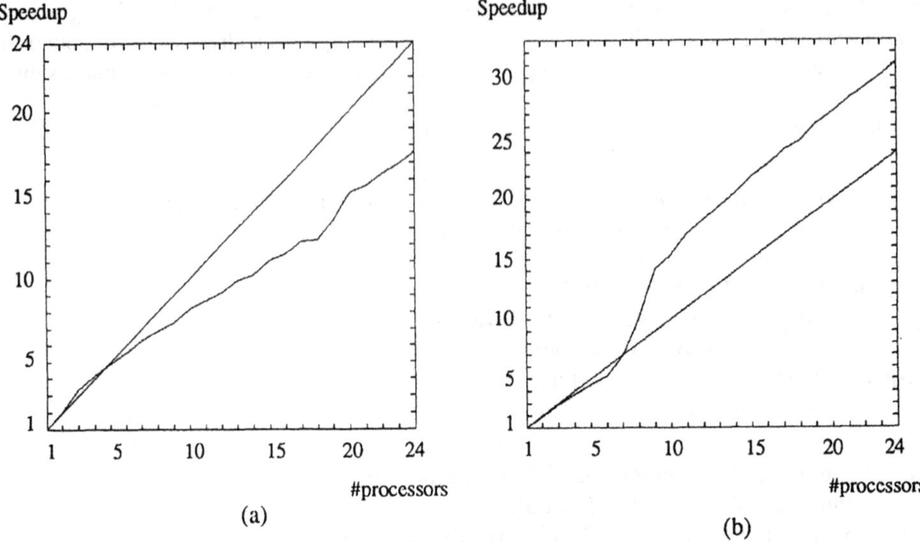

Figure 4 : The Travelling Salesman Problem solved by b&b algorithm. Speedups obtained by 1-24 processors on 2 randomly generated distribution of nodes.

rightmost branches. These are two extreme cases for b&b. In general, the speedups are close to linear, and sometimes super-linear, depending randomly on the problem. This is intuitively reasonable for a truly blind search, where the parallel b&b algorithm has a better chance of hitting a good bound early. Others have demonstrated in practice that parallel b&b can produce superlinear speed-ups [12] [5]. These results indicate good scope for parallelism, and a successful algorithm. In this problem, global publishing of lemmas represents a small burden because a bound is simply a number, and bounds are only likely to be updated infrequently. The cost of checking with the bound before expanding subtrees is the same as for the sequential algorithm, provided it is held locally. This does depend on a reasonable implementation of the distribution of the lemma-base.

5.2 The minimax algorithm and alpha-beta optimisation

Our next programming example comes from the application area of game-tree searches.The minimax algorithm performs an exhaustive search to a given depth of the nodes that form the state space of a partial game tree, with the aim of finding the best reachable terminal node for **max**. An evaluation function returns a heuristic estimation of the state of the game at any given terminal node (more positive values favour **max** and more negative values favour **min**). Establishing the best reachable terminal node is achieved by maximising the value of the descendent nodes whenever **max** is to play or minimising it otherwise, then passing this to the parent node. Finally, the best value together with the best continuation of the game will be available at the root node.

The drawback of the minimax algorithm is that the evaluation function needs to be applied to all terminal nodes. The evaluation function for non-trivial games like chess is complicated and computationally expensive. The alpha-beta cut-off is an optimisation of the minimax algorithm which prunes some branches of the search tree if they are guaranteed to produce a worse solution than the solution found so far. One could imagine alpha-beta pruning to be an augmented b&b algorithm that works at two levels, one trying to maximise the beta bound, the other trying to minimise the alpha bound.

5.2.1 A simple parallelisation of alpha-beta

Parallelising the minimax algorithm is straight-forward, since every processor can independently perform a search in a different subtree. The source of difficulty in parallelising alpha-beta is the inherently sequential nature of the algorithm; a cut-off for one subtree comes from a better value found in a previously explored subtree. However we can use the same mechanism for a Brave parallel solution as we used for b&b: processors use the global lemma-base as a way of communicating cut-offs from already explored subtrees. First, here is a Brave program for the minimax algorithm (the notion of node type mentioned in the program is used later):

```
/* Pos-Type represents the tuple <Current Position,Position Type>  */
/* arc/2 represents the tuple <Current Position,Children Positions> */
/* D is the depth of the game tree to be searched                  */

minimax(D,PlayA,Pos-Type,[arc(Pos-Type,Children)|Path],BestSucc,BestVal):-
        ( D = 0 ->
                BestSucc = Pos-Type,
                evaluate(Player,BestSucc,BestVal)
        ;
                generate(D,Pos-Type,Children),
                switch(PlayA,PlayB),
                NewDepth is D-1,
                bagof( (Successor,Value),
                            proceed(NewDepth,PlayB,Type, Children,
                                    [arc(Pos-Type,Children)|Path],Successor, Value),
                        ListOfSuccessors),
                best(PlayA,ListOfSuccessors,BestSucc,BestVal)
        ).

proceed(D,Player,AnyType,Children,PathSoFar,Succ, Value):-
        expand(X,Children),
        minimax(D,Player,X,[arc(X,C)|PathSoFar],Succ,Value).

switch(min,max).
switch(max,min).
```

Adding the following meta-directives transforms the minimax search into a parallel alpha-beta algorithm in which cut-offs are made globally available:

:-lemmapattern(minimax(_,Player,Pos,_,_,Value),garbage-collect,both) if true.
:-prune(minimax(_,_,_,[_|Ancestors],_,_)) if cut_off(Ancestors).

where **cut_off/1** is defined in the meta-program:

```
cut_off( [arc(Node1, Children1), arc(Node2,Children2) | _] ):-
        member(X1, Children1),
        lemma(minimax(max, X1, V1)),
        member(X2, Children2),
        lemma(minimax(min, X2, V2)),
        V1 =< V2.
cut_off( [arc(Node1, Children1),arc(Node2,Children2) | _] ):-
        member(X1, Children1),
        lemma(minimax(min, X1, V1)),
        member(X2, Children2),
        lemma(minimax(max, X2, V2)),
        V1 >= V2.
cut_off( [ P,G | Rest] ):-
        cut_off(Rest).
```

The first two clauses for **cut_off/1** describe the criterion for shallow cut-offs whereas the third recursive clause gives the criterion for deep cut-offs, as illustrated in figure 5. This distinction arises because in parallel alpha-beta subtrees which are destined to be pruned may be already being explored. Thus search at some depth in a subtree may have to be terminated. These deep cut-offs are a potential worry as the test for pruning involves a search back up through ancestor nodes. This is not a significant worry for fat, high branching factor trees, as they do not reach great depths. Deeper, thinner trees, might however be a problem.

A further improvement on the alpha-beta algorithm is to use the lookup meta-directive in order to avoid re-evaluating positions which have been evaluated before:

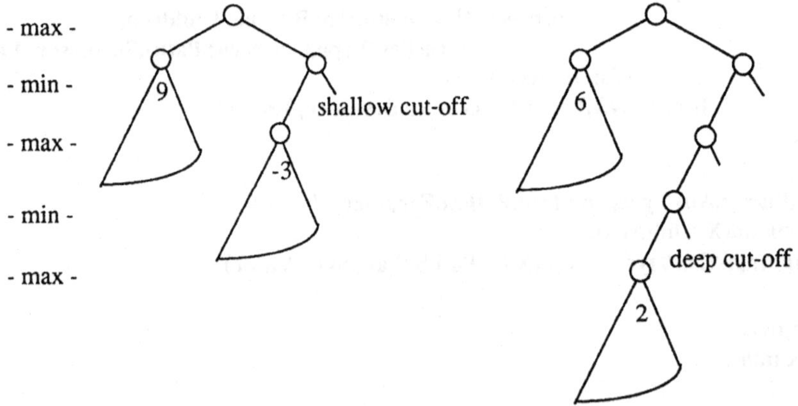

Figure 5 : Examples for Shallow and Deep alpha-beta Cut-offs, as described by the Brave meta-program.

:-lookup(minimax/6).

This is a commonly used technique in games like chess (transposition tables) which avoids redundant computation. However, we do not take this into account when we present the results since our intention is to show the OR-parallelism available in the alpha-beta algorithm itself.

There is a potential problem with the lemma-base. If it grows too large we may pay a penalty in space and an even larger penalty in lookup time. This is mostly under programmer control but the implementation of Brave can help by performing garbage collection on lemmas that have become redundant. The lemma-base may be used to hold information which persists after exploration of a subtree collapses, in which case lemmas will be made permanent (the flag in the lemmapattern directive). However lemmas may not be needed when a subtree is finished and can be labelled garbage-collect. Figure 6 illustrates the garbage-collection mechanism for the alpha-beta program. When all nodes below P are evaluated then the lemma for P will garbage collect all lemmas created for each of its children. We have also monitored the dynamic size of the lemma-base for simulator runs of the alpha-beta search programs to evaluate the effectiveness of this approach. The results have shown us that garbage collection of lemmas is a highly desirable feature. The average level of lemmas is greater for larger numbers of processors, but does not grow as the programs run. Rather it exhibits a quasi-cyclic behaviour as subtrees expand and contract.

In practice game-trees are not often truly random and the moves which expand from any particular node may be more or less strongly ordered in their likelihood of leading to a good position. In order

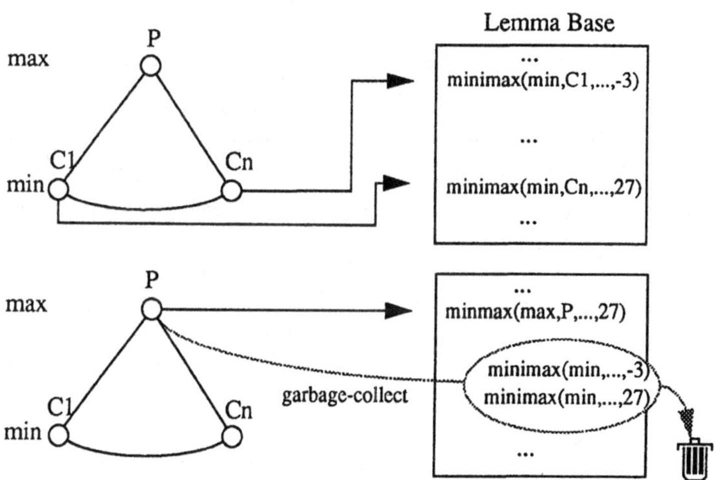

Figure 6 : Garbage-Collection of Lemmas in alpha-beta program. If all nodes Cl to Cn have produced a minimax value then node P evaluates its children and garbage-collects the lemmas made by Cl to Cn.

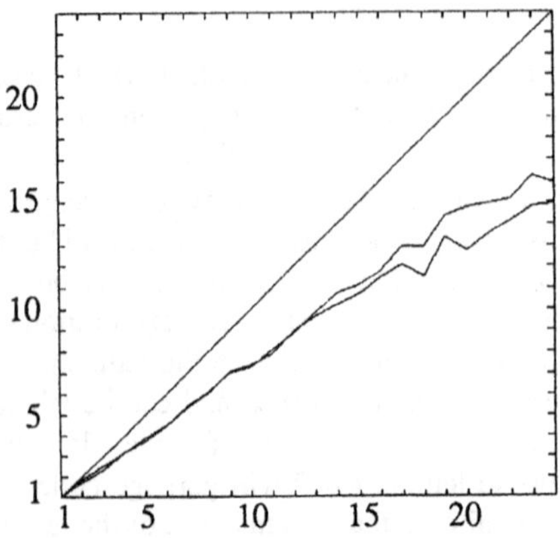

Figure 7 : alpha-beta on random trees.

Application of alpha-beta to three kinds of trees: un-ordered, moderately-ordered and perfectly ordered. The trees searched are of Branching Factor 10 and Depth 3. The vertical axis is the speedup obtained and the horizontal axis is the number of simulated processors (1-24). The two lines in each diagram correspond to different values a=1,b=1 and a=1000 and b=10000 of node generation and node evaluation costs.

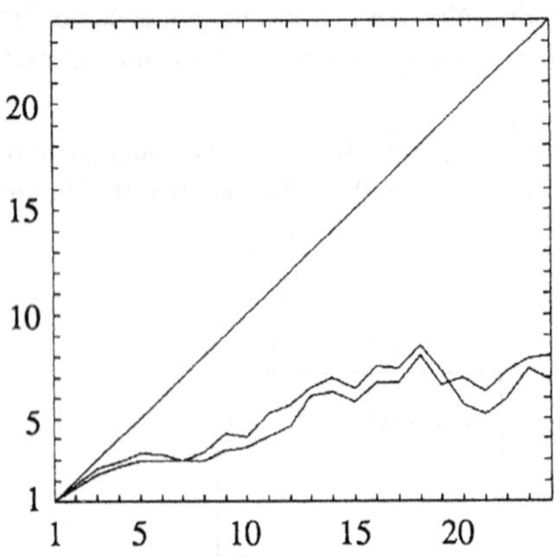

Figure 8 : Ordered Trees

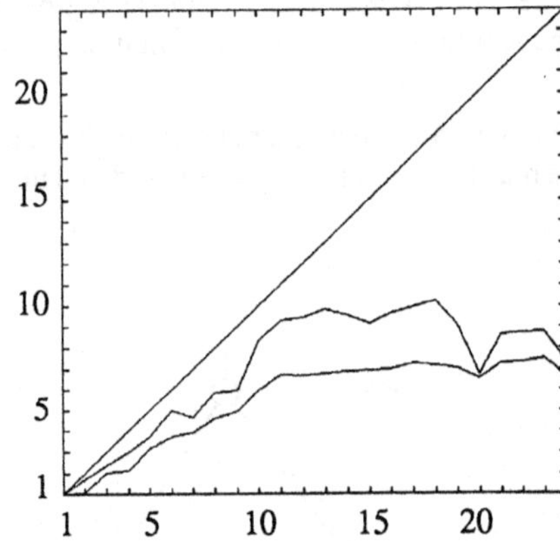

Figure 9 : Ordered Trees with Types

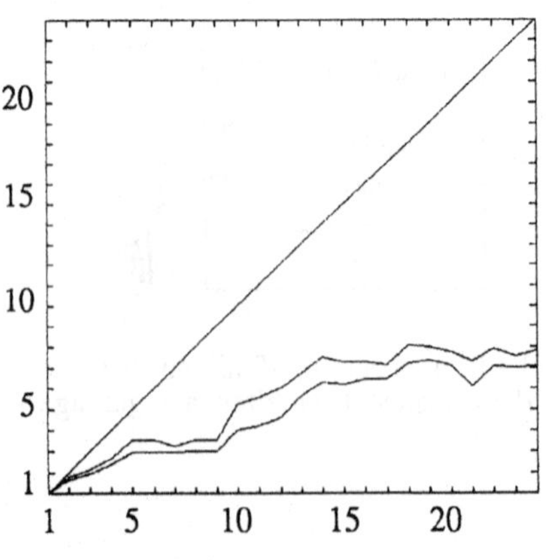

Figure 10 : Perfectly Ordered Trees

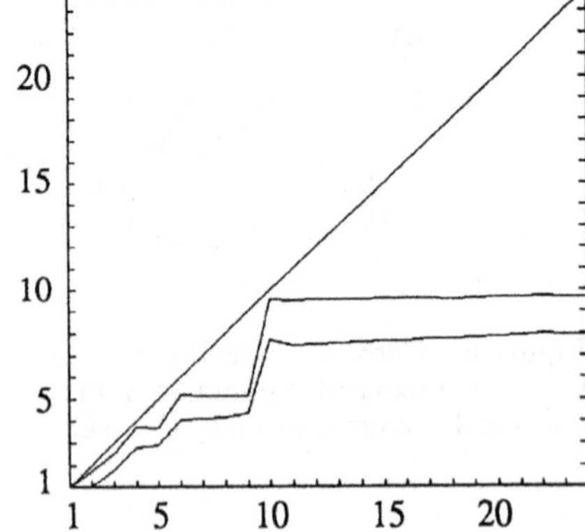

Figure 11 : Perfectly Ordered Trees with Types

to evaluate the simple alpha-beta algorithm we generated game-trees randomly, but with a varying degree of bias in the ordering.

Figures 7,8,10 show data taken for a search tree with a branching factor of 10 and depth 3. The alpha-beta program is general in the sense that it can be applied to a range of two-persons games. In order to estimate the expected speedup for any particular game we make some guesses as to the relative costs (a,b) of the position generation and the position evaluation operations. The cost is relative to one simulator cycle and varies from a=1,b=1 for trivial games up to a=1,000, b=10,000 for more complicated games like chess. The work performed by parallel execution now becomes:

$$\mathbf{Cycles_N + 1.5{\times}LC_N + 2{\times}LR_N + 50{\times}TS_N + a{\times}G_N + b{\times}E_N}$$

where G_N and E_N are the numbers of node generations and node evaluations. Figures 7,8 and 10 show results taken for trees which are, respectively, un-ordered, moderately-ordered and perfectly ordered. Clearly searching random trees can easily employ more processors. This is to be expected as alpha-beta pruning is least effective for these trees, leaving more searching to be done. The moderately ordered case is intermediate. We approach the perfectly ordered case where the algorithm improves performance only up to a speedup of about 8 with 20 processors. This represents the point of useful parallelisation. More processors simply explore tree that will be pruned. As with branch and bound, this will be determined by problem size. A chess playing program which searches a game-tree with branching factor 30, to a depth of 15 moves will evaluate 2000 million terminal positions with ideal alpha-beta pruning. There is thus plenty of scope for employing thousands of processors.

5.2.2 An improved parallelisation of alpha-beta for ordered trees

The algorithm of 5.2.1 can be improved for game-trees where it is known that moves can be well-ordered. We start by classifying nodes into three types [4].

type1: the root node is type1, and all the leftmost children of type1 nodes are type1

type2: other children of type1 nodes are type2 and the children of type3 nodes are type2

type3: the children of type2 nodes are type3

If we assume perfect ordering, all but one child of node type 2 can be pruned whereas the children of the other two types can be searched in parallel. In this situation pruning would be ideal and favours the sequential alpha-beta algorithm. In our program above the definition of **proceed/7** does not take into account node type. We can modify it as follows:

```
proceed(D,Player, 1 ,[First|_],PathSoFar,Succ,Value):-
        minimax(D,Player,First,[arc(Pos,_)|PathSoFar],Succ,Value).
proceed(D,Player, 1 ,[_|RestChildren],PathSoFar,Succ,Value):-
    expand(Pos,RestChildren),
        minimax(D,Player,Pos,[arc(Pos,_)|PathSoFar],Succ,Value):
proceed(D,Player, 2 ,Children,PathSoFar,Succ,Value):-
    expand(Pos,Children),
        minimax(D,Player,Pos,[arc(Pos,_)|PathSoFar],Succ,Value):
proceed(D,Player, 3 ,Children,PathSoFar,Succ,Value):-
    expand(Pos,Children),
    minimax(D,Player,Pos,[arc(Pos,_)|PathSoFar],Succ,Value):
```

The first two clauses for **proceed** are tried sequentially, ie. the leftmost child of node type 1 is searched first. If the tree is perfectly ordered, the value found at the leftmost branch will ideally prune the rest of the tree. Perfectly ordered trees rarely occur and it is the case that if a tree is perfectly ordered, search would have been unecessary. Figures 9 and 11 show results from the simulator for running the alpha-beta algorithm with types.

Conclusion

We have presented a language based on Prolog, but designed for parallel execution. Clearly we hope that users in the AI application domain will find the features we provide sufficient to express good algorithms. We also hope that reasonably clear programs will result. We accept that meta-directives have the same attendant problems as meta-predicates in Prolog. It is possible to write programs in Brave for which no consistent logical reading can be given. We hope though, that the separation of the meta-directives from the program proper is a clean and natural division, and that further work on meta-programming [6] will give us a consistent logical framework in which to combine target and meta-language. In previous work on Brave [11] we have made full speed implementations for sequential and parallel hardware. We have followed the general approach of the Aurora computation model of [7]. Having designed the additional features described in this paper and tested them with a simulator we are now testing them in full speed implementation on the GRIP parallel architecture [10].

Acknowledgement

This work is funded by the Science and Engineering Research Council of Great Britain through an IED grant and a studentship for the second author. We would like to acknowledge two other members of the Brave team: George Gourbalis and Frankiskos Ieromnimon for their work on aspects of the Brave simulator.

References

[1] R.A. Finkel, J.P. Fishburn. "Parallelism in alpha-beta Search". *Artificial Intelligence 19*, pages 89-106, 1982

[2] S. Haridi, P. Brand. "ANDORRA Prolog - An Integration of Prolog and Committed Choice Languages". *Proceedings of the FGCS '88 Conference*, ICOT, pages 745-754, 1988

[3] B. Hausman, A. Ciepielewski, A. Calderwood. "Cut and Side Effects in Or-Parallel Prolog". *Proceedings of the FGCS '88 Conference*, ICOT, pages 831-840, 1988

[4] D.E. Knuth, R.W. Moore. "An Analysis of alpha-beta Pruning". *Artificial Intelligence 6*, pages 293-326, 1975

[5] T.H. Lai, S. Sahni. "Anomalies in Parallel Branch and Bound Algorithms". *Communications of the ACM*, Vol.27, pages 594-602, 1984

[6] J.W. Lloyd. "Directions for Meta-Programming". *Proceedings of the FGCS '88 Conference*, ICOT, pages 609-617, 1988

[7] E. Lusk et al., D.H.D. Warren et. al, S. Haridi et al. "The Aurora OR-Parallel Prolog System". *Proceedings of the FGCS '88 Conference*, ICOT, pages 819-830, 1988

[8] L.Naish. "Negation and Control in Prolog". *G. Goos and J.Hartmanis (ed.), LNCS 238,* Springer-Verlag, 1985

[9] M. Newborn. "A Parallel Search Chess Program". *Proceedings of the ACM Annual Conference,* ACM, New York, pages 272-277, 1985

[10] S.L. Peyton Jones, C. Clack, J. Salkild, M. Hardie. "GRIP - a high-performance architecture for parallel graph reduction". *Proceedings of the IFIP Conference on Functional Programming Languages and Computer Architecture, Portland, G. Kahn (ed.), LNCS 274,* Springer-Verlag, pages 98-112, 1987

[11] T.J. Reynolds, A.J. Beaumont, A.S.K. Cheng, S.A. Delgado-Rannauro, L.A. Spacek. "BRAVE - A Parallel Logic Language for Artificial Intelligence". *Future Generations Computer Systems 4,* North-Holland, pages 69-75, 1988

[12] P. Van Hentenryck. "Parallel Constraint Satisfaction in Logic Programming: Preliminary results of CHIP within PEPSys". In *Logic Programming, Proceedings of the 6th International Conference on Logic Programming,* Lisbon, MIT Press, pages 165-180, 1989

[13] D.H.D.Warren. "An Abstract Prolog Instruction Set", *Technical Report TN-309,* SRI, October 1983

[14] D.H.D.Warren. "The SRI model for OR-Parallel Execution of Prolog. Abstract Design and Implementation Issues". In *the IEEE Computer Society Press (ed.), Proceedings of the 1987 Symposium on Logic Programming,* pages 46-53, IEEE, September 1987

A Declarative Debugging Environment for DATALOG

Francesco Russo, Mirko Sancassani *
V.le Silvani, 1
Bologna, Italy

Abstract

In this paper we discuss a declarative approach to debugging for DATALOG, a logic programming language based on a bottom-up evaluation strategy. We focus on the DATALOG implementation provided by the ALPE programming environment. In ALPE, a DATALOG program is translated by the Logic Query Compiler - LQC - into a Prolog program whose top-down evaluation corresponds to the bottom-up evaluation of the source DATALOG program. It follows then that the computational models of the source and the compiled programs are completely different. This makes the use of a tracer practically pointless.

Conversely, the definition of a declarative debugger in this framework requires that an adeguate representation of the computation steps be stored in order to rebuild the evaluation tree and to allow a conversation with the user to occur at source program level. In this paper we show how to produce such information from the DATALOG program.

1 Introduction

In this paper we discuss a declarative approach to debugging for DATALOG, a logic programming language based on a bottom-up evaluation strategy. More precisely, we consider the development of a declarative debugger [Sha82] within the framework of the DATALOG implementation provided by the ALPE environment [Mo90]. ALPE is a logic programming environment for the development of advanced database applications. It provides a user-friendly knowledge-based framework for querying a multi-database system and supports a compilative implementation for DATALOG.

*This work has been supported by "Progetto Finalizzato Sistemi Informatici e Calcolo Parallelo" o[f] CNR. undergrant n. 90.00757.69

A DATALOG program is traslated by the Logic Query Compiler - LQC - into a PROLOG program whose top-down evaluation corresponds to the bottom-up evaluation of the source DATALOG program. The transformation is achieved through the application of new and adavanced logic query optimization methods and execution algorithms (Mini Magic Set [BeRa87] and Mini Magic Counting [SaZa87]).

The choice of ALPE as the underlying architecture doesn't involve any loss of generality since our method does not rely on any specific feature of the ALPE compiler and is actually well-suited for any class of compiled DATALOG programs.

The pure declarative semantics of DATALOG and the nature of its implementation, which is tipically based on a compilative approach, have a number of consequences for the design of the architecture of a debugging system.

In fact, it looks clear that a conventional debugging tool, that is a tool which essentially follows step by step the execution of the program, would be almost useless if applied to a DATALOG environment. Indeed, with such a tool, while being able to *develop* a program keeping in mind only its declarative interpretation, the user would be required to know everything about the transformation algorithms and the computational behaviour of his system to *debug* it!

It follows then, that the only acceptable choice for a debugging tool well-suited for DATALOG is that of a declarative tool capable of filtering out all the procedural aspects of the computation and of presenting the user with a declarative view of its program.

2 Query Evaluation Methods in DATALOG

Generally, given a query Q=(G,LP), where LP is the set of *horn clauses* and G is the *query goal* if G has some bound variables a simple bottom-up evaluation (*naive* or its optimized *semi-naive* version, by the *query-subquery approach* [BaRa86]) is highly inefficient since the selection of the interesting answers is made only when all the possible answers for the unbound query goal G are found [Mo90]. Then, when G has the *binding passing property* i.e. it has the ability of propagating top-down its bindings through the rules at some level of recursion, it is possible to take advantage of several evaluation methods to improve the performance of the system. The most famous methods found in the literature are: The *Magic Set* method [BeRa87], the *Minimagic Set* method [SaZa87b] , the *Counting* method [SaZa86], the *Minimagic Counting* method [SaZa87].

In ALPE the DATALOG compiler LQC receives as input a pair (DATALOG program,query),determines the subprograms that are independently computed, then it determines which arguments in the recursive predicates will be instantiated during the top-down phase. In a furthe phase, it adds rules to the source program (using the *Minimagic* or *Counting* Methods) tha actually implement the top-down propagation of bound rules. Finally,LQC executes the **translation** of these rules into a PROLOG program whose top-down execution corresponds to the optimized bottom-up computation of the source program [Mo90].

To give you an informal overview on the underlying ideas of these methods, we describe in the following the *Magic Set Method.* Actually, the mostly used methods by LQC are the *Minimagic Set* and the *Minimagic Counting*, which are optimizations of the *Magic Set* method. We don't describe in this paper these tecniques. The interested reader may refer to the appropriate literature.

2.1 Magic Set Method

The Magic Set Method is a method that provides an efficient treatment of logic queries having the *binding passing property*. The objective of the *Magic Set* method is to trasform at compile time the query Q=(G,LP) into a query $Q^* = (G^*, LP^*)$ which has the same answers as Q but is more efficient. The underlying idea is to evaluate top-down the bindings for the variables occurring in the rules of LP during the computation, to produce only the answers which we are interested in. The information about binding propagation is stored in special predicates called *magic-predicates* that are computed by special rules called *magic-rules*. These rule are generated at compile time. Now, every rule of LP is transformed into a modified rule with the addition of a *magic-predicate* in the body of the rule. This make it possible to advantage of the above information by restricting the generation and evaluation of useless facts. The answer can be finally efficiently built by means of two fixpoint computation. The run time binding propagation is accomplished by evaluating the *magic-rules* (first fixpoint). The next *bottom-up* evaluation of the modified rules computes the answer (second fixpoint), having some variables constrained to range on sets generated by the previous phase. Let us consider the well known *same generation* example:

r1: **sg(X,X) :- h(X).**
r2: **sg(X,Y) :- p(X,X1),sg(X1,Y1),p(Y,Y1).**

The predicate **sg(X,Y)** means that X is in the same generation of Y, **p(X,Y)** means that Y is parent of X and finally, **h(X)** says that X is a human. The semantic of the program says that X is in the same generation of itself if X is a human. X is in the same generation of Y if their parent are in the same generation.
The *Magic Set* method produces the following modified query program:

rm1: **ms(a).**
rm2: **ms(X1) :- ms(X),p(X,X1).**
rm3: **sgm(X,X) :- ms(X),h(X).**
rm4: **sgm(X,Y) :- ms(X),p(X,X1),sgm(X1,Y1),p(Y,Y1).**

The query goal is rewritten as G^*: **?- sgm(a,Y).**
We see that the *Magic Set* Method is more efficient than a simple bottom-up computation since **ms/1** restricts the set of facts to be considered only to those that can produce the answers which we are interested in.

2.2 The Translation Step

The DATALOG programs implementing the query optimization methods which are described in the previous section are traslated by the Logic Query Compiler - LQC - into a PROLOG program whose top-down evaluation corresponds to the bottom-up evaluation of the source DATALOG programs.

In the following, we sketch the basic traslation mechanism:

Let **r** be a rule **P :- E.** in the source code, where $\mathbf{E} = E_1,...,E_k$. Then, if **r** is a non recursive rule, **r** is modified into:

query :- E,not(P),assert(P),fail.

and if **r** is a recursive rule, **r** is modified into:

query :- E,not(P),assert(P),query.

3 Declarative Error Diagnosis in DATALOG

We report here a brief informal description of the foundational notions of the declarative approach to program diagnosis. For a detailed description, the interested reader may refer to [Ll87b], [PC88a], [Sha82].

definition: let P be a program and I an intended interpretation for it. We say that a program P is correct wrt an interpretation I, if any answer $G\theta$ computed by P for a goal G is valid in I and any atom W satisfiable in I is computed by P.

Two are then the types of errors a declarative diagnoser can detect: the existence of an *uncovered atom* A and the presence of a *wrong clause* $A : -B$ in the program.

Intuitively, an atom A is uncovered either if there is no clause in P whose head matches A, or if for any such clause $A' : -B$ the execution of the its body $B\theta$ *legitimately fails*, i.e. each failing goal in B is indeed unsatisfiable in I. We also say that the predicate definition for the predicate symbol in A is *incomplete*. Furthermore, the intuitive meaning of the notion of *wrong clause* is that if an incorrect atom A is produced by clause $A' : -B$ even when each atom in the body of $B\theta$ is correct, then the clause must be incorrect.

The next proposition connects the concepts introduced above:

proposition: Let P be a program and I an intended interpretation for P. Then P is incorrect wrt I iff there is an uncovered atom for P wrt I or there is an incorrect clause instance for P wrt I.

These ideas can be already formalized as a logic program. However, this declarative program needs to be informed about the definitions of *validity* and *satisfiability* and needs a procedural component whose role is to guide and try to optimize the search for an error using either an algorithm or some heuristics. These funcionalities are usually obtained by querying an *oracle*, which provides the intended interpretation of the program. The oracle is usually the user himself, who is asked questions about validity, satisfiability and also admissibility of an atomic goal. The diagnoser avoids querying the oracle whenever possible: this is achieved by recording the oracle's answers (avoiding asking again for any instance of an atom already known as valid or unsatisfiable) and using type information or other kinds of computable constraints whenever available. Furthermore, the whole search strategy of the diagnoser should be directed to minimize the number of queries needed to detect an error.

All the concepts described above can be applied to a Datalog system. The main problem in this case is the choice of an adeguate representation of the SLD derivation tree which is needed by the diagnoser.

In general, a diagnostic tool can either perform its search *while executing* the program or by exploring the *stored trace* of a preceding execution.

The latter approach is more powerful, since it allows for more general and flexible ways of searching the derivation tree to be defined and for different heuristics to be applied in the choice of query nodes. Furthermore, it makes it easier the integration with graphical browsing tools and advanced user interfaces in general.

The classical technique for obtaining a stored trace of the execution is to produce an instrumented version of the program, capable of recording, during the evaluation of a query, the information that is needed to rebuild the execution tree. This task is complicated in the ALPE environment by the presence of the compiler since the execution model of the source DATALOG program is different from that of the Prolog compiled one.

The type of trace information needed is essentially a set of recorded facts each describing a node of the derivation tree. The level of detail of this description can vary determining different degrees of complexity in the reconstruction phase performed by the diagnoser.

A minimal description should include at least the following information:

- a set of assertions of the form $derived(G, R)$ for each atom G (in the form in which it appears in the source program) which has been derived from the rule R during program execution;

- a set of assertions of the form $query(G)$ where for G is the top query or a subquery invoked in the top-down part of program evaluation.

A more detailed description could also include the following additional information:

- a set of node representations of the form $derived(G, R, Sub)$, where G is a derived atom (in the form in which it appears in the source program), R the rule used for deriving it and Sub a list of pointers to the subnodes (of the form $derived/3$ or $db\_fact/1$) for the goals in rule R's body;

- a set of assertions of the form *query*($G, Sols$) where G is a queried goal (either the top query or a subquery) and $Sols$ is a list (maybe empty) of pointers to *derived*/3 nodes, one for each solution found to the query.
- an information of the form *reordered*(H, B) which reports the reordering of a source program clause body B upon a specific call instantiation of its head H.

Both these sets, together with the original program clauses, will be sufficient to reconstruct dinamically the structure of the whole SLD search tree, and thus the justification for any incorrect program behaviour. The differences consists essentially in the following points:

- With the minimal trace, we don't have stored informations about the instantiation of goals in the body of a clause which has led to the derivation of an instantiation of its head. Hence this information has to be derived by matching the body of the clause with the set of *known* facts (either derived or data-base ones). Notice that, althoug there can be ambiguities in this process, (since there can be multiple such instantiations), this doesn't affect the declarative behaviour of the diagnoser: what can change is just the order in which bugs would be detected, if more than one is present.
- If we have only informations on full instanciated atoms, rather than the pair (Goal instantiation at the call, derived Goal instantiation), it is not always possible to detect the case of inadmissible goal call, and thus the diagnosis could be less specific.
- To explore the full search tree for a given goal (not just the success paths, which are well described by recursion on the *derived* node information), the diagnoser will use a top-down process where all clauses defining the predicate in the goal are matched with it and the body of each matched clause is recursively examined in left-to-right order. For each success goal in the body, we will find a *derived* node or a data-base fact in the stored trace; otherwise the goal is a failed one, and the search can recur on it. For this process to be successful and efficient, we need that clause bodies are explored in the same order which has been produced by the compiler (which is dictated by clause head instantiation at the call and can be different from the source clause one). If this information is not stored in the trace, the diagnoser will have to perform a reordering of the clause by itself.

The following section describes in detail how to extend the compiler to produce instrumented Prolog programs capable of generating the required representation of the execution of the source DATALOG programs.

4 Towards a Post-Mortem Evaluation Trace

The term *post mortem* means that the evaluation trace is *rebuilded* once the computation of the program has been completed. As already mentioned at the beginning of the previous section, the problem of producing an instrumented version of the compiled programs is a non trivial one if the interaction between the user and the diagnoser is to occur at the level of the source program. In fact, the representation of the execution tree should be adequate for the diagnoser to present the user with a set of questions which are consistent with the declarative view of the source program. Yet, the program which is actually executed is a Prolog program which is evaluated via a top-down strategy. Moreover, the use of MMS and MMC introduces further complications as some rules are splitted up in several pieces.

We approach the problem in two different ways. In both cases the idea is to extend the compiler so that it automatically generates the instrumented programs.

The first method - the *Analytic Method* - produces a detailed representation of the execution tree in order to make it easy to rebuild the evaluation trace during the diagnostic process. To obtain such an information is not staightforward since the optimization methods split up the recursive rules in the object code, furthermore the names and the arguments of predicates are changed.

The second method - the *Synthetic Method* - aims at minimizing the complexity of the representation of the evaluation tree thus symplifying the task of instrumenting the object programs accomplished by a DATALOG compiler. As an obvious consequence, the rebuilding phase carried out by the diagnoser gets more complex and inefficient.

In the following section we give a general overview of the *Synthetic Method*. Detailed information about the two methods can be found in [RuSa91].

4.1 The Synthetic Method

The underlying idea of the *Synthetic Method* is the following: during the bottom-up evaluation of the DATALOG program, all the instantiated heads of the succesful rules are first recorded. Then, if we know which source rule produced a certain fact, we can easily rebuild the instanciated body of such rule. In fact, all the *base* and *derived* predicates verified during the bottom-up computation are still available in the database. Therefore, the reconstruction of the execution trace can be carried out in a top-down evaluation of the source code by exploiting Prolog's backtraking mechanism. Notice that the literals in the body of a source rule that are instanciated in this way might be different from those which would have been actually used in a *bottom-up* evaluation. Fortunately, from a declarative point of views this is not a problem. In fact, what matters is indeed the fact that a certain fact was succesfully evaluated rather than execution path that led to the success of the given fact.

A schematic description of how the *Synthetic Method* operates in the different cases of simple bottom-up, MMS and MMC query optimization methods can be given as follows.
Every time a new *derived predicate* is asserted during the evaluation of the program, we assert piece of information about the rule used and the original name of such a predicate. This information is encoded in term of the predicate **derived(Or_der_pred,Num_rule** where **Or_der_pred** is the fact which has been derived and **Num_rule** is the rule whose

evaluation led to the success of **Or_der_pred**.

Let's consider an example. Let **P** be a derived predicate symbol written in **Minimagic** form and let **p** be the original name. Let also **A** be the set of arguments for **p** and **r** be the index to the source rule. Then, in the **Semi-Naive** and in the *Minimagic* methods, the sequence:

...,not(P(A)),assert(P(A)),...

will be expanded to include the assertion of the new fact **derived(p(A),r))**.
In the MMC method, the generation of the predicate **derived/2** is more complex since only the initially unbound arguments of the verified predicate get asserted. Then in order to rebuild the *original predicate*, we need to know the **counting** predicate with the same recursion index as the newly verified *derived predicate.*
Assume that **P** is a derived predicate symbol written in **Counting** form and **p** is the original name. Let then **A** be the set of initially unbound arguments for **p** and **W** be the set of initially bound arguments for **p**. Finally, let **r** be the index of the source rule and **J1** be the current level of recursion and let' s assume that, in **p((W,A),r)** the predicates in the conjunction **W,A** are appropriately reordered.
Then, in the MMC Method, the sequence:

...,not(P(_,A)),assert(P(J1,A)),...

will be expanded as follows:

...,not(P(_,A)),assert(P(J1,A)),cs(J1,W),assert(derived(p((W,A),r)),...

Accordingly, in the *same generation* example, the assertion of the fact **derived(sg(a,b),2** means that the source rule **2** produces the *intensional predicate* **sg(a,b)**.

5 Conclusions

In this paper we have discussed a declarative approach to debugging for DATALOG. The idea is to produce object programs that are instrumented to record an image of the evaluation trace. Such trace is then used by the diagnoser to establish a dialogue with the user at the source program level. The instrumented programs are obtained by a suitable extension to the compiler.
We argue that declarative diagnosis algorithms and graphical browsing tools could be integrated in a powerful *debugging environment* for Datalog.
Such a diagnosis environment can be seen as a top-level shell which prompts the user for a query to solve, compiles the query in an executable prolog program using "debugging-mode" transformations, executes the query and shows the list of results to the user. The user could then ask for a graphical display of the SLD search tree for the query, focusing either on the derivation tree of a given solution or inspecting the failing branches or a missing one he could provide. In alternative, the user could invoke the diagnosis algorithms and let them do the work, providing just answers to their queries. Both approaches could be used in an integrated way, switching from one to the other, either with the invocation of declarative diagnosis on a given node visualized by the browser,

or asking the display of a node about which the diagnoser is querying.
Finally, we remark that the described methods for a post-mortem rappresentation of the derivation tree can be sucessfully used for the realization of both the declarative and graphical tools.

References

[BaRa86] Bancilhon, F., Ramakrishnan, R., "An amateur's introduction to Recursive Query Processing Strategies", ACM Proc. on Database Systems, 1986.

[BeRa87] Beeri, C., Ramakrishnan, R., "On the Power of Magic" , ACM Proc. on Database Systems, 1987.

[DoSi] Donato, C., Sicilia, M., "The Logic Query Compiler of Alpe" CRAI (Cosenza), Italy, Internal Report.

[Ll87b] J.W. Lloyd, "Declarative Error Diagnosis", New Generation Computing 5, 2 (1987), pp. 133-154.

[Mo90] Mostardi, T., " An Introduction to the Advanced Logic Programming Environment", Internal Tech. Report for the ALPE Project, CRAI (Cosenza), Italy.

[PC88a] L.M. Pereira, M. Calejo, "A Framework for Prolog Debugging", Procs. 5th Int. Conf. on Logic Programming, MIT Press, 1988.

[RuSa91] F.Russo, M. Sancassani, "Debugging Tools for the ALPE environment: A Feasibility Study", Internal Tech. Report for the ALPE Project, V. Silvani 1, DS Logics s.r.l., Bologna, Italy.

[SaZa86] Sacca', D., Zaniolo, C., "The generalized counting method for recursive logic queries", Proc. 1st International Conf. on Database Theory, Rome, 1986, to appear in TCS.

[SaZa87] Sacca', D., Zaniolo, C., "Magic Counting Methods", Proc. ACM SIGMOD Conf. San Francisco, 1987.

[SaZa87b] Sacca', D., Zaniolo, C., "Implementation of Recursive Queries for a Data Language based on Pure Horn Logic", Proc. Fourth Int. Conference on Logic Programming . Melbourne, Australia, 1987.

[Sha82] E.Y. Shapiro, "Algorithmic Program Debugging", MIT Press, 1982.

A SEQUENT CALCULUS FOR A FIRST ORDER LINEAR TEMPORAL LOGIC WITH EXPLICIT TIME

Jūratė SAKALAUSKAITĖ

Institute of Mathematics and Informatics,
Lithuanian Academy of Sciences,
Akademijos 4, Vilnius, 2600, Lithuania
email: logica @ ma – mii.lt.su

1. Introduction

We consider a first order logic of linear discrete time, denoted FTL. FTL is based on first order language with equality which is extended with temporal modalities ○ ("next in the future"), □ ("always in the future"). A set of functional symbols of the underlying first order language is restricted to constant symbols. As in [1] some constants have time-independent meaning which we call rigid constants, other constants have time-dependent meaning which we call non-rigid. It must be noticed that variables have time-independent meaning in FTL. Non-rigid constants are functions on time and can be used to formulate explicitely conditions on time in temporal logic. These conditions are of a great importance in specifying the so called real-time reactive systems ([4]). It would be desirable to have a proof-search procedure for FTL which can serve as a basis for executable temporal languages ([2]). As in the case of first order predicate logic ([3]) a Gentzen-type cut-free sequent calculus can be used to construct such a proof-search procedure. Unfortunatelly, there is no finite sound and complete axiomatic system w.r.t. the intended semantics for FTL with function symbols as it is proved in [1]. We present an infinite Gentzen-type sequent calculus for FTL. The calculus presented is cut-free and enjoys some kind of subformula property. This calculus is constructed as an extension of the sequent calculus for the first order temporal logic without equality and function symbols from [5]. It must be noticed that in LFTL there are some rules of inference for equality which have no counterparts in first order logic with equality. These are rules 3.3.2 for non-rigid constants and the last rule in 3.3.1 for variables and rigid constants below. We prove that the calculus considered is sound and complete for the intended semantics of FTL.

Recently, there has been an interesting related work on proof-theoretic aspects of first order temporal logic ([6]), in which method to convert infinite sequent calculus into finite one is elaborated.

In section 2. we rewiew the syntax and semantics of FTL. In section 3 we describe a sequent calculus for FTL and prove the soundness of it. In this calculus we give a proof of the valid formula which cannot be proved in the finite axiomatic system proposed in [1]. In section 4. the proof of completeness of the sequent calculus is presented.

2. Temporal logic FTL

A language of FTL consists of countable collections of

- individual variables $x, y, z, \dots$;
- m-ary relation symbols R_1^m, $R_2^m, \dots$ $(m = 0, 1, 2, \dots)$;
- constant symbols $a, b, c, \dots$;

and the following operators:

- equality $=$;
- logical connectives $\urcorner, \vee, \wedge, \supset$;
- quantifiers $\forall$, $\exists$;
- temporal modalities $\bigcirc$, $\Box$, $\Diamond$.

Terms are variables and constant symbols.

Formulas are defined inductively as follows:

- if R is a relational symbol of arity m and $t_1, \dots, t_m$ are terms, then $R(t_1, \dots, t_m)$ is an (atomic) formula;
- if t_1, t_2 are terms, then $t_1 = t_2$ is an (atomic) formula;
- if p, q are formulas, then $\urcorner p$, $p \wedge q$, $p \vee q$, $p \supset q$, $\bigcirc p$, $\Box p$, $\Diamond p$ are formulas;
- if p is a formula and x is a variable then $\forall x p, \exists x p$ are formulas.

Let w denote the set of natural numbers. Let p be a formula, then $\bigcirc^0 p$ denotes p and $\bigcirc^i p$ denotes $\underbrace{\bigcirc \cdots \bigcirc p}_{i \text{ times}}$, $i \in w$, $i \neq 0$. $\bigcirc^i p$ means "p holds at a time moment i".

Informally FTL formulas are evaluated over a sequence of states which extends infinitely towards the future. We call a constant of FTL rigid if its value does not vary all over each sequence of states; otherwise a constant is non-rigid. Thus, in a given language of FTL each constant is rigid or non rigid. We assume that for each non-rigid constant, say a, there exists an infinite sequence of rigid constants $a_0, \dots, a_i, \dots$ such that values of a and a_i coincide at the i-th state $i \in w$. We call a_i the i-th *constant associated with* a. The formula $\bigcirc^i(a = a_i)$ holds in FLT for a and a_i.

Formally, standard semantics is defined as follows. A standard structure N is a quadruple $< D,\ w,\ ',\ \leqslant >$, where

- D is a nonempty set of individuals;
- $'$ is the successor function on ω;
- $\leqslant$ is the usual order relation on ω.

A model M is a triple $< N,\ I,\ \alpha >$, where

- N is a standard structure;
- α is an assignment of values from D to variables, i.e. $\alpha(x) \in D$ for each variable x;
- I is an interpretation of relation and constant symbols defined as follows:

for each $n \in w$ and each relation symbol p of arity m, $I(n, p) \in \underbrace{D \times \cdots \times D}_{m}$;

for each $n \in w$ and each constant symbol a $I(n, a) \in D$ and the following condition is satisfied:

- if a is a non-rigid constant and $a_0, \dots, a_i, \dots$ is a sequence of rigid constants associated with a, then

$$I(i, a) = I(i, a_i), \quad i \in \omega.$$

We define *the evaluation function* τ for terms, which evaluates terms in a model M at a time moment $n \in \omega$ as follows:

- for a variable x $\tau(M,n,x) = \alpha(x)$;
- for a constant symbol a $\tau(M,n,a) = I(n,a)$.

If $d \in D$, then $M(x \leftarrow d)$ denotes the model obtained from M by modifying its assignment function α to map the variable x to d.

We inductively define the satisfaction relation "a formula p is true in a model $M = (N, I, \alpha)$ at a time moment $n \in \omega$" as follows:

• for atomic formulas:
$M,n \models R(t_1,\ldots,t_m)$ iff $(\tau(M,n,t_1),\ldots,\tau(M,n,t_m)) \in I(n,R)$;
$M,n \models t_1 = t_2$ iff $\tau(M,n,t_1) = \tau(M,n,t_2)$;
• for logical connectives:
$M,n \models \neg p$ iff not $M,n \models p$;
$M,n \models p \wedge q$ iff $M,n \models p$ and $M,n \models q$;
$M,n \models p \vee q$ iff $M,n \models p$ or $M,n \models q$;
$M,n \models p \supset q$ iff not $M,n \models p$ or $M,n \models q$;
• for quantifiers:
$M,n \models \forall xp$ iff for each $d \in D$ $M(x \leftarrow d), n \models p$;
$M,n \models \exists xp$ iff there exists $d \in D$ $M(x \leftarrow d), n \models p$;

for temporal modalities:
$M,n \models \circ p$ iff $M, n+1 \models p$
$M,n \models \Box p$ iff $M, n+m \models p$ for each $m \in \omega$;
$M,n \models \Diamond p$ iff $M, n+m \models p$ for some $m \in \omega$.

We say that p is true in a model M if $M, 0 \models p$. A formula p is universally valid if p is true in each model.

3. A sequent calculus LFTL

We use capital Greek letters Γ, Δ to denote finite (possibly empty) sets of formulas of FTL. An expression of the form $\Gamma \rightarrow \Delta$ is called a seguent. Let it be $\Gamma = \{p_1,\ldots,p_m\}$, $\Delta = \{q_1,\ldots,q_n\}$. A sequent $\Gamma \rightarrow \Delta$ is true in a model M, if there exists $p_i \in \Gamma$, which is false in a model M or there exists $q_i \in \Delta$, which is true in a model M. A sequent $\Gamma \rightarrow \Delta$ is universally valid if $\Gamma \rightarrow \Delta$ is true in each model.

If p is a formula, r and t are terms, then $p(t/r)$ denotes the result of substituting t in p for each occurrence of r.

LFTL has the following *axioms*:
$\Gamma \rightarrow \Delta$, where $\Gamma \cap \Delta \neq \emptyset$;
$\Gamma \rightarrow \Delta,\ \circ^k(t = t),\ k \in \omega$.

Rules of inference of LFTL. The rules of inference of LFTL are divided into three groups: rules for logical connectives and quantifiers, rules for temporal modalities and rules for equality. All the rules except the rules for equality are analogous to the corresponding rules in [5].

3.1 *rules for logical connectives and quantifiers:*

$$\frac{\bigcirc^n p, \Gamma \to \Delta}{\Gamma \to \Delta, \bigcirc^n \neg p} \ (\neg(n)\text{right}) \qquad \frac{\Gamma \to \Delta, \bigcirc^n p}{\bigcirc^n \neg p, \Gamma \to \Delta} \ (\neg(n)\text{left});$$

$$\frac{\Gamma \to \Delta, \bigcirc^n p; \Gamma \to \Delta, \bigcirc^n q}{\Gamma \to \Delta, \bigcirc^n (p \wedge q)} \ (\wedge(n)\text{right}) \qquad \frac{\bigcirc^n p, \bigcirc^n q, \Gamma \to \Delta}{\bigcirc^n (p \wedge q), \Gamma \to \Delta} \ (\wedge(n)\text{left});$$

$$\frac{\Gamma \to \Delta, \bigcirc^n p, \bigcirc^n q}{\Gamma \to \Delta, \bigcirc^n p \vee \bigcirc^n q} \ (\vee(n)\text{right}) \qquad \frac{\bigcirc^n p, \Gamma \to \Delta; \bigcirc^n q, \Gamma \to \Delta}{\bigcirc^n (p \vee q), \Gamma \to \Delta} \ (\vee(n)\text{left});$$

$$\frac{\bigcirc^n p, \Gamma \to \Delta, \bigcirc^n q}{\Gamma \to \Delta, \bigcirc^n (p \supset q)} \ (\supset (n)\text{right}) \qquad \frac{\Gamma \to \Delta, \bigcirc^n p; \bigcirc^n q, \Gamma \to \Delta}{\bigcirc^n (p \supset q), \Gamma \to \Delta} \ (\supset (n)\text{left});$$

$$\frac{\Gamma \to \Delta, \bigcirc^n p(y/x)}{\Gamma \to \Delta, \bigcirc^n \forall x p} \ (\forall(n)\text{right}) \qquad \frac{\bigcirc^n p(t/x), \bigcirc^n \forall x p, \Gamma \to \Delta}{\bigcirc^n \forall x p, \Gamma \to \Delta} \ (\forall(n)\text{left});$$

$$\frac{\Gamma \to \Delta, \bigcirc^n \exists x p, \bigcirc^n p(t/x)}{\Gamma \to \Delta, \bigcirc^n \exists x p} \ (\exists(n)\text{right}) \qquad \frac{\bigcirc^n p(y/x), \Gamma \to \Delta}{\bigcirc^n \exists x p, \Gamma \to \Delta} \ (\exists(n)\text{left});$$

with the following restrictions:

- for ($\forall(n)$ right) and ($\exists(n)$ left) the variable y does not occur in the lower sequent;
- for ($\forall(n)$ left) and ($\exists(n)$ right) the substitution (t/x) does not create any new bound occurrence of variable and any new occurrence of non-rigid constant in the scope of temporal modalities.

3.2 *rules for temporal modalities:*

$$\frac{\{\Gamma \to \Delta, \bigcirc^{n+m} p\} m \in \omega}{\Gamma \to \Delta, \bigcirc^n \Box p} \ (\Box(n)\text{right}) \qquad \frac{\bigcirc^{n+m} p, \bigcirc^n \Box p, \Gamma \to \Delta}{\bigcirc^n \Box p, \Gamma \to \Delta} \ (\Box(n)\text{left});$$

$$\frac{\Gamma \to \Delta, \bigcirc^n \Diamond p, \bigcirc^{n+m} p}{\Gamma \to \Delta, \bigcirc^n \Diamond p} \ (\Diamond(n)\text{right}) \qquad \frac{\{\bigcirc^{n+m} p, \Gamma \to \Delta\} m \in \omega}{\bigcirc^n \Diamond p, \Gamma \to \Delta} \ (\Diamond(n)\text{left}).$$

3.3 *rules for equality:*
3.3.1

$$\frac{\bigcirc^n (s = r), \Gamma \to \Delta}{\bigcirc^n (r = s), \Gamma \to \Delta}$$

$$\frac{\bigcirc^n p(s/r), \bigcirc^n p, \bigcirc^n (r = s), \Gamma \to \Delta}{\bigcirc^n p, \bigcirc^n (r = s), \Gamma \to \Delta} \ (= (n)\text{left})$$

$$\frac{\bigcirc^n (r = s), \Gamma \to \Delta, \bigcirc^n p, \bigcirc^n p(s/r)}{\bigcirc^n (r = s), \Gamma \to \Delta, \bigcirc^n p} \ (= (n)\text{right})$$

$$\frac{\bigcirc^m(r=s), \bigcirc^n(r=s), \Gamma \to \Delta}{\bigcirc^n(r=s), \Gamma \to \Delta} \ (=(n)m), \ m \in \omega;$$

with the restriction: p is an atomic formula, r, s are not non-rigid constants.

3.3.2

$$\frac{\bigcirc^n p(s/r), \Gamma \to \Delta}{\bigcirc^n p, \Gamma \to \Delta} \ (=(n)(\text{non}-\text{rigid})\text{left});$$

$$\frac{\Gamma \to \Delta, \bigcirc^n p(s/r)}{\Gamma \to \Delta, \bigcirc^n p} \ (=(n)(\text{non}-\text{rigid})\text{right})$$

with the following restriction: p is an atomic formula, r is a non-rigid constant and s is the n-th rigid constant associated with r.

Note: It can be verified by analysis of the proof of completeness theorem that it is sufficient to have the simplified inference rules

$$\frac{\bigcirc^n p(s/r), \bigcirc^n(r=s), \Gamma \to \Delta}{\bigcirc^n p, \bigcirc^n(r=s), \Gamma \to \Delta}$$

$$\frac{\bigcirc^n(r=s), \Gamma \to \Delta, \bigcirc^n p(s/r)}{\bigcirc^n(r=s), \Gamma \to \Delta, \bigcirc^n p}$$

instead of the rules $(=(n)\text{left})$, $(=(n)\text{right})$ for the case p is of the form $r=t$.

Theorem 1. (soundness of LFTL). If a sequent $\Gamma \to \Delta$ is provable in LFTL, then $\Gamma \to \Delta$ is universally valid.

Proof is carried out by induction on the depth of the sequent in the proof and it is omitted here.

Let p be an atomic formula and let a be a non-rigid constant. Let φ be the formula

$$\forall x \Diamond(a=x) \wedge p(a) \wedge \forall x \forall y \ \varphi_0(x,y) \supset \forall x p(x),$$

where $\varphi_0(x,y)$ denotes $p(x) \wedge \Diamond(a = x \wedge \bigcirc(a = y)) \supset p(y)$. We give a proof of the sequent $\to \varphi$ to illustrate the utility of associated constants. φ is taken from [1] where it is shown that φ is not provable in the finite axiomatic system proposed in [1]. To prove φ in [1] an auxilliary definition of a *new relation symbol* not occuring in φ is used. On the contrary, we have not "to guess" any new definitions as the proof of $\to \varphi$ below is cut-free.

To prove $\to \varphi$ it is sufficient to prove the sequent

$$\forall x \Diamond(a=x), \ p(a), \ \forall x \forall y \varphi_0(x,y) \to \forall x p(x). \tag{1}$$

Let z be a variable not occurring in φ. The figure (I) below shows that to prove (1) it is sufficient to prove the sequent

$$\bigcirc^n(a=z), \ p(a), \ \forall x \forall y \ \varphi_0(x,y) \to p(z), \tag{2_n}$$

for $n \in \omega$.

(I):

$$\cfrac{\cfrac{\cfrac{\{\bigcirc^n(a=z),\ \Gamma \rightarrow p(z)\} n \in w}{\Diamond(a=z),\ \Gamma \rightarrow p(z)}\ (\Diamond(0)\ \text{left})}{\forall x \Diamond(a=x),\ \Gamma \rightarrow p(z)}\ (\forall(0)\ \text{left})}{\forall x \Diamond(a=x),\ \Gamma \rightarrow \forall x p(x)}\ (\forall(0)\ \text{right})$$

where Γ denotes $\{p(a),\ \forall x \forall y\ \varphi_0(x,y)\}$.

The figure (II) below shows that to prove (2_n) for fixed n it is sufficient to prove

$$\bigcirc^0(a_n=z), \bigcirc^n(a_n=z), p(a_0),\ \forall x \forall y\ \varphi_0(x,y) \rightarrow p(a_n), p(z) \tag{3_n}$$

where a_n is the n-th rigid constant associated with a, $n \in w$.

(II):

$$\cfrac{\cfrac{\cfrac{\cfrac{\bigcirc^0(a_n=z), \bigcirc^n(a_n=z),\ p(a_0),\ \gamma \rightarrow p(a_n), p(z)}{\bigcirc^0(a_n=z), \bigcirc^n(a_n=z),\ p(a),\ \gamma \rightarrow p(a_n), p(z)}\ (=(0)\ (\text{non}-\text{rigid})\ \text{left})}{\bigcirc^0(a_n=z), \bigcirc^n(a_n=z),\ p(a),\ \gamma \rightarrow p(z)}\ (=(0)\ \text{right})}{\bigcirc^n(a_n=z),\ p(a),\ \gamma \rightarrow p(z)}\ (=(n)0)}{\bigcirc^n(a=z),\ p(a),\ \gamma \rightarrow p(z)}\ (=(n)\ (\text{non}-\text{rigid})\ \text{left})$$

where γ denotes $\forall x \forall y\ \varphi_0(x,y)$.

Let Γ_1 be an arbitrary set of formulas. We prove

$$\Gamma_1,\ p(a_0),\ \forall x \forall y \varphi_0(x,y) \rightarrow p(a_n), p(z) \tag{4_n}$$

for $n \geq 0$. Provability of (3_n) follows from provability of (4_n).

If $n = 0$, then (4_n) is an axiom of LFTL.

Let it be $n > 0$. Let Γ_2 be the sequent $\Gamma_1 \cup \{p(a_0),\ \forall x \forall y \varphi_0(x,y).\}$ The figure (III) below shows that to prove (4_n) it is sufficient to prove (4_{n-1}) and

$$\Gamma_2 \rightarrow \Diamond((a = a_{n-1} \wedge \bigcirc(a = a_n)) \tag{5_n}$$

(III):

$$\cfrac{\cfrac{\cfrac{\cfrac{\Gamma_2 \rightarrow p(a_{n-1});\ \Gamma_2 \rightarrow \Diamond((a=a_{n-1}) \wedge \bigcirc(a=a_n))}{\Gamma_2 \rightarrow p(a_{n-1}) \wedge \Diamond((a=a_{n-1}) \wedge \bigcirc(a=a_n));\ \Gamma_2,\ p(a_n) \rightarrow p(a_n), p(z)}\ (\wedge(0)\ \text{right})}{\Gamma_2,\ \varphi_0(a_{n-1}/x, a_n/y) \rightarrow p(a_n), p(z)}\ (\supset(0)\ \text{left})}{\Gamma_2, \forall y \varphi_0(a_{n-1}/x, y) \rightarrow p(a_n), p(z)}\ (\forall(0)\ \text{left})}{\Gamma_2, \forall x \forall y \varphi_0(x,y) \rightarrow p(a_n), p(z)}\ (\forall(0)\ \text{left})$$

Thus, to end the proof of (4_n) and also the proof of $\rightarrow \varphi$ we have to prove (5_n). The proof of (5_n) is as follows.

$$\dfrac{\dfrac{\dfrac{\Gamma_2 \to \bigcirc^{n-1}(a_{n-1} = a_{n-1})}{\Gamma_2 \to \bigcirc^{n-1}(a = a_{n-1});}\,(=(n-1)\,(\text{non-rigid})\text{right}) \qquad \dfrac{\Gamma_2 \to \bigcirc^{n}(a_n = a_n)}{\Gamma_2 \to \bigcirc^{n-1}\bigcirc(a = a_n)}\,(=(n)(\text{non-rigid})\ \text{right})}{\Gamma_2 \to \bigcirc^{n-1}((a = a_{n-1}) \wedge \bigcirc(a = a_n))}\,(\wedge(n-1)\text{right})}{\Gamma_2 \to \Diamond((a = a_{n-1}) \wedge \bigcirc(a = a_n))}\,(\Diamond(0)\text{right})$$

This ends the proof of $\to \varphi$ in LFTL.

4. The proof of completeness of LFTL

Theorem 2. (completeness theorem). If a sequent S is universally valid, then S is LFTL provable.

The completeness theorem follows directly from the following:

let S be a sequent which is not provable in LFTL, then there is a model M in which S is not true.

To construct M a canonical reduction tree T_S for S is used. T_S is constructed inductively beginning from S and applying the rules of inference in a bottom up fashion similarly to the case of first order predicate logic (e.g. [7]). We construct T_S analogously to the case of first order temporal logic without equality, in [5].

T_S is constructed as follows.

Stage 0: We write down S at the root of the tree.

Stage k :

Case 1. Each topmost sequent is an axiom of LFTL. Then procedure stops.

Case 2. Otherwise. This case is divided into $k \equiv 1\ (mod\ 21), \ldots, k \equiv 20\ (mod\ 21)$. $k \equiv 1, 2$ concern to formulas $\bigcirc^n \neg p$; $k \equiv 3, 4$ concern to formulas $\bigcirc^n(p \wedge q)$; $k \equiv 5, 6$ concern to formulas $\bigcirc^n(p \vee q)$; $k \equiv 7, 8$ concern to formulas $\bigcirc^n \forall x p$; $k \equiv 9, 10$ concern to formulas $\bigcirc^n \exists x p$; $k \equiv 11, 12$ concern to formulas $\bigcirc^n \Box p$; $k \equiv 13, 14$ concern to formulas $\bigcirc^n \Diamond p$. These subcases are defined as in [5] and it is omitted here.

$k \equiv 15$ concern to formulas $\bigcirc^n(r = s)$; $k \equiv 16, 17$ concern to formulas $\bigcirc^n p$, where p is an atomic formula; $k \equiv 18$ concern to formulas $\bigcirc^n(r = s)$, where r, s are not non-rigid constants; $k \equiv 19, 20$ concern to $\bigcirc^n p$, where p is an atomic formula, which has an occurrence of a non-rigid constant. These subcases are defined as follows.

Let $\Gamma \to \Delta$ be any topmost sequent obtained at stage $k - 1$ such that $\Gamma \to \Delta$ is not an axiom of LFTL.

$k \equiv 15$: Let $\bigcirc^{n_1}(r_1 = s_1), \ldots, \bigcirc^{n_m}(r_m = s_m)$ be all the formulas such that $\bigcirc^{n_i}(r_i = s_i) \in \Gamma$. and r_i, s_i are not non-rigid constants, $1 \leqslant i \leqslant m$. Then write down

$$\bigcirc^{n_1}(s_1 = r_1), \ldots, \bigcirc^{n_m}(s_m = r_m), \quad \Gamma \to \Delta$$

above $\Gamma \to \Delta$.

$k \equiv 16$: Let $\bigcirc^{n_1}p_1, \bigcirc^{n_1}(r_1 = s_1), \ldots, \bigcirc^{n_m}p_m, \bigcirc^{n_m}(r_m = s_m)$ be all the pairs such that $\bigcirc^{n_i}p_i \in \Gamma$, p_i is an atomic formula, r_i occurs in p_i and $\bigcirc^{n}(r_i = s_i) \in \Gamma$, r_i, s_i are not non-rigid constants $1 \leqslant i \leqslant m$. We write down

$$\bigcirc^{n}p_1(s_1/r_1), \ldots, \bigcirc^{n}(s_m/r_m), \quad \Gamma \to \Delta$$

above $\Gamma \to \Delta$.

$k \equiv 17$: Let $\bigcirc^{n_1}p_1, \bigcirc^{n_1}(r_1 = s_1), \ldots, \bigcirc^{n_m}p_m, \bigcirc^{n_m}(r_m = s_m)$ be all the pairs such that $\bigcirc^{n_i}p_i \in \Delta$, p_i is an atomic formula, r_i occurs in p_i, $\bigcirc^{n_i}(r_i = s_i) \in \Delta$, r_i, s_i are not non-rigid constants $1 \leqslant i \leqslant m$. We write down

$$\Gamma \to \Delta, \; p_1(s_1/r_1), \ldots, p_m(s_m/r_m)$$

above $\Gamma \to \Delta$.

$k \equiv 18$: Let $\bigcirc^{n_1}(r_1 = s_1), \ldots, \bigcirc^{n_m}(r_m = s_m)$ be all the formulas of the form $\bigcirc^{n}(r = s)$ in Γ, where r, s are not non-rigid constants. Let Γ' be the result of omitting $\bigcirc^{n_i}(r_i = s_i)$, $1 \leqslant i \leqslant m$ in Γ. We write down

$$\bigcirc^{0}(r_1 = s_1), \ldots, \bigcirc^{k}(r_1 = s_1), \ldots, \bigcirc^{0}(r_m = s_m), \ldots, \bigcirc^{k}(r_m = s_m), \quad \Gamma' \to \Delta$$

above $\Gamma \to \Delta$.

$k \equiv 19$: Let $\bigcirc^{n_1}p_1, \ldots, \bigcirc^{n_m}p_m$ be all the formulas in Γ such that p_i is an atomic formula which has an occurrence of a non-rigid constant r_i, $1 \leqslant i \leqslant m$. Let s_i be the n-th rigid constant associated with r_i. Let Γ' be the result of omitting $p_1, \ldots, p_m$ in Γ. Then we write down

$$p_1(s_1/r_1), \ldots, p_m(s_m/r_m), \quad \Gamma' \to \Delta$$

above $\Gamma \to \Delta$.

$k \equiv 20$ is defined simetrically to $k = 19$ considering Δ instead of Γ.

Let T_S be a reduction tree. As in the case of first order predicate logic we have two cases: i) each branch of T_S is finite and the topmost sequent of it is an axiom or ii) there is an infinite branch W such that all sequents in W are not axioms. If i) holds then we obtain the proof of S (in LFTL) from T_S. This contradicts our assumption that S is not provable in LFTL. Thus, ii) holds.

A model M is obtained from W in a similar way as in the case of first order predicate logic with equality in [3]. By W_l and W_r we denote all the formulas which occur in antecendents of sequents and succedents of sequents from W, respectively.

By the construction of T_S it follows that W_l, W_r satisfy the following properties:

Property 1: let r, s be not non-rigid constants, if $\bigcirc^{i}(r = s) \in W_l$ for some $i \in \omega$ then $\bigcirc^{j}(r = s) \in W_l$ for each $j \in \omega$.

Property 2: if $\bigcirc^{i}(r = s) \in W_l$, then $\bigcirc^{i}(s = r) \in W_l$ and if $\bigcirc^{i}(r = s) \in W_l$ and $\bigcirc^{i}(s = t) \in W_l$, then $\bigcirc^{i}(r = t) \in W_l$.

Property 3: If $\bigcirc^n R(t_1,\ldots,t_m) \in W_l(W_r)$ for a m-ary relation symbol R and $s \equiv t$ for $s,t \in H$, then there exists $\Gamma \to \Delta$ from W such that $\bigcirc^n R(t_1,\ldots,t_m) \in \Gamma(\Delta)$ and $\bigcirc^n(s=t) \in \Gamma(\Delta)$.

Let H be all free variables and rigid constants from W. We define a binary relation $\underset{n}{=}$ on H as follows:

$$t \underset{n}{=} t \qquad \text{for each } t \in H$$

and

$$t \underset{n}{=} s, \qquad \text{if } \bigcirc^n(t=s) \in W_l.$$

We define a binary relation $\equiv$ on H as follows:

$$t \equiv s \text{ iff for each } n \in \omega \;\; t \underset{n}{=} s.$$

It can be verified using properties 1, 2 that $\equiv$ is an equivalence relation.

It can be shown using property 3 and construction of T_S

Property 4. Let p be an atomic formula, s be a term occurring in p and $\bigcirc^n p \in W_r(W_l)$.
(i) If p is of the form $R(t_1,\ldots,t_m)$ and $s \in H$, then $\bigcirc^n p(t/s) \in W_l(W_r)$ for each $t \equiv s$, $t \in H$.
(ii) If $s \notin H$, then $\bigcirc^n p(t/s) \in W_l(W_r)$, where t is the n-th rigid constant associated with s.

We define the grade $g(p)$ of a formula p as follows:

1) for each formula $\bigcirc^n p$, where p is an atomic formula $g(p)=0$,

2)

$$g(\bigcirc^n \neg p) = g(p)+1,$$
$$g(\bigcirc^n(p \wedge q)) = g(\bigcirc^n(p \vee q)) = g(\bigcirc^n(p \supset q)) = g(p)+g(q)+1,$$
$$g(\bigcirc^n \forall x p) = g(\bigcirc^n \exists x p) = g(p)+1,$$
$$g(\bigcirc^n \Box p) = g(\bigcirc^n \Diamond p) = g(p)+1.$$

Let D be the set of equivalence classes of terms from H modulo $\equiv$. We put D as a set of individuals of our model M. Let t be a term from H, then $\bar{t}$ denotes an equivalence class from D such that $t \in \bar{t}$.

We define the assignment α and interpretation I in our model M as follows:

- $\alpha(x) = \overline{x}$, where x is a variable from H;
- $I(n,a) = \begin{cases} \overline{a}, & \text{if } a \text{ is a rigid constant from } H; \\ \overline{a}_n, & \text{if } a \text{ is non}-\text{rigid constant and} \\ & a_n \text{ is the } n-\text{th} \\ & \text{constant associated with } a; \end{cases}$
- $(\bar{t}_1,\ldots,\bar{t}_m,n) \in I(n,R)$ iff
  $\bigcirc^n R(t_1,\ldots,t_m) \in W_l$ or $(\bigcirc^n R(t_1,\ldots,t_m) \notin W_l$ and $\bigcirc^n R(t_1,\ldots,t_m) \notin W_r)$.

It can be verified using (i) of property 4 that each relation in M is well defined, i.e., that the definition of the $I(n, R)$ doesn't depend on the particular choice of represantatives $t_1, \ldots, t_m$ in the equivalence classes of terms (in H).

We have the following lemma, from which it follows that M is a model in which the sequent S is not true.

Lemma Let φ be any formula from W. If $\varphi \in W_l$, then $M, 0 \models \varphi$; if $\varphi \in W_r$, then not $M, 0 \models \varphi$.

Proof is made by induction on the grade $g(\varphi)$. Let it be $g(\varphi) = 0$. If φ is of the form $R(t_1, \ldots, t_m)$, t_i is a term, $1 \leqslant i \leqslant m$, then the lemma follows using (ii) of property 4 by definition of I and the fact that W doesn't contain axioms; if φ is of the form $\circ^n(t_1 = t_2)$, then the lemma follows using (ii) of property 4 by definition of $\equiv$, property 2 and the mentioned fact.

The proof of the step of induction is based on the fact that W contains subformulas of the formulas which occur in W. It is straightforward.

REFERENCES

1. M.Abadi, The power of temporal proofs, Theoret. Comput. Sci. 65 (1989) 35–83.
2. D.Gabbay, Modal and temporal logic programming In: Temporal logics and their applications (ed. A.Galton) (Academic Press, 1987) 197–237.
3. J.H.Gallier, Logic for computer science (Harper & Row, Publishers, New York, 1986).
4. E.Harel, O. Lichtenstein, A.Pnueli, Explicit clock temporal logic, Proc. fifth Annual IEEE Symp. on Logic in Comput. Sci., June 4–7, 1990, 402–413.
5. H.Kawai, Sequential calculus for a first order infinitary temporal logic, Zeitschr. f. math. Logic und Grundlagen d.Math. 33 (1987), 423–432.
6. R.Pliuškevičius, Investigation of finitary calculus for a discrete linear time logic by means of finitary calculus, LNCS 50 2 (1991) 504–528.
7. G. Takeuti, Proof theory (North-Holland, 1975).

A logical-based language for feature specification and transmission control

P. Sébillot
Irisa, Campus de Beaulieu, F-35042 Rennes Cédex, France
e-mail: sebillot@irisa.fr

Abstract

In this paper, we present a logic-programming modelling of *feature filtering mechanisms.* Feature filtering is an important topic in natural language processing. We give a logical-based specification, valid for different theories of features (including their evolution from the lexicon, expressed in terms of axioms, and including control tools). We define the logical foundations of these specifications, which have been shown useful for natural-language grammar programmers. The logical-based language and our formalization are very general and independent from any parsing strategy. They permit to represent feature and control systems proposed by different linguistic theories.

Area: logic programming, natural language processing

1. INTRODUCTION

Our long-term aim is to develop effective syntactic parsers for natural language written texts. In order to obtain a general and clear understanding of these texts, we study different linguistic theories and we try to transfer their results into the domain of automatic natural language processing. Therefore, our work consists in modelling specific linguistic problems and translating these studies into their computational representation. We select *logic* as a formal means to achieve this linguistic-to-computer-science transfer because logic exhibits a clear and precise semantics and permits a description of the different aspects of a selected linguistic model, i.e. the lexicon, the syntactic rules for parsers, ... Moreover, through *logic programming*, such a logical-oriented approach provides us with computational means, also.

Here we focus on the way the different linguistic schools *represent* phonologic, morphologic, syntactic and semantic *knowledge* in lexicons of natural language parsers and on the way they describe the propagation and the control of this information in grammar rules. Most linguistic theories use a kind of *feature* notation to represent the knowledge that is associated with the words in the lexicons. A feature is a distinctive and relevant mark [1], often called attribute, with which a value is associated (e.g. gender: masculine). Features are often viewed as attribute-value pairs [2], but most current linguistic theories use *complex features* and construct a hierarchy of features. Features represent characteristics that are specific to each word while giving a precise description of the environments allowed for this word. In addition to this *descriptive role*, they are used to *restrict the applicability* of grammar rules containing them - we shall refer to that latter role as *feature filtering* - and they must be propagated from the words in the lexicon into the rules. They are used, for example, to treat agreement or sub-categorization problems.

In this paper, we present a modelling, within a logic context, of feature representation and of the management of these features. Let us emphasize that we do not propose a logical-based expression of a new general theory of features, but we rather give an advanced logical-based specification of features and of their propagation in rules, valid for different theories. This specification permits to treat *feature inheritance* phenomena, from the lexicon and from rules to rules, to express the computation and the control of feature values and the transfer of features between the right and the left parts of rules. We propose a *set of control tools*, relevant in a logic programming context, in order to build the elements of a *linguistic environment*, intended to help a grammar expert to describe the useful features for a given application and to manage their evolution.

The main advantages of our approach over the previous published ones lie in that it is simpler, more modular and declarative. Moreover, our approach benefits by the clear semantics of logic programming languages. We represent features in a simple and general way (we use a unique complex feature in which all the information required to address an application appears). Our solution to express features and constraints operating on them is not bound to a specific language or linguistic theory. The primitives defined in this way will be inserted into natural language parsers, which are also modelled in terms of logic programs. Note that the resulting global description will also be independent of any parsing strategy.

The structure of the paper is as follows: first, we present the logical-based language for feature expression and control that we propose; then we focus on the semantics of its elements.

2. THE LOGICAL-BASED LANGUAGE

Let us describe the structure we choose to represent feature description. We present the basic operation on this structure: *the augmented unification*. We define the elements of the language that permit to express feature transmission and control.

2.1. Feature representation

We present the notion of *feature descriptor*, a structure adapted from [3], whose generality seems quite relevant to our problem. We describe the different allowed feature values in such a structure.

2.1.1 Feature descriptor

Every feature formalism or theory works on quite a large set of features. In applications yet, only a subset of these features is used. It is possible to represent this subset by a tree (or a hierarchy). The basic idea to express features by a feature descriptor is to determine a feature tree and the allowed values for those features in a given application - P. Saint-Dizier [3] calls it a *prototype of feature descriptor* - and to affect a more or less instanciated version of this prototype, called a *feature descriptor*, to each terminal and non-terminal symbol of the grammar. The prototype of feature descriptor depends on the considered language and on the degree of precision we need. Once defined, it is fixed for all the elements of the lexicon and the symbols of the rules of the grammar (it permits global assignments). From a computational point of view, it can be considered as an abstract data type.

More formally:

A prototype of feature descriptor

Definition 1: A *feature value* is an element of a finite Set of the Allowed Values (SAV) for that feature.

Definition 2: A *preterminal feature* in a tree is a feature whose immediate daughters are feature values.

Definition 3: A *prototype of feature descriptor* is a tree whose root node is F (feature), in which each node is labelled with a feature and where each preterminal node is bound to a set of its own allowed values (SAV). Figure 1 shows an example of such a prototype for a simple English application. We just consider morpho-syntactic features in all the examples we present, even if our model can deal with other feature types.

A feature descriptor

The descriptor assigned to each symbol of the grammar contains "precise" values for each feature.

We explain now what the allowed values are.

2.1.2 Feature values

We have tried to get an expression of the values which verify two types of properties. It must be easy to use and precise; moreover it must be general enough to permit the expression of principles of different linguistic theories.

We have chosen 5 possible values:

an atomic value val: val must be one of the elements of the SAV of the considered feature. Only this value is possible.

a constant av: the feature may take All the Values of his SAV. All the values are "equi-possible".

a disjunction of values OR(val_1, ..., val_n): each val_i, $i \in [1,n]$ must be in the SAV of the feature and only these val_i are possible.

a negation of values NOT(val_1, ..., val_m): each val_i, i $\in [1,m]$, must be in the SAV of the feature. It is equivalent to the disjunction of the other values of the SAV, i.e. to the complement of $\{val_1, ..., val_m\}$ in the SAV of the feature.

a logical variable X or _: it means that the feature is not relevant to the considered phenomenon or that the feature value is unknown or does not present any interest now.

Figure 2 shows an example of a feature descriptor which may be considered as a type instance of the prototype of feature descriptor of figure 1.

2.2 Augmented unification mechanism of two feature descriptors

The basic operation on two descriptors is the *augmented unification* which builds a resulting descriptor from two descriptors. The augmented unification of two descriptors is the unification of the feature values. For atomic values and logical variables, the unification mechanism is quite simple and usual. We must yet explain the way it works for disjunction and negation of values (this is why we name it an augmented unification or an *A-unification*). First, we present the formal semantics of this operation and then we explain how it works on descriptors. This section uses notions and definitions that are

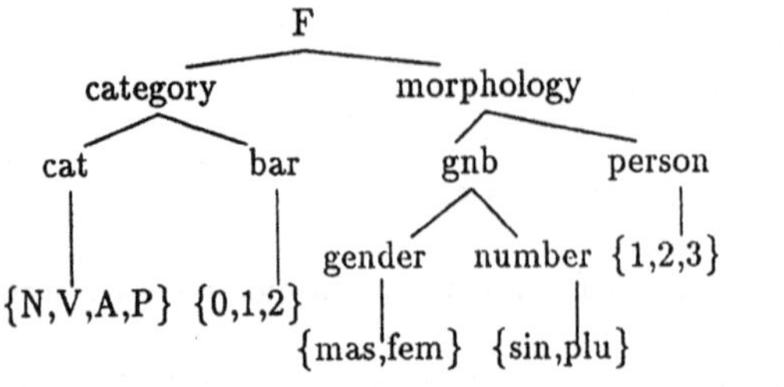

Figure 1: Example of a prototype of feature descriptor

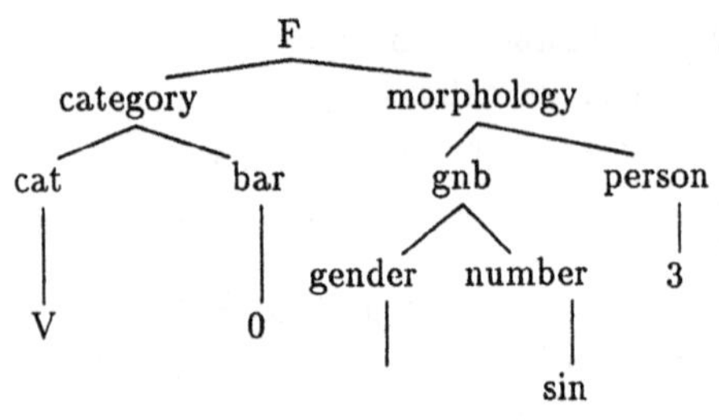

Figure 2: Example of a feature descriptor of the word "walks"

explained in [4], [5] or [6].

2.2.1 Formal semantics of the A-unification of feature values

Traditional unification algorithms address simple expression sets (atoms or terms) and allow one to determine a unique mgu (i.e. most general unifier) of these expressions when it exists. It seems quite interesting that our definition of A-unification preserves this unicity for the calculated unifier. We first present the A-unification mechanism on values of features.

We are working in closed domains; therefore a feature has only a finite set of possible values (shown in the prototype). Let *t be a feature and E his finite SAV.*

We can use the following notations and interpretations for the values.

Values in the descriptor	Set of considered values
t(v_m)	$\{v_m\}$
t(OR(v_i, ..., v_k))	$\{v_i, \ldots, v_k\}$
t(NOT(v_j, ..., v_l))	$C_E^{\{v_j,\ldots,v_l\}}$ (complement)
t(av), t(X), t(_)	E

We present the augmented unification of two sets of feature values in terms of a calculation of the greatest lower bound (glb) of these two sets under the partial order $\subseteq$.

Let us give a few definitions.

Definition 1: A relation R on a set S is a *partial order* if the following conditions are satisfied:

(a) xRx, for all x $\in$ S.

(b) xRy and yRx imply x=y, for all x, y $\in$ S.

(c) xRy and yRz imply xRz, for all x, y, z $\in$ S.

We note $\leq$ a partial order.

Definition 2: Let S be a set with a partial order $\leq$. Then a $\in$ S is an *upper bound* of a subset X of S if x $\leq$ a, for all x $\in$ X. Similarly, b $\in$ S is a *lower bound* of X if b $\leq$x, for all x $\in$ X.

Definition 3: Let S be a set with a partial order $\leq$. Then a $\in$ S is the *least upper bound* of a subset X of S if a is an upper bound of X and, for all upper bounds a' of X, we have a $\leq$ a'. Similarly, b $\in$ S is the *greatest lower bound* of a subset X of S if b is a lower bound of X and, for all lower bounds b' of X, we have $b' \leq$ b.

The least upper bound of X is unique, if it exists, and is denoted by lub(X). Similarly, the greatest lower bound of X is unique, if it exists, and is denoted by glb(X).

Definition 4: A partially ordered set L is a *complete lattice* if lub(X) and glb(X) exist for every subset X of L.

Let us come back to our specific case where E is the SAV of the feature t. Let 2^E be the set of all subsets of E; then set inclusion $\subseteq$ is a *partial order* on 2^E. 2^E under $\subseteq$ is a *complete lattice.* In fact, the least upper bound of a collection of subsets of E is their union and the greatest lower bound (glb) is their intersection.

We choose to represent A-unification by this lattice under $\subseteq$ and the calculated general a-unifier of a set of subsets of E is the glb of this set.

Consequence: The calculated general a-unifier is *unique.*

Examples:

Let us consider the feature person which has three possible values: 1, 2, 3.

Example 1: Determination of the a-unifier of the 4 following feature subdescriptors: person(X), person(OR(1,3)), person(NOT(3)) and person(1).

Notation: X is represented by {1,2,3}, OR(1,3) by {1,3}, NOT(3) by {1,2} and 1 by {1}.

The following figure represents the lattice :

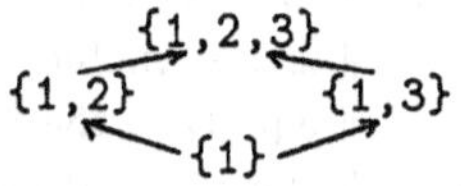

```
           {1,2,3}
{1,2}                {1,3}
           {1}
```

The glb is {1} and is the general a-unifier (in fact, person(1)) for the 4 subdescriptors.

Example 2: Let us now consider person(1) and person(OR(2,3)).

The lattice is {1} {2,3}. The intersection is $\emptyset$ = glb({1},{2,3}). It is viewed as an A-unification failure.

2.2.2. Extension to the A-unification of two feature descriptors

The unification of each set of feature values of two descriptors may be modelled by a complete lattice of the set of these values under the partial order $\subseteq$. The *unification of two descriptors* may be modelled by a *cartesian product* of complete lattices under $\subseteq$. According to well-known results from lattice theory, the cartesian product of complete lattices is a *complete lattice.*

2.2.3 What "really" happens during the A-unification of two descriptors

Let D_i and D_j be two feature descriptors. We give, for the same feature, the condition of success for the A-unification and the result. These results come from the semantics of the elements presented in the previous sections. Let E be the SAV of the feature.

Feature value in D_i	Feature value in D_j	Condition	Result
atomic value val1	atomic value val2	val1 = val2	the value val1
atomic value val1	disjunction of values OR(val1,val2,val3)	val1 of D_i belongs to the disjunction in D_j	the value val1
atomic value val1	negation of values NOT(val3,val4)	val1 of D_i does not belong to the negation, i.e. belongs to $C_E^{\{val3,val4\}}$	the value val1
disjunction of values OR(val1,val2,val3)	disjunction of values OR(val2,val3,val4)	the intersection between the two lists is not empty	the intersec tion of the disjunc tions OR(val2,val3
negation of values NOT(val2,val3)	negation of values NOT(val1,val2)	the union of all the negated values is not equal to E, i.e. $C_E^{\{val2,val3\}} \cup C_E^{\{val1,val2\}} \not\subseteq \{val1, val2, val3\}$	the negation of th union of the value NOT(val1, val2,val3) = $C_E^{\{val1,val2,val3\}}$
negation of values NOT(val1,val2)	disjunction of values OR(val2,val3)	the list of values in the disjunction is not included (or equal) in the list of the negated values, i.e. $C_E^{\{val1,val2\}} \cap \{val2, val3\} \neq \emptyset$	the va ues in the disjun tion which do no belong to the list the negated value val3 (it is the no empty intersectio of the condition)

We present now the elements of the language that we have developed to permit the expression of feature transmission and of the control of their values. All these tools are based on the A-unification.

2.3. Feature transmission and control expression

We have already presented a first control element: *the prototype of feature descriptor* may be seen as a tool that controls the coherence of features of any lexical entry. The SAV of each feature may be stored in a database (list) of allowed preterminal feature values.

When a lexical entry is built, we must test if the given values match the conditions listed in the database, i.e. if the values associated with each preterminal feature of the entry belong to the list of the possible values for that feature in the database. The negation of feature values is also controlled and interpreted with this list, that may be automatically constructed, given a prototype.

We define two classes of tools: one at the lexical level, the other one in the rules of the grammars.

2.3.1. In the lexicon

Given a language and a degree of precision for the processing, we can express feature Value Coexistence Incompatibilities (VCI). For example, an intransitive verb cannot be in the passive voice. These incompatibilities are the second way to control the lexicon coherence. We choose to treat these two types in the same way and we give them the same name VCI.

VCI may be viewed as *integrity constraints* of two types:

**positive constraints*: they express what a feature descriptor must always verify; they state to which set the values of a preterminal feature must belong. Example: gender (mas, fem)

**negative constraints*: they express feature value incompatibilities in a given feature descriptor. Example: (transitivity(no), voice(passive))

These controls of the coherence of the lexicon are inserted by a *VCI predicate* added to each axiom that forms the lexicon and represents a lexical entry. It has a single argument: the feature descriptor of the considered entry. It "consults" the database of negative and positive constraints of the application to accept or reject the lexical entry. The verification of coherence is based on the previously described A-unification; therefore, in the database, the two types of constraints (negative and positive) are stored in feature descriptors. For example, for positive constraints, the whole descriptor is empty (_ as feature values) except for a specific feature to which its SAV is attached. For negative constraints, only the relevant feature values are stored in the descriptor. To be correct, the feature descriptor of a lexical entry must a-unify (augmented unification) with all the descriptors that model positive constraints and with no descriptor that models a negative constraint. During lexical insertion, a non-terminal inherits feature values from the lexical word which is attached to it; the inherited features have been filtered by the VCI. They percolate up in the derivation tree in the grammar rules. So we present now the tools that allow and control this evolution.

2.3.2 In the grammar rules

A grammar rule has three parts: a *derivation* part, a *feature control* part (constraints on feature values) and a *construction and inheritance* part (assignment of values to features, propagation of values in the parsing tree and control). The two last parts are realized by *predicates* that work on the feature descriptors associated with the terminal and non-terminal symbols of the grammar. These parts are based on the *augmented unification mechanism.*

So a rule form is: Left part $\rightarrow$ Right part, {Pred}

where *Pred* is a predicate whose first argument is a list, that may be empty, which offers an eventual choice of *actions*; the other arguments are the different possible actions.

Pred(List, ..., ..., ...): if List is empty, there is a single other argument which is executed. Otherwise the ith argument of the list leads to the execution of the $(i+1)^{th}$ argument of Pred.

The *actions* are conjunctions of *logic tools* (predicates). We present them now. For each tool, we give a definition and a simplified example. In the examples, D_i represents the feature descriptor associated with each symbol of the grammar.

AUNIF(t_1, D_1, t_2, D_2): true if the subdescriptor of D_1 whose root node is t_1 a-unifies with the subdescriptor of D_2 whose root node is t_2. It treats the agreement problems and the construction of non-instanciated descriptors. AUNIF is directly based on the A-unification.

Example: Agreement checking between two categories and feature transmission
$N^2(D_0) \leftarrow det(D_1),\ N^1(D_2),\ \{Pred([],$
$(AUNIF(gnb,\ D_1,\ gnb,\ D_2),\ AUNIF(F,\ D_2,\ F, D_0)))\}$

The first AUNIF states that the values of the features "gender" and "number" in D_1 and D_2 must agree; the second AUNIF instanciates D_0 with the values of the descriptor of N^1.

ASSIGN$(val,\ t_1,\ D_1)$: true; binds the value *val* to the feature t_1 of D_1.

ASSIGN$(t_1,\ D_1,\ t_2,\ D_2)$: true; binds the subdescriptor of D_1 whose root node is t_1 to the subdescriptor of D_2 whose root node is t_2.

So it binds or assigns a new value to a feature or a subdescriptor.

Example: During a passivization operation, the verb gets the passive voice.
$V^0(D_0) \leftarrow V^0(D_1),\ [by],\ N^2(D_2),\ \{Pred([],\ ASSIGN(passive,\ voice,\ D_1))\}$

VERIF$(arg,\ t_1,\ D_1,\ val)$: checks if a specific feature t_1 of D_1 has the value *val*.

* If arg = s: *strict verification*; true if t_1 has the value val, av or a disjunction of values which contains val.

* If arg = e: *existential verification*; true in the same cases as the strict verification plus the possibility of having a negation of values without val.

* If arg = b: *broad verification*; true in the same cases as the existential verification plus the possibility of having a logical variable as value.

Example: A subject noun phrase must have the nominative case.
$S(D_0) \leftarrow N^2(D_1),\ V^2(D_2),\ \{Pred([],\ (VERIF(e,\ case,\ D_1,\ nominative)))\}$

CALCUL$(t_3,\ D_3,\ t_1,\ D_1,\ t_2,\ D_2)$: true; the subdescriptor of D_3 whose root node is t_3 is deduced from the subdescriptor of D_1 whose root node is t_1 and from the subdescriptor of D_2 whose root node is t_2. It is used to build a resulting descriptor from parts of two others.

We have also defined another tool *CALC* which is used for generalization purpose to modify the value of a feature or to calculate it from values of other features (e.g. number in coordination). As it is quite specific, we just mention it here.

3. CONTROL AND TRANSMISSION TOOL SEMANTICS

We briefly present the basic theoretical foundations of feature filtering, reformulating the semantics of logic programs in order to take constraints on features into account. Some of the material presented here has emerged from a reformulation of definitions and theorems of [4].

3.1. VCI semantics

In this section, we consider lexicons as a collection of lexical entries. We define the notion of *VCI-filtered lexicons and logic programs.* We present the concept of logical consequence of a VCI-filtered logic program.

Definition 1: A *non-filtered lexicon* is a definite program P that only contains a finite set of unit clauses.

Definition 2: A *VCI-filtered lexicon* is a non-filtered lexicon P whose unit clauses $(A\ \ \leftarrow)$ are transformed into definite clauses by addition of an antecedent VCI; this antecedent states two types of conditions on the arguments of A: coexistence and membership.

A logic program which models a grammar generally contains unit clauses (corresponding to the lexicon or to a part of it) and conditional definite clauses (corresponding to derivation rules).

Definition 3: A *VCI-filtered definite program* is a definite program P whose unit clauses (corresponding to the lexicon) have been transformed into conditional clauses by addition of an antecedent VCI.

Definition 4: Let P be a definite program and F a formula. F is a *logical consequence of VCI-filtered P* iff - F is a logical consequence of P.

- In the proof tree for F, the successive goals do not show, in their arguments, any of the incompatibilities defined by VCI.

We will not say more on the declarative and the procedural semantics of this type of filtering. The reader may consult [7] and adapt it to the case of feature filtering.

3.2. TRANSMISSION AND CONTROL TOOL SEMANTICS

We just give here the basic ideas of the semantics of feature filtered logic programs. [7] gives a more detailled presentation even if it presents another type of filtering.

Definition 5: Let P be a definite program. Transforming P in *a feature filtered logic program* consists of the addition of an antecedent *Pred* at the end of each clause of P. This antecedent states conditions on the arguments of the elements of the clauses. These conditions are numerous: verification of values, assignment ,...

Definition 6: Let P be a definite program and F a formula. F is a *logical consequence of feature-filtered P* iff - F is a logical consequence of P.

- In the proof tree for F, the goals that contain the predicate Pred are verified, i.e. the arguments of the other goals verify every condition of Pred that constrain them.

4. CONCLUSIONS

In this paper, we have proposed an original logic-programming expression of feature filtering. The interest of our expression of features lies mainly in its simplicity and generality, thanks to the use of a prototype of feature descriptor that can be considered as an abstract data type.

Our logical feature filtering modelling of lexicons is based on a predicate that consults a database of positive and negative constraints, specific to each application. We have also specified a set of logical tools that permit to express feature transmission and control in the rules of logical-based grammars. The logical language defined in this way exhibits a clear semantics. Obtained from linguistic motivations and specifications, it is defined in a precise aim of filtering expression and must therefore be clear and easy enough to be used by a linguist. The proposed language permits to create elements of a linguistic environment that will help a natural-language grammar expert to express and manage features (linguistic engineering).

We sketched the nature of the semantics of logic programs with which a feature filtering is associated. The filtering consists firstly in transforming the unit clauses (axioms) that model the lexicon into definite clauses, by adding an antecedent VCI. Secondly, it consists in introducing an antecedent Pred into the definite clauses of the grammar that manages the control and the construction of descriptors that are arguments of the elements of these clauses. We have also introduced how to adapt the semantics of logic programs when we need to solve problems in a constrained universe.

This approach has been shown useful in the context of a computational application of syntactic feature filtering in a CDG (Contextual Discontinuous Grammars [8]) parser that models concepts of the Government and Binding theory [9, 10].

The main advantages of our approach over the previous published ones lie in that it is simpler, more modular and declarative. The logic programming-based language is also more natural, adequate and expressively efficient with respect to the considered linguistic problem. More than a static tool, we have presented a flexible model with a clear semantics, valid for different feature filtering problems. This method can be undertaken for every type of features and provides us with a partial solution to semantical problems, partial solution which we plan to extend in future works.

References

[1] J. Nivette. *Principes de grammaire générative.* Fernand Nathan, Labor, 1974.

[2] L. Karttunen. Features and values. In *Proc. Int. Conf. on Computational Linguistics*, Stanford, CA, 1984.

[3] P. Saint-Dizier. Expression of syntactic and semantic features in logic-based grammars. *Computational Intelligence, 2*, 1–8, 1986.

[4] J.W. Lloyd. *Foundations of Logic Programming.* Springer-Verlag, second extended edition, 1987.

[5] M.H. Van Emden and R.A. Kowalski. The semantics of predicate logic as a programming language. *Journal of the ACM, 23, 4*, 733–742, 1976.

[6] H. Ait-Kaci and R. Nasr. Login : a logic programming language with built-in inheritance. *The Journal of Logic Programming, 3*, 185–215, 1986.

[7] P. Sébillot. Logical modelling of constraints on acceptable syntactic tree forms for natural language analysis. In *Proc.* 7th *IAICV*, Ramat Gan, Israel, 1990.

[8] P. St-Dizier. Contextual discontinuous grammars. In V. Dahl and P. St-Dizier, editors, *Natural Language Understanding and Logic Programming II*, North Holland Pub., 1988.

[9] N. Chomsky. *Lectures on Government and Binding.* Foris Pub., Dordrecht, Holland, 1982.

[10] N. Chomsky. Barriers. *Linguistic Inquiry Monograph No. 13, MIT Press*, 1986.

PROGRAM TRANSFORMATIONS AND WAM-SUPPORT FOR THE COMPILATION OF DEFINITE METAPROGRAMS

Paul Tarau

Département d'Informatique,
Université de Moncton, Moncton, Canada, E1A 3E9,
E-mail: taraup@umoncton.ca

ABSTRACT. *By combining binarization and elimination of metavariables we compile* **definite metaprograms** *to equivalent* **definite binary programs**, *while preserving first argument indexing. The transformation gives a* ***faithful embedding*** *of the essential ingredient of full Prolog to the more restricted class of binary definite programs while preserving a strong operational equivalence with the original program. The resulting binary programs can be executed efficiently on a considerably simplified WAM. We describe WAM-support that avoids increasing code size by* **virtualizing** *links between predicate and functor occurrences that result in the binarization process. To improve the space-efficiency of our run-time system we give a program transformation that simulates* **resource-driven failure** *while keeping* **conditional answers**. *Then we describe the WAM-support needed to implement the transformation efficiently. We also discuss its applications to* **parallel execution of binary programs** *and as a surprisingly simple* **garbage collection** *technique that works in time proportional to the size of useful data.*

1. Introduction

We suppose the reader is familiar with basic logic programming theory (see [LL87]) and compilation of logic programs (see [WA83]).

The architectural elegancy and the efficiency of the WAM [WA83] have made it the unchallenged implementation paradigm for logic programs. Much work has been reported (see for example [JD88], [ME90], [UM90]) on various interesting optimizations. Those layers of successive optimizations make an efficient, full implementation of the WAM a huge and fairly complex software engineering project. This has some drawbacks when we focus on formal verification or hardware implementation. Although optimizations perform impressively on popular benchmarks, large code size and lack of *locality of reference* can induce page faults that often backfire on user-level performances in unexpected ways.

In previous work (see [TA91a], [TA91b]) we have observed that, through program transformations, a small subset of the WAM can replace the full abstract machine and still support execution of definite programs. This keeps the software simpler than a C-Prolog style interpreter, while performances are close to emulator based compilers like Sicstus-Prolog and SB-Prolog[1].

The basic idea is to restrict the WAM to binary definite programs, as in this case there is no environment and all variables are temporary (see [WA83]).

This approach is comparable with using PROPLOG and DATALOG as intermediate (interpretative) steps towards the WAM (see [MW88]). However, those restricted logic languages are strictly less

---

[1] Our current WAM emulator runs at 42000 LIPS on a Sparcstation IPC. The compiler supports full Prolog and works on a clause by clause basis. The system has been bootstrapped on various machines from PCs to Sparcs.

expressive than Prolog, while focussing on binary definite programs gives, from the beginning, the full power of the language for a comparable conceptual complexity. A similar approach, using successive refinements of an abstract machine, is given in [AK90], although binary definite programs are not singled out as an intermediate step towards the full WAM. Independent work by B.Demoen and A.Marien [DM91] shows that complex builtins can be implemented efficiently in the framework of binary logic programs.

This paper extends our previous work with a 3-step program transformation from definite metaprograms to definite binary programs that preserves first argument indexing and therefore is within a constant factor in space and time efficiency when compared with general WAM-code. We show that, with suitable WAM-support, it is possible to avoid undesirable code expansion and improve performances.

Garbage collection of the Prolog heap is one of the most intricate parts of a WAM implementation. To address it in a comparably simple way we propose a program transformation that simulates *resource-driven failure* while keeping *conditional answers*. After giving a meta-interpreter based specification, we describe a partial evaluation based transformation technique. With suitable WAM support, the transformation can be implemented efficiently and introduces an interesting form of parallel execution of logic programs.

2. Efficient binarization of definite metaprograms

First we describe how a fairly general class of logic programs can be reduced to equivalent definite binary programs, while preserving the indexing capabilities of the original code. This improves on a similar sequence of transformations presented in [TB90].

Definition 1. A *definite metaclause* has the form

B_0 :- B_1,B_2,...B_n.

where each B_k for k=0 to n, is either an atom, or a variable. A *definite metaprogram* is a set of definite metaclauses.

Definition 2. Let P be a definite metaprogram and *\$demo/1* a new predicate symbol. For each clause

C: B_0 :- B_1,B_2,...B_n.

of P, we construct the *definite closure* of C, denoted CL(C):

\$demo(B_0) :- *\$demo*(B_1),*\$demo*(B_2),...*\$demo*(B_n).

Following [TB90], the logical meaning of a definite metaprogram is the minimal model of its definite closure, therefore definite metaprograms are basically first order objects despite variables occurring as atoms.

On the other hand, to preserve first argument indexing, a more efficient (but apparently more space consuming) technique can be used, as shown in the following transformation.

Definition 3 Let P be a definite metaprogram. We call the *Warren definite closure*[2] of P (denoted META2DEF(P)) the program obtained from P by means of the following transformations:

a) replace each *metavariable* V of P by *$demo*(V)

b) add to the program obtained at **a)**, for every *predicate p/n* of P, a clause:

$demo(p(V1,V2,...,Vn)) :- p(V1,...,Vn)

where *$demo/1* is a new predicate symbol and *V1,...,Vn* are of new distinct variables.

Let us remark that this definition also covers the case when the heads of clauses are variables.

The semantic equivalence of META2DEF(P) and CL(P) is obvious as they perform essentially the same unifications in a SLD-refutation. Moreover, if we consider as *observables* the answer substitutions and the order in which the answers are generated, META2DEF and CL(P) provide *faithful embeddings* of the class of definite metaprograms into the class of definite programs.[3]

Definition 4. Let P be a definite program and *Cont* a new variable. Let T and E=p(T1,...Tn) be two expressions. We denote ϕ(E,T) the expression p(T_1,...,T_n,T). Starting with the clause

(C) A :- B_1,B_2,...,B_n.

we construct the binary metaclause

(C') ϕ(A,Cont) :- ϕ(B_1,ϕ(B_2,...ϕ(B_n,Cont)...).

Let us denote DEF2MBIN(P) the metaprogram obtained from the (implicit) unit clause "*true*" and the set of clauses C'. The following example shows the result of the transformations DEF2MBIN.

% definite program P

```
perm([],[]).
perm([X/Xs],[Y/Zs]):-del(Y,[X/Xs],Ys), perm(Ys,Zs).

del(X,[X/Xs],Xs).
del(X,[Y/Xs],[Y/Ys]):-del(X,Xs,Ys).
```

% binary metaprogram DEF2MBIN(P)

```
perm([],[],Cont):-Cont.
perm([X/Xs],[Y/Zs],Cont):-del(Y,[X/Xs],Ys,perm(Ys,Zs,Cont)).

del(X,[X/Xs],Xs,Cont):-Cont.
del(X,[Y/Xs],[Y/Ys],Cont):-del(X,Xs,Ys,Cont).
```

A more general binarization technique of Sato and Tamaki [ST89], based on continuations and input-output argument classification, can be used for the same purpose. Variations of this technique can be obtained by folding some of the goals into new predicates before applying the transformation

---

2 This technique is similar to that used by Warren in [WA82] to define a clause "apply(F,X_1,...,X_n):-F(X_1,...,X_n)" for each predicate symbol F of arity N.

3 These powerful language comparison tools introduced by Shapiro [SH91] are based on the existence of a compilation scheme from one language to the another.

DEF2MBIN. Variables occurring only locally in the new clauses need not to appear in the heads of the new predicates as in the following:

Example 1. Starting with the clause

a(X1,X5):-b(X1,X2),c(X2,X3),d(X3,X4),e(X4,X5).

we get after folding:

a(X1,X5):-$new1(X1,X3),$new2(X3,X5).
$new1(X1,X3):-b(X1,X2),c(X2,X3).
$new2(X3,X5):-d(X3,X4),e(X4,X5).

and after DEF2MBIN:

a(X1,X5,Cont):-$new1(X1,X3,$new2(X3,X5,Cont)).
$new1(X1,X3,Cont):-b(X1,X2,c(X2,X3,Cont)).
$new2(X3,X5,Cont):-d(X3,X4,e(X4,X5,Cont)).

This is slightly more flexible than the alternative transformation proposed in [DM91] that simulates the WAM's environment trimming.

Definition 5. Let P be a definite metaprogram. Let us denote META2BIN(P) = META2DEF(DEF2MBIN(META2DEF(P))). We shall refer to the result of META2BIN simply as the *binarization* of the definite metaprogram P.

Clearly, as both META2DEF and DEF2MBIN preserve first argument indexing, this property also holds for META2BIN. Therefore WAM-executions of P and META2DEF(P) have the same properties with respect to determinism detection and choice point creation on the local stack.

Example 2. If P =

eq(X,X).
or(X,_):-X.
or(_,Y):-Y.
memb(X,[Y|Ys]):-or(eq(X,Y),memb(X,Ys)).

then META2BIN(P) =

$demo(true).
$demo($demo(A,B)):-$demo(A,B).
$demo(eq(A,B,C)):-eq(A,B,C).
$demo(memb(A,B,C)):-memb(A,B,C).
$demo(or(A,B,C)):-or(A,B,C).
$demo(true,A):-$demo(A).
$demo(eq(A,B),C):-eq(A,B,C).
$demo(memb(A,B),C):-memb(A,B,C).
$demo(or(A,B),C):-or(A,B,C).
eq(A,A,B):-$demo(B).
or(A,B,C):-$demo(A,C).
or(A,B,C):-$demo(B,C).
memb(A,[B|C],D):-or(eq(A,B),memb(A,C),D).

As DEF2MBIN provides a *faithful embedding* from the class of definite programs to the class of binary metaprograms, META2BIN is also a *faithful embedding* from the class of definite metaprograms to the class of binary definite programs. As a practical consequence we have the *reversibility* of the transformation, useful in decompilation or while tracing transformed clauses in a debugger. This strong operational embedding of *definite metaprograms* (that are the basic computational ingredient of full Prolog) into the much smaller class of binary definite programs is the main motivation of our approach to Prolog compilation.

3. Virtualizing demo-predicates at WAM-level

In order to reduce the implementation complexity of the WAM we designed a simplified abstract machine working on binary definite programs (see [TA91b]). Our current implementation covers full Prolog and runs at about half of the speed of Sicstus-Prolog 0.7. The 17K Prolog compiler uses META2BIN as a preprocessor and is now able to compile itself on various machines from PCs to Sparcs. We will summarize here its principles.

The basic function of environments in standard WAM can be seen as a form of compile time garbage collection capitalizing on the fact that arguments of the current goal (thanks to a new-to-old pointing policy) can never be referenced by non-local pointers so that they can be on the rewritable local stack instead of the monotonically growing heap. Combined with environment trimming, their use is accountable for impressive space savings. Warren rejected goal-stacking in his [WA83] paper for imposing run-time checking on unsafe variables, although he recognized its implementation simplicity among other advantages. Clearly, after binarization unsafe variables simply vanish (together with other permanent variables) and there's no danger of having dangling references.

In the framework of [TB90], SLD-resolution can be described using the ⊕ "arrow composition" operator (basically unfolding of binary clauses) as the following (nondeterministic) algorithm:

while (Goal-arrow ≠ A:-true) {
Select(Clause-arrow)
Goal-arrow:=Goal-arrow ⊕ Clause-arrow
}

where(P:-Q_1) ⊕ (Q_2:-R) is (P:-R) mgu(Q_1,Q_2).

The process starts with an arrow of the form G:-G or \$answer(X_1,...,X_n):-G if we are interested only in variables X_1,...,X_n occurring in the atomic goal :-G.

At a given time, the content of the *heap* can be seen as representing the current state of *Goal-arrow*. The **S** and **H** pointers (see [WA83]) follow locations on the *heap* used to pass from the state *Goal-arrow* to the state *Goal-arrow* ⊕ *Clause-arrow*. The **S** pointer walks through existing fragments of *Goal-arrow* which is destructively updated when composed with a *Clause-arrow*. The **H** pointer creates (lazily) new structures coming from (virtual) copies of *Clause-arrows*.

Our run-time system follows closely this abstract interpreter. Remark that we do not need environment frames as there are neither explicit continuations (the CP register of the WAM) nor permanent variables. However, as in standard WAM (unless a call is deterministic), we have to push to the stack *choice point* frames containing:

- the argument registers $A_1 \ldots A_n$
- the current value of the top of the heap **H**
- the current value of the top of the trail **TR**
- the current value of the instruction pointer **P**

As in standard WAM this choice point is created by a TRY_ME_ELSE instruction, updated by RETRY_ME_ELSE instructions and freed by a TRUST_ME instruction.

The management of continuations (the variables "*Cont*" introduced by the transformation DEF2MBIN) is also implicit in the unification instructions.

As it can be seen in the example 2, we can avoid generating code for the predicates $demo/1 and $demo/2 as they have a quite obvious structure. This is important if we want to avoid any code size increase with respect to standard WAM and to be able to work on a clause by clause basis. Remark also that this ensures reversibility of the transformation on a clause by clause basis and can be useful if we want to implement dynamic decompilation as in ALS-Prolog.

The easiest way to "*virtualize*" the functionality of $demo/1 and $demo/2 is by introducing two builtins *DEMO1* and *DEMO2* which simply transfer the control to the first instruction of the predicate, after using its functor as a key in a hash table (similarly to the SWITCH_ON_FUNCTOR instruction in standard WAM).

Remark that metavariables occurring as arguments of *$demo/1* have (exactly) one occurrence in the head, which is guaranteed to instantiate them. All we have to do is to dereference the metavariable to a predicate name, copy its arguments from the heap to the registers and then execute the first instruction of the predicate, as shown in the following fragment of our WAM emulator:

```
case DEMO1:
{   register term pred=regs+1;
    DEREF2(pred,functor);
    P=(instr) hget(PREDMARK,functor);
    arity=ARITY(FUNC2SYM(functor));
    for(k=1; k<=arity; k++)  SETREF(regs+k,pred+k);
}
continue;
```

We can also "virtualize" calls to $demo/2 via the builtin DEMO2 that links the term occurrence $p(V_1,\ldots,V_n)$ obtained by dereferencing the argument register X_1 to the predicate occurrence $p(V_1,\ldots,V_n,X_2)$ where X_2 holds the continuation argument obtained by the binarization process.

An easy way to implement this link is by using a hashing function that associates instruction addresses to predicate functors and a little word-surgery to go from p/n to p/n+1 before searching in the hash table.

After locating the code for the predicate *p/n+1*, we copy from the heap the *n* arguments of *p/n* to the first *n* argument registers and X_2 (implemented as regs[2]) to the *n+1*th argument register:

```
case DEMO2:
{    sym s; term continuation;
     register term pred=regs+1;
     DEREF2(pred,functor);
     s=FUNC2SYM(functor); arity=ARITY(s);
     functor=SYM2FUNC(SETARITY(s,++arity));
     P=(instr) hget(PREDMARK,functor);
     continuation=(term)regs[2];
     for(k=1; k<arity; k++) SETREF(regs+k,pred+k);
         SETREF(regs+arity,continuation);
}
continue;
```

After the execution of DEMO2 the argument registers are in a state similar to what would have happened if p/n+1 was called. Obviously no choice point is created on the stack as $demo/1 and $demo/2 are deterministic predicates. No special care has to be taken to "virtualize" the functionality of clauses like

```
$demo(true).
$demo($demo(A,B)):-$demo(A,B).
$demo(true,A):-$demo(A).
```

as they are dealt with automatically by an appropriate combination of DEMO1 and DEMO2.

4. Space management by resource-driven failure

Our optimized design of a simplified WAM (see [TA91b]) can be combined with a marking garbage collector as is usually done in Prolog systems. However, the complexity (and the code-size) of such a garbage collector contradicts our primary objective to compensate for the constant (and small) loss of performance compared to standard WAM by a very compact run-time system that entirely fits in on-processor caches and/or has a cost efficient hardware implementation.

The program transformation that leads to binary definite programs highly facilitates an alternative approach that will be described here.

Let's start with a simple but effective programming hack that overcomes the absence of a garbage collector in Prolog.

In the Craft of Prolog, R.A. O'Keefe (see [OK90] - p.85) uses "findall" to reduce the space used for the evaluation of a goal to the total size of its successful instances:

```
gc_call(Goal):-
    findall(Goal,Goal,Instances),
    member(Goal,Instances).
```

Remark that O'Keefe's program is limited by the space requirements of the longest OR-branch of the SLD-search tree and that the presence of an infinite AND branch will push the whole findall in an infinite loop. *Gc_call* is also sensible to resource limitations within execution of *Goal*. Just think about what happens if calling Goal generates a heap-overflow.

The solution we retained is to "distribute" the work of findall over the SLD-tree, so that it will work even if individual branches exceed the resources. A good partial evaluator that deals with *assert* and *retract* can eventually infer our technique by "interleaving" the work of *findall* with the evaluation of *Goal* in *gc_collect*.

The basic idea is to *send* a conditional answer consisting in the *original goal* and a *resolvent* (see [CV88], [CV90], [TB90] for the theory behind conditional answers and some their applications) to a failure-driven top-level loop while simulating failure inside the SLD-search process. This failure occurs on an "abstract" resource limitation (a heap overflow, a depth limitation of the SLD-tree or some other control mechanism).

The following metainterpreter gives a prototype implementation of the approach, by simulating randomly occurring resource-limitations (see the predicate *overflow/0*).

```
% toplevel call to goal G
run(G):-
    send(G),
    repeat,iterate,!.

% initiates execution of one process with limited resources
iterate:-finished,!.
iterate:-
    receive(ToDo),!,
    solve(ToDo).

finished :- \+ todo(_).

% limited resource meta-interpreter with failure on overflow
% conditional answers are "sent" to iterate/0 to be restarted later as new processes

solve((A:-true)):-!,answer(A).
solve((A:-H)):-overflow,!,send((A:-H)),fail.
solve((A:-call(G))):-!,solve((A:-G)).
solve((A:-H)):-
    clause(H,B),
    solve((A:-B)).

% simulates some resource limitation (for example a heap overflow)

overflow:-X is random, X < 0.10.
send(A):-assertz(todo(A)).
receive(CondAnswer):-retract(todo(CondAnswer)),!.
answer(A):-write(A),nl,fail.
```

We isolated the actual implementation of *overflow/0*, *send/1*, *receive/1*, and *answer/1* to emphasize that they are more abstract that the underlying sequential Prolog primitives. Remark that *assert* and *retract* are used simply to implement a *queue of conditional answers* that survive backtracking. The next section will discuss WAM-support and alternative implementations of these operations.

The following example runs on top of the metainterpreter. By including calls to *statistics/0* in the *receive/1* predicate, the reader can test space savings due to the implicit garbage collection on various Prolog systems.

Example 3.

```
:-run((answer(A):-nrev([a1,a2,a3],A,true))).

app([],Ys,Ys,Cont):-call(Cont).
app([A|Xs],Ys,[A|Zs],Cont):-app(Xs,Ys,Zs,Cont).

nrev([],[],Cont):-call(Cont).
nrev([X|Xs],Zs,Cont):-nrev(Xs,Ys,app(Ys,[X],Zs,Cont)).
```

The overhead of metainterpretation can easily be eliminated by a well-known program transformation technique (see [TA86], [TF86]). We will give here only the transformed program for our example, together with the predicates *solve/1* and *$call/2* that jointly implement the fail-on-overflow-and-send mechanism, while the predicate *iterate/0* implements the receive-and-solve mechanism that restarts the computation from a conditional answer.

Example 4.

```
:-run(nrev([a1,a2,a3],Answer,true(Answer))).

run(G):- send(G),repeat,iterate,!.

iterate:-finished,!.
iterate:-receive(ToDo),!,solve(ToDo).

finished :- \+ todo(_).

$call(true(Answer)):-answer(Answer).
$call(append(A,B,C,D)):-append(A,B,C,D).
$call(nrev(A,B,C)):-nrev(A,B,C).

solve(A):-overflow,!,send(A),fail.
solve(A):-$call(A).

append([],A,A,B):-$call(B).
append([A|B],C,[A|D],E):-solve(append(B,C,D,E)).

nrev([],[],A):-$call(A).
nrev([A|B],C,D):-solve(nrev(B,E,append(E,[A],C,D))).
```

Remark that instead of working explicitly on a conditional answer of the form (Answer:-Goal) we can incorporate Answer in the terminator *true* that becomes *true(Answer)*. The following example can be used to test the technique on nondeterministic predicates.

Example 5.

:-run(subset([a,b,c,d,e,f],[A,B,C,D],true([A,B,C,D]))).

```
$call(true(Answer)):-answer(Answer).
$call(subset(A,B,C)):-subset(A,B,C).

subset([],[],true(A)):-answer(A).
subset([_|Xs],Ys,A):-solve(subset(Xs,Ys,A)).
subset([X|Xs],[X|Ys],A):-solve(subset(Xs,Ys,A)).
```

Remark that the transformed code preserves first argument indexing and therefore it can be used jointly with META2BIN as an efficient binarization technique that also ensures automatic garbage collection. Compared with "gc_call" which is essentially programmer controlled, our technique has the advantage that it can be system controlled at WAM-level.

5. WAM support and parallelization

Clearly, *assert* and *retract* are an overkill in the implementation of *send* and *receive*. Let us focus on what is happening inside "assert" and "retract" on a Prolog system.

In the worst case, the system compiles dynamically when it asserts a clause and then decompiles it when retracting it (ALS-Prolog). This is clearly expensive for *send* and *receive* as it generates code (used only once) to create on the new heap a resolvent. *Copying it directly from the old heap at WAM-emulator level does the same work far more efficiently.* Even in Prolog implementations with special purpose, optimized dynamic code, copying the answer and the resolvent to the new heap is simpler when taking into account the hidden overhead in the implementation of assert and retract, like compaction of code-space or relinking of indexing chains.

Within a copying garbage collector that uses two or more heaps we can simply put the conditional answer

Answer:-Resolvent

on the new heap, load the argument registers from *Resolvent* and then jump to the predicate of *Resolvent*.

Remark that we get (implicit) garbage collection with an effort proportional to the size of useful data (the resolvent). This is clearly better than marking algorithms that use space and time proportional to the size of heap used so far.

Starting a new WAM can be done in *parallel* while continuing work on the parent WAM. For the parent, every conditional answer generates a failure. Thus space will be also recovered in the parent WAM by backtracking. Execution with a limited set of WAM-workers can be done by suspending parent processes that try to detach conditional answers when no accessible WAM-worker is free.

The condition that triggers detaching a conditional answer can be a variation of iterative deepening (iterative increment of the depth of SLD derivations) or, at a more abstract level, some other *measure* of the structural complexity of the resolvent.

For example, in the case of "nrev" on a list of length N, copying only useful data takes O(N) while marking garbage collection of the heap takes $O(N^2)$ as "nrev" uses $O(N^2)$ heap-space on the WAM (although the stack and the trail are O(1)).

We can also ensure fairness of the SLD-resolution (Prolog depth-first search rule is known as unfair) with a priority queue implementation of send-receive, where the priority of a conditional answer that is used to restart a computation depends on the length of the SLD-branch.

6. Conclusion

The program transformations discussed in this paper are a step towards extending our simplified WAM-system to support a fairly large subset of Prolog. By using these techniques, the implementor can focus on an optimized WAM emulator that covers only binary definite programs and still be able to execute clauses containing an arbitrary combination of conjunctions, disjunctions and metavariables in their body. Binarization allows to add space management by resource-driven failure in a way that is very simple both semantically and at implementation level. Combined with WAM support it gives cheap garbage collection and allows parallel evaluation. Resource-driven failure can be implemented with the same semantics as in Prolog or, alternatively, by a fair search rule.

As source-to-source transformations, even without WAM-support, the techniques presented here can be used to add metaprogramming to restricted implementations of Prolog like PDC-Prolog and garbage collection to systems like C-Prolog.

Future work will focus on optimizations of our abstract machine and on parallel implementations. An interesting direction is to explore the use of alternative binarization techniques and various forms of compile time garbage collection. Recent work by B.Demoen [DE90] and our preliminary results using the Mixtus partial evaluator at binary clause level show dramatic improvements in heap consumption and speed not attainable by partial evaluation of the original program. The key point is that access to continuations gives more information to the partial evaluator while considerably simplifying the structure of the program. Run-time access to a term representation of continuations can also be useful in experimenting with new control mechanisms for logic languages.

Acknowledgments

We gratefully acknowledge support from the Canadian Natural Sciences and Engineering Research Council (grant OGP0107411) and the FESR of the Université de Moncton.

References

[AK90] H. Ait-Kaci: *"The WAM: A (Real) Tutorial"*, Digital PRL, January 1990.

[CV88] M.H.M. Cheng, M.H. van Emden, P.A. Strooper, "Complete Sets of Frontiers in Logic-based Program Transformation", J.W. Lloyd (ed.), *Proceedings of the Workshop on Meta-Programming in Logic Programming*, p. 213-225, Bristol, 1988.

[CV90] M.H.M. Cheng, M.H. van Emden, B.E. Richards: "On Warren's Method for Functional Programming in Logic", D.H.D. Warren, P. Szeredi (eds.), *Logic Programming Proceedings of The Seventh International Conference*, p.546-561, MIT Press 1990.

[DE90] B. Demoen, *"On the transformation of a Prolog Program to a more efficient binary Program"*, Report CW 130, December 1990, K.U.Leuven.

[DM91] B. Demoen, A. Marien, "Implementation of Prolog as binary definite Programs", *Russian Conference of Logic Programming*, September 1991.

[HM89] T. Hickey, S. Mudambi: "Global Compilation of Prolog", *Journal of Logic Programming*, 1989:7, p.193-230.

[JD88] G. Janssens, B. Demoen, A. Marien: "Improving the Register Allocation of WAM by Reordering Unification", R. Kowalski, K.A. Bowen (eds.), *Logic Programming: Proceedings of the Fifth International Conference and Symposium*, p. 1388-1402, MIT Press, 1988".

[LL87] J.W. Lloyd, *"Foundations of Logic Programming"*,(2nd ed.), Springer-Verlag,1987.

[ME90] M.Meier: "Compilation of Compound Terms in Prolog", S. Debray, M. Hermenegildo (eds.) *Proceedings of the 1990 North American Conference*, p.63-79, MIT Press 1990.

[MW88] D. Maier, D.S. Warren, "Computing with Logic", Benjamin/Cummings Publishing, 1988.

[OK90] R.A. O'Keefe, *"The Craft of Prolog"*, MIT Press, 1990.

[SH91] E. Shapiro, *"Separating Concurrent Languages with Categories of Language Embeddings"*, Tech. Report CS91-05, The Weizmann Institute of Science, March 1991.

[ST89] T. Sato, H. Tamaki: "Existential Continuation", *New Generation Computing* 6, p. 421-438, 1989.

[TA86] A. Takeuchi, *"Affinity between Meta Interpreters and Partial Evaluation"*, ICOT Tech. Report TR-166, Institute for New Generation Computer Technology, Tokyo, 1986.

[TA91a] P. Tarau: *"A Compact Definite Metaprogram Compiler Based on Program Transformations"*, Tech. Report TR-91-1, Dept. d'Informatique, Université de Moncton, January 1991.

[TA91b] P. Tarau: "A Simplified Abstract Machine for the Execution of Binary Metaprograms", *Preceedings of the Logic Programming Conference'91*, p. 119-128, ICOT, Tokyo 1991.

[TB89] P. Tarau, M. Boyer: "Prolog Meta-Programming with Soft Databases", Harvey Abramson and M.H. Rogers (eds.) *Meta-Programming in Logic Programming*, p. 365-382, MIT Press, 1989.

[TB90] P. Tarau, M. Boyer: "Elementary Logic Programs", P. Deransart, J. Maluszynski, (eds), *Programming Language Implementation and Logic Programming*, p.159-173, LNCS 456, Springer Verlag, 1990.

[TF86] A. Takeuchi, K. Furukawa, *"Partial Evaluation of Prolog Programs and its Application to Meta Programming"*, ICOT Tech. Memorandum TM-0164, Institute for New Generation Computer Technology, Tokyo, 1986.

[UM90] Z.D. Umrigar: "Finding Advantageous Orders for Argument Unification for the Prolog WAM", S. Debray, M. Hermenegildo (eds.) *Logic Programming: Proceedings of the 1990 North American Conference*, p.80-96, MIT Press 1990.

[WA82] D.H.D. Warren, "Higher order extensions to PROLOG: are they needed", *Machine Intelligence 10*, Ellis Horwood, 1982.

[WA83] D.H.D. Warren: *"An Abstract Prolog Instruction Set"*, SRI International, 1983.

Some Considerations on the Logic P_FD

A Logic Combining Modality and Probability

Wiebe van der Hoek
Free University Amsterdam
Department of Mathematics and Computer Science
De Boelelaan 1081, 1081 HV Amsterdam, The Netherlands.

Abstract

We investigate a logic P_FD, as introduced in [FA]. In our notation, this logic is enriched with operators $P_r^>$, ($r \in [0,1]$) where the intended meaning of $P_r^> \varphi$ is 'the probability of φ (at a given world) is strictly greater than r'. We also adopt the semantics of [FA]: a class of 'F-restricted probabilistic Kripkean models'. We give a completeness proof that essentially differs from that in [FA]: our 'peremptory lemma' (a lemma *in* P_FD rather than *about* it) facilitates the construction of a canonical model for P_FD considerably. We show that this construction can be carried out using only finitary means, and also give a filtration-technique for the intended models. We then define an alternative (in some sense more natural) semantics for the logic, and derive some of its properties. Finally, we prove decidability of the logic.

1. Introduction

The division into 'necessary', 'actual', and 'possible facts' is almost standard, witness the traditional inference rule *"Ab oportere ad esse, ab esse ad posse, valer illatio"* (necessary being implies actual being implies possible being). Many proposals have been given that refine this tripartition, like the recent developments in so called non-monotonic logics, that try to add a fourth component, like 'default-' or 'preferred facts'. Obviously, reasoning with this kind of 'defaults' has a strong relation with reasoning with probabilities, which is clear if one compares the several approaches in the literature that try to model those entities in modal logic.

It appears that logics that try to combine modalities and probabilities show some variety in the way these probabilities are made explicit. They are rather implicitly present in an application in the field of epistemic logic ([Le]), where φ is defined to be believed if 'φ is more likely than $\neg\varphi$'. In the seventies, Segerberg introduced his 'qualitative modality' ($\varphi \geq \psi$; φ is at least as likely as ψ), which also finds a natural interpretation on probabilistic Kripke models - although we showed in [H1] that probability measures are not really necessary here, so that '$\geq$' may as well be interpreted as 'at least as many'.

Whereas probability theory on itself is a well-understood area in mathematics, its applications in AI and computer science in general justify the study and analyses of our reasoning about probabilities. In spite of the existing (and fastly growing) literature on probabilistic reasoning, we know of few endeavours at defining a logic enabling explicit reasoning *about* probabilities.

Our logic P_FD, as introduced in [FA], is essentially modal. Instead of interpreting the modal (possibility) operator as a 'likelihood operator' (cf. [HR]), it is a logic designed for reasoning with probabilities. Formally, the logic contains operators $P_r^>$ for each $r \in [0,1]$, with meaning of $P_r^> \varphi$ "the probability of φ is greater than r". $P_0^>$ is identified with the classical possibility operator M (we assume familiarity with modal logic and its semantics - cf. [Ch]). This is a bridge between probability and modal logic, but as presented in [FA], only a narrow one: for $P_{0.5}^> \varphi$ to be true at w, there is no corresponding semantic requirement (that the measure of worlds verifying φ, exceeds 0.5, for example). This mirrors a choice between assigning probabilities to sets or to formulas, a distinction that is typically ignored. We show, after having presented an alternative semantics (enabling us to interpret the logics dealing with modalities and probabilities in one framework), that for our logic this choice is not essential.

A peculiar property of P_FD is that it only allows probability measures that have a *finite range* (F). This assumption restores compactness for our logic of probabilities, as we will indicate in section 3. Although restricting ourselves to a finite set F of granted measures (not eliminating the possibility to *reason* about and *express* properties of arbitrary probabilities) is a serious logical restriction, we think it allows for interesting applications. Taking F = {0,1} gives us ordinary modal logic. If $0.5 \in F$, we might represent Lentzen's logic of Belief ([Le]). Letting F have 9 elements might be suitable to model Driankov's linguistic estimates *impossible, extremely unlikely, very low chance, small chance, it may, meaningful chance, most likely, extremely likely, certain* (cf. [Dr]. Also more general applications to fuzzy reasoning seem interesting. We think that in many occasions, to let an agent reason about probabilistic events, the granularity of F can be chosen conveniently.

In the area of AI, [FH] presents a profound investigation and a framework to systematically reason with probabilities. In their formalism, a formula is typically a Boolean combination of expressions of the form $a_1 w(\varphi_1) + \ldots a_k w(\varphi_k) \geq c$, where $a_1, \ldots a_k$, c are integers. P_FD is at least as expressive, it would represent the above formula by

$$(P^{=}_{w(\varphi_1)}\varphi_1 \wedge \ldots P^{=}_{w(\varphi_{n-1})}\varphi_{n-1}) \rightarrow P^{\geq}_r \varphi_n, \text{ where } r = \frac{c - (a_1 w(\varphi_1) + \ldots a_{n-1} w(\varphi_{n-1}))}{a_n}$$

However, in this formalism, the φ's must be purely propositional, thus omitting 'higher order weight formulas'. Moreover, they do not allow 'w-free formulas' unabling expressions like "p is true, although its probability is less than 0.1". Also, their system explicitly postulates as axioms all valid formulas of linear inequalities, and the completeness proof heavily relies on results in the are of linear programming. Finally, their (and all related systems we know of) system suffers from a logical complication: it is not (logically) compact (cf. section 3).

The rest of this paper is organized as follows. In section 2, we introduce the formal system P_FD, and prove some of its theorems. In particular, we prove, purely syntactically, so *within the system*, that each formula has a probability in the set F, (which is given at forehand). This is the backbone of the completeness proof that we give for P_FD with respect to the semantics presented in section 3. In the same section, we develop a filtration technique for the proposed models, guaranteeing the finite model property for P_FD. Then we show that, although P_FD is not finitely axiomatised, it *is* decidable. Finally, we introduce an alternative semantics for the system, proving completeness again. In the last section, we end with some conclusions and mention directions for further research.

2. The logic P_FD

The (uncountable) language L of P_FD consists of a countable set P of propositional symbols, parentheses, the logical connectives ¬ and ∨ (with standard definitions for ⊥, T, ∧, → and ↔, and $(\varphi \nabla \psi) \equiv (\varphi \vee \psi) \wedge \neg(\varphi \wedge \psi))$, and, for each real number $r \in [0,1]$, a unary operator $P_r^>$.

2.1. Definition. A set F is a base (for a logic P_FD) if it satisfies:

i. F is finite

ii. $\{0,1\} \subseteq F \subseteq [0,1]$

iii. F is 'quasi-closed under addition': $r, s \in F$ & $(r + s) \leq 1 \Rightarrow (r + s) \in F$.

iv. F is 'closed under taking complements': $r \in F \Rightarrow (1 - r) \in F$.

P_FD is given relative to a fixed base $F = \{r_0, \dots, r_n\} \subseteq [0,1]$. We assume that $r_i < r_{i+1}$, if $i < n$ (implying $0 = r_0$, $r_n = 1$). We will now give the axioms for P_FD. Primitive notion, 'φ has a probability strictly greater than r' is denoted by $P_r^>\varphi$. Moreover, we will use the following abbreviations (for all $s, r \in [0,1]$):

def1 $P^{\geq}_{r}\varphi \equiv \neg P^{>}_{1-r}\neg\varphi$

def2 $P^{<}_{r}\varphi \equiv P^{>}_{1-r}\neg\varphi$

def3 $P^{\leq}_{r}\varphi \equiv \neg P^{<}_{1-r}\neg\varphi$

def4 $P^{=}_{r}\varphi \equiv \neg P^{>}_{r}\varphi \wedge \neg P^{<}_{r}\varphi$

P_FD has inference rules R1 and R2 and axioms A1 - A7:

R1 Modus ponens: From φ and $\varphi \rightarrow \psi$ derive ψ.

R2 Necessitation: From φ derive $P^{\geq}_{1}\varphi$.

A1 Propositional tautologies.

A2 $P^{\geq}_{1}(\varphi \rightarrow \psi) \rightarrow [(P^{>}_{r}\varphi \rightarrow P^{>}_{r}\psi) \wedge (P^{>}_{r}\varphi \rightarrow P^{\geq}_{r}\psi) \wedge (P^{\geq}_{r}\varphi \rightarrow P^{\geq}_{r}\psi)]$ $r \in [0,1]$

A3 $P^{\geq}_{1}(\varphi \rightarrow \psi) \rightarrow (P^{\geq}_{r}\varphi \rightarrow P^{>}_{s}\psi)$ $r, s \in [0,1]$ and $s < r$

A4 $P^{\geq}_{0}\varphi$

A5 $P^{>}_{r+s}(\varphi \vee \psi) \rightarrow (P^{>}_{r}\varphi \vee P^{>}_{s}\psi)$ $r, s \in [0,1]$ such that $r + s \in [0,1]$

A6 $P^{\geq}_{1}\neg(\varphi \wedge \psi) \rightarrow ((P^{>}_{r}\varphi \wedge P^{\geq}_{s}\psi) \rightarrow P^{>}_{r+s}(\varphi \vee \psi))$ $r, s \in [0,1]$ such that $r + s \in [0,1]$

A7 $P^{>}_{r_i}\varphi \rightarrow P^{\geq}_{r_{i+1}}\varphi$ $i < n$

2.2. Lemma. The following are derivable in P_FD:

a. $P^{\leq}_{1}\varphi$ b. $P^{\geq}_{1}\varphi \leftrightarrow P^{=}_{1}\varphi$

Proof.

a. $\neg P^{\leq}_{1}\varphi \Rightarrow_{def3} P^{<}_{0}\neg\varphi \Rightarrow_{def2} P^{>}_{1}\varphi \Rightarrow_{def1} \neg P^{\geq}_{0}\neg\varphi \Rightarrow_{A4} \perp$.

b. $P^{=}_{1}\varphi \Rightarrow_{def4} \neg P^{>}_{1}\varphi \wedge \neg P^{<}_{1}\varphi \Rightarrow_{A1} \neg P^{<}_{1}\varphi \Rightarrow_{def2} \neg P^{>}_{0}\neg\varphi \Rightarrow_{def1} P^{\geq}_{1}\varphi$

$P^{\geq}_{1}\varphi \Rightarrow_{a} P^{\geq}_{1}\varphi \wedge P^{\leq}_{1}\varphi \Rightarrow_{def1,\, def3} \neg P^{>}_{0}\neg\varphi \wedge \neg P^{<}_{0}\neg\varphi \Rightarrow_{def2} \neg P^{<}_{1}\varphi \wedge \neg P^{>}_{1}\varphi \Rightarrow_{def4} P^{=}_{1}\varphi$.

2.3. Remark. Although definitions def1 - def4 are very straightforward in our notation, we only ensure in 2.6. that they have the desired properties; for instance, that $P^{\geq}_{r}\varphi$ equals $(P^{>}_{r}\varphi \vee P^{=}_{r}\varphi)$. By 2.2.b, R2 is equivalent to $(\vdash\varphi \Rightarrow \vdash P^{=}_{1}\varphi)$, which is perhaps a better motivation for the name 'Necessitation'. Axiom 7 is the axiom where our finite set F comes into play. The axiom forbids any formula to have a probability which is not in F: if the probability for φ is greater than some $r_i \in F$, it must be at least the next value of F, viz. r_{i+1}. Axiom A6 resembles a natural property of probability measures ('additivity'). A5 assures that the probability of a disjunction is distributed over the disjunctives. A4, together with lemma 2., guarantees that probabilities are in a proper range: $\vdash (P^{\geq}_{0}\varphi \wedge P^{\leq}_{1}\varphi)$. Axiom A2 implies that the probability of the consequent of a necessary implication is at least the probability of the antecedent. It also records that 'greater than' implies 'at least'. A3 is the axiom allowing one to jump from one probability to another (a smaller, to be more precise). Although A3 is, like A2, stated for necessary implications, we can weaken it to A3': $P^{\geq}_{r}\varphi \rightarrow P^{>}_{s}\varphi$, $r, s \in [0,1]$ and $s < r$.

The name 'Necessitation rule' for R2 is motivated by the following.

2.4. Proposition. Let, by definition, $\Box\varphi = P^{\geq}_{1}\varphi$. Then:

a. (1) $\vdash \varphi \Rightarrow \vdash \Box\varphi$

(2) $\vdash \Box(\varphi \rightarrow \psi) \rightarrow (\Box\varphi \rightarrow \Box\psi)$

(3) $\vdash \neg\Box\bot$

b. We say that a formula $\in$ L is *modal* if it is build from atomic propositions, the logical connectives and the operator $\Box$. We allege, for all modal formulas φ: $P_FD \vdash \varphi$ iff $KD \vdash \varphi$.

2.5. Definition. A logic $\mathcal{L} \supseteq (P_FD \setminus A7)$ is *permitting a probability-assignment* (ppa) if for each $\mathcal{L}$-consistent set Γ and formula φ there is an $r \in [0,1]$, such that $\Gamma \cup \{P^{=}_{r}\varphi\}$ is $\mathcal{L}$-consistent.

Obviously, the property of ppa is necessary to allow models that assign probabilities to formulas. We will show in the sequel that P_FD is ppa. It guarantees completeness with respect to the semantics of section 3, but first we give some properties of P_FD (for proofs, cf [H2]), showing that def1 - def4 are 'right', and in preparation of our peremptory lemma.

2.6. Theorem. The following are P_FD-theorems for all $\varphi, \psi \in$ L and all r, s $\in [0,1]$.

a. $P^{\geq}_{r}\varphi \leftrightarrow (P^{>}_{r}\varphi \vee P^{=}_{r}\varphi) \wedge P^{\leq}_{r}\varphi \leftrightarrow (P^{<}_{r}\varphi \vee P^{=}_{r}\varphi)$

b. $P^{>}_{r}\varphi \mathbin{\nabla} P^{=}_{r}\varphi \mathbin{\nabla} P^{<}_{r}\varphi$

c. $\neg(P^{=}_{r}\varphi \wedge P^{=}_{s}\varphi)$ $\qquad (r \neq s)$

d. $(\neg P^{<}_{r}\varphi \leftrightarrow P^{\geq}_{r}\varphi) \wedge (\neg P^{>}_{r}\varphi \leftrightarrow P^{\leq}_{r}\varphi)$

e. $P^{=}_{r}\varphi \leftrightarrow (P^{\geq}_{r}\varphi \wedge P^{\leq}_{r}\varphi)$

f. $(P^{>}_{r}\varphi \rightarrow P^{>}_{s}\varphi)$ $\qquad (s \leq r)$

g. $P^{=}_{r}\varphi \leftrightarrow P^{=}_{1-r}\neg\varphi$

Proof. We prove the '$\rightarrow$'-part of the first conjunct of a:

$P^{\geq}_{r}\varphi =_{def1} \neg P^{>}_{1-r}\neg\varphi =_{def2} \neg P^{<}_{r}\varphi \Rightarrow_{A1} (\neg P^{<}_{r}\varphi \wedge P^{>}_{r}\varphi) \vee (\neg P^{<}_{r}\varphi \wedge \neg P^{>}_{r}\varphi) \Rightarrow_{A1} P^{>}_{r}\varphi \vee (\neg P^{<}_{r}\varphi \wedge \neg P^{>}_{r}\varphi) \Rightarrow_{def4} P^{>}_{r}\varphi \vee P^{=}_{r}\varphi$.

Now we state our 'peremptory-lemma'. In the section 3 we will see that this, with theorem 2.6 and 2.9, are the properties that we use when building a model from maximal consistent sets.

2.7. (Peremptory) Lemma. $\vdash P^{=}_{r_0}\varphi \mathbin{\nabla} P^{=}_{r_1}\varphi \mathbin{\nabla} \ldots \mathbin{\nabla} P^{=}_{r_n}\varphi$ (recall $F = \{0 = r_0, \ldots, r_n = 1\}$).

Proof. By theorem 2.6.c, it is sufficient to show that $\vdash P^{=}_{r_0}\varphi \vee P^{=}_{r_1}\varphi \vee \ldots \vee P^{=}_{r_n}\varphi$.

If $i < n$, $P^{\geq}_{r_i}\varphi \Rightarrow_{2.6.a} P^{=}_{r_i}\varphi \vee P^{>}_{r_i}\varphi \Rightarrow_{A7} P^{=}_{r_i}\varphi \vee P^{\geq}_{r_{i+1}}\varphi$ (1).

Also, $P^{\geq}_{1}\varphi \Rightarrow_{2.6.a} P^{>}_{1}\varphi \vee P^{=}_{1}\varphi \Rightarrow_{def2} P^{<}_{0}\neg\varphi \vee P^{=}_{1}\varphi \Rightarrow_{2.6.d} \neg P^{\geq}_{0}\neg\varphi \vee P^{=}_{1}\varphi \Rightarrow_{A4,A1} P^{=}_{1}\varphi$ (2).

Now we apply (1) n times to A4, and then (2) to its result:

$\vdash P^{\geq}_{0}\varphi \Rightarrow_{\text{def F}} P^{\geq}_{r_0}\varphi \Rightarrow_{(1)} P^{=}_{0}\varphi \vee P^{\geq}_{r_1}\varphi \Rightarrow_{(1)} P^{=}_{0}\varphi \vee P^{=}_{r_1}\varphi \vee P^{\geq}_{r_2}\varphi \Rightarrow_{(1)} \ldots \Rightarrow_{(1)} P^{=}_{r_0}\varphi \vee P^{=}_{r_1}\varphi \vee \ldots \vee P^{\geq}_{r_n}\varphi \Rightarrow_{(2)} P^{=}_{r_0}\varphi \vee P^{=}_{r_1}\varphi \vee \ldots \vee P^{=}_{r_n}\varphi$.

2.8. Definition.

$r\Uparrow = \min \{s \in F \mid s > r\}$ $\qquad r \in [0,1)$

$r\Downarrow = \max \{s \in F \mid s < r\}$ $\qquad r \in (0,1]$

2.9. Corollary. The following are $P_F D$-theorems ($r, s \in [0,1]$):

a. $P^{>}_{r}\varphi \rightarrow P^{=}_{r_i}\varphi \,\nabla\, P^{=}_{r_{i+1}}\varphi \,\nabla \ldots \nabla\, P^{=}_{r_n}\varphi$, with $r_i = r\Uparrow$

b. $P^{<}_{r}\varphi \rightarrow P^{=}_{r_0}\varphi \,\nabla\, P^{=}_{r_1}\varphi \,\nabla \ldots \nabla\, P^{=}_{r_j}\varphi$, with $r_j = r\Downarrow$

c. $(P^{>}_{r}\varphi \leftrightarrow P^{\geq}_{r\Uparrow}\varphi) \wedge (P^{<}_{s}\varphi \leftrightarrow P^{\leq}_{s\Downarrow}\varphi)$ $\qquad r \in [0,1), s \in (0,1]$

d. $[P^{\geq}_{1}\neg(\varphi \wedge \psi) \wedge P^{=}_{r}\varphi] \rightarrow [P^{=}_{s}\psi \leftrightarrow P^{=}_{r+s}(\varphi \vee \psi)]$ $\qquad r, s, r+s \in [0,1]$

2.10. Corollary. $P_F D$ is ppa.

3. A semantics for $P_F D$ and some of its properties

3.1. Definition. For a base F $\mathcal{QPK}_F$ is the class of models $\mathcal{M} = \langle W, R, P_F, V\rangle$ for which;

a. W is a non-empty set (of 'worlds')

b. R is a binary relation on W that is serial, i.e: $\forall w \exists v Rwv$.

c. P_F is a function P_F: $W \times L \rightarrow F$, satisfying:

- $P_F(w,\varphi \wedge \psi) = 0 \Rightarrow P_F(w,\varphi \vee \psi) = P_F(w,\varphi) + P_F(w,\psi)$
- $P_F(w,\varphi) = 1$ iff for all v such that Rwv, $(\mathcal{M},v) \models \varphi$

d. V is a valuation: $W \times L \rightarrow \{\textit{true}, \textit{false}\}$.

e. The truth-definition for formulas is defined inductively, with interesting cases:

$(\mathcal{M},w) \models P^{\geq}_{1}\varphi$ iff for all v such that Rwv, $(\mathcal{M},v) \models \varphi$

$(\mathcal{M},w) \models P^{>}_{r}\varphi$ iff $P_F(w,\varphi) > r$

What is peculiar in this semantics is, that to verify $P^{\geq}_{1}\varphi$ at w, a condition has to be fulfilled both for R and P_F. For the other operators, ($P^{>}_{r}$, $r < 1$), only P_F plays a role. Where R is defined upon the set of worlds (i.e. interpretations of formulas), P_F is defined purely syntactically. As a consequence, verifying the truth of a formula at a world w is typically not recursively an occupation. For instance, to verify the truth of $P^{>}_{r}P^{<}_{s}P^{>}_{t}p$ at world w, one only compares $P_F(w,P^{<}_{s}P^{>}_{t}p)$ with the value r.

3.2 Proposition. P_FD is sound with respect to $\mathcal{QPK}_F$, i.e. $P_FD \vdash \varphi \Rightarrow \mathcal{QPK}_F \models \varphi$.

The following theorem ("completeness"), stating that for each P_FD-consistent formula there is a $\mathcal{QPK}_F$-model $\mathcal{M}$, is also to be found in [FA], but our proof differs in essence. Where the proof in [FA] uses infinite maximal consistent sets Γ of formulas, computing for each Γ the sup of $P^{\geq}_{r}\varphi$ for which φ in Γ and using properties of converging sequences $<s_n : n \in \omega>$ for which $P^{\geq}_{s_n}\varphi \in \Gamma$, our construction of $\mathcal{M}$ entirely rests upon our peremptory-lemma. In some sense, we can prove the existence of an assignment P_F to each formula φ at each world (i.e., maximal consistent set) *within the logic P_FD!*

3.3. Theorem. Let ψ be any P_FD-consistent formula. Then there is a $\mathcal{QPK}_F$-model $\mathcal{M}$ and a world w such that $(\mathcal{M},w) \models \psi$.
Proof. We construct a *canonical model* $\mathcal{M}^c = <W^c, R^c, P_F^c, V^c>$ as follows:
(1) $W^c = \{\Gamma \mid \Gamma \text{ is maximal consistent in } P_FD\}$
(2) $R^c\Gamma\Delta$ iff for all $P^{\geq}_{1}\varphi \in \Gamma$ it holds that $\varphi \in \Delta$.
(3) $P_F^c(\Gamma,\varphi) = r$ iff $P^{=}_{r}\varphi \in \Gamma$.
(4) $V^c(\Gamma,p) = \mathit{true}$ iff $p \in \Gamma$.
Definition (3) is correct because of our peremptory-lemma. It also guarantees that $P_F^c : W \times L \rightarrow F$. An easy induction, which we omit here, shows:
(*) $(\mathcal{M}^c,\Gamma) \models \varphi$ iff $\varphi \in \Gamma$.

3.4. Corollary. For all $\varphi \in L$, $P_FD \vdash \varphi \Leftrightarrow \mathcal{QPK}_F \models \varphi$.

Now that we know that each consistent formula is satisfiable, some natural questions come to surface with respect to the nature of this satisfiability. For instance, is there always a *finite* model? If so, can this model be *obtained constructively* (i.e., using only finite sets and finitary means)? What does this mean for the matter of *decidability* of P_FD? We will present answers to this questions in (3.8, 3.9 and 3.10, respectively).

Key-idea for the proof of the finite model property (f.m.p.) is obtained from a technique called *filtration* (cf. [HC]). However, we need a major adjustment. Suppose $(\mathcal{M},w) \models \varphi$. Let Φ be the set of subformulas of φ. In standard filtration-techniques, each equivalence class [v] of worlds $u \in W$ verifying the same Φ-formulas defines a world v^* in W^*. There are finitely many of those classes, together defining W^*. Then, V^* is defined, for $p \in \Phi$, as $V^*([w],p) = V(w,p)$. However, we can not do the same thing for P_F^*, since two worlds u and v may have the same Φ-theory without satisfying $P_F(u,p) = P_F(v,p)$. One might try to define:

(Δ) $[u] = [v] \Leftrightarrow$ u and v verify the same Φ-formulas *and* $P_F(u,\phi) = P_F(v,\phi)$ for all $\phi \in \Phi$,

However, (Δ) would not help us here because it does not guarantee a finite set of equivalence classes. Therefore we will not use these equivalence classes as our worlds in W^*, but rather choose a representative in each class. We define this construction in 3.5.

3.5. Definition. Let Φ be a set of formulas closed under subformulas, and $\mathcal{M} = \langle W, R, P_F, V\rangle$ a $\mathcal{QPK}_F$-model. We define $\mathcal{M}^* = \langle W^*, R^*, P_F^*, V^*\rangle$, the *filtration of* $\mathcal{M}$ *through* Φ as follows. Starting point is the equivalence relation induced by Φ on W: $[u] = [v]$ iff for all $\varphi \in \Phi$, $(\mathcal{M},u) \models \varphi \Leftrightarrow (\mathcal{M},v) \models \varphi$. Let $W+ = \{x \mid x = [w]$ for some $w \in W\}$ denote the set of those equivalence classes. Let f: $W+ \rightarrow W$ associate one unique world with each equivalence class. Let !: $W \rightarrow W$ be a function that associates with each $w \in W$ the unique representative of its class: $w! = f([w])$. Note that $x! \in [x]$, and that $x! = y! \Leftrightarrow [x] = [y]$. We define $\mathcal{M}^*$:

(1) $W^* = \{w \in W \mid w = f([x])$ for some $[x] \in W+\} = \{w \in W \mid w = w!\}$

(2) R^*: (i) for any $x, y \in W^*$, if there is some $z \in W$ such that Rxz and $z \in [y]$, then R^*xy

(ii) if R^*xy, then for any $P_1^{\geq}\varphi \in \Phi$: $(\mathcal{M},x) \models P_1^{\geq}\varphi \Rightarrow (\mathcal{M},y) \models \varphi$

(3) $P_F^*(x,\varphi) = P_F(x,\varphi)$, for all $\varphi \in \Phi$, $x \in W^*$.

(4) $V^*(x,p) = V(x,p)$, for all $p \in \Phi$, $x \in W^*$.

3.6. Lemma. For all $w \in W$, $\varphi \in \Phi$, $r \in [0,1]$: $P_F^*(w!,\varphi) > r \Leftrightarrow P_F(w,\varphi) > r$.

3.7. (Filtration-) lemma. Let $\mathcal{M}^* = \langle W^*, R^*, P_F^*, V^*\rangle$ be a filtration of $\mathcal{M} = \langle W, R, P_F, V\rangle$ through Φ. Then:

(i) for all $\phi \in \Phi$: $(\mathcal{M},w) \models \phi \Leftrightarrow (\mathcal{M}^*,w!) \models \phi$

(ii) $\mathcal{M}^* \in \mathcal{QPK}_F$, if $\mathcal{M} \in \mathcal{QPK}_F$

Proof.

(i) We prove (i) with induction on φ. The interesting cases are $P_r^{>}\psi$ and $P_1^{\geq}\psi$, the former not relying on any induction hypothesis. $(\mathcal{M},w) \models P_r^{>}\psi \Leftrightarrow_{3.1} P_F(w,\psi) > r \Leftrightarrow_{3.6} P_F^*(w!,\psi) > r \Leftrightarrow_{3.1} (\mathcal{M}^*,w!) \models P_F(w!,\psi)$. Next, suppose $(\mathcal{M},w) \models P_1^{\geq}\psi$. Then, for all v with Rwv: $(\mathcal{M},v) \models \psi$ & $P_F(w,\psi) = 1$. We have to show that for all y with $R^*w!y$, $(\mathcal{M}^*,y) \models \psi$ and $P_F^*(w!,\psi) = 1$. So, suppose $R^*w!y$. Since $P_r^{>}\psi \in \Phi$ and $P_F(w,\psi) = 1$, we know from 3.5(2)(ii) that $(\mathcal{M},y!) \models \psi$. Next, since $P_F(w,\psi) = 1$ and $P_1^{\geq}\psi \in \Phi$, we derive that $P_F(w!,\psi) = 1$(w and w! verify the same Φ-formulas). Then, using 3.5(3), we get $P_F^*(w^*,\psi) = 1$. Conclusion: $(\mathcal{M}^*,w!) \models P_1^{\geq}\psi$. Conversely, suppose $(\mathcal{M},w!) \models P_1^{\geq}\psi$. Then for all y! such that $R^*w!y!$, $(\mathcal{M},y!) \models \psi$ and also $P_F^*(w!,\psi) = 1$. From 3.5.(3) immediately follows $P_F(w,\psi) = 1$. Suppose Rwv. By 3.5(2)(i), we know that $R^*w!v!$, so $(\mathcal{M},v!) \models \psi$. Our induction hypothesis says that $(\mathcal{M},v) \models \psi$ and thus $(\mathcal{M},w) \models P_1^{\geq}\psi$.

(ii) Note that all the clauses of definition 3.1 are fulfilled for $\mathcal{M}^*$, since $W^* \subseteq W$ and V^* and R^* are just the proper restriction of V and R to W^*, respectively. We only have to show that $\mathcal{M}^*$ is serial. Choose $w \in W^*$. Since $w \in W$, we there is some $v \in W$ with Rwv. Now $v! \in W^*$ and $v \in [v!]$, so that, by 3.5(3), $R^*wv!$.

Now the f.m.p. follows from the filtration lemma by observing that if $(\mathcal{M},w) \models \varphi$, the set of subformulas Φ of φ is finite, and one can make a filtration $\mathcal{M}^*$ (which is finite and a $\mathcal{QPK}_F$-model) of $\mathcal{M}$ through Φ.

3.8. Corollary. P_FD has the *finite model property*, in the sense that every P_FD consistent formula is satisfiable in a finite model.

Now we know a way to distil a model for φ from an arbitrary model for φ. However, if we now want to give a finite model for any consistent formula, we firstly have to construct our canonical model (cf. 3.3) and then filtrate this model through φ (cf. 3.5) to show the formula's finite satisfiability. In [H2] a method is given to directly construct a finite model for a given formula, using only finite sets, assigning a truth-value to a finite number of propositional atoms and a probability value to a finite number of formulas. So:

3.9. Theorem. For any P_FD-consistent formula φ, a finite model $\mathcal{M}^c_{fin} = \langle W', R', P_F', V' \rangle$ can be directly build for φ using only finitary means: W' is a finite set of finite consistent sets, P_F' and V' need only be specified for a finite set of formulas, in each world.

Since P_FD is not a finite axiomatisation (A3 reflects an infinite set of axioms: for each $r > 0$, there are (uncountable) infinitely many $s < r$) 3.8 does not guarantee decidability of P_FD. Indeed, the fact that there is a finite model for each consistent formula, does not imply that there are only finitely many possible models that refute $\neg\varphi$. For, if P_F is defined for all propositional atoms in φ, it is not the case that $P_F\psi$ is determined for all ψ that have no other atoms than φ has (see our comment following definition 3.1). However, there is this 'narrow escape'.

3.10. Theorem. P_FD is decidable.

Proof. Let φ be any formula $\in$ L. From 3.8, we deduce that if φ is not a theorem, there is a finite model $\mathcal{M}$ that refutes φ. There are only finitely many candidates for this $\mathcal{M}$ as is seen as follows. Let Φ be the set of subformulas of $\neg\varphi$, and $P^{-1}(\Phi) = \{\psi \mid$ there is some $r \in [0,1]$ and ~

$\in \{<, >, \leq, \geq, =\}$ such that $P_r^{\sim}\psi \in \Phi\}$. Then an upper bound of the magnitude of the models we have to consider (as a candidate for refuting φ) is $2^{|\Phi|} \times |P^{-1}(\Phi)|^{|F|}$: for each non-equivalent world (a priori there are at most $2^{|\Phi|}$ many) we have to assign probabilities to each formula of $P^{-1}(\Phi)$. For each of its members, there are only $|F|$ many choices for such a probability.

Now we make precise, that for P_FD, the difference between assigning probabilities to sets (of situations, or worlds) and to formulas is not a crucial one.

3.11. Definition. A Probability Kripke model $\mathcal{M}$ over F is a tuple $<W, R, P_F, V>$ where W, R and V are as in a Quasi Probability Kripke model (definition 3.1) but now P_F: $W \times \mathcal{P}(W) \to F$ is a function from the powerset of W to F, for each $w \in W$, satisfying:

$$X \cap Y = \emptyset \Rightarrow P_F(w,X \cup Y) = P_F(w,X) + P_F(w,Y) \qquad X, Y \in \mathcal{P}(W)$$

$$P_F(w,\{v \mid Rwv\}) = 1$$

The truth definition for P_FD formulas is obtained straightforwardly, with the modal case

$$(\mathcal{M},w) \models P_r^{>}\varphi \text{ iff } P_F(w,\{v \mid Rwv \text{ and } (\mathcal{M},v) \models \varphi\}) > r$$

Note that now we have a truly recursive truth-definition, and that we no longer need a separate clause for $P_r^{\geq}\varphi$ (cf. definition 3.9). We denote the class of all these models by $\mathcal{PK}_F$.

We might allow some more generality here. For instance, we might leave out the accessibility relation R as a primitive and define $Rwv \Leftrightarrow P_F(w,\{v\}) > 0$ (cf. [H1]). Also, instead of forcing the range of P_F to be F, we might weaken this to

$$P_F(w,X) \in F \text{ for each X that is the denotation of a formula } \varphi$$

However, for our purposes, definition 3.11 is sufficient.

Given this definition of Probability model, we elaborate a bit on the problem of compactness for P_FD-like systems, and the role of A7 in this. Axiom A7 is a logical compromise. On one hand, it restricts us to probability measures with finite range, on the other hand, it guards us against some serious logical complications. To be more precise, consider $P_FD \setminus A7$, and let

$$\Gamma = \{\neg P_r^{=}q \mid r \in [0,1]\}.$$

In [H2] we claim is that Γ is consistent. However, we can also use Γ to show that $P_FD \setminus A7$ is not compact. For, although Γ is not satisfiable on General Probabilistic Kripke Models ($\mathcal{GPK}$): (models similar to those of $\mathcal{PK}_F$, but allowing F to be $\mathbb{R}$) each finite subset of Γ is (in fact each proper subset Γ' of Γ is). Stated equivalently: although (for each $s \in [0,1]$), $\Gamma \setminus \{P_s^{=}\varphi\} \models P_s^{=}\varphi$, there is no finite subset $\Gamma' \subseteq \Gamma \setminus \{P_s^{=}\varphi\}$ such that $\Gamma' \models P_s^{=}\varphi$. It seems that the only way to resolve this (at least to avoid consistency of Γ) is to admit an infinitary logic, for which $\sum_{r \in [01]} P_r^{=}\varphi$ holds, i.e., guaranteeing that each assertion has a probability. We only know of one attempt of allowing an infinite rule that guarantees such a property (in a logic for '$P_s^{=}$'), (cf.[Al]).

3.12. Definition. We define a canonical model $\mathcal{M}^c$, given a consistent formula φ. Let Ψ be as defined in 3.9: the set of subformulas of φ closed under single negation and satisfying

$$(P_r^{\sim}\varphi \in \Phi \text{ for any } \sim \in \{<, >, \leq, \geq, =\} \Rightarrow \{P_{r_i}^{=}\varphi \mid r_i \in F\} \subseteq \Psi).$$

Ψ is finite, say $|\Psi| = k$. Let $\Gamma_1, \ldots \Gamma_n$ be the Ψ-maximal consistent sets, $n \leq 2^k$, and γ_i the conjunction of formulas in Γ_i, $i \leq n$. Let $\Psi' = \Psi \cup \{P_r^{=}\gamma_i \mid r \in F,\ i \leq n\}$. Using standard arguments, we have:

i. $\vdash \neg(\gamma_i \wedge \gamma_j),\ i \neq j$

ii. $\vdash (\gamma_1 \vee \ldots \vee \gamma_n)$

iii. $\vdash \psi \leftrightarrow (\gamma_{\psi 1} \vee \ldots \vee \gamma_{\psi r})$, where $\gamma_{\psi 1}, \ldots, \gamma_{\psi r}$ are exactly those γ's which contain ψ as a conjunct, for each $\psi \in \Psi$.

Now let $\Delta_1, \ldots \Delta_n$ be 'saturated' extensions of $\Gamma_1, \ldots, \Gamma_n$:

a. $\Gamma_i \subseteq \Delta_i,\ i \leq n$,

b. $\gamma_i \in \Delta_i,\ i \leq n$,

c. for each Δ_i and γ_j, $j,i \leq n$, there is a $r_j \in F$ such that $P_{r_j}^{=}\gamma_j \in \Delta_i$ (possible, by 2.7).

Our canonical Probability model $\mathcal{M}^c = \langle W^c, R^c, P_F{}^c, V^c \rangle$ is defined as follows:

(1) $W^c = \{\Delta_1, \ldots \Delta_n \mid \Delta_i \text{ defined as above}\}$

(2) $R^c\Delta_i\Delta_j$ iff for all $P_1^{\geq}\varphi \in \Delta_i$, it holds that $\varphi \in \Delta_j$, $i, j \leq n$.

(3) $P_F{}^c(\Delta_i,X) = r_{i1} + \ldots r_{ini}$, if $X = \{\Delta_{i1}, \ldots \Delta_{ini}\} \subseteq W^c$ and $\{P_{r_{i1}}^{=}\gamma_{i1}, \ldots P_{r_{ini}}^{=}\gamma_{ini}\} \subseteq \Delta_i$, $i, i_1, \ldots, i_{ni} \leq n$.

(4) $V'(\Delta_i,p) = \mathit{true}$ iff $p \in \Delta_i$, for all $p \in \Psi'$.

Note that in contrast to our Quasi Probabilistic canonical models, the truth of any formula $\vec{O}\varphi$, where $\vec{O}$ is a sequence of probability operators and $\psi \in \Psi'$, is determined. Note also that $P_F(\Delta_i,X)$ is determined by the values it assigns to singletons $\{x\} \subseteq X$.

3.13. Lemma. Let $\mathcal{M}^c$ be as defined in 3.12, for a given consistent formula φ. Then:

for all $\psi \in \Psi'$, $\Delta_i \in W^c$, $(\mathcal{M}^c,\Delta_i) \models \psi$ iff $\psi \in \Delta_i$ $(i \leq n)$.

3.14. Corollary. For all P_FD-formulas φ, $\mathcal{PK}_F \models \varphi$ iff $\mathcal{QPK}_F \models \varphi$.

3.15. Corollary. P_F' of 3.9 extends straightforwardly to formulas not occurring in Ψ.

4. Conclusion and directions for further research

We were able to use some elementary P_FD-theorems for constructing its canonical model and also provided a method to obtain a model for any consistent formula using only finitary means. Moreover, we showed that P_FD has the finite model property by developing a filtration method on the models for P_FD and obtained decidability. We indicated that P_FD as a system combining modal- and probability logic has two major drawbacks. Firstly, the semantics introduced in [FA] and used here initially, have a crucial syntactic component. We showed that when interpreting the system on Probability Kripke models with range F, how methods developed here can be applied, thus providing a bridge to other approaches that study probabilistic Kripke models ([Se], [H1]).

Secondly, a challenge P_FD leaves us with, as also mentioned in [FA], is the question whether the restriction to a finite F can be abandonned (resulting in a logic that *is* compact), in favour of a regular probability measure (on the set of worlds). Although this seems to be a difficult task, we think that under some restrictions, this is possible. One approach is to allow for infinitary rules (cf. [Al]), but we think there are perhaps other ways to tackle this problem.

Acknowledgements. I like to thank Professor J.-J.Ch. Meyer for his very profound reading of an earlier version of this paper, and also for his stimulating discussions on the subject.

References.

[Al] N.A. Alyoshina, *Probabilistic Logic 1;* report Inst. of Philosophy, Moscow (1991).
[Ch] B. Chellas, *Modal Logic, an Introduction;* Cambridge University Press (1980).
[Dr] D. Driankov, *Reasoning under Uncertainty: Towards a Many-valued Logic of Belief;* IDA research report, Linköping University & Institute of Technology (1987) 113-120.
[FA] M. Fattorosi-Barnaba & G. Amati, *Modal operators With Probabilistic Interpretations, I;* Studia Logica 4, 1989 383 - 393.
[FH] R.F. Fagin & J.Y. Halpern, *Reasoning about Knowledge and Probability*; Proc. of the 2nd. Conf. on Theoretical Aspects of Reasoning About Knowledge (1988) 277-295.
[HC] G.E. Hughes & M.J. Creswell, *A Companion to Modal Logic;* Methuen, London (1984).
[H1] W. van der Hoek, *Qualitative Modalities,* proceedings of the Scandinavian Conference on Artificial Intelligence 91, B. Mayoh (ed.) IOS Press, Amsterdam (1991), 322-327.
[H2] W. van der Hoek, *Some Considerations on the Logic P_FD*, Report Free University IR-227, Amsterdam (1990).
[HR] J.Y. Halpern and Rabin, *A Logic to Reason about Likelihood;* Artificial Intelligence 32:3 (1987) 379 - 405.
[Le] W. Lenzen, *Glauben, Wissen und Warscheinlichkeit;* Wien, Springer Verlag (1980).
[Se] K. Segerberg, *Qualitative Probability in a Modal Setting;* Proc. of the 2nd Scand. Log. Symp., Fenstad (ed.), Amsterdam (1971).

Logic Programming with Bounded Quantifiers

Andrei Voronkov*
ECRC
Arabellastr.17
8000 Munich 81
Germany
voronkov@ecrc.de

Abstract

This paper describes an extension of Horn clause logic programs by bounded quantifiers. Bounded quantifiers had been extensively used in a part of mathematical logic called theory of admissible sets [2]. Later some variants of bounded quantifiers had been introduced in logic programming languages [12, 19, 21, 9, 6, 7]. We show that an extension of logic programs by bounded quantifiers has several equivalent logical semantics and is efficiently implementable using a variant of SLD-resolution, which we call SLDB-resolution. We give examples showing that introduction of bounded quantifiers results in a high level logical specification language. An expressive power of subsets of Horn clauses and subsets of logic programs with bounded quantifiers is compared. We also show that the use of bounded quantifiers sheds new light on classical negation in logic programming.

1 Introduction

Although Horn clause programs are sufficiently rich to express all computable predicates, they are not sufficiently expressive to naturally represent some relations which are easily expressed in richer languages, for example in full first order logic. Since the publication of early papers on logic programming and on Prolog there has been many attempts to extend Horn clauses programs in various ways. Examples of practical extensions are numerous in different implementations of Prolog. These are usually control primitives like cut or built-in primitives. Among more theoretical extensions are programming with full first order logic, higher order logic programming, etc.

There are many foundational problems with extensions of Horn clauses. The practical solutions are mostly non-logical which means that they have no natural logical semantics. As a consequence, such programs are more difficult to understand and to verify. The problem with more theoretical extensions is that most of them cannot be efficiently implemented. In most cases the inefficiency is inherent – for example negation combined

*On leave from the International Laboratory of Intelligent Systems (SINTEL), 630090, Universitetski Prospect 4, Novosibirsk 90, Russia.

with recursion leads to non-computable predicates. Similar problem arise when using universal quantifiers.

Here we present an extension of logic programming with *bounded quantifiers* – i.e. quantifiers over finite domains. We prove that this extension can be efficiently implemented. Moreover we show on some practical examples that bounded quantifiers can be used in practical cases to express iterative algorithms and to specify the exhaustive search over finite domains. Our extension is logical in the sense that it enjoys a complete and sound model-theoretical semantics while still being efficiently implementable.

In our presentation we also handle built-in domains, so our semantics can in principle be easily generalized to a semantics of constraint logic programming as well. The motivation to consider built-in domains with built-in predicates defined on these domains was to handle correctly non-toy examples (and in particular not to define natural numbers via the constant 0 and the function symbol s).

In fact we extend the usual model-theoretical semantics of logic programming in three ways. The first extension consists of considering built-in predicates as well. The second extension is the introduction of sorts. And the third extension is the introduction of bounded quantifiers.

Intuitively, bounded quantifiers are quantifiers ranging over some finite domains, in particular over finite lists or sets. Expressions containing bounded quantifiers give natural and elegant examples of executable specifications. Consider, for example, the specification of disjoint sets

$$\mathtt{disjoint}(S_1, S_2) \text{ iff } (\forall x_1 \in S_1)(\forall x_2 \in S_2)\, x_1 \neq x_2$$

This specification implies the obvious way to check if two given sets are disjoint. Similar uses of the bounded quantifiers can be found already in the language SETL [19] in which the concept of a set is a first-class concept. It seems very natural to use such expressions in logic programs as well. Combining the technique of finite search with the logic programming techniques (a variant of SLD-resolution and unification) allows one to use such specifications for constructing sets with given properties. The use of such specifications also makes logic programming more logical.

Although the semantics of the above expression is quite clear, its usual representation in Horn clause logic programming

```
disjoint([],S).
disjoint([A|As],S) :-
   nonmember(A,S),
   disjoint(As,S).

nonmember(A,[]).
nonmember(A,[B|Bs]) :-
   A≠B,
   nonmember(A,Bs).
```

is in contrast not so easy to understand. The Prolog program for `disjoint` also lacks in elegance compared to the specification using bounded quantifiers.

This paper is concerned with the logical justification of logic programming with bounded quantifiers. We restricted our attention only to two types of the bounded quantifiers, originally introduced in [12], but our techniques is quite general and is also applicable to another kinds of bounded quantifiers.

In this section we give the basic definitions of sorts, lists and generalized logic programs. Some natural examples of specifications using bounded quantifiers are given in section 2.4 In section 2 we introduce several semantics of logic programs with bounded quantifiers and prove their equivalence. Section 3 describes the procedural semantics of the language, which generalizes SLD-resolution. A unification algorithm for sorted expressions is also given. In section 5 we discuss the expressive power of the language with bounded quantifiers. In general it is equivalent to the expressive power of the Horn clauses – both kinds of languages can express exactly all computable predicates in the least model semantics . First in section 5.1 we show a natural translation from the language with bounded quantifiers to Horn clauses. This translation can for example be used to automatically obtain from the above definition of `disjoint` with bounded quantifiers a definition of `disjoint` in the language of Horn clauses. Then we construct a non-recursive metainterpreter for Horn clause programs, written in the language with bounded quantifiers. Section 6 is concerned with the use of negation in logic programs. Finally, in section 7 we discuss some other possible applications of logic programming with bounded quantifiers.

1.1 Related work

Bounded quantifiers were considered among others in the following papers[1] [12, 13, 14, 9, 10, 6, 7]. Also related are papers on introducing set constructs in logic programming [3, 9, 10, 8]. For us the main motivation was the series of papers on Σ-programming [12, 13, 14] and some of our results on semantics of Σ-programs and on a translation of Σ-programs into logic programs [22, 26, 27]. Bounded quantifiers were already introduced in the first of the above-mentioned papers [12] which was inspired by the Kripke-Platek formalization of the theory of admissible sets [2]. However the authors could not find a satisfiable procedural semantics for their language. Later ideas for defining such a semantics were introduced in our paper [26] based on the translation of Σ-programs to Horn clause programs described in [22, 27] (some ideas could even be found in even earlier preprints [24, 25], written in 1985).

Later bounded quantifier were introduced in [9, 10] with the purpose of enriching logic programming languages with sets. But the absence of the set constructor in Kuper's language leads to problems with the procedural semantics of his language. Set constructor was introduced later [6] where a more satisfiable procedural semantics for a logic programming language with finite sets had been constructed. However sets are not easy to handle: the unification problem for finite sets is NP-complete [6]. The language presented in our paper has a procedural semantics comparable with SLD-resolution for Horn clauses. As we try to show, most of the applications of logic programming with sets are easily expressed in our language.

[1]Some kind of bounded quantifiers had been implemented in the seventies in a Prolog-like language developed in Hungary (We do not have any exact references).

Bounded quantifiers were also considered in constraint logic programming [7]. In most of the programming literature they are called restricted quantifiers, but bounded quantifiers introduced earlier in the mathematical literature (e.g. [2]) seem to capture better the idea of the search on finite domains.

Bounded quantifiers are usually quantifiers over finite domains represented by lists or sets. In the papers [12, 13, 14, 9, 10] lists and sets are considered as a superstructure of the usual Herbrand universe, which prohibits using terms like f([]), since [] is a list. In our paper we introduce a flexible sort structure which allows to treat lists as an ordinary sort. In [6] function symbols are allowed also to have set arguments. In that paper sorts are not introduced but they are used in the unification algorithm, and without sorts the procedural semantics from [6] becomes incorrect.

The proof theory is different in different approaches. In [12] an analog of Kripke-Platek theory for admissible sets [2] called GES is used for proving properties of Σ-programs. In [9] the underlying proof system is the calculus with the extensionality axiom `∀z(z∈x≡z∈y)⊃x=y`. We can use at least two types of calculi to provide a proof theory for our language. The first is an analog of GES, which treats lists as a special kind of objects with induction axioms for lists. The second is a theory of inductive definitions, which seems more flexible for proving properties of programs. However to prove completeness of our semantics any of the two theories is convenient.

Our approach to defining a procedural semantics for our language, which we call SLDB-resolution, is more efficient compared to the cited papers. In [6] an exponential unification algorithm for sets is used. In [9] no satisfiable procedural semantics is provided. For example given the program

```
q :- (∀x∈Y)fail.
```

and the query `?-q`. Kuper's system has to make a substitution `[Y←{}]` during unification of `q` from the query with `q` from the head of the clause. In [12] the proposed procedural semantic comprises the exhaustive search over infinite universes. According to their first papers, the answer to the query `?-X=5`, where `X` ranges over rational numbers, should be found by the exhaustive search for the substitution for `X` over all rational numbers. In the papers following [13, 14] it was noted that in some cases the search procedure can use other strategies, but in general case their strategy includes the exhaustive search over infinite domains.

However both [9] and [6] have other motivations and for some applications finite sets could be more appropriate than lists used in our paper.

Also similar to bounded quantifiers are some of Zermelo-Fraenkel set theory expressions of the functional programming language Miranda [21].

2 Logic programs with bounded quantifiers

In this section we introduce the main notions of the paper. In our presentation of generalized logic programs sorts are assigned to terms. In section 2.1 we introduce sort definitions and their semantics. In section 2.2 we define an instance of sorts called lists. Lists are used in defining bounded quantifiers and generalized logic programs in section 2.3. We give some examples in section 2.4 to show the practical importance of bounded quantifiers.

2.1 Sorts

Consider a simple expression $(\forall x \in l)\varphi(x)$ containing a bounded quantifier. Intuitively this expressions means that for every element x of thc list l $\varphi(x)$ holds. If such an expression occurs in a query, we need to check that l is a list. However according to the logic programming philosophy l may be any term, for example a variable. Of course we can call a predicate stating that l is a list or a predicate generating all lists each time when such a query is posed. But it would be extremely inefficient and it would obscure the semantics of our language. So we need to distinguish lists from all other elements. To this end we introduced sorts in our language. Sorts are also needed to correctly handle built-in predicates.

The use of non-sorted structures for our purposes is inefficient and leads to some semantic problems. For example, in [12, 14] it is not clear what version of a (many sorted) predicate calculus is used in the list theory GES introduced there, which makes some considerations quite obscure. The operational semantics of the extended logic programming language introduced in [12, 13] is based on model theory and therefore comprises the exhaustive search over an infinite universe. The sort structure introduced here helps to provide an efficient operational semantics for the language with bounded quantifiers. The possibility of unification-based operational semantics was noted in earlier papers [22, 27], but in these papers we used an algorithm verifying if the terms from the binding expressions are lists, which is inefficient in general.

Lets us now give formal definitions. The set of sorts consists of predefined sorts and defined sorts. Defined sorts can be introduced via sort definitions. Suppose that we have two alphabets **Symb** and **Sort** for symbols and sorts and **P** is a distinguished subset of **Sort** (the subset of predefined sorts). Let **Dsort** (defined sorts) stands for **Sort****P**. Then the sort definitions **Def** are defined as

$$\mathbf{Def} := \mathbf{Symb} \in \mathbf{Dsort} \mid \mathbf{Symb} \in \mathbf{Sort} \times \ldots \times \mathbf{Sort} \rightarrow \mathbf{Dsort} \mid \mathbf{Sort} \subseteq \mathbf{Dsort}$$

The intuitive semantics of the sort definitions is the following. A sort expression of the form `Symb∈Sort` means that `Symb` is a constant of the sort `Sort`. The expression $\texttt{Symb}\in\texttt{Sort}_1\times\ldots\times\texttt{Sort}_n\rightarrow\texttt{Sort}$ means that `Symb` is a function symbol of the sort $\texttt{Sort}_1\times\ldots\times\texttt{Sort}_n\rightarrow\texttt{Sort}$ and $\texttt{Sort}_1\subseteq\texttt{Sort}_2$ means that each term of the sort $\texttt{Sort}_1$ is also a term of the sort $\texttt{Sort}_2$.

Below are some examples of the sort definitions:

```
nil∈Atom
Integer⊂Atom
String⊂Atom
nil∈List
cons∈List×List→List
cons∈Atom×List→List
nil∈Words
cons∈Atom×Words→Words.
```

Here `Integer`, `Strings` are predefined sorts.

From now on we assume that a finite set **P** of predefined sorts and a finite set **D** of sort definitions are fixed.

The idea to restrict variables to sorts has also some semantic advantages. If the sorts are fixed then the semantics of programs is not sensitive to extensions of the Herbrand Universe. Such sensitivity may cause problems, e.g. in the semantics of negation as failure in logic programming.

The definition of terms of a sort α is straightforward from the definition of sorts. We assume that for each sort α we have a countable set C_α of variables of this sort. Let for each predefined sort α C_α denotes the set of all constants of this sort. The set $Term_\alpha$ of terms of the sort α is defined as follows:

1. If the sort α is predefined then $Term_\alpha = V_\alpha \cup C_\alpha$;
2. $V_\alpha \subseteq Term_\alpha$;
3. For every sort definition in **D** of the form $c \in \alpha$, we have $c \in Term_\alpha$;
4. For each definition of the form

   $$f \in \alpha_1 \times \ldots \times \alpha_n \rightarrow \alpha$$

   and for each $t_1 \in Term_{\alpha_1}, \ldots, t_n \in Term_{\alpha_n}$, we have $f(t_1, \ldots, t_n) \in Term_\alpha$;
5. For each definition of the form $\alpha \subseteq \beta$, if $t \in Term_\alpha$, then we have $t \in Term_\beta$.

A term t or formula φ is ground iff it contains no free variables.

If a set D of sort definitions contains a definition of one of the following forms: $f \in \alpha_1 \times \ldots \times \alpha_n \rightarrow \alpha$ or $c \in \alpha$, or $\beta \subseteq \alpha$, we shall say that α is defined in D.

We assume that for each predefined sort α there is at least one constant in C_α.

Let $HU(\alpha)$ (the Herbrand universe for the sort α) be the set of all ground terms of the sort α.

From now on we impose a restriction on sort definitions: for any sort α, $HU(\alpha)$ must be non-empty. This property of the sort definitions is easy to check.

The notion of sorts introduced here is sufficiently rich to express many interesting domains. A term can belong to several sorts. However this sort structure admits an efficient unification algorithm. The set of the most general unifiers can be infinite. If the set of basic (predefined) sorts is closed under intersection, then two terms can have only finite (up to renaming) number of the most general unifiers.

The domain definitions of PDC-Prolog [17] are very similar to the sort definitions from this paper. The main difference is that we allow function symbols and constants to belong to several sorts.

2.2 Lists

In this paper we shall use the notion of a (many-sorted) *model* in the model-theoretical sense (see e.g. [5]), i.e. as sets with some constants, functions and predicates on this sets. The idea of an interpretation for built-in predicates is introduced via basic models. A basic model $\mathcal{M}$ is a model with the following restrictions:

1. There are no function symbols in the signature of $\mathcal{M}$;

2. Each element of $\mathcal{M}$ is the value of one and only one constant from the signature of $\mathcal{M}$.

The basic models are needed to define interpretations for built-in domains. Many practical built-in domains satisfy this properties, for example integers and strings in programming languages may have a unique representation as suitable constants. In what follows we shall refer to the sorts, predicates etc. of basic models as predefined sorts, predefined predicates etc.

Among all possible sorts we shall distinguish the sort of lists in the following way. Let $\mathcal{M}$ be a basic model, D a set of sort definitions such that `List`, `Atom` and `Univ` are not defined in D and there is no definition of the binary function symbol `cons`. (Note. Each element of the model $\mathcal{M}$ can belong to several sorts). Let $\beta_1, \ldots, \beta_n$ are all sorts which are in the model $\mathcal{M}$ or are defined by D. Then $L(\mathcal{M}, D)$ will denote the following set of sort definitions:

nil∈Atom
β_1 ⊆Atom
...
β_n ⊆Atom
nil∈List
Atom⊆Univ
List⊆Univ
cons∈Univ×List→List

Thus we allow all built-in and defined elements to be atoms and we add the constant `nil` to atoms. `Univ` is a universal sort comprising both atoms and lists. Lists are built from atoms and lists by applying the list constructor `cons`.

Instead of writing `nil` and `cons` we shall adopt the standard Prolog notation:

[]	stands for	nil;
[s|t]	stands for	cons(s,t);
[s_1,...,s_n|t]	stands for	cons(s_1,...,cons(...,cons(s_n,t)...));
[s_1,...,s_n]	stands for	cons(s_1,...,cons(...,cons(s_n,nil)...)).

Lists are elements of HU(`List`). We do not define here lists as a superstructure of the ordinary terms as it has been done in [Goncharov 85] following the theory of admissible sets [Barwise 75]. Such a superstructure is convenient for more theoretical purposes (to distinguish sets from urelements in the theory of admissible sets), but from the viewpoint of programming it has some disadvantages. In particular, it forbids to use terms with subterms containing lists, e.g. f([],[a]). Our sorts allow to use such terms. For example, we can define the sort of such a function f via e.g.

f∈List×List→Pairs

We can introduce the function `cons` on lists in a natural way. We define also two relations ∈, ⊆ on lists as follows:

x∈[y_1,...,y_n] iff for some $y = 1, \ldots, n$ we have x=y_i;
x⊆[y_1,...,y_n] iff x=[] or for some $i = 1, \ldots, n$ we have x=[y_i,...,y_n].

2.3 Bounded quantifiers and generalized logic programs

Here we introduce bounded quantifiers. Apart from sorts they are basically equivalent to bounded quantifiers introduced in [12].

Bounded quantifiers are expressions of the form $(\forall x \in t)$, $(\exists x \in t)$, $(\forall x \sqsubseteq t)$, $(\exists x \sqsubseteq t)$, where the variable x is of the universal sort Univ and does not occur in the term t of the sort List.

A Σ-formula is any formula constructed from atoms using $\wedge$, $\vee$, $\exists$, and bounded quantifiers. To distinguish the ordinary existential quantifier $\exists$ from bounded quantifiers we shall call the former unrestricted existential quantifier. Δ_0-formula is a Σ-formula containing no occurences of the unrestricted quantifier $\exists$.

A generalized logic program $\mathcal{P}$ is a triple $\langle \mathcal{M}, \mathbf{D}, \mathbf{C} \rangle$ where $\mathcal{M}$ is a basic model with a signature σ, $\mathbf{D}$ a set of sort definitions, $\mathbf{C}$ a set of clauses of the form

$$P_i(\bar{x}_i) \colon - \varphi_i(\bar{x}_i),$$

where $i = 0, \ldots, n$, P_i are predicate symbols which are not in the signature σ, φ_i are Σ-formulas with free variables from $\bar{x}_i$ in the following signature δ:

the constants of δ consist of nil, of the constants from $\mathcal{M}$, and of the constants defined in $\mathbf{D}$;

the function symbols of δ are cons and the function symbols defined in $\mathbf{D}$;

the predicate symbols of δ are $P_0, \ldots, P_n$,= and the predicate symbols of σ.

To make the generalized logic programs shorter and to make the syntax closer to that of the ordinary logic programs we shall introduce some notations. The set of expressions

$P(\bar{t}_1)$:- φ_1

...

$P(\bar{t}_n)$:- φ_n

will denote the clause

$$P(\bar{x}) \text{ :- } \exists \bar{y}_1(\bar{x} = \bar{t}_1 \wedge \varphi_1) \vee \ldots \vee \exists \bar{y}_n(\bar{x} = \bar{t}_n \wedge \varphi_n),$$

where $\bar{x}$ are new variables, $\bar{y}_i$ are all the variables of t_i.

Query to a generalized logic program $\mathcal{P}$ is any Σ-formula.

In what follows we assume that a generalized logic program $\mathcal{P} = \langle \mathcal{M}, \mathbf{D}, \mathbf{C} \rangle$ is fixed.

In the literature on logic programming with sets only one kind of bounded quantifiers is used, namely $\forall\in$. The quantifier $\forall\sqsubseteq$ was introduced in [12]. This quantifier is very expressive. For example in section 5.2 we present a non-recursive metainterpreter for Horn clause logic programs, which uses only unrestricted existential quantifiers and the bounded quantifier $\forall\sqsubseteq$. It is difficult to introduce such bounded quantifier for sets, because there is no analog of the relation $\sqsubseteq$ on sets.

2.4 Examples

To explain the use of the bounded quantifiers we give some examples below. The interesting property of these examples is that they are not recursive.

Note. In all examples of this paper we use the standard Prolog notation: variable names start from the upper-case letters, while all other symbols start from the lower-case letters.

Example 2.1 *A program computing whether a list L is ordered:*

```
ordered(L) :-
   (∀X⊑L)(X=[]∨singleton(X)∨ordered2(X)).

singleton([X]).

ordered2([X,Y|Z]) :-
   X≤Y.
```

We assume that ≤ is a built-in relation.

Example 2.2 *A program computing whether all elements of a list* L *satisfy a property* p*:*

```
allp(L) :-
   (∀X∈L)p(X).
```

Example 2.3 *A program computing the subset relation:*

```
subset(L1,L2) :-
   (∀X1∈L1)(∃X2∈L2)X1=X2.
```

Example 2.4 *A program verifying whether a given list* L *has no repetitions:*

```
norep(L) :-
   (∀X⊑L)norep1(X).

norep1([A|As]) :-
   (∀X∈As)A≠X.
```

Here X1≠X2 is a built-in relation which is true iff X1 and X2 are not equal.

Example 2.5 *A program finding a route in a graph. We assume that vertices* a *and* b *are connected in the graph iff the fact* arc(a,b) *is in the program.*

```
route(A,B,C) :-
   path(C)∧start(A,C)∧finish(B,C).

path(P) :-
   (∀S⊑P)(S=[]∨singleton(S)∨connected(S)).

singleton([X]).
```

```
connected([X,Y|Z]) :-
   arc(X,Y).

start(X, [X|Xs]).

finish(X,Xs) :-
   (∃S⊑Xs)(S=[X]).
```

3 Semantics of generalized logic programs

The theory of logic programming is based on the fundamental fact that the (declarative) model-theoretical semantics coincides with the provability by SLD-resolution and some other kinds of provabilities, e.g. those of intuitionistic and classical logics.

In this section we shall adapt these semantics to generalized logic programs. The features specific to generalized logic programs are sorts, built-in predicates and bounded quantifiers. The semantics of Horn clause programs must be modified so as to accommodate these features.

In section 3.1 we introduce a model theoretical semantics for generalized logic programs, which simply expresses the intended declarative meaning of programs with bounded quantifiers. In section 3.2 a least fixedpoint semantics is introduced which serves as a bridge between the declarative model-theoretical semantics and the procedural interpretation introduced later. Then, in sections 3.3 and 3.4, we define two types of classical and constructive provabilities — one with axioms for lists from [12], and another one which considers lists as elements generated by inductive definitions.

Section 3.5 presents so called natural semantics of the generalized logic programs which was originally introduced in [26]. The calculus $\mathbf{Nat}(\mathcal{P})$ from this section represents in a declarative way ideas from the procedural semantics introduced in section 4.

3.1 Model-theoretical semantics

The main semantics of generalized logic programs is the model theoretical semantics which allows a declarative reading of programs. Our semantics is similar to the semantics introduced in the other papers [12, 9, 6]. The main difference between our semantics and that of [12] is that we allow function symbols to be constructors, which means that we can define new terms, whereas in [12] the model is fixed - the set of all lists with atoms from the basic model. Our semantics also treats built-in sorts, while both [9] and [6] do not consider built-in sorts.

To introduce the model-theoretical semantics of our programs we have to define the meaning of quantified expressions because we use a non-standard sort structure and because we introduce bounded quantifiers as primitives.

Let us note that we did not introduce any restrictions on usage of sorts for the arguments of predicates. In other words, we can use any terms as arguments. However terms have sorts, which leads to the definitions given below.

Now we define the notion of truth for the quantified formulas. Let x be a variable of the sort α, y a variable of the sort Univ, $\mathcal{N}$ a model. Then

1. $\mathcal{N} \models \forall x \varphi(x)$ iff for all elements a of $HU(\alpha)$ we have $\mathcal{N} \models \varphi(a)$;
2. $\mathcal{N} \models \forall x \varphi(x)$ iff for some element a of $HU(\alpha)$ we have $\mathcal{N} \models \varphi(a)$;
3. $\mathcal{N} \models (\forall y \in b) \varphi(y)$ iff for every element a of b we have $\mathcal{N} \models \varphi(a)$;
4. $\mathcal{N} \models (\exists y \in b) \varphi(y)$ iff for some element a of b we have $\mathcal{N} \models \varphi(a)$;
5. $\mathcal{N} \models (\forall y \sqsubseteq b) \varphi(y)$ iff for every $a \sqsubseteq b$ we have $\mathcal{N} \models \varphi(a)$;
6. $\mathcal{N} \models (\forall y \sqsubseteq b) \varphi(y)$ iff for some $a \sqsubseteq b$ we have $\mathcal{N} \models \varphi(a)$.

A clause $A(\bar{x})$:- $\varphi(\bar{x})$ is true on a model $\mathcal{N}$ iff such is the formula $\varphi(\bar{x}) \supset A(\bar{x})$.

Let $\mathcal{N}$ be a many-sorted model, $\langle \mathcal{M}, \mathbf{D}, \mathbf{C} \rangle$ be a generalized logic program. Then $\mathcal{N}$ is a model of $\langle \mathcal{M}, \mathbf{D}, \mathbf{C} \rangle$ iff

1. The sorts of $\mathcal{N}$ are the sorts of $\mathcal{M}$ and the sorts defined in $L(\mathcal{M}, \mathbf{D})$;
2. The interpretation of any predefined sort is the set of all constants of this sort;
3. The predicates of $\mathcal{N}$ are those of $\mathcal{M}$ and those of $\langle P_0, \ldots, P_n, = \rangle$;
4. The interpretation of any defined sort α is $HU(\alpha)$;
5. The interpretation of any predefined predicate symbol in $\mathcal{N}$ is the same as in $\mathcal{M}$;
6. The interpretation of $=$ is equality;
7. All clauses from $\mathbf{C}$ are true on $\mathcal{N}$.

As usual we introduce the relation $\subseteq$ between models of a generalized logic program $\mathcal{P}$ in the following way: $\mathcal{N}_1 \subseteq \mathcal{N}_2$ iff for any ground atomic formula φ, $\mathcal{N}_1 \models \varphi$ implies $\mathcal{N}_2 \models \varphi$.

Lemma 3.1 *Let $\mathcal{N}_1, \mathcal{N}_2$ be models of a generalized logic program $\mathcal{P}$. Then $\mathcal{N}_1 \subseteq \mathcal{N}_2$ iff for any ground Σ-formula φ, from $\mathcal{N}_1 \models \varphi$ follows $\mathcal{N}_2 \models \varphi$.* ■

We omit proofs of all lemmas and theorems in this paper.

Theorem 3.2 *For every generalized logic program $\mathcal{P}$ there exists a model $\mathcal{N}$ of $\mathcal{P}$ which is minimal w.r.t. $\subseteq$.* ■

Given a generalized logic program $\mathcal{P}$, let $\mathbf{Mod}(\mathcal{P})$ denotes the set of all ground Σ-formulas that are true in the minimal model of $\mathcal{P}$.

3.2 The least fixedpoint semantics

The definition of the least fixedpoint semantics is similar to the definitions given in [12] and to the formalization of logic programming proposed in [1]: a (generalized) logic program is considered as a monotonic mapping from interpretations to interpretations.

An interpretation $\Im$ is any set of ground Σ-formulas satisfying the following properties:

1. $\varphi \wedge \psi \in \Im$ iff $\varphi \in \Im$ and $\psi \in \Im$;
2. $\varphi \vee \psi \in \Im$ iff $\varphi \in \Im$ or $\psi \in \Im$;
3. $\exists x \varphi(x) \in \Im$ iff for some term t of the same sort as x, $\varphi(t) \in \Im$;
4. $(\exists x \in s)\varphi(x) \in \Im$ iff for some term $t \in s$, $\varphi(t) \in \Im$;
5. $(\forall x \in s)\varphi(x) \in \Im$ iff for every term $t \in s$, $\varphi(t) \in \Im$;
6. $(\exists x \sqsubseteq s)\varphi(x) \in \Im$ iff for some term $t \sqsubseteq s$, $\varphi(t) \in \Im$;
7. $(\forall x \sqsubseteq s)\varphi(x) \in \Im$ iff for every term $t \sqsubseteq s$, $\varphi(t) \in \Im$;

An interpretation $\Im$ is an interpretation for the model $\mathcal{M}$ iff

1. For every predefined predicate symbol P and any tuple of terms $\bar{t}$, $P(\bar{t}) \in \Im$ iff $\mathcal{M} \models P(\bar{t})$.
2. For every terms s, t, $s = t \in \Im$ iff s is identical to t.

Note. There is a minimal interpretation for $\mathcal{M}$ which coincides with the set of all Σ-formulas true on $\mathcal{M}$ extended by equality. We shall denote it by $\Im_0$.

Lemma 3.3 *Any interpretation is uniquely characterized by the set of its atomic formulas.* ■

Let $\mathbf{C}^*$ be the set of all ground instances of clauses in $\mathbf{C}$. The immediate consequence operator defined by a program $\mathcal{P}$ is the function $\mathbf{I}_\mathcal{P}$ on the set of all interpretations defined as follows: for an interpretation $\Im$ and a ground atomic formula ψ, $\psi \in \mathbf{I}_\mathcal{P}(\Im)$ if $\mathbf{C}^*$ contains a clause $\varphi \supset \psi$ such that $\varphi \in \Im$.

Lemma 3.4 *The operator $\mathbf{I}_\mathcal{P}$ is monotonic.* ■

Theorem 3.5 *There is the least fixed point* $\mathbf{Lfp}(\mathcal{P})$ *of the operator $\mathbf{I}_\mathcal{P}$ among all interpretations containing $\Im_0$. It can be computed as*

$$\mathbf{Lfp}(\mathcal{P}) = \bigcup_{i=0}^{\infty} \Im_i,$$

where $\Im_0$ is as defined, and $\Im_{i+1} = \Im_i \bigcup \mathbf{I}_\mathcal{P}(\Im_i)$. Moreover $\mathbf{Lfp}(\mathcal{P})$ *coincides with* $\mathbf{Mod}(\mathcal{P})$. ■

3.3 Classical provability

Classical proof systems for lists are obtained from the classical predicate calculus by adding axioms expressing properties of lists. There are two ways to define appropriate extensions of the predicate calculus. The first approach is similar to the approach used in [12] for lists and in [2] for hereditarily finite sets. According to this approach all elements but lists are considered as urelements and some axioms expressing properties of list are added (the calculus Clt($\mathcal{P}$) below). The second possibility is to treat sort definitions (including the definition of lists) as inductive definitions. Inductive definitions for sorts define universes for these sorts and also give induction rules for proving properties of elements of the sorts. These induction rules are given in the calculus **CInd**($\mathcal{P}$) below.

The many-sorted predicate calculus we use differs from the ordinary non-sorted predicate calculus in the following restrictions on the axioms for the quantifiers. In the axioms $\forall x\varphi(x)\supset\varphi(t)$ and $\varphi(t)\supset\exists x\varphi(x)$, if the variable x is of the sort α, then t must be a term of the same sort.

We also do not consider in this section bounded quantifiers as primitives, but as notations:

$$\begin{array}{lll}(\forall x\lambda t)\varphi(x) & \text{stands for} & \forall x(x\lambda t\supset\varphi(x)),\\ (\exists x\lambda t)\varphi(x) & \text{stands for} & \exists x(x\lambda t\wedge\varphi(x)),\end{array}$$

where λ denotes $\in$ or $\sqsubseteq$.

3.3.1 The calculus Clt($\mathcal{P}$)

The calculus Clt($\mathcal{P}$) (the classical list theory) is obtained from the classical predicate calculus with equality by adding the following axioms:

1. Axioms for $\mathcal{M}$: all ground atomic formulas true on $\mathcal{M}$;

2. Axioms for **C**. If a clause $P(\bar{x})$:- $\varphi(x)$ belongs to **C**, then the formula $\forall\bar{x}(\varphi(\bar{x})\supset P(\bar{x}))$ is an axiom of Clt($\mathcal{P}$);

3. Axioms for lists:

$$\begin{array}{l}[\mathtt{s}_1|\mathtt{t}_1]=[\mathtt{s}_2|\mathtt{t}_2]\supset\mathtt{s}_1=\mathtt{t}_1\wedge\mathtt{s}_2=\mathtt{t}_2\\ \neg\mathtt{s}=[\,]\\ \mathtt{r}\in[\mathtt{s}|\mathtt{t}]\equiv(\mathtt{r}=\mathtt{s}\vee\mathtt{r}\in\mathtt{t})\\ \mathtt{t}\sqsubseteq[\,]\equiv\mathtt{t}=[\,]\\ \mathtt{t}_1\sqsubseteq[\mathtt{s}|\mathtt{t}_2]\equiv\mathtt{t}_1=[\mathtt{s}|\mathtt{t}_2]\vee\mathtt{t}_1\sqsubseteq\mathtt{t}_2\end{array}$$

   where $\mathtt{s}_i,\mathtt{r}_i$ are terms of the sort `Univ`, $\mathtt{t}_i$ terms of the sort `List`.

4. Induction axioms:

$$\varphi([\,])\wedge\forall\mathtt{x}\forall\mathtt{y}(\varphi(\mathtt{y})\supset\varphi([\mathtt{x}|\mathtt{y}]))\supset\forall\mathtt{y}\varphi(\mathtt{y}),$$

   where φ is any formula, x a variable of the sort `Univ`, y a variable of the sort `List`.

5. Foundation axiom:

$\forall$y(($\forall$x$\in$y)φ(x)$\supset\varphi$(y))$\supset\forall$yφ(y)

where φ is any formula, x a variable of the sort Univ, y a variable of the sort List.

This theory is almost identical to GES defined in [12].

Theorem 3.6 *A ground Σ-formula φ is provable in* Clt($\mathcal{P}$) *iff* φ $\in$Mod*($\mathcal{P}$).* ■

3.3.2 The calculus CInd($\mathcal{P}$)

The calculus **CInd($\mathcal{P}$)** treats lists not as a special sort, but simply as an instance of a sort definition. A set of sort definitions gives an induction scheme to prove the properties of the defined sorts, including lists. A somewhat similar approach to the semantics of logic programming had been used in [15]. But in this article inductive definitions are not sorts definitions, but clauses of logic programs.

The calculus **CInd($\mathcal{P}$)** is obtained from the classical predicate calculus with equality by adding the following axioms and rules. The first three groups of axioms are identical to those of **Clt($\mathcal{P}$)**. The other axioms are

4. Induction rules. Let $\alpha_1, \ldots, \alpha_n$ be all built-in and $\alpha_{n+1}, \ldots, \alpha_k$ all defined sorts from $L(\mathcal{M}, \mathbf{D})$, $\varphi_1(x), \ldots, \varphi_k(x)$ be formulas. Then the following is a rule of **Clt($\mathcal{P}$)**.

$$\frac{\forall x_1 \varphi_1(x_1) \ldots \forall x_n \varphi_n(x_n) \quad \Gamma_{n+1} \ldots \Gamma_k}{\forall z_i \varphi_i(z_i)}$$

where x_j are variables of the sorts α_j, the variable z_i may be of any sort. The sets Γ_i of formulas are formed as follows (below z_i are variables of the sorts α_i).

If $c \in \alpha_i$ is a definition in $L(\mathcal{M}, \mathbf{D})$, then $\varphi_i(c) \in \Gamma_i$;

If $f \in \alpha_m \times \ldots \times \alpha_p \to \alpha_i$ is a definition in $L(\mathcal{M}, \mathbf{D})$, then $\forall z_m \ldots \forall z_p(\varphi_m(z_m) \wedge \ldots \wedge \varphi_p(z_p) \supset \varphi_i(f(z_m, \ldots, z_p))) \in \Gamma_i$;

If $\alpha_m \subseteq \alpha_i$ is a definition in $L(\mathcal{M}, \mathbf{D})$, then $\forall z_m(\varphi_m(z_m) \supset \varphi_i(z_m)) \in \Gamma_i$;

There are no other formulas in Γ_i.

As we can easily see, if we have the following set of sort definitions, describing natural numbers

0 $\in$Nat
s $\in$Nat$\to$Nat

then the usual induction scheme for natural numbers

$$\frac{\varphi(0) \qquad \forall n(\varphi(n) \supset \varphi(s(n)))}{\forall n \varphi(n)},$$

where n is a variable of the sort Nat, can be derived from our rules. However we prefer to consider natural numbers as a built-in sort.

Theorem 3.7 *A ground Σ-formula φ is provable in* Clt($\mathcal{P}$) *iff* φ $\in$Mod*($\mathcal{P}$).* ■

3.4 Constructive provability

The systems **Ilt**($\mathcal{P}$) (intuitionistic list theory) and **IInd**($\mathcal{P}$) have the same axioms and rules as **Clt**($\mathcal{P}$) and **CInd**($\mathcal{P}$) but both are based on the intuitionistic predicate logic instead of the classical one. The intuitionistic variant of list theory was introduced in [23]. In [28] we proved that it is constructive from the viewpoint of a constructive semantics, which in particular means that it has a variant of the existential property: if a ground formula $\exists x\varphi(x)$ is provable then it is possible to effectively find a term t such that $\varphi(t)$ holds.

Theorem 3.8 *For a ground atomic* Σ*-formula* φ*,* **Ilt**$(\mathcal{P})\models\varphi$ *iff* $\varphi \in$ Mod$(\mathcal{P})$*.* ∎

Theorem 3.9 *For a ground atomic* Σ*-formula* φ*,* **IInd**$(\mathcal{P})\models\varphi$ *iff* $\varphi \in$ Mod$(\mathcal{P})$*.* ∎

3.5 The natural calculus

We call this calculus natural because it gives a natural semantics to formulas with bounded quantifiers. The rules of the natural calculus treat these formulas in a very natural and elegant way. The calculus **Nat**($\mathcal{P}$) introduced below is similar to the ground positive hyperresolution on Horn clauses. At the same time the natural calculus serves as a basis for the procedural semantics of generalized logic programs. The natural semantics for Σ-programs [12] was introduced in [26] with the aim of showing that Σ-programs can be efficiently executed using unification instead of the exhaustive search.

The calculus **Nat**($\mathcal{P}$) consists of the following axioms and inference rules:

1. Axioms.
   (a) All ground atomic formulas true on $\mathcal{M}$;
   (b) All axioms of the form `t=t`, where `t` is a ground term.
2. Rules for **C**:

   $$\frac{\varphi}{A},$$

   if A:-φ belongs to **C***.
3. Rules for the logical connectives:

   $$\frac{\varphi}{\varphi\vee\psi} \qquad \frac{\psi}{\varphi\vee\psi} \qquad \frac{\varphi \quad \psi}{\varphi\wedge\psi}$$

4. A rule for the existential quantifier:

   $$\frac{\varphi(t)}{\exists x\varphi(x)}$$

   where x is a variable of a sort α, t a term of the same sort.

5. Rules for bounded quantifiers (here t is a term of the sort List, s an arbitrary term):

$$\frac{\varphi(\mathtt{s})}{(\exists \mathtt{x}\in [\mathtt{s}|\mathtt{t}])\varphi(\mathtt{x})} \qquad \frac{(\exists \mathtt{x}\in \mathtt{t})\varphi(\mathtt{x})}{(\exists \mathtt{x}\in [\mathtt{s}|\mathtt{t}])\varphi(\mathtt{x})}$$

$$\frac{}{(\forall \mathtt{x}\in [\,])\varphi(\mathtt{x})} \qquad \frac{\varphi(\mathtt{s}) \quad (\forall \mathtt{x}\sqsubseteq \mathtt{t})\varphi(\mathtt{x})}{(\forall \mathtt{x}\in [\mathtt{s}|\mathtt{t}])\varphi(\mathtt{x})}$$

$$\frac{\varphi(\mathtt{t})}{(\exists \mathtt{x}\sqsubseteq \mathtt{t})\varphi(\mathtt{x})} \qquad \frac{(\exists \mathtt{x}\sqsubseteq \mathtt{t})\varphi(\mathtt{x})}{(\exists \mathtt{x}\sqsubseteq [\mathtt{s}|\mathtt{t}])\varphi(\mathtt{x})}$$

$$\frac{\varphi([\,])}{(\forall \mathtt{x}\sqsubseteq [\,])\varphi(\mathtt{x})} \qquad \frac{\varphi([\mathtt{s}|\mathtt{t}]) \quad (\forall \mathtt{x}\sqsubseteq \mathtt{t})\varphi(\mathtt{x})}{(\forall \mathtt{x}\sqsubseteq [\mathtt{s}|\mathtt{t}])\varphi(\mathtt{x})}$$

Theorem 3.10 *For any ground Σ-formula φ, $\varphi \in$*Mod*$(\mathcal{P})$ iff* Nat$(\mathcal{P})\vdash \varphi$. ∎

4 Procedural semantics: SLDB-resolution

The natural calculus of section 3.5 represents in a declarative way ideas of the operational semantics. This calculus treats only ground formulas. To produce the operational semantics from the calculus it is sufficient to show how to treat non-ground formulas and how to formalize the top-down search. To this end we introduce a unification algorithm for our terms in section 4.1 and SLDB-resolution in section 4.2.

4.1 Unification

The sorts introduced here require a special unification algorithm. Such an algorithm is possible if we impose some restrictions on the set of all predefined (or basic) sorts. The restrictions are the following:

1. There is an algorithm verifying whether a given constant c belongs to a given basic sort α;

2. There is an algorithm which by given predefined sorts $\alpha, \ldots, \beta$ decides if $C_\alpha \bigcap \ldots \bigcap C_\beta = \emptyset$.

Note. An algorithm for 2 always exists if there is only a finite number of the predefined sorts.

The first restriction is needed to unify a variable of a predefined sort with a constant. The second restriction is necessary to unify variables of different predefined sorts.

For simplicity we assume that the set of all predefined sorts is closed under intersection, i.e. for each two predefined sorts α, β if $C_\alpha \bigcap C_\beta \neq \emptyset$, then there is a predefined sort γ such that $C_\gamma = C_\alpha \bigcap C_\beta$.

The following theorem gives a theoretical basis for the unification algorithm.

Theorem 4.1 *The following statements are true:*

1. *There exists an algorithm verifying whether two sorts have a nonempty intersection.*
2. *If the set of all predefined sorts is closed under intersection, then the set of all sorts is closed under intersection. More precisely, if D is a set of sort definitions, and α, β are sorts with $HU(\alpha) \bigcap HU(\beta) \neq \emptyset$, then there is a set $D' \supseteq D$ of sort definitions which defines a sort γ such that*
   (a) *The Herbrand universe $HU(\xi)$ for any sort ξ defined in D is identical in D and in D';*
   (b) *$HU(\alpha) \bigcap HU(\beta) = HU(\gamma)$ in D'.* ∎

According to theorem 4.1 we can assume that the set of all sorts is closed under intersection. Before defining the unification algorithm for our sorts we introduce some definitions.

A substitution is any expression of the form $[t_1/x_1, \ldots, t_n/x_n]$, $n \geq 0$, such that for every i, the variable x_i has the same sort as the term t_i. Such a substitution with $n = 0$ is the empty substitution. The application $t\theta$ of a substitution θ to a term or formula t is the result of the simultaneous replacement of all free occurences of x_i in t by t_i. We denote by $\theta_1 \circ \theta_2$ the substitution θ such that for each term t, $t\theta = (t\theta_2)\theta_1$. A substitution θ is a unifier of terms t_1 and t_2 iff $t_1\theta = t_2\theta$. Two terms are unifiable iff they have a unifier. A covering set of unifiers of t_1 and t_2 is a set $\{\theta_1, \ldots, \theta_n\}$ of unifiers of these terms such that

1. for any other unifier θ of t_1 and t_2 there exist $i \in \{0, \ldots, n\}$ and a substitution θ' with $\theta = \theta' \circ \theta_i$;
2. Any proper subset of $\{\theta_1, \ldots, \theta_n\}$ does not satisfy 1.

A substitution θ is called a most general unifier of t_1 and t_2 iff it belongs to some covering set of unifiers of t_1 and t_2.

Theorem 4.2 *If the set of all sorts is closed under intersection then there is an algorithm verifying whether two given terms t_1 and t_2 are unifiable, and if they are unifiable giving a finite covering set of unifiers of t_1 and t_2.* ∎

Now we shall give the unification algorithm. The algorithm consists of repeated transformations of pairs $\langle S, \theta \rangle$, consisting of a set of equations S and a substitution θ. The initial set of equations consists of one equation $t_1 = t_2$, and the initial substitution is the empty substitution ε. The algorithm is as follows:

If the set of equations is empty then the algorithm stops and gives the substitution as the result. Otherwise, let $s_1 = s_2$ be the first equation, S be the rest of the equations and θ be the current substitution. Consider the following cases.

1. s_1 is a constant.
   (a) s_2 is a constant. Then if s_1 and s_2 are identical, we change the current state to S, θ.

(b) s_2 is a variable x of a sort α. Then if s_1 is of the sort α then we change the current state to $\langle S, \mathit{0}\rangle[s_1/x]$, otherwise the algorithm fails on this state.

(c) Otherwise the algorithm fails on this state.

2. s_1 is a variable x of a predefined sort α.

   (a) s_2 is a constant. This case is symmetrical to case 1b.

   (b) s_2 is a variable y of a sort β. Then if α and β have a non-empty intersection γ, we change the state to $\langle S, \mathit{0}\rangle[z/x, z/y]$, where z is a new variable of the sort γ, otherwise the algorithm fails on this state.

   (c) Otherwise the algorithm fails on this state.

3. s_1 is a variable x of a defined sort α.

   (a) s_2 is a constant. This case is symmetrical to case 1b.

   (b) s_2 is a variable y of a sort β. This case is similar to case 2b.

   (c) s_2 is a term $f(t_1, \ldots, t_n)$. Then if there is a definition $f \in \alpha_1 \times \ldots \times \alpha_n \rightarrow \alpha$ then we take n new variables $x_1, \ldots, x_n$ of the sorts $\alpha_1, \ldots, \alpha_n$ and change the state to $\langle S \bigcup \{x_1 = t_1, \ldots, x_n = t_n\}, \mathit{0}\rangle[f(x_1, \ldots, x_n)/x]$, otherwise the algorithm fails on this state.

4. s_1 is a term $f(t_1, \ldots, t_n)$.

   (a) s_2 is a variable x of a defined sort α. This case is symmetrical to case 3c.

   (b) s_2 is a term $f(r_1, \ldots, r_n)$. Let α be a sort such that there is a sort definition of the form $f \in \alpha_1 \times \ldots \times \alpha_n \rightarrow \alpha$ and let $x_1, \ldots, x_n$ be new variables of the sorts $\alpha_1, \ldots, \alpha_n$. Then we change the state to the states of the form $\langle S \bigcup \{t_1 = r_1, \ldots, t_n = r_n, x_1 = r_1, \ldots, x_n = r_n\}, \mathit{0}\rangle$ for all such sorts α.

   (c) Otherwise the algorithm fails on this state.

The only inefficient part of the algorithm is case 4b, where the algorithm can branch and give several unifiers. If we do not allow definitions of the same function symbol as belonging to different sorts, then the most general unifier (if there is one) is unique w.r.t renaming. The domain definitions of PDC-Prolog [17] do not allow multiple declarations of the same function symbol.

4.2 SLDB-resolution

In this section we give a procedural semantics of the language which generalizes SLD-resolution for Horn clause programs. We call it SLDB-resolution (SLD-resolution with Bounded quantifiers). There is no difference between SLDB-resolution and SLD-resolution in the treatment of program clauses, but there are special features in processing built-in predicates and complex formulas.

Goal is any list of Σ-formulas. We assume that the reader is familiar with the notion of SLD-resolution (see e.g. [11]. As usually, the computation rule is a function from the set of all non-empty goals to the set of Σ-formulas such that the value of the function on a goal is a formula, called <u>selected</u> formula, in that goal. Let $\varphi_1,\dots,\varphi_n$ be a goal and φ_i is the selected formula. Then the goal $(\varphi_1,\dots,\varphi_{i-1},\Gamma,\varphi_{i+1},\dots,\varphi_n)\theta$, where θ is a substitution, Γ is a list of Σ-formulas, is a <u>successor</u> of the goal $\varphi_1,\dots,\varphi_n$ with the substitution θ iff one of the following conditions holds:

1. φ_i is a built-in atom (an atom of the form $P(\bar{t})$, where P is from the signature of the basic model), Γ the empty list, and θ a substitution which makes φ_i true on $\mathcal{M}$;

2. φ_i takes the form $\psi_1 \vee \psi_2$, Γ is ψ_i, $i = 1,2$, θ is the empty substitution.

3. φ_i takes the form $\psi_1 \wedge \psi_2$, Γ is ψ_1, ψ_2, θ the empty substitution.

4. φ_i takes the form $\exists x \psi(x)$, Γ is $\psi(y)$, θ the empty substitution, y a new variable of the same sort as x.

   In the following y is a new (not occuring in the original goal) variable of the sort `List`, z a new variable of the sort `Univ`.

5. φ_i takes the form $(\exists x \in t)\psi(x)$, Γ is $\psi(z)$, θ the most general unifier of t and $[z \mid y]$.

6. φ_i takes the form $(\exists x \in t)\psi(x)$, Γ is $(\exists x \in y)\psi(x)$, θ is the most general unifier of t and $[z \mid y]$.

7. φ_i takes the form $(\forall x \in t)\psi(x)$, Γ is $\psi(z)$,$(\forall x \in y)\psi(x)$, θ is the most general unifier of (x,t) and $(z,[z \mid y])$

8. φ_i takes the form $(\forall x \in t)\psi(x)$, Γ is empty, θ is the most general unifier of t and $[]$.

9. φ_i takes the form $(\exists x \sqsubseteq t)\psi(x)$, Γ is $\psi(y)$, θ is the subsitution $[t/y]$.

10. φ_i takes the form $(\exists x \sqsubseteq t)\psi(x)$, Γ is $(\exists x \sqsubseteq y)\psi(x)$, θ is the most general unifier of t and $[z \mid y]$.

11. φ_i takes the form $(\forall x \sqsubseteq t)\psi(x)$, Γ is $\psi([z \mid y])$,$(\forall x \sqsubseteq y)\psi(x)$, θ is the most general unifier of (x,t) and $([z \mid y],[z \mid y])$

12. φ_i takes the form $(\forall x \sqsubseteq t)\psi(x)$, Γ is $\psi([])$, θ is the most general unifier of t and $[]$.

13. φ_i is an atom $P(\bar{t})$, Γ is $\varphi(\bar{t})$, θ is the empty substitution, where there is a clause of the form $P(\bar{x})$:-$\varphi(\bar{x})$ in C.

14. φ_i is an atom t_1=t_2, Γ is empty, θ is a most general unifier of t_1 and t_2.

<u>SLDB-derivation</u> is any sequence of pairs $\langle\Gamma_0,\theta_0\rangle,\dots,\langle\Gamma_n,\theta_n\rangle$ of goals and substitutions such that for every $i \in \{1,\dots,n\}$ there exists a substitution θ such that

1. Γ_i is a successor of Γ_{i-1} with the substitution θ;

2. $\theta_i = \theta \circ \theta_{i-1}$.

We say that a goal Γ is provable with the substitution θ iff there is an SLDB-derivation starting from $\langle \Gamma, \varepsilon \rangle$ and finishing at $\langle \Lambda, \theta \rangle$, where ε is the empty substitution and Λ the empty goal.

The following theorem states that SLDB-resolution is independent of the computation rule.

Theorem 4.3 *If a goal Γ is provable with a substitution θ under one computation rule, than Γ is provable with θ under any other computation rule.* ■

Let for any formula φ, $\forall\varphi$ denotes the formula $\forall\bar{x}\varphi$, where $\bar{x}$ is the sequence of all free variables of φ. The following theorem states completeness and correctness of SLDB-derivations:

Theorem 4.4 *The following statements are true:*

1. *(Correctness.) If a Σ-formula φ is provable with a substitution θ, then the formula $\forall(\varphi\theta)$ is true in* Mod*($\mathcal{P}$).*

2. *(Completeness.)If φ is a Σ-formula and ψ is its ground instance true in* Mod*($\mathcal{P}$), then there are substitutions θ, θ_1, such that φ is provable with the substitution θ and $\varphi\theta\theta_1$ is identical to ψ.* ■

Corollary 4.5 *If φ is a ground Σ-formula, then the following conditions are equivalent.*

1. *φ is provable (with the empty substitution ε);*

2. *φ is true in* Mod*($\mathcal{P}$).* ■

5 Expressive power

In this section we prove some results about the expressive power of generalized logic programs. In section 5.1 we show a natural translation of generalized logic programs into Horn clause programs. During the translation some new predicates may be defined in the programs. However the obtained Horn clause programs can be recursive, while the initial generalized programs are not recursive. In section 5.2 we construct a non-recursive metainterpreter for Horn clauses in the language of generalized logic programs. It shows the expressive power of generalized logic programs — every computable predicate can be expressed by a non-recursive generalized programs, which means that it can be expressed by a generalized program consisting of one nonrecursive definition. We also prove that this can not be achieved using only bounded quantifiers or only the unbounded existential quantifier.

5.1 Translation to Horn clauses

In this section we will show that the generalized logic programs can be naturally translated to Horn clause programs by adding new predicate symbols. A similar translation can be done for the {log} language of [6], but in this language the only allowed bounded quantifiers are those over elements of a set, which corresponds to our $(\forall x \in t)$. We prove correctness and completeness of such a translation. The existence of the translation is not surprising, because Horn clauses form a universal programming language (a language in which all the computable predicates can be expressed). The interesting features of our translation are that it is quite natural and that non-recursive programs with the bounded quantifiers may be translated into recursive Horn clause programs. In section 5.2 we will show that it can not be avoided in general.

Since our programs are sorted, then we shall assume that the corresponding Horn clause programs are sorted in the same way, and that their semantics is a restriction of our semantics when we omit the bounded quantifiers.

The details of the translation are well known in cases of disjunction, conjunction and the existential quantifier. Suppose that we have a generalized logic program with the set of clauses C. We shall define its translation - a Horn clause program $\widehat{C}$ in the following way. If there is a non-Horn definition in C, we change it to one or more definitions according to the rules given below in the table, until we get a Horn clause program.

Sentence:	**Its translation:**

`A(x̄) :- B(x̄)∧C(x̄)`	`A(x̄) :- D(x̄),E(x̄)`
	`D(x̄) :- B(x̄)`
	`E(x̄) :- C(x̄)`

`A(x̄) :- ∃vB(x̄,v)`	`A(x̄) :- B(x̄,v)`

`A(x̄) :- (∃y∈t)B(x̄,y)`	`A(x̄) :- D(x̄,t)`
	`D(x̄,[y|z]) :- B(x̄,y)`
	`D(x̄,[y|z]) :- D(x̄,z)`

`A(x̄) :- (∀y∈t)B(x̄,y)`	`A(x̄) :- D(x̄,t)`
	`D(x̄,[])`
	`D(x̄,[y|z]) :- B(x̄,y),D(x̄,z)`

`A(x̄) :- (∃y⊑t)B(x̄,y)`	`A(x̄) :- D(x̄,t)`
	`D(x̄,z) :- B(x̄,z)`
	`D(x̄,[y|z]) :- D(x̄,z)`

`A(x̄) :- (∀y⊑t)B(x̄,y)`	`A(x̄) :- D(x̄,t)`
	`D(x̄,[]) :- B(x̄,[])`
	`D(x̄,[y|z]) :- B(x̄,[y|z]),D(x̄,z)`

Here `D`,`E` are new predicate symbols, $\bar{x}$ are all free variables of the clause in the left part of the table. In the right part of the table `y` is a new variable of the sort `Univ`, `z` a new variable of the sort `List`.

Theorem 5.1 *For any generalized logic program* C, C *is equivalent to* $\widehat{C}$ *in the sense that for any atomic formula* φ, φ *is provable by SLDB-resolution from* C *with a substitution* θ *iff it is provable by SLDB-resolution from* $\widehat{C}$ *with the same substitution.* ∎

Let us note that this equivalence implies the semantic equivalence - the least models of the two programs are identical. It is also easy to see that if to omit sorts and built-in predicates, then SLDB-resolution on Horn clauses is identical to SLD resolution.

5.2 A metainterpreter for Horn clause programs

In the examples from section 2.4 we already had shown that many iterative programs, which are usually expressed in Prolog via recursion, have simple non-recursive definitions using bounded quantifiers. Here we give a more interesting example: a metainterpreter for Horn clause programs. We assume that the clauses of the object level Horn clause program of the form `A:-B1,...,Bn` are represented as facts of the form `rule(A,[B1,...,Bn])`, and the facts of the form `A` are represented as `rule(A,[])`. The definition of the metainterpreter is given below:

```
call(Goal) :-
   (∃List)(trace_of_execution(List,[])∧starts(List,[Goal])).

starts([X|Xs],X).
trace_of_execution(List,Last_element) :-
   (∀Sublist⊑List)(Sublist=[]
                   ∨Sublist=[Last_element]
                   ∨step_of_execution(Sublist)).

step_of_execution([State1,State2|States]) :-
   transition(State1,State2).

transition([Atom|Atoms],NewAtoms) :-
   rule(Atom,Tail),
   append(Tail,Atoms,NewAtoms).

append(L1,L2,L3) :-
   (∃List)trace_of_append(L1,L2,L3,List).

trace_of_append(L1,L2,L3,List) :-
   starts(List,[L3,L1]),
   (∀L⊑List))(L=[]∨L=[[[],L2]]∨step_of_append(L)).

step_of_append([[X|Xs],[X|Ys]],[Xs,Ys]|States]).
```

Theorem 5.2 *Let* C *be a Horn clause logic program (possibly with sort declarations). Let* R *is obtained from* C *by replacing each rule* `A:-B1,...,Bn` *(each fact* `A` *resp.) with the facts* `rule(A,[B1,...,Bn])` *(facts* `rule(A,[])` *resp.) and by adding the above clauses (with the same sort declarations). Then for any atomic φ, φ is provable by SLDB-resolution from* C *with a substitution θ iff the goal* `call(φ)` *is provable from* R *with the same substitution.*
■

From this theorem we can easily infer that any computable predicate can be expressed by a non-recursive generalized logic program (and hence by a non-recursive generalized logic program consisting of only one definition).

Theorem 5.3 *Let* C *be a non-recursive Horn clause program with sort declarations. Then the minimal model computed by* C *is decidable (i.e. there is an algorithm verifying if a given ground atomic formula belongs to the model).* ■

Corollary 5.4 *The class of predicates that are computable by the non-recursive generalized logic programs is strictly larger then the class of predicates computable by non-recursive Horn clause programs.* ■

We can also note that the use of bounded quantifiers is essential:

Theorem 5.5 *The class of predicates computable by non-recursive Horn clause programs coincides with the class of programs computable by non-recursive generalized logic programs without bounded quantifiers.* ■

6 Negation

Throughout this section negation means classical negation, unless the inverse is explicitly stated. By classical negation we mean the following: the negation of a formula is true on some elements, if the formula is not true on this elements.

The traditional approaches to handling negation in logic programming are not satisfactory. The main reason is very easy - there can not be a complete and correct implementation of classical negation in logic programs. In the literature concerning negation usually some conditions are given which show when the negation of a predicate defined by Horn clauses satisfies some desirable properties. There are two aspects of using negation: the first concerns computability and the second concerns semantic issues. Let us briefly consider the two aspects.

1. Negation is very hostile to computability. The main reason is the universality of the Horn clause language. Any computable predicate can be represented as a Horn clause program, which means that the negation of predicates defined by Horn clauses is not computable. Even if the negation of a predicate is computable in the Herbrand universe of a program, then the program computing the negation for this Herbrand universe is not correct for extensions of the universe. The usual solution of the computability problem is negation as failure, which is incomplete. It is also difficult to find satisfiable general cases for which negation as failure is complete.

2. As for the semantic aspects, one of the usual solutions is to restrict the class of admissible programs to so called stratified programs or some other classes. These programs have a (stratified) least model, but this model is not computable in general.

Let us informally call predicates with the computable negation negatable predicates. The desirable solution of the two aspects can be summarized as follows: to find a class of programs which is sufficiently rich, but which defines only negatable predicates. We show one such class of generalized logic programs. This class of programs is sufficiently rich, for instance, all the examples of section 2.4 are in this class.

The bounded quantifiers can easily be negated using the following equivalencies:

$$\neg(\forall x \in t)\varphi \equiv (\exists x \in t)\neg\varphi$$
$$\neg(\exists x \in t)\varphi \equiv (\forall x \in t)\neg\varphi$$
$$\neg(\forall x \sqsubseteq t)\varphi \equiv (\exists x \sqsubseteq t)\neg\varphi$$
$$\neg(\exists x \sqsubseteq t)\varphi \equiv (\forall x \sqsubseteq t)\neg\varphi$$

Let us note, that this property is already sufficient to use negation for a wide class of predicates. For instance, all examples of section 2.4 can easily be transformed into non-recursive programs without unrestricted quantifiers. Some of the clauses of the examples contain unrestricted quantifiers, e.g.

```
ordered2(A[X,Y|Z]) :-
   X≤Y.
```

which is a notation for

```
ordered2(L) :-
   ∃X∃Y∃Z(L=[X,Y|Z]∧X≤Y).
```

However this clause is equivalent to the clause

```
ordered2(L) :-
   (∃X∈L)(∃Y∈L)(∃Z⊑L)(L=[X,Y|Z]∧X≤Y).
```

which contains only bounded quantifiers.

It is interesting, that we can validate such a use of patterns in the head of clauses without rewriting them into expressions with bounded quantifiers, using one general theorem proved below. Let us first give precise definitions. A predicate P defined by a generalized logic program $\mathcal{P}$ is called negatable iff there is a generalized logic program $\mathcal{R}$ computing the negation R of P, i.e. the predicate R such that for any tuple t of ground terms from the Herbrand universe of $\mathcal{P}$, $P(t)$ is true in the minimal model for $\mathcal{P}$ iff $R(t)$ is false in the minimal model for $\mathcal{R}$.

Theorem 6.1 *Let* N *be the class of the generalized logic programs with the following properties:*

1. *No recursion is used in the programs from* N*;*
2. *All occurences of the unrestricted existential quantifier in the programs from* N *take the form*

   $$\exists y(\varphi(\bar{x},y)\wedge\psi(\bar{x},y)),$$

   where $\bar{x}$ *are all free variables of* φ, ψ *and*

   (a) φ *is negatable;*

   (b) For any tuple of ground terms s *there is at most one term* t *such that* φ(t,s)

Then every defined in N *predicate is negatable.* ■

Following [12], we shall call the generalized logic programs without unrestricted quantifiers Δ_0-programs.

Corollary 6.2 *If $\mathcal{P}$ is a Δ_0-program without recursion, all built-in predicates are negatable and the equality = is negatable, then any predicate defined in $\mathcal{P}$, is negatable.* ■

Now we apply theorem 6.2 to extensions of Δ_0-programs:

Theorem 6.3 *Let the equality = and all built-in predicates be negatable. Let $\mathcal{P}$be a non-recursive generalized logic program with clauses of the form*

$P(\bar{t})$:- φ,

where φ is a Δ_0-formula, and all free variables of φ occur it $\bar{t}$. Then for every predicate P defined in $\mathcal{P}$, P is negatable. ■

Note. If there are no infinite predefined sorts, then the equality is negatable.

Now we shall consider the case of recursive Δ_0-programs. The first result about negation in such programs is negative.

Theorem 6.4 [2] *For every computable set S of ground terms there is a Δ_0-program $\mathcal{P}$ defining a predicate P such that for every ground term t, $P(t)$ belongs to* Mod*($\mathcal{P}$) iff $t \in S$.*

This result shows that there can be no correct and complete implementations of negation for Δ_0-programs in general. However the (recursive) Δ_0-programs have interesting properties related to negation as failure. There are approaches to solve the problem of negation in logic programs by constructing programs which computes the finite failure state of a given Horn clause program (see e.g.[18]). However the programs generated in such a way can be very complicated even when the original programs are very simple. The programs with bounded quantifiers admit a very elegant solution for constructing such a dual program.

Let every built-in predicate P is negatable and the predicate $\widehat{P}$ represents its negation. Let = be also negatable and $\neq$ is the negation of the equality. Let C be a set of clauses which do not contain unrestricted quantifiers and which defines predicates $P_1, \ldots, P_n$ We introduce new dual predicate symbols $\widehat{P}_1, \ldots, \widehat{P}_n$. For each clause $C \in$ C the dual clause $\widehat{C}$ is constructed as follows: all atomic formulas $R(\bar{t})$ are replaced by $\widehat{R}(\bar{t})$, $t_1 = t_2$ by $t_1 \neq t_2$, $\wedge$ by $\vee$, $\vee$ by $\wedge$, all occurences of $\exists$ by $\forall$, and all occurences of $\forall$ by $\exists$. For example, the dual clause to

```
ordered(L) :-
  (∀X⊑L)(X=[]∨singleton(X)∨ordered2(X)).
```

s

[2]This theorem had been proved by Starchenko and myself

```
not_ordered(L) :-
   (∃X⊑L)(X≠[]∨not_singleton(X)∨not_ordered2(X)).
```

where we denoted the dual predicate symbols with the prefix not_. The dual program $\widehat{\mathrm{C}}$ to C consist of all such dual clauses together with the definitions of predicates that are dual to built-in predicates and to equality.

Theorem 6.5 *For any predicate P defined in the program* C *the predicate $\hat{P}$ from the program $\widehat{\mathrm{C}}$ computes the finite failure set for P.* ■

7 Concluding remarks

There are other aspects of the programming with bounded quantifiers, which are not considered in this paper. In this section we briefly mention some possible research issues on logic programming with bounded quantifiers.

7.1 Other kinds of bounded quantifiers

Intuitively, from the viewpoint of programming bounded quantifiers represent the idea of the iterative search over finite domains. There are domains of different sorts that have not been considered in this paper. For example, the quantification over subsets of a finite set may be needed. If we represent sets by lists, then all our results can be easily generalized for bounded quantifiers over subsets. For example, the step of the translation of the following expression, containing $\forall\subseteq$-quantifier, where $\subseteq$ is the subset relation

```
A(x̄) :- (∀y⊆t)φ(x̄,y).
```

to Horn clauses gives

```
A(x̄) :- D(x̄,[],t).
D(x,l,[]) :- φ(x̄,l).
D(x̄,l,[y|z]) :- D(x,l,z), D(x,[y|l],z).
```

Bounded quantification over an integer interval [k..n] consisting of the numbers from k to n:

```
A(x̄) :- (∀m∈[k..n])φ(x̄,m).
```

can be translated to

```
A(x̄) :- D(x̄,k,n).
D(x̄,k,n) :- k>n.
D(x̄,k,n) :- k≤n,
             φ(x̄,k),
             k1 is k+1,
             D(x̄,k1,n).
```

The other results of this paper can be formulated for the integer intervals as well, with the difference that integers are considered as a predefined sort.

7.2 Other applications

There are many other applications of bounded quantifiers. It had been noted that they can be used in the constrained logic programming over finite domains [7]. Constraints using the bounded quantifiers can be used in constraint logic programming to keep the set of constraints in a smaller size. To this end it is interesting to develop resolution-like calculus for the formulas with bounded quantifiers to resolve upon constraints similar to the theorem proving technique developed in [4].

The most obvious application of bounded quantifiers is parallel and concurrent logic programming, as was also noted in [14]. The bounded universal quantifier captures AND-parallelism, while the bounded existential quantifier - OR-parallelism. The kind of AND-parallelism inherent to bounded universal quantifiers is similar to FOR-ALL-parallelism from [16]. As for bounded existential quantifiers, their procedural treatment is completely different from the interpretation of unrestricted existential quantifier, which serves only for unification purposes. At the same time the bounded existential quantifiers are easily expressed via the unrestricted quantifier, e.g.

```
(∃x∈y)G :- ∃x(member(x,y)∧G)
```

which has a different procedural interpretation. It shows that the bounded quantifiers can also be used for expressing in a declarative way the control of program execution.

In our opinion bounded quantifiers can also be applied in deductive and relational databases. If we consider databases as finite objects, then bounded quantifiers seems to capture the intuitive semantics of databases better then unrestricted quantifiers. Variants of SLDB-resolution can be also used to formalize different kinds of finite search in databases.

8 Acknowledgements

I am grateful to Francois Bry, who had made many comments on a preliminary version of this paper.

References

[1] K.Apt and M.van Emden. Contributions to the theory of logic programming. *Jornal of the Association for Computing Machinery*, 29(3), 1982.

[2] J.Barwise. *Admissible Sets and Structures.* Springer Verlag, 1975.

[3] C.Beeri, Sh.Naqvi, R.Ramakrishnan, O.Shmueli, and Sh.Tsur. Sets and negation in a logic database language (LDL1). In *Proc. 6th ACM SIGACT-SIGMOD-SIGART Symposium on Principles of Database Systems*, pages 21–36. ACM Press, 1987.

[4] H.-J.Bürkert. A resolution principle for clauses with constraints. In M.E.Stickel, editor, *Proc. 10th CADE*, volume 449 of *Lecture Notes in Artificial Intelligence*, pages 178–192, 1990.

[5] C.C.Chang and H.J.Keisler. *Model theory.* North Holland, 1977.

[6] A.Dovier, E.G.Omodeo, E.Pontelli, and G.Rossi. {log}: A logic programming language with finite sets. In *Proc. ICLP'91*, pages 109–124. MIT Press, 1991.

[7] P.van Hentenryck, V.Saraswat, and Y.Deville. Constraint processing in cc(fd). Technical report, Brown University, December 1991.

[8] B.Jayaraman and D.A.Plaisted. Programming with equations, subsets and relations. In *Proc. NACLP'89*, Cleveland, 1989. MIT Press.

[9] G.M.Kuper. Logic programming with sets. In *Proc. 6th ACM SIGACT-SIGMOD-SIGART Symposium on Principles of Database Systems*, pages 11–20. ACM Press, 1987.

[10] G.M.Kuper. On the expressive power of logic programming languages with sets. In *Proc. 7th ACM SIGACT-SIGMOD-SIGART Symposium on Principles of Database Systems*, pages 10–14. ACM Press, 1988.

[11] J.W.Lloyd. *Foundations of Logic Programming.* Springer Verlag, 1984.

[12] S.S.Goncharov and D.I.Sviridenko. *Σ-programming (in Russian)*, volume 120 of *Vychislitelnye Systemy*, pages 3–29. Novosibirsk, 1985.

[13] S.S.Goncharov and D.I.Sviridenko. Theoretical aspects of σ-programming. In *Mathematical Methods of Specification and Synthesis of Software Systems'85*, volume 215 of *Lecture Notes in Computer Science*, pages 169–179, 1986.

[14] S.S.Goncharov, Yu.L.Ershov, and D.I.Sviridenko. Semantic programming. In *IFIP'86*, pages 1093–1100. Elsevier Science, 1986.

[15] M.Hagiya and T.Sakurai. Foundation of logic programming based on inductive definition. *New Generation Computing*, 2(1):59–77, 1984.

[16] R.Kowalski. Logic programming. In *Proc.IFIP'83*, pages 133–145. Elsevier Science, 1983.

[17] M.Alexander, P.Bilse, L.Jensen, and e.a. *PDC Prolog User's Guide.* Prolog Development Center, 1990.

[18] T.Sato and H.Tamaki. Transformational logic program synthesis. In *Proc. of the Conference on Fifth Generation Computer Systems*, pages 195–201. ICOT, 1984.

[19] J.T.Schwartz, R.B.K.Devar, E.Dubinski, and E.Schonberg. *Programming with sets. an Introduction to SETL.* Springer Verlag, 1986.

[20] Sh.Tsur and C.Zaniolo. LDL: A logic-based language. In *Proc. 12th International Conference on Very Large Databases*, pages 33–40, Kyoto, Japan, 1986.

[21] D.A.Turner. An overview of Miranda. *ACM SIGPLAN Notices*, 21(12):158–166, 1986.

[22] A.Voronkov. Program execution methods in σ-programming (in Russian). In *Proc 4th Soviet Conf. on Applications of Mathematical Logic*, pages 51–53, Tallinn, 1986.

[23] A.Voronkov. Intuitionistic list theory (in Russian). In *Proc. 8th Soviet Conf. on Mathematical Logic*, page 32, Moscow, 1986.

[24] A.Voronkov. Logic programs and their synthesis (in Russian). Technical Report 23, Institute of Mathematics, Novosibirsk, 1986.

[25] A.Voronkov. Synthesis of logic programs (in Russian). Technical Report 24, Institute of Mathematics, Novosibirsk, 1986.

[26] A.Voronkov. *A natural calculus for Σ-programs (in Russian)*, volume 120 of *Vychislitelnye Systemy*, pages 14–23. Novosibirsk, 1987.

[27] A.Voronkov. Logic programming and Σ-programming (in Russian). *Kibernetika*, (1):67–72, 1989.

[28] A.Voronkov. N-realizability: one more constructive semantics. Technical Report 71, Monash University, Department of Computer Science, Clayton, Australia, 1991.

Lecture Notes in Artificial Intelligence (LNAI)

Vol. 345: R.T. Nossum (Ed.), Advanced Topics in Artificial Intelligence. VII, 233 pages. 1988.

Vol. 346: M. Reinfrank, J. de Kleer, M. L. Ginsberg, E. Sandewall (Eds.), Non-Monotonic Reasoning. Proceedings, 1988. XIV, 237 pages. 1989.

Vol. 347: K. Morik (Ed.), Knowledge Representation and Organization in Machine Learning. XV, 319 pages. 1989.

Vol. 353: S. Hölldobler, Foundations of Equational Logic Programming. X, 250 pages. 1989.

Vol. 383: K. Furukawa, H. Tanaka, T. Fujisaki (Eds.), Logic Programming '88. Proceedings, 1988. IX, 251 pages. 1989.

Vol. 390: J.P. Martins, E.M. Morgado (Eds.), EPIA 89. Proceedings. 1989. XII, 400 pages. 1989.

Vol. 395: M. Schmidt-Schauß, Computational Aspects of an Order-Sorted Logic with Term Declarations. VIII, 171 pages. 1989.

Vol. 397: K.P. Jantke (Ed.), Analogical and Inductive Inference. Proceedings, 1989. IX, 338 pages. 1989.

Vol. 406: C.J. Barter, M.J. Brooks (Eds.), Al '88. Proceedings, 1988. VIII, 463 pages. 1990.

Vol. 418: K.H. Bläsius, U. Hedtstück, C.-R. Rollinger (Eds.), Sorts and Types in Artificial Intelligence. Proceedings, 1989. VIII, 307 pages. 1990.

Vol. 419: K. Weichselberger, S. Pöhlmann, A Methodology for Uncertainty in Knowledge-Based Systems. VIII, 132 pages. 1990.

Vol. 422: B. Nebel, Reasoning and Revision in Hybrid Representation Systems. XII, 270 pages. 1990.

Vol. 437: D. Kumar (Ed.), Current Trends in SNePS – Semantic Network Processing System. Proceedings, 1989. VII, 162 pages. 1990.

Vol. 444: S. Ramani, R. Chandrasekar, K.S.R. Anjaneyulu (Eds.), Knowledge Based Computer Systems. Proceedings, 1989. X, 546 pages. 1990.

Vol. 446: L. Plümer, Termination Proofs for Logic Programs. VIII, 142 pages. 1990.

Vol. 449: M.E. Stickel (Ed.), 10th International Conference on Automated Deduction. Proceedings, 1990. XVI, 688 pages. 1990.

Vol. 451: V. Marík, O. Stepánková, Z. Zdráhal (Eds.), Artificial Intelligence in Higher Education. Proceedings, 1989. IX, 247 pages. 1990.

Vol. 459: R. Studer (Ed.), Natural Language and Logic. Proceedings, 1989. VII, 252 pages. 1990.

Vol. 462: G. Gottlob, W. Nejdl (Eds.), Expert Systems in Engineering. Proceedings, 1990. IX, 260 pages. 1990.

Vol. 465: A. Fuhrmann, M. Morreau (Eds.), The Logic of Theory Change. Proceedings, 1989. X, 334 pages. 1991.

Vol. 475: P. Schroeder-Heister (Ed.), Extensions of Logic Programming. Proceedings, 1989. VIII, 364 pages. 1991.

Vol. 476: M. Filgueiras, L. Damas, N. Moreira, A.P. Tomás (Eds.), Natural Language Processing. Proceedings, 1990. VII, 253 pages. 1991.

Vol. 478: J. van Eijck (Ed.), Logics in Al. Proceedings. 1990. IX, 562 pages. 1991.

Vol. 481: E. Lang, K.-U. Carstensen, G. Simmons, Modelling Spatial Knowledge on a Linguistic Basis. IX, 138 pages. 1991.

Vol. 482: Y. Kodratoff (Ed.), Machine Learning – EWSL-91. Proceedings, 1991. XI, 537 pages. 1991.

Vol. 513: N. M. Mattos, An Approach to Knowledge Base Management. IX, 247 pages. 1991.

Vol. 515: J. P. Martins, M. Reinfrank (Eds.), Truth Maintenance Systems. Proceedings, 1990. VII, 177 pages. 1991.

Vol. 517: K. Nökel, Temporally Distributed Symptoms in Technical Diagnosis. IX, 164 pages. 1991.

Vol. 518: J. G. Williams, Instantiation Theory. VIII, 133 pages. 1991.

Vol. 522: J. Hertzberg (Ed.), European Workshop on Planning. Proceedings, 1991. VII, 121 pages. 1991.

Vol. 535: P. Jorrand, J. Kelemen (Eds.), Fundamentals of Artificial Intelligence Research. Proceedings, 1991. VIII, 255 pages. 1991.

Vol. 541: P. Barahona, L. Moniz Pereira, A. Porto (Eds.), EPIA '91. Proceedings, 1991. VIII, 292 pages. 1991.

Vol. 542: Z. W. Ras, M. Zemankova (Eds.), Methodologies for Intelligent Systems. Proceedings, 1991. X, 644 pages. 1991.

Vol. 543: J. Dix, K. P. Jantke, P. H. Schmitt (Eds.), Nonmonotonic and Inductive Logic. Proceedings, 1990. X, 243 pages. 1991.

Vol. 546: O. Herzog, C.-R. Rollinger (Eds.), Text Understanding in LILOG. XI, 738 pages. 1991.

Vol. 549: E. Ardizzone, S. Gaglio, F. Sorbello (Eds.), Trends in Artificial Intelligence. Proceedings, 1991. XIV, 479 pages. 1991.

Vol. 565: J. D. Becker, I. Eisele, F. W. Mündemann (Eds.), Parallelism, Learning, Evolution. Proceedings, 1989. VIII, 525 pages. 1991.

Vol. 567: H. Boley, M. M. Richter (Eds.), Processing Declarative Kowledge. Proceedings, 1991. XII, 427 pages. 1991.

Vol. 568: H.-J. Bürckert, A Resolution Principle for a Logic with Restricted Quantifiers. X, 116 pages. 1991.

Vol. 587: R. Dale, E. Hovy, D. Rösner, O. Stock (Eds.), Aspects of Automated Natural Language Generation. Proceedings, 1992. VIII, 311 pages. 1992.

Vol. 590: B. Fronhöfer, G. Wrightson (Eds.), Parallelization in Inference Systems. Proceedings, 1990. VIII, 372 pages. 1992.

Vol. 592: A. Voronkov (Ed.), Logic Programming. Proceedings, 1991. IX, 514 pages. 1992.

Lecture Notes in Computer Science